# HUMAN COLONIC BACTERIA: ROLE IN NUTRITION, PHYSIOLOGY, AND PATHOLOGY

*Edited by*

**Glenn R. Gibson, Ph.D.**
Medical Research Council Scientist
Department of Microbiology
Dunn Clinical Nutrition Centre
Cambridge, England

**George T. Macfarlane, Ph.D.**
Medical Research Council Senior Scientist
Department of Microbiology
Dunn Clinical Nutrition Centre
Cambridge, England

**CRC Press**
**Boca Raton Ann Arbor London Tokyo**

**Library of Congress Cataloging-in-Publication Data**

Human colonic bacteria: role in nutrition, physiology, and pathology / edited by Glenn R. Gibson
  and George T. Macfarlane.
      p.    cm.
    Includes bibliographical references and index.
    ISBN 0-8493-4524-3
    1. Colon (Anatomy)—Microbiology.    I. Gibson, Glenn R.    II. Macfarlane, George T.
    [DNLM:    1. Colon—microbiology.    2. Bacteria—metabolism.    3. Nutrition.
    4. Colonic Diseases—pathology.  WI 520 H918 1995]
    QR171.A43H85  1995
    612.3′6—dc20
    DNLM/DLC
    for Library of Congress                                                               94-28242
                                                                                               CIP

*NWST*
*IAFQ 0479*

# THE EDITORS

**Glenn R. Gibson, Ph.D.**, has been a Medical Research Council scientist at the Dunn Clinical Nutrition Centre, Cambridge, England since 1987. He also teaches in the Department of Pathology at the University of Cambridge. Dr. Gibson obtained a B.Sc. in microbiology from the Department of Biological Sciences at the University of Dundee, Scotland in 1983 and his Ph.D. from the same department and the Scottish Marine Biological Association, Oban in 1986. He is a member of the Society for General Microbiology, Society for Applied Bacteriology, American Society for Microbiology, and currently serves as an editor for the Journal of Applied Bacteriology. Dr. Gibson is an author or co-author of over 60 papers in the scientific literature. He received the W. H. Pierce Memorial Prize in 1991 which was awarded by the Society for Applied Bacteriology and Unipath.

**George T. Macfarlane, Ph.D.**, is a senior scientist with the Medical Research Council at the Dunn Clinical Nutrition Centre in Cambridge, where he has worked since 1984. He also teaches in the Department of Pathology at the University of Cambridge. Dr. Macfarlane received a B.Sc. in microbiology in 1980 from the University of Dundee, where he later obtained a Ph.D. in 1984. He is a member of several learned societies including the Society for General Microbiology, Society for Applied Bacteriology, and American Society for Microbiology. Dr. Macfarlane is a Senior Editor of the Journal of Applied Bacteriology, and has published more than 80 research articles dealing with various aspects of the microbiology of marine and estuarine sediments and the large intestine.

# PREFACE

When we consider that over 90% of all cells associated with the human body are bacteria living in the large intestine, it is surprising that so little is known about what these microorganisms actually do. While there is a great deal of information available on the relative numbers and different types of bacteria that can be isolated from stool specimens, we feel that it is now time to consider the gut microbiota from a more functional perspective. The last decade has seen a number of significant advances in our understanding of certain aspects of colonic microbiology and one of the purposes of this book is to summarize some of these findings in the contexts of nutrition, physiology, and pathology. To this end, we invited a number of research workers who are active in the field to participate in the project. Our original purpose was that whenever possible, a critical assessment be given that reflected the authors own, possibly idiosyncratic opinions.

By definition, a handbook is a reference work for research and practice, and this is one of the main objectives of this volume. However, it is also hoped that the contents of this book will be of interest to the undergraduate, and that it will assist in some way towards increasing the awareness of human intestinal microbiology in institutions of higher learning. With the exceptions of the more common bacterial enteric diseases, such as salmonellosis and shigellosis, large intestinal microbial pathology usually receives only a cursory mention in undergraduate level courses. The early work of Metchnikoff, Arbuthnott Lane, and others, at the beginning of this century, alluded to a wider role played by bacteria in the large intestine in affecting health and the quality of life. Today, many serious gut diseases such as ulcerative colitis, large intestinal cancer (in western populations the colon is the second most common site for carcinomas), and Crohn's disease are major sources of morbidity and mortality, yet still have unknown etiologies. However, it is almost certain that microorganisms are important agents involved in the initiation, promotion, and/or development of these disorders.

The colonic microbiota should also be considered in terms of human health as well as disease. For example, increased energy gain for the host may be obtained from fermentation, while more research is being directed at manipulating the composition and activities of the gut microflora towards what is vaguely termed an improved "balance". Although this description is poorly defined, the goals of some of this work are credible. In the not too distant future, it may be envisaged that subtle changes in diet, genetic manipulation, and careful selection of probiotic microorganisms will, on an individual level, all be perceived as being of importance for health.

This book has attempted to demonstrate the complexity of the gut ecosystem such that the colon should not merely be considered as an appendage situated at the end of the digestive tract whose sole functions are the absorption of water and storage of waste materials. In this volume, individual contributions on microbial ecology, carbohydrate, protein, short chain fatty acid, host glycoprotein, and gas metabolism may be found, together with texts dealing with newer areas of research such as molecular biology and probiotics. The remaining sections discuss toxicology, infections, and inflammatory bowel disease.

In conclusion, we thank each of the authors involved in this project for their efforts and acknowledge the help given by CRC Press, in particular by Marsha Baker and Jim Labeots. Last but not least, we would like to express our gratitude to each of our young children Gemma, Cara, Benjamin, and Laurie for occasionally allowing us the use of "their" home computers during the preparation of this book.

**Glenn R. Gibson**
**George T. Macfarlane**

# CONTRIBUTORS

**Clive Allison, Ph.D.**
Department of Microbiology
Unilever Research
Port Sunlight Laboratory
Bebington, England

**Robert P. Anderson, Ph.D.**
Wellcome Medical Research Institute
University of Otago Medical School
Dunedin, New Zealand

**Vinton S. Chadwick, M.A., M.Sc., M.D.**
Wellcome Medical Research Institute
University of Otago Medical School
Dunedin, New Zealand

**Stefan U. Christl, M.D.**
Medical Department
University of Wurzburg
Wurzburg, Germany

**Patricia L. Conway, M.Sc., Ph.D.**
School of Microbiology and Immunology
University of New South Wales
Kensington, New South Wales, Australia

**John H. Cummings, M.A., M.Sc., M.B., Ch.B.**
MRC Dunn Clinical Nutrition Centre
Cambridge, England

**Glenn R. Gibson, Ph.D.**
Department of Microbiology
MRC Dunn Clinical Nutrition Centre
Cambridge, England

**Michael J. Hudson, B.Sc.**
Department of Microbial Pathogenicity
PHLS Centre for Applied Microbiology and
    Research
Porton Down, England

**Sean M. Kelly, MRCP, M.D.**
Department of Gastroenterology
Addenbrooke's Hospital
Cambridge, England

**Michael D. Levitt, M.D.**
Department of Medicine
Minneapolis VA Medical Center
Minneapolis, Minnesota

**George T. Macfarlane, Ph.D.**
Department of Microbiology
MRC Dunn Clinical Nutrition Centre
Cambridge, England

**Sandra Macfarlane, B.Sc.**
Department of Microbiology
MRC Dunn Clinical Nutrition Centre
Cambridge, England

**Philip D. Marsh, Ph.D.**
Department of Pathogenicity
PHLS Centre for Applied Microbiology and
    Research
Porton Down, England

**Michael E. Quigley, B.Sc.**
Carbohydrate Laboratory
MRC Dunn Clinical Nutrition Centre
Cambridge, England

**Ian R. Rowland, Ph.D.**
Department of Microbiology and Nutrition
BIBRA International
Carshalton, England

**Gerald W. Tannock, Ph.D.**
Department of Microbiology
University of Otago Medical School
Dunedin, New Zealand

**Jerry M. Wells, Ph.D.**
Department of Pathology
University of Cambridge
Cambridge, England

# CONTENTS

Chapter 1    Microbial Ecology of the Human Large Intestine ................................................ 1
*Patricia L. Conway*

Chapter 2    Molecular Genetics of Intestinal Anaerobes ..................................... 25
*Jerry M. Wells and Clive Allison*

Chapter 3    Carbohydrate Metabolism in the Colon.............................................61
*Michael J. Hudson and Philip D. Marsh*

Chapter 4    Proteolysis and Amino Acid Fermentation ......................................... 75
*Sandra Macfarlane and George T. Macfarlane*

Chapter 5    Short Chain Fatty Acids ............................................................... 101
*John H. Cummings*

Chapter 6    Gas Metabolism in the Large Intestine ......................................... 131
*Michael D. Levitt, Glenn R. Gibson, and Stefan U. Christl*

Chapter 7    Toxicology of the Colon — Role of the Intestinal Microflora ...................................... 155
*Ian R. Rowland*

Chapter 8    Structure, Function, and Metabolism of Host Mucus Glycoproteins ............................ 175
*Michael E. Quigley and Sean M. Kelly*

Chapter 9    Bacterial Infections and Diarrhea ................................................. 201
*George T. Macfarlane and Glenn R. Gibson*

Chapter 10 The Role of Intestinal Bacteria in Etiology and Maintenance
of Inflammatory Bowel Disease ....................................................227
*Vinton S. Chadwick and Robert P. Anderson*

Chapter 11 Role of Probiotics ........................................................................257
*Gerald W. Tannock*

Index ........................................................................................ 273

*Chapter 1*

# Microbial Ecology of the Human Large Intestine

*Patricia L. Conway*

## CONTENTS

I. Introduction .................................................................................................................... 1
II. Human Intestinal Ecosystem ........................................................................................ 2
    A. Ecological Principles ............................................................................................. 2
    B. Description of the Ecosystem ............................................................................... 3
    C. Methods for Study ................................................................................................. 3
        1. Experimental Design ....................................................................................... 4
            a. *In Situ* Studies ....................................................................................... 4
            b. *In Vitro* and *In Vivo* Models .............................................................. 4
        2. Sample Analysis .............................................................................................. 5
            a. Structure .................................................................................................. 6
                i. Cultivation of Major Bacterial Groups ........................................... 6
                ii. Molecular Genetic Techniques ....................................................... 7
            b. Functional State ....................................................................................... 8
III. Individual Components of the Ecosystem ..................................................................... 9
    A. Description of the Components ............................................................................. 9
        1. Host Factors ..................................................................................................... 9
        2. Dietary Contributions ...................................................................................... 9
        3. Microbiota ..................................................................................................... 10
    B. Diversity Within the Ecosystem ......................................................................... 10
    C. Interactions Within the Ecosystem ..................................................................... 10
        1. Interactions Between the Microbiota and Other Constituents ....................... 11
            a. Diet ........................................................................................................ 11
            b. Host Factors .......................................................................................... 11
            c. Susceptibility to Clinical Conditions ................................................... 11
        2. Interactions Between Components of the Microbiota ................................... 12
IV. The Microbiota ........................................................................................................... 12
    A. Acquisition .......................................................................................................... 12
    B. Adult Profile ........................................................................................................ 13
        1. Fecal Populations ........................................................................................... 13
        2. Luminal Populations ...................................................................................... 14
        3. Mucosal-Associated Populations ................................................................. 15
    C. Unusual Components ........................................................................................... 16
V. Stability of the Ecosystem .......................................................................................... 16
    A. Ecological Principles Governing Stability ......................................................... 16
    B. Disturbances ........................................................................................................ 17
    C. Manipulation ....................................................................................................... 17
VI. Summary ..................................................................................................................... 18
References ........................................................................................................................... 18

## I. INTRODUCTION

The term "ecology" originates from the Greek words *oikos* (household or dwelling) and *logos* (law) and is therefore, by definition, the law of the household. It was first defined in 1866 by the German biologist Ernst Häeckel and refers to the interrelationship between organisms and their living and nonliving surroundings.[1] As such, microbial ecology relates to interactions between microorganisms and their environment.

In reflecting on the boundaries of microbial ecology, Lynch and Hobbie[2] proposed that "it is necessary to understand the behavior of microorganisms under defined and controlled conditions in the laboratory, in order to understand behavior in natural environments." A microbial ecologist views the intestinal tract as the ideal ecosystem because it can be studied in its entirety and the whole ecosystem can be treated as a unit, which lends itself to be controlled at a range of levels. This represents a considerable advantage when compared to aquatic or terrestrial ecosystems that may be studied in the laboratory by sampling and using collected material in an *in vitro* microcosm. Unfortunately, sampling of these environmental ecosystems inherently perturbs the system and one is always concerned about how representative a single sample in fact is. For studies involving the gastrointestinal tract, the entire ecosystem can be exposed to an altered diet, stress, or drug regime and effects on the ecosystem and components thereof can be monitored. Furthermore, the intestinal tract is sterile at birth and subsequently successively colonized during the first year of life until a stable complex microbial population is established. The developing population is valuable for studies directed toward understanding the ecology of the system and the role of different bacterial populations.[3-5] The ecosystem can be studied in a very controlled way by using hosts with a developing microbiota, or germfree animal models.[6]

Although the intestinal tract is theoretically an ideal ecosystem, it has been relatively difficult to study. Progress in our understanding of interactions within the intestinal tract has been markedly hindered by difficulties of studying the system *in situ*, its enormous complexity, the dense diverse indigenous bacterial populations, the fact that most of the microbes were strict anaerobes requiring specific culture conditions, and the ethical and practical considerations of human studies. These aspects are discussed in detail below.

Since Dubos and colleagues[7] first applied ecological theory to the gastrointestinal microbiota of non-ruminants, there have been numerous reviews addressing the interactions between components.[8-15] Over the last few years there has been an increasing trend to use the term "ecology of the intestine" for published studies of the bacterial profile of the human colon. Microbial ecology is a complex discipline which addresses both individual constituents of an ecosystem and their influence on the unit. The main goal of this chapter is therefore to use ecological principles in order to develop an understanding of both the composition and function of the microbial community of the human colon. The importance of function in relation to compositional analysis will be highlighted. For example, the profile of the major bacterial groups has been shown to be relatively insensitive to dietary change while alterations in the metabolism of the population have been more evident.[8]

## II. HUMAN INTESTINAL ECOSYSTEM

### A. ECOLOGICAL PRINCIPLES

The concept of modern ecological theory as presented by Alexander[9] defines an ecosystem as being composed of habitats and niches. Habitats are physical spaces in the system that are occupied by communities of autochthonous organisms. Niche is described as the way an organism makes a living in its habitat. The composition of these communities is maintained "reasonably constant with the passage of time."[9] Transient, or allochthonous, organisms are also found in a habitat, however, they normally contribute little to the economy of the ecosystem, implying no colonization or growth in the healthy ("normal") host.[10] Disturbance of an ecosystem could result in an autochthonous species vacating a habitat which could then be colonized by allochthonous microbes. The critical distinction here is that autochthonous organisms colonize the normal habitat whereas allochthonous species cannot colonize except under abnormal conditions.

In 1965, Dubos and co-workers[7] first applied ecological theory to the gut and hypothesized that experimental perturbations of the system would give reproducible results. They proposed that microbes of the gastrointestinal tract which they called "indigenous," comprised those which have evolved with the host (autochthonous), those that are ubiquitous in the community so that they are established in practically all members (normal), and true pathogens that are accidently acquired. Their designation of whether an organism was autochthonous or normal was based on population levels. Accordingly, the indigenous microbes would include resident nonpathogenic species as well as resident and nonresident pathogens. Population levels of some microbes which are undoubtedly autochthonous, are normally low in adults, e.g., *Escherichia coli* and *Enterococcus faecalis*.[11] Consequently, Savage[10,12] revised these concepts so that indigenous and autochthonous are now considered synonyms which are applied to microbial systems. Furthermore, the designation by Dubos and colleagues of pathogens as a separate entity has some limitations. Some pathogens can be autochthonous to the ecosystem and normally live in harmony with the host, except when the system is disturbed.

The gastrointestinal tract is an integrated unit composed of numerous microbial habitats. Each habitat is normally colonized by one or more autochthonous microbes that may be allochthonous in another habitat. The entire mass of microbes colonizing the intestinal tract is generally referred to as "microflora" while to many microbial ecologists this microbial mass is called the "microbiota,"[12] which is more correct.[13] The two terms have different origins and implications. Knowledge gained by understanding the ecology of the system can be useful in understanding why the popular term "microflora" is in such general use instead of "microbiota." Basically, the term microflora initially established itself in the scientific community and has successfully competed with the more recently introduced description, microbiota. The term microbiota cannot easily colonize all habitats in the scientific community despite support by certain microbial ecologists.[12,13] Perhaps, one will need to oust the term microflora before the use of microbiota can successfully establish itself.

## B. DESCRIPTION OF THE ECOSYSTEM

The entire length of the colon in adult humans is about 150 cm. It consists of the cecum (to which the appendix is attached), the ascending colon, transverse colon, descending colon, and the sigmoid rectum.[14] Simplistically, the entire organ can perhaps be considered as a large continuous culture fermentor with an approximate volume of 500 ml and containing an average of 220 g of contents.[15] Plate 1 shows the colon, removed at autopsy, of a 59-year-old male who was healthy up until the time of death. In excess of $10^{11}$ bacterial cells per gram of contents have been reported,[16-18] which is about ten times higher than the number of host cells.[12] Complex fermentations carried out by bacteria in the colon produce a large number of metabolites which can be beneficial or detrimental to the health and nutritional status of the host. For example, some bacterial metabolites have been implicated in a number of colonic disorders.[19] In contrast, colonic bacteria also produce short chain fatty acids which are absorbed by the host as an energy source (see Chapter 5).[20]

The suggestion that the colon can be thought of as a continuous culture fermentor is certainly an oversimplification of this ecosystem. One reason for its complexity is the number of microbes (approximately $10^{13}$ in total) belonging to hundreds of different species.[16] Many colonic microbes are in steady state at the same time. Study of a limited number of microbes in steady state is not straightforward, so one can but muse at the number of doctoral programs that may be needed to investigate complex steady state conditions of colonic microorganisms. Not only does the microbiota influence the host, but the culture vessel of the colon is composed of living animal cells[13] that actively secrete and take up components. These secretions, together with the diet of the host, form nutrients of the ecosystem with the host immune system also contributing. These aspects, which are discussed more extensively in the next section, are cited here to give some perspective of the system complexity.

The colonic ecosystem can be roughly divided into three influential aspects: diet, host physiology, and the microbiota. In the healthy adult, a fine balance exists between these parameters. An understanding of these interactions helps predict the response and sensitivity of the host to various environmental pressures. This chapter will emphasize the acquisition and development of the microbiota and how it interacts with other components of the ecosystem.

## C. METHODS FOR STUDY

The colonic microbiota can be studied either *in situ*, *in vitro*, or by using animal models. In all cases, the system must be sampled either for direct analysis or for obtaining inocula for the model studies. Because of the inaccessibility of the organ, difficulties are encountered in trying to sample from the colon. Much of the pioneer work used fecal rather than colonic samples,[19] even though fecal samples are not entirely representative of the colonic microbiota. One must sample the exact site of interest in the colon. While initial studies were directed toward analyzing the composition or structure of the microbiota, it is now generally accepted that the function of the population is also a major consideration.[21] In fact, Miller and Hoskins[22] proposed that one could study functional activity that affects human health and thus estimate the population density.

## 1. Experimental Design

### a. In Situ Studies

Studies of the colonic microbiota can be carried out *in situ* using human volunteers or donors. Non-invasive studies can be performed relatively simply by monitoring the fecal microbiota of subjects exposed to various experimental regimes, such as dietary alteration. The advantage of such studies is that freshly void samples can be collected and theoretically can be processed without much delay. Furthermore, these studies can be performed on healthy subjects who are not exposed to conditions which may affect the gut ecosystem, as will be the case for surgery patients. Although these studies do shed some light on the responses of the "normal" microbiota to environmental change, results must be interpreted with caution. There is increasing evidence that the fecal microbiota does not mirror the colonic situation. Ideally, the colon of healthy subjects should be sampled directly; however, unless accident victims are used, there are several difficulties with this approach. First, the ethical ramifications of such studies need to be assessed in terms of the information gained and effects on the subject. The second major consideration is the type of equipment used for sampling and transportation. Unless feces are used, any sampling of the colon is invasive and involves intubation or collection during operation. Both methods can have undesirable effects. Intubation may result in contamination with bacteria from another region, while sampling at operation will inevitably mean that the host is anesthetized and fasted. These factors will obviously have an effect on the microbes, and in many cases, the patients may have also undergone pre-operative antibiotic treatment. The colon is an anaerobic environment and consequently a sample collected from the colon by intubation, or at operation, must be introduced into an $O_2$-free environment as rapidly as possible. In addition, the time and temperature of storage as well as the nature of the transport medium will greatly influence the outcome of subsequent culture work. Qualitative and quantitative changes in the microbiota have been seen to occur in samples stored for a relatively short amount of time.[23]

### b. In Vitro and In Vivo Models

Because of the complexity of the intestinal ecosystem, there is a real need to utilize *in vitro* and *in vivo* models to allow controlled studies of individual components. The use of models for studying the human colonic microbiota has been recently reviewed by Rumney and Rowland.[24] Numerous models have been developed and they vary considerably in complexity ranging from simple batch cultures[25] to semi-continuous systems[26] and multistage continuous-culture fermentors.[27] While some useful information can be obtained from each type of model, the limitations of the method being used must be understood and examined in terms of the questions being investigated. For example, when studying interactions between pairs of strains, inhibition of growth of one microbe by another may be interpreted as the action of an antagonistic metabolite produced by the other. However, it could equally be attributed to exhaustion of nutrients or competitive processes.

The anaerobic continuous-flow culture model has been extensively used[28-33] and it has been shown that the *in vivo* numerical relationships of the major bacterial groups can be duplicated. Moreover, bacterial interactions demonstrated *in vivo* can be mimicked. It can be argued that if one can maintain the natural balance among the numerous species populating the large intestine *in vitro*, one may conclude that the ecological control mechanisms in the model must be similar to those operating *in vivo*. It is difficult to imagine two different sets of mechanisms that fortuitously bring about similar equilibria in populations as complex as those of the indigenous large intestinal microbiota.[34] One criticism that can be made of anaerobic continuous-culture fermentors is that growth media do not closely resemble substrates available to the microbes *in vivo*. This fact may not be especially important, as not only do profiles of the bacterial populations reflect the *in vivo* situation, but the concentrations of major metabolic end products in a continuous culture of mouse cecal flora were consistent with those detected in the animal.[35] In a very rigorous attempt to apply ecological principles to investigate intestinal microbes, Freter and co-workers combined anaerobic continuous culture conditions and mathematical modeling for the rodent system.[34-36] In these studies, the concept of surface colonization was addressed in terms of growth of microbes on the walls of the fermentor. While bacterial adhesion to small intestinal epithelial mucosa is important for colonization,[37] no conclusive evidence exists that adhesion to colonic epithelium is a prerequisite for establishment in this region, since transit time is markedly longer in the colon than in the small intestine. The concept that invaders of the intestine need to adhere to the epithelium in order to avoid being washed out of the system may not therefore be so relevant in the large intestine.

Some aspects of *in vivo* conditions are difficult to create in continuous-culture fermentors. In particular, the mucosal surface with the overlying mucus gel is not included in such systems. While surfaces can be included in fermentors, it is difficult to maintain the type of surface area to volume ratio that occurs in the intestine, since the microvilli present a vast surface area. It is also difficult to include host factors and the mucus gel in *in vitro* models, especially in continuous-culture fermentors. As a way of addressing these issues, the epithelial mucosa has been reconstructed *in vitro* by immobilizing epithelial cells in microtiter wells and then overlaying with mucus, scraped from the mucosa. This technique was originally developed for studying the colonization potential of *Salmonella typhimurium* and has subsequently been used to study mechanisms of pathogenicity of enterotoxigenic *E. coli* strains.[38] Unfortunately, this methodology is largely limited to batch culture, a major limitation is the amount of material available and therefore the volume of the reactor, unless larger animals are used as the source of mucosal material. This is not a totally unsurmountable problem for the dedicated researcher as exemplified by the recent comment by Charles Kurland:"Use an elephant!" As discussed earlier in this chapter, it can be difficult to acquire sufficient material of human origin for such studies.

Because of difficulties of working with *in vitro* models, researchers frequently turn to the germ-free animal or streptomycin-treated mouse for studying microbial interactions. While the streptomycin-treated mouse is sometimes referred to as an inexpensive but imperfect substitute for the germ-free animal, the latter also differs markedly from its conventional counterpart.[6] As discussed later in this chapter, the indigenous microbiota contributes to the ecosystem and consequently a host lacking this microbiota may differ physiologically from the normal host. Perhaps the most correct model is the ex-germ-free mouse colonized with human colonic microbiota.[39] An important consideration of this technique is the origin and complexity of the inoculum, since simple mixes of facultative anaerobes such as those containing *Enterococcus faecalis*, *Enterobacter aerogenes*, *Proteus vulgaris*, and two strains of *E. coli*[40] would not function as a complex indigenous population. Various laboratory animals, e.g., rat, mouse, hamster, guinea pig, pig, and marmoset, have been examined as possible animal models. None of these species had a functional microbiota consistent with humans as determined by fecal enzyme activities.[41] However, the pig is reported to show good correlation on other parameters.[42] All models have limitations but valuable information may still be obtained. Results obtained from model studies can only be meaningful within the framework or context of the model being used. One could conclude, from the data listed above, that all studies directed at understanding the *in vivo* situation ought to employ continuous-culture fermentors. However, we have successfully used a simple, small volume batch culture system and shown excellent correlation with *in vivo* studies for the parameter being investigated.[43] Here, gene transfer was studied in (a) *in vivo* using germ-free mice, (b) *in vitro* using intestinal extracts from germ-free mice and batch conditions, and (c) in a range of laboratory media and batch conditions. Batch culturing of intestinal extracts gave the same high frequency of transfer of plasmid RP1 as noted *in vivo*. Very low transfer occurred in the range of laboratory media tested for one set of *E. coli* strain, with another set of strain showing lower transfer both *in vivo* and in batch culturing of gut extracts, when compared to laboratory media.

## 2. Sample Analysis

Early work analyzing the colonic microbiota has focused on enumeration and identification of microbial isolates. However, these approaches only examined dominant strains of the major bacterial groups in this diverse community. Today, it is recognized that in addition to determining the type and numbers of individual isolates and groups of microbes, the metabolic state is also of major importance. Although total numbers of groups of bacteria may remain constant under various conditions, metabolism of the subpopulations can vary. For example, no major differences in the fecal microbiota were noted in people who initially ate a high beef diet and subsequently shifted to a nonmeat diet, but a decrease in fecal β-glucuronidase was noted.[18] Borriello[44] studied the weekly variation in fecal bacteria and while all major genera of bacteria, including *Clostridium*, were stable in terms of total numbers per gram of feces, numbers of *Clostridium perfringens* varied considerably from week to week. In addition, the major serotype of *C. perfringens* was also changed.

As discussed by Freter,[34] detection of microbes in the colon does not confirm colonization of the niche. It may reflect colonization of a niche higher up in the intestinal tract that results in microbes merely existing on a transient basis. Under these circumstances, the transient species may not be actively growing in the colon and as such will not release metabolites that play a role in the ecological balance of the habitat. Furthermore, bacteria detected in the lumin may not actually be multiplying therein, but instead could be accumulated inactive daughter cells sloughed from mucosal-associated populations. This hypothesis originated from a study[45] in which the populations of *E. coli* were drastically reduced in the stomach but not in the large intestine with oral administration of lactobacilli, although viable counts of lactobacilli were slightly higher in the intestine. The authors proposed that lactobacilli were colonizing the stomach and were metabolically active, but only transient in the intestine where the concentration of metabolites would be significantly lower.

Analyses of colonic specimens and samples from model systems cannot be limited to enumeration and identification of the major bacterial groups. In addition, function must also be examined. This is best studied in terms of profiling metabolic end products, as discussed below.

## a. Structure

Analyses of the major bacterial groups in the large bowel can be fraught with difficulty. Many species have very exacting nutritional requirements which, even if identified, can be very difficult to include in selective media. While pure isolates of various representatives of many bacterial groups in the colon can be cultured in laboratory media, there may not be a suitable selective medium which allows preferential growth of particular species that occurs in the colonic microbiota. If one succeeds with enumeration and isolation of colonic microbes, the next major hurdle is identification. Traditionally, identification methodology is based on phenotypic characteristics, as measured by biochemical profiling of metabolites and fermentation patterns. This technique will not be further elaborated on in this chapter. Recently, taxonomic studies have utilized genotypic and molecular biological tools for identification also. These are briefly discussed below and in Chapter 2.

## i. Cultivation of major bacterial groups

These can be studied using the classical technique of enumerating colony forming units of a particular group or genus of microorganisms in a sample and is referred to as viable counting. It is interesting to note the successive accumulation of knowledge as new culture techniques have become available and are applied to enumerating viable bacterial populations of the intestine. For example, with the development of considerably improved anaerobic techniques in the late 1960s, the human intestinal flora could be more satisfactorily analyzed.[46] Finegold and colleagues[47] concluded in 1973 that the common technique in use at that time (i.e., the anaerobic jar), was unsuitable for the isolation of strict anaerobes from the mouse cecum. In 1974, Moore and Holdeman[16] cultured anaerobes in stools from 20 healthy Japanese-Hawaiians using a roll-tube technique. These workers concluded that as techniques improve, new species will continue to be identified. Today, a detailed description of the individual components of the microbiota is not hindered by limitations in the degree of anaerobiosis that can be achieved. However, total anaerobiosis may not be the requirement for some microbes. There is a need for analyzing the redox-potential and composition of the microhabitat under investigation. As an example, one can cite the growth requirements of *Campylobacter* spp. and *Helicobacter pylori*. As could be anticipated from the micro-habitat colonized by these bacteria, total anaerobiosis is not required, but rather enhanced levels of $CO_2$. These pathogens colonize the epithelial mucosa which presents a very different environment for the microbe than luminal contents.

Today, compositional analyses of the microbiota are hindered, or limited, by the range of media available. For example, numbers of bacteroides are enumerated by spreading serial dilutions of the sample on a nutritionally rich medium containing the antibiotic vancomycin because Gram-positive bacteria are inhibited by vancomycin. Unfortunately, some species of *Bacteroides* are also sensitive to this antibiotic[48] possibly causing an underestimate of viable counts.

In 1983, Finegold et al.[49] published comprehensive tables of the composition of the human intestinal microbiota. Even with the use of selective media and strict anaerobic techniques, it has been estimated that only about 10% of human cecal organisms are cultivated *in vitro*.[50] Others workers[51] report that approximately 60% of bacteria observed in a direct microscopic count could be recovered. The undetected microbes are largely those existing in smaller numbers in the sample, and without detailed knowledge of their function, it is difficult, if not impossible, to develop an enrichment process. Attempts have been made to manufacture laboratory media based on gut extracts, in order to simulate intestinal conditions, that allow culturing of suspected pathogens from a fecal or intestinal sample. This approach is usually unsuccessful without including selective agents or some sort of enrichment process. In one instance (Conway and Adams, unpublished results), a toxin producing clinical isolate was successfully enriched in the small intestine of the infant mouse using the model developed for studying infantile botulism.[52]

Quantitative analyses of colonic samples also needs to be addressed. In some studies, results are expressed as viable counts per gram dry weight. Determining the dry weight of samples will exclude variation due to differing water content, but it is not always possible to determine the dry mass of biological material especially with small samples. Counts of microbes are usually expressed as colony forming units per gram wet weight of sample. This unit is reasonable for fecal samples and for the analysis of colonic mucosa in small animal models. It does not allow, however, for comparison of results for viable counts per mucosal sample. The amount of epithelium will vary between samples without the actual mucosal surface area, with associated microbes, differing since the tissue can vary in thickness. The problem is real for human mucosal samples, since the epithelium is much thicker than that of small animals. Counts of mucosal-associated microbes are frequently expressed per unit surface area, provided a standardized sample can be collected. This allows comparison of results from one study to another. One is tempted, however, to compare numbers of microbes associated with the epithelium with numbers of microbes in luminal contents. As the former is per unit area and the latter per gram wet weight, this comparison cannot be made.

Quantification of mucosal-associated populations is greatly influenced by the washing procedure used to remove luminal contents and reversibly bound bacteria. Unfortunately, most studies fail to give any details of washing procedures. Our approach to this problem has been to insert a uniformly sized epithelial piece in a glass rod plugged with glass-wool and connected to a vacuum pump.[53] The washing procedure was standardized by sucking a uniform volume of wash buffer through the tube at a constant flow rate. This procedure was applicable in regions of the digestive tract with fast transit times, e.g., the small intestine, where microbes need to adhere. Colonization of the colonic mucosa may not necessarily involve adhesion to the epithelium but to the mucus overlying the epithelial cells. This mucus layer is relatively easily removed by washing and consequently, a carefully designed washing procedure is needed to distinguish mucosal-associated populations, including those within the mucus, from luminal populations. Technically, this may be very difficult to achieve.

## ii. Molecular genetic techniques

These techniques provide new tools for investigating the microbial ecology of the intestinal tract.[54] Gut microbes can be studied *in situ* using nucleic acid probes for detecting specific groups or species of microbes.[55] An oligonucleotide probe targeting ribosomal RNA (rRNA) can be used for direct hybridization of whole cells. The specificity of the probe can often be tested by using the CheckProbe program at the Ribosomal Database Project.[56] The probe can be labeled with a fluorescent dye that allows visualization by epifluorescent microscopy.[57] For example, fluorescent oligonucleotide probing of whole cells has been used for determinative, phylogenetic, and environmental microbiology of biofilms.[58] Because such probes target ribosomes, one can estimate the number of ribosomes in a single cell *in situ*. This information can be used to calculate the growth rate of the cell, since ribosomal content and growth show correlation.[59] This has been performed by comparing data gained from *in situ* hybridization of single cells with similar hybridizations of cells from pure cultures with known generation times and thereby estimating growth rate.[60] Previously, *in vivo* growth rate has been estimated using radioisotopes[61] or by dilution of a nonreplicating genetic marker. These techniques give an estimate of overall growth rather than that of individual cells, which may vary from one site to another.

One may well ask if this technology is directly applicable to some of the more esoteric bacteria of the colon. At present, various laboratories are developing rRNA probes for some, but by no means all colonic microbes. Alternatively, indigenous microorganisms can be studied in terms of SDS-PAGE protein profile of outer membrane proteins[62,63] or by restriction enzyme mapping.[62] Both of these methods compare the profile of the study isolate with that of known reference strains. Unfortunately, these techniques are only valid for pure cultures and therefore are not applicable to *in situ* measurements. Because many colonic microbes are nutritionally demanding or fastidious with long incubation times being required, genetic detection systems without culturing are favored. The technique using polymerase chain reaction (PCR) has facilitated detection and confirmation of microbes *in situ*. The PCR assay can be used to detect a cryptic piece of DNA: the gene encoding 16S rRNA, genes coding for species specific antigens, and fimbrial subunits.[64] Because the assay involves amplifying a portion of a gene, it has a very

**Table 1**  Selected Microflora-Associated Characteristics (MAC) that are Used to Study the Functional State of the Colonic Microbiota. The Microorganism Involved in the Various Parameters are Included Where Possible

| Parameter | Bacteria |
| --- | --- |
| Conversion of cholesterol to coprostanol | Eubacteria |
| Dehydroxylation of bile acids | Anaerobic, nonsporing Gram +ve rods |
| Transformation of conjugated bilirubin | Several spp., e.g., *Clostridium ramosum* |
| Inactivation of tryptic activity | Unknown |
| Breakdown of mucin | Several spp., e.g., peptostreptococci, ruminococci, bifidobacteria |
| Absence of some dipeptides such as β-aspartyl-glycine | Unknown |
| Presence of short-chain fatty acids | Several spp. |
| Analysis of intestinally produced gases in the respiratory air | Unknown |

From Midtvedt, T., *Recent Advances in Microbial Ecology*, Hattori, T., et al., Eds., Japan Scientific Societies Press, Tokyo, 1989, 515.

low detection limit and theoretically can be used for quantification if the process is sufficiently controlled to allow enumeration of the number of cycles required to yield a signal.

### b. Functional State

Considerable effort has been made in quantifying and characterizing the colonic microbiota in both health and disease. Interest often focuses on predicting the biochemical function of the microorganisms. However, an alternative approach addresses the question of functional status by determining the microbes' previous processes.[8,21,41,65-67] For this approach to be meaningful, it is important to clarify those mechanisms that are host related and those which are attributable to the microbiota itself.[21] Host-related mechanisms have been well established in a range of animals by using germ-free species and conventional counterparts. The newborn and the developing infant have been used for establishing host-related functions in humans. Consequently, a series of parameters such as anatomical structure as well as physiological and biochemical functions have been associated with the host itself and baseline values established. Some normal functions of microbes have been determined by Midtvedt and co-workers[68] who coined the term microflora associated characteristics (MAC) to refer to any anatomical structure, or physiological or biochemical function of the host which has been influenced by the microbiota. Accepting that not all such functions can be evaluated, these workers have selected a number of MACs (Table 1) that do not involve complicated sample collection or microbiological cultivation but which can be carried out by any major clinical laboratory. This table also lists the microbes known to be involved in the various parameters. The conversion of cholesterol to coprostanol, dehydroxylation of bile acids, and transformation of conjugated bilirubin are included in order to elucidate some aspects of hepatic and intestinal interaction. Pancreatic intestinal co-function can be examined by studying the inactivation of tryptic activity, while the integrity and function of the intestinal mucosa can be monitored by determining the breakdown of the mucin layer which overlies the epithelium. Mucin is an important source of carbon and energy for the microbes and therefore the extent of its degradation can be used as a measure of the degree of gut colonization.[66] Microbial interaction with compounds mainly derived from the diet can be examined by studying levels of dipeptides such as β-aspartyl-glycine and the presence of short chain fatty acids (SCFA). These fatty acids, which mostly include acetate, propionate, and butyrate, are the principal end products of fermentation and are the predominant anions in the colon with concentrations ranging from 20–200 mmol/l.[65] Enzyme activities have also been used to accumulate a metabolic profile of the microbiota.[41] The enzymes most frequently studied are those involved in the generation of toxic, mutagenic, or carcinogenic metabolites in the gut, e.g., β-glucosidase, β-glucuronidase, nitroreductase, nitrate reductase, and azoreductase.[67] Because these enzymes have a very broad specificity, it is possible to predict the metabolism of a wide variety of foreign and endogenously produced compounds.

This approach of monitoring the functional state of the microbiota has been applied to studying the effects of antibiotics[68] and diet[8,65] on the microbiota, as well as to following the development of the functional microbiota in children during the first five years.[69,70] A status of the microbiota of mice, in both conventional and *Lactobacillus* free animals, has been extensively studied by monitoring 26 MACs.[71]

## III. INDIVIDUAL COMPONENTS OF THE ECOSYSTEM

## A. DESCRIPTION OF THE COMPONENTS

The individual components of the ecosystem can be broadly divided into host-related factors, factors attributable to diet, and the microbiota itself. While it is easy to delineate between these groups, a dynamic equilibrium exists within the ecosystem.[10] Although there is a fine balance between these parameters in the healthy host, a disturbance in one factor can create a domino effect with ramifications for all components in the ecosystem. Each component will be briefly described in this section which also contains some dicussion of the types of interactions.

### 1. Host Factors

The colonic epithelial mucosa consists of columnar epithelial cells covered by a layer of mucus which is secreted by specialized goblet cells.[72,73] The mucus layer consists predominantly of mucin which is a 2000 kDa glycoprotein responsible for the viscosity of mucus as well as many smaller proteins, glycoproteins, lipids, and glycolipids.[72,74,75] Anatomically, microhabitats exist since the environment deep within the crypts of the mucosa would not be consistent with that exposed to luminal contents, and would be less influenced by peristalsis. Although this latter factor plays a major role in microbial colonization of the small intestine, it is generally less critical in the colon where, as mentioned earlier, transit time is slower. Both the host and the microbiota contribute to enzymic digestion in the gut.[76]

Secretory IgA is the major immunoglobulin in human exocrine secretions.[77] It is synthesized and released as a dimer by local plasma cells in the lamina propria. The dimer complexes with polypeptides on the surface of the secretory cells and the complex transported across the cells and released at the apical surface into the intestinal lumin.[78] The secreted complex is relatively resistant to proteolysis and protects the mucosal membrane by coating potential pathogens and thereby sterically hindering attachment.[79,80] This protective effect is enhanced by lactoferrin.[81] Another arm of the immune response includes cytokines which can be thought of as hormones of the immune system. Cells of the immune system can communicate with each other during immune responses by secreting cytokines that bind specifically to membrane receptors on the same or different cell. Cytokines can act alone or in combination to stimulate or inhibit immune responses.[82]

### 2. Dietary Contributions

The diet of the host is one source of nutrients for the microbiota. In addition, some host secretions are also utilized by the microbiota and *in vitro* studies by Macfarlane and co-workers[27,83] show that indigenous microbes can rapidly metabolize these compounds. The host secretes fermentable carbohydrates including glycoproteins such as mucins and polysaccharides such as chondroitin sulfate which can be utilized predominantly by bifidobacteria, some bacteroides, and ruminococci. Host pancreatic enzymes can also be metabolized as a source of nitrogen since many gut species are proteolytic (see Chapter 4).

The effects of ingested nutrients on the fecal microbiota have been extensively studied, however few direct effects have been reported. Early studies on the influence of diet on the human fecal microbiota showed no major differences in microbial numbers,[84,85] except after administrating a synthetic chemically defined diet.[86,87] There has been controversy as to the reliability of reported minor differences due to sampling difficulties, and the complexity of the microbiota and its growth requirements.[88] Studies on interactions between dietary components and the colonic microbiota have been largely restricted to either determinations of the major constituents of diet, e.g., protein and fat levels, or biotransformations of potential carcinogens.[89] Detailed compositional data for human feces from special dietary groups within hospitalized subjects have been presented by Finegold and co-workers.[49] In reviewing the effects of diet on fecal microbes, Borriello[90] concluded that high fiber and high protein diets did not alter the composition of the microbiota. However, elevated levels of carbohydrates increased numbers of bifidobacteria, while more fat increased bacteroides and an elemental diet decreased enterococci and lactobacilli. This is consistent with the conclusion by Tannock[91] that from published studies, enteric bacteria and lactobacilli emerge as indicators of stress including dietary stress, with the lactobacilli decreasing and the other bacteria increasing relative to the normal profile.

The main constituents of the host diet which would evade digestion in the upper regions of the human digestive tract include ingested carbohydrates. In fact, up to 60 g of carbohydrate reaches the colon per day,[20] of which the main fermentable proportion, of up to 40 g/d is resistant starch, i.e., resistant to pancreatic amylases but not necessarily bacterial enzymes, and up to 18 g/d is nonstarch polysaccharides. The endogenous carbohydrate contribution to the colon is about 2–3 g/d. Protein can also be used by the colonic microbiota with up to 9 g/d of dietary origin and up to 6 g/d of endogenous enzymes.[14] Most simple sugars are absorbed in the small intestine[92] with some exceptions, e.g., lactose and raffinose.[20] Substrate availability results in different carbohydrate fermentation rates in the proximal and distal colon, the former being carbohydrate rich.[14] Conversely, protein fermentation increases distally (see Chapters 3 and 4).[93] This varied distribution of nutrients is reflected in the composition of the microbiota in the various regions.

### 3. Microbiota

The large bowel is the most densely colonized region of the digestive tract of humans and it is proposed that there are at least 400 to 500 different species of bacteria, as well as yeast, fungi, and protozoa. The bacterial microbiota has been extensively studied and is discussed in detail in the next section. It includes a very complex population of aerobes, facultative anaerobes, and strictly anaerobic species with the nonsporing anaerobes predominating. It is generally recognized that the structure and function of the microbiota is regulated by environmental conditions such as nutrient availability, pH, redox potential, and microbial interactions. In addition, Gorbach and co-workers presented evidence that the microbiota is altered in the presence of chemical or pathogen induced diarrhea. In particular, chemically induced diarrhea resulted in an increase in fecal coliforms[94] while patients with cholera had a $10^5$-fold decrease in bacteroides.[95] It can be generalized that the composition and function of the microbiota is environmentally controlled and varies between individuals. This generalization is supported by the findings of van der Merwe and colleagues[96] who report that the aerobic and anaerobic intestinal flora differs between individuals. These workers note an exception for identical twins and suggest that genetic identity plays a role in the composition of the anaerobic flora of humans which in turn can have implications for Crohn's disease patients.

The yeast, fungi, and protozoa are most frequently overlooked and poorly quantified. Many bacteriologists consider that protozoa are not part of the normal microbiota and only expect to find them during disease.[90] However, in many instances parasitic infections are asymptomatic. Since protozoa can colonize and replicate in humans without disease, according to ecological definitions, they can be called indigenous. A number of commensal protozoa have been reported.[97]

### B. DIVERSITY WITHIN THE ECOSYSTEM

There is an extremely diverse microbial population within the colon of humans as well as a vast range of specialized ecological niches to fill. When one tabulates the variety of foods the average individual consumes over a short period, the type of nutrients and nonnutrients available for the microbes is itself diverse. In addition, secretions from the host, as well as sloughed off intestinal cells, provide a mixture of secretory and membrane glycoproteins. Colonic bacteria possess glycosidases which degrade glycoproteins releasing from the oligosaccharide chains, sugars that may be utilized as carbon sources for other bacteria.[98] In fact, colonic microbes degrade a large number of nutrients thereby widening the range of substrates available within the colon. The diverse range of biochemical activities of gut microorganisms has been established for some time.[99]

Innumerable microbial interrelationships have been found in the colon where one species utilizes materials excreted by another. These interactions can be divided into two classes: the metabolism of primary substrates digested by extracellular enzymes and waste product utilization.[100] This complexity and cross feeding make it feasible for many species to co-exist in the intestine thereby maintaining microbial diversity.[14]

Freter[101] proposed that four microhabitats can be identified in the colon, namely, the surface of the epithelial cells, the mucus gel overlying the villi, the mucus gel within crypts, and luminal contents. The role of each microhabitat varies significantly. Interest in the mucus gel both overlying the epithelial cells and within the crypts has received considerable attention over the last decade with discussions as to whether it serves a protective role and/or is a site of colonization (see Chapter 8). For example, receptors in mucus may competitively inhibit bacterial adhesion to analogous receptor sequences on the surface of the underlying epithelial cells.[102,103]

### C. INTERACTIONS WITHIN THE ECOSYSTEM

The interactions between constituents of the ecosystem can have very significant ramifications as far as host welfare is concerned. Interactions can be divided into those between the microbiota and other constituents of the ecosystem, and between components of the microbiota.

## 1. Interactions Between the Microbiota and Other Constituents

### a. Diet

As discussed earlier, host secretions and sloughed off epithelial cells can be utilized by some bacteria which in turn provide breakdown products that are available for metabolism by other microbes. This reverse interaction has also been demonstrated to contribute to the well being of the host. For example, it has been proposed that fatty acids produced during microbial fermentation are major sources of energy for the colonic mucosa[104] and regulation of mucosal cell turnover and carbohydrate metabolism.[21] This is a good example of mutualism with evolution by the colonocytes of ketogenic enzymes. Locally, SCFA directly or indirectly contribute to the regulation of water and electrolyte absorption with an antidiarrheic effect.[21] The impact of the microbiota on intestinal structure and function has been addressed in the review by Abrams.[105] In addition, the functional relationship between bacteria and colonic mucosa, with bacteria contributing to the mucosal cell growth, has also been discussed.[76] Likewise, the interrelationship between the colon flora, fermentation, and large bowel digestive function has been reviewed.[14]

### b. Host Factors

Host functions such as intestinal peristalsis, mucus secretion, and immunological defense are well known to be stimulated by the indigenous microbiota.[12,101,106] Consequently, only a few examples are cited here in order to highlight some types of interactions that can be anticipated between the microbiota and host factors that contribute to the ecosytem.

It is well established that local immune responses in the intestine protect the host from invasion by enteric pathogens. The question is frequently raised as to what effect this local immunity has on the indigenous microbiota, because the immune system has a capacity to respond to the antigens of the normal microbiota as well as pathogens. Low levels of antibody to many intestinal microbes can be found in serum.[107] It is interesting that the parenteral inoculation of anaerobes isolated from an animal's own fecal flora provoked only a very minor immunological response,[108] which suggests that a state of suppression of the immune response or immunological tolerance may exist. It has been suggested that synergy exists between the ecological control and immunological control of the microbiota. For example, *Vibrio cholerae* colonized in greater numbers in immunized germ-free animals as compared to animals inoculated with a biota of antagonistic bacteria.[40] It was speculated that indigenous bacteria, which form large populations in the intestine, might be affected little by local immune mechanisms. From microscopic examination of freshly void fecal microbiota of humans, it was found that fluorescent anti-IgA antibodies did not coat anaerobic microorganisms from the intestine, but did appear as a paint on enterobacterial strains.[109] This coating of IgA may lead to rejection of the strain from the mucosal surface. Consequently, only strains that are immunologically selected can colonize for extended periods.[110] Consistently, evidence was presented that the immune response to some indigenous bacterial species was impaired,[111] although this was not entirely consistent for all reports.[111,112] Evidence for antigen induced nonresponsiveness of the gut associated immune system may be related to the presence of suppressor T cells in the Peyer's patches.[113] It may also play a crucial role in acquisition of the microbiota. For example, acquisition of an *E. coli* strain soon after birth could occur because the strain originates from the mother and has therefore already been immunologically selected. After critically evaluating the literature, Lee[13] found that from the data available there was no evidence that the immune system had any influence on fluctuations in the microbial populations of the gut. Although categorically disregarding the role of the immune response, it was acknowledged that data from immunodeficient states could provide alternative information, since numbers of aerobic and anaerobic microbes increased in the proximal jejunum of patients with agammaglobulinemia.[114] Recently, Freter[34] concluded that there was little evidence that the immune system had a significant effect on indigenous populations, since it is difficult to imagine how this microbiota can exist at all if it were to be affected by the immune response. Although the immune system may well play a significant role in the acquisition and development of the indigenous microbiota, it probably has little effect on the stable developed microbial ecosystem of the healthy adult.

### c. Susceptibility to Clinical Conditions

The interactions between the microbiota and the host in relation to infection, irritable bowel disorders, and colon cancer are discussed elsewhere in Chapters 7, 9, and 10 of this book. It is interesting to note that about 20% of patients with bowel inflammation, e.g., Crohn's disease and ulcerative colitis, also suffer joint inflammation. Bowel infection by *Salmonella*, *Shigella*, and *Yersinia* can provoke joint inflammation which suggests an etiological link between colonic bacteria and arthritis. It has been proposed that gut contents contain soluble bacterial cell wall products and are arthropathic in an animal model. Bacterial cell wall fragments of dominant intestinal microbes,[115] in particular *Eubacterium aerofaciens*,[116] and peptidoglycan-polysaccharide complexes isolated from human feces,[117] have been shown to induce joint inflammation in rats.

## 2. Interactions Between Components of the Microbiota

The interactions between constituents of the microbiota can be discussed at two levels: the effects of the indigenous microbes on potential pathogens and sensitivity towards pathogens. Colonization of the intestine and growth *in situ* may be influenced by competition for nutrients and by metabolites of indigenous microbes that are antagonistic to invaders. These effects are often referred to as colonization resistance, a subject which has been reviewed elsewhere[118] and therefore will not be discussed in detail here.

Conditions reflecting the sensitivity of the intestinal ecosystem to pathogens and the balance between individual components can be illustrated by examining *Clostridium botulinum* and *Clostridium difficile* infections. For example, the peak of infant botulism is uniquely age related. The incidence of infant botulism frequently correlates with the introduction of solid food, a subsequent decrease in breast feeding and therefore a decrease in protective maternal antibodies, and an immature host immune response.[119] In one case of infant botulism in a breast-fed baby,[120] onset of symptoms coincided with an alteration in ascorbic acid dosage. It was subsequently shown in laboratory studies, using the infant mouse animal model, that cessation of high doses of ascorbic acid predisposed the host to the botulinum toxin.[121] There is also a fine balance between the indigenous microbiota and *C. difficile*[94] infections which include both antibiotic associated diarrhea and pseudomembranous colitis. The suitability of conditions in the intestine for clostridial toxin production and/or absorption can depend on a number of factors, including bacterial-, host-, and diet-related parameters. It is reported that colostral cell secretions neutralize the activity of *C. difficile* toxin and that breast milk can contain secretory immunoglobulin type A antibody against this toxin.[122] It has been proposed that the presence of one microbe can enhance absorption of toxin produced by another,[123] while in another report[124] it has been shown that one toxin can influence receptor sites of another, thereby affecting the permeability of the intestinal mucosa to toxins.

# IV. THE MICROBIOTA

Most work on the microbes of the large bowel has concentrated on the analysis of fecal samples. There is much less information available on the microbiota of luminal populations or those associated with the mucosa. Another consideration when quantifying the colonic microbiota is to distinguish between mucosal-associated and luminal populations. This distinction can be difficult, if not impossible, since mucosal-associated populations will also be detected in the lumin as sloughed cells or daughter cells of attached populations. The procedure used for sample collection will also influence whether loosely associated populations are removed by a washing step.

## A. ACQUISITION

It is generally accepted that the gastrointestinal tract is sterile at birth and is then successively colonized until the complex microbiota develops, although one report[125] presents data showing that newborn babies delivered by cesarean section and with intact membranes carried viable microbes in gastric aspirates. The time taken for bacteria to establish in the mouth and subsequently the digestive tract and feces varies considerably, depending on the type of delivery and exposure to bacteria during the birth process. The newborn is colonized with microbes originating from the anal and genital tracts of the mother during delivery,[126] or cutaneous contamination from the mother as can occur during breast feeding. In the case of cesarean or premature babies, microbes have also been traced from the environment, i.e., from other infants via the air, equipment, and nursing staff. The acquisition of the microbiota of the newborn has been extensively reviewed[42] and will not be discussed further. Bacteria have been isolated during the first 24 h of life and include facultative anaerobes, micrococci, streptococci, enterobacteria, lactobacilli, bacteroides, and clostridia.[127] Many individual studies exist regarding the fecal microbiota of newborns fed different diets and varying in age, as reviewed by Moughan et al.[42] with some prospective studies of the successive development in both normal delivery and cesarean born infants.[4,127,128] It is interesting to note that the summary of all individual studies of specific age groups reflects a pattern similar to that observed by Stark and Lee[4] who studied one group of normal healthy breast-fed infants and one group of formula fed infants for the first 12 months. The general pattern that emerged was that facultative organisms appear first and are then followed by a limited number of anaerobes during the first two weeks. By two days, streptococci and enterobacteria are present in both breast and formula fed infants and generally have peaked before the end of Week 1. The anaerobes are largely bifidobacteria during the first week of life, with bacteroides, clostridia, and anaerobic streptococci beginning to gradually appear over the first two weeks. *Escherichia coli* and streptococci occur at approximately $10^9$ organisms per gram of feces. In breast fed infants, there is subsequently a major decrease in *E. coli* and *Enterococcus* counts, as well as an almost complete disappearance of clostridia and bacteroides. These reductions do not occur in formula fed infants which maintain the more complex flora. By four weeks, the fecal microbiota of breast fed infants consists predominantly of bifidobacteria (about log 10.6 per gram) until dietary supplementation occurs. There is no consistency in reports as to which species of *Bifidobacterium* is most dominant. The most common bacteroides found in both breast and formula fed infants are *B. distasonis*, *B. vulgatus*, and *B. fragilis*.[128] The difference between breast fed and formula fed infants disappears with the introduction of solid food. By 12 months, the anaerobic fecal populations begin to resemble that of adults in number and composition as the facultative anaerobes decrease. By two years of age the profile resembles that of the adult.

The development of the bifidobacteria in cesarean-sectioned infants can be delayed beyond the two week period and facultative bacteria are often enterobacteria other than *E. coli*.[129] In addition, when studying colonization during the first 14 d in newborns delivered by cesarean section, it was noted that *C. perfringens* seems to be a precursor for installation of bacteroides and other clostridial species.[130] These levels of *C. perfringens* could be lowered in bottle fed infants by oral supplementation of *Bifidobacterium bifidum*.[131]

## B. ADULT PROFILE

The composition of the fecal microbiota has been the subject of many investigations and has been summarized in numerous reviews.[15,49,90] Most studies focus on enumeration of the major groups of microbes with some studies characterizing to the genus level.[132,133] Few workers have been as thorough as Moore and Holdeman[16] and Finegold and co-workers,[17,49,85,134,135] who identified the species level wherever possible.

### 1. Fecal Populations

Bacteria constitute a major proportion (approximately 50%) of feces. Figure 1 shows a scanning electron micrograph of bacteria in gut contents. A summary of the fecal microbiota as reported by Finegold and his colleagues[17,49,134,135] is presented in Table 2 and discussed in detail below. The findings are consistent with those present by Moore and Holdeman[16] for 20 subjects. It is interesting to note that despite the fact

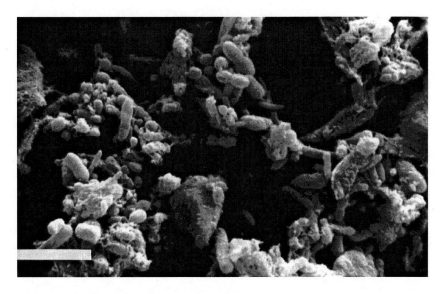

**Figure 1** Scanning electron micrograph of human feces showing various bacterial morphologies and the presence of particulate material. Bar marker = 5 μm.

**Table 2**  Fecal Microbiota in Various Dietary Groups Including Seventh-Day Adventists who Were Strictly Vegetarian, Japanese who Consumed an Oriental Diet that Included Fish but no Beef, and Healthy Subjects who Consumed a Western Diet with Relatively Large Quantities of Beef[49]

| Microorganisms | Strict Vegetarian (13) %[a] | Strict Vegetarian (13) Mean[b] | Japanese (15)[c] % | Japanese (15)[c] Mean | Western (62) % | Western (62) Mean | Total[d] (141) % | Total[d] (141) Mean |
|---|---|---|---|---|---|---|---|---|
| *Bacteroides* | 100 | 11.7 | 93 | 10.8 | 100 | 11.3 | 99 | 11.3 |
| *Fusobacterium* | 0 | — | 40 | 8.1 | 24 | 8.6 | 18 | 8.4 |
| Anaerobic streptococci | 8 | 11.4 | 60 | 9.5 | 32 | 10.5 | 34 | 10.3 |
| *Peptococcus* | 8 | 11.2 | 47 | 9.4 | 37 | 10.1 | 33 | 10.0 |
| *Peptostreptococcus* | 23 | 11.1 | 80 | 10.2 | 35 | 10.2 | 45 | 10.1 |
| *Ruminococcus* | 54 | 10.2 | 60 | 10.3 | 45 | 10.0 | 45 | 10.2 |
| Anaerobic cocci | 85 | 10.3 | 100 | 10.7 | 98 | 10.6 | 94 | 10.7 |
| *Actinomyces* | 31 | 10.5 | 0 | — | 2 | 5.7 | 7.8 | 9.2 |
| *Arachnia-propionibacterium* | 38 | 10.0 | 0 | — | 2 | 5.5 | 9.2 | 8.9 |
| *Bifidobacterium* | 69 | 10.9 | 80 | 9.7 | 79 | 10.4 | 74 | 10.2 |
| *Eubacterium* | 92 | 11.0 | 93 | 10.6 | 95 | 10.6 | 94 | 10.7 |
| *Lactobacillus* | 85 | 11.1 | 73 | 9.0 | 73 | 9.3 | 78 | 9.6 |
| *Clostridium* | 92 | 9.4 | 100 | 9.7 | 100 | 10.2 | 100 | 9.8 |
| *Streptococcus* | 100 | 8.6 | 100 | 8.7 | 100 | 9.1 | 99 | 8.9 |
| Gram-negative facultatives | 100 | 8.2 | 100 | 9.2 | 98 | 8.9 | 98 | 8.7 |
| *Candida albicans* | 15 | 4.9 | 47 | 5.6 | 14 | 5.4 | 14.2 | 5.4 |
| Other yeasts | 23 | 5.6 | 53 | 5.8 | 31 | 5.2 | 36.2 | 5.6 |
| Filamentous fungi | 0 | — | 0 | — | 3 | 3.8 | 3.5 | 5.9 |
| *Bacillus* spp. | 69 | 4.2 | 80 | 6.2 | 82 | 5.0 | 82.3 | 5.2 |
| TOTAL[e] | 100 | 12.6 | 100 | 11.8 | 100 | 12.2 | 100 | 12.2 |

a  % Positive. [b]Mean count expressed as organisms log10/g dry weight feces. [c]Number of subjects per dietary group. [d]Total for all 141 subjects including colonic polyp, colonic cancer, and vegetarians who consume some meat. [e]Total of all microbes detected (including other genera and groups not listed above).

that the groups studied by Finegold and co-workers varied considerably in diet and health, numbers of the major bacterial groups in all 141 subjects were remarkably similar. Bacteroides were found in highest numbers in all fecal samples, with *B. thetaiotaomicron* being the most common species. High counts of *B. vulgatus*, *B. distasonis*, *B. fragilis*, and *B. ovatus* were also noted, which is consistent with other published studies.[136]

The most frequently isolated anaerobic cocci were peptostreptococci, ruminococci, viellonella, and anaerobic streptococci, with *Peptostreptococcus productus* being the most commonly isolated species (log 9.7 per gram in 29% of subjects). Of the anaerobic Gram-positive nonsporing rods, eubacteria were found in highest numbers (log 10.7 per gram of feces in 94% of patients), with the most commonly isolated species being *E. aerofaciens*, *E. contortum*, *E. cylindroides*, *E. lentum*, and *E. rectale*. Lactobacilli were isolated from 78% of subjects, at a level of log 9.6 per gram, with *L. acidophilus* being the most common species. Frequently isolated bifidobacteria include *B. adolescentis*, *B. infantis*, and *B. longum*. *Clostridium* species were found in 100% of subjects at a level of log 9.8 per gram, with *Clostridium ramosum* being the most common (53% of subjects) with a mean count of log 9.1 per gram. *Clostridium perfringens* varied between dietary groups, being high in Japanese subjects (log 7.6 in 73% of subjects) and absent in strict vegetarians. Other studies have reported *C. perfringens* to be a common member.[137,138]

Of the facultative anaerobes, enterococci were present in almost all patients (99.3% of subjects) in relatively high numbers (log 8.9 per gram), with *Ent. faecalis* predominating. Consistent with another study,[132] coliforms were the most common of the Gram-negative facultative anaerobes, with *E. coli* being ubiquitous and present in 92.9% of subjects at a mean count of log 8.6 per gram. *Candida albicans* was present in all dietary groups but most common in those on a Japanese diet (47% of subjects, mean count log 5.6) and only in 14% of those consuming a western diet.

## 2. Luminal Populations

Most representatives of the fecal populations were found in the luminal contents.[90] Initially, it was proposed that there were 100-fold less colonic microbes than in feces.[132] Subsequently, Bentley et al.[139]

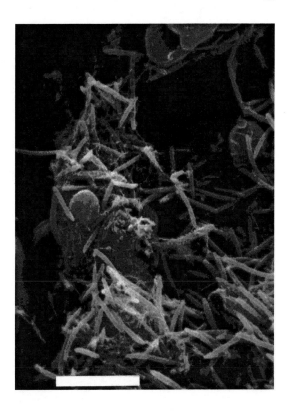

**Figure 2** Scanning electron micrograph of microbiota associated with the colonic epithelial mucosa of the rodent. The microorganisms are located in the mucus overlying the epithelial cells and, due to sample preparation, this mucus dries and appears as strands between bacterial cells. Bar marker = 5 μm.

compared microbes in the transverse colon and cecum with stool samples using improved culture techniques. The stool samples contained higher numbers of bacteria with log 9.6 anaerobes and log 7.5 coliforms per gram. The transverse colon averaged 2–3 log values lower than the fecal samples. Despite these quantitative differences, the major groups of bacteria were similar for all samples. This is also supported by the findings in rodents[53] that no major differences existed between cecal and fecal counts for the major bacterial groups, i.e., enterobacteria, bacteroides, fusiforms, and lactobacilli.

## 3. Mucosal-Associated Populations

In many animal systems, there is marked delineation between luminal and mucosal-associated populations.[140] Defined populations are detectable in the various microhabitats of the mucosal surface and living within the mucus layer overlying the epithelial cells. For example, in the colon of the rodent, fusiform-like microbes can be seen on the mucosal surface actually located within the mucus layer.[141] A typical population is illustrated in Figure 2 where such bacteria can be seen "trapped" within the mucus. They appear as strands because the mucus is dehydrated during preparation for scanning electron microscopy. There is no definite evidence that the bacteria actually associate with the epithelial surface in the colon,[101] as is demonstrated for other regions of the digestive tract in a range of hosts.[140,142] In fact, they may remain entirely within the mucus layer, perhaps by chemotaxis.[143] If the mucus blanket with associated microbes is removed, the bacteria-free epithelial surface is revealed, as shown in Figure 4. This mucus blanket was shown to be affected by antibiotics and diarrhea,[144] some dietary components,[53] and by nutrient deprivation.[145] These conditions resulted in a depleted mucus layer and showed that in the absence of the overlying bacterial populations, spiral shaped microbes came out of the mucosal crypts (Figure 3).

The human colonic mucosa has been less well studied due to sampling difficulties. It has been shown to be colonized by spirochetes,[146,147] mixed Gram-positive and negative rods and cocci[148] and extremely diverse populations.[62,149-152] While there is disagreement between these reports on the anaerobe to aerobe ratio of the populations, there is agreement for the total numbers per gram of colonic tissue (max log 8 per gram) and the fact that bacteroides are the most frequently isolated organism. Anaerobic microbes on the mucosa have been reported to reduce from $3.4 \times 10^7$ to $1.8 \times 10^2$ cfu/g by antibiotic dosage.[153]

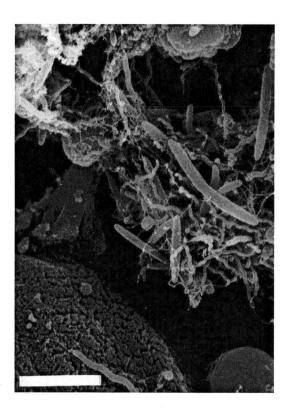

**Figure 3** Scanning electron micrograph of colonic epithelial mucosa of rodents after treatment to remove the overlying mucus. Spiral microorganisms, which normally remain in the crypts, can be detected at the surface. Bar marker = 5 μm.

## C. UNUSUAL COMPONENTS

Detailed taxonomic studies of the human colonic microbiota promise to reveal new species. For example, the work of Moore and Holdeman[16] resulted in the description of twelve new species of bacteria and one new genus, *Coprococcus*. Finegold and his co-workers[49] found 105 isolates of *Eubacterium* which could not be speciated and accounted for a significant percentage of all isolates of this genus. A similar difficulty exists with clostridia which frequently cannot be speciated, despite the use of accepted identification schemes. For example, 67 isolates could not be speciated in one study[49] while in another, 29% of clostridial isolates were unidentified.[138]

A number of interesting microbes are now being recovered from fecal samples. Methanogens have been shown to be the most predominant bacteria in some individuals, with levels reaching log 10 per gram dry weight of feces.[154,155] Spirochetes have been isolated from patients with rectal proctitis[156,157] and *Anaerobiospirillum* species from subjects with diarrhea,[158] although little is known of levels in healthy humans. Spirochetes have been detected on colonic mucosa but they have defied attempts at culture.[146,147]

## V. STABILITY OF THE ECOSYSTEM

### A. ECOLOGICAL PRINCIPLES GOVERNING STABILITY

As the year 2000 approaches, we are increasingly aware of the impact of stress on the environment. The gut ecosystem is similarly exposed to many factors which can affect stability of the unit. The subject has been addressed by examining factors which contribute to the control of colonic microbiota. These factors are reviewed by Drasar and Roberts,[159] and include studying microbial selection in continuous culture[160] and mechanisms that control the microflora in the large intestine[101] as well as trying to understand control and consequences of fermentation in the colon.[20]

The complexity and diversity of the colonic ecosystem is a major obstacle to understanding control mechanisms within it. At any given time, each species is affected by a large number of factors, which if studied in isolation, do not lead to an understanding of the whole system. While this is a daunting task for the researcher, it is this diversity which protects the stability of the ecosystem. As proposed by Alexander,[9] "it is likely that the tendency to maintain a balance among the resident organisms is enforced with increasing species diversity."

## B. DISTURBANCES

The colon of healthy adults is a very stable ecosystem in which invading or allochthonous microbes have great difficulty surviving. Radical changes in diet seem to have little influence on the composition of the microbiota,[91] which is rather surprising when one considers the incidence of traveler's diarrhea. Susceptibility to enteric pathogens is often increased in travelers who would inherently experience a dietary change at the same time as they are exposed to altered pathogen levels and types. It is tempting to propose that susceptibility is enhanced due to the inevitable dietary change as well as increased exposure to pathogens. Tannock[91] proposes that "any external influence that adversely affects an animal is likely to change the gastrointestinal ecosystem through changes produced in host physiology in response to that condition. Any dietary or environmental condition that adversely affects the optimal functioning of the host can thus influence the gastrointestinal ecosystem in a direct or indirect manner."

One agent which can dramatically disturb the ecosystem is chemotherapy. In fact, the importance of gut microorganisms in the healthy individual was not fully realized until Smith[161] showed that pertubations of the colonic microbiota resulting from antibiotic dosage could induce diarrhea and enteritis. Antibiotic-induced suppression of the microbiota predisposes the host to infection by enteric pathogens. A classic case study[162] exemplifies this fact. Three children with cystic fibrosis were dosed with a broad-spectrum antibiotic as a prophylaxis against chest infection, but at the same time received another pharmaceutical agent which unfortunately contained 44 salmonellae. While the infective dose in the undisturbed gut is recognized to be at least $10^5$, these antibiotic-treated children were infected with only 44 cells.

The ecological impact of antibiotics on the human colonic microbiota has been investigated by Nord and co-workers who presented disturbances in the profile of the microbiota as a result of antibiotic dosage.[163,164] Ingestion of antimicrobial agents has been shown to influence the ecosystem not only in terms of the composition of the microbiota, but also the antibiotic resistance profile of indigenous microbes and physiological and biochemical characteristics of the ecosystem.[165] In this study of quinolones, it was shown that in contrast to all the other drugs tested, ofloxacin exerted almost no influence on the characteristics investigated.

From animal studies, it has been conclusively shown that nutrient deprivation can cause alterations in the microbiota, as reviewed by Tannock.[91] Alterations have not only been noted in luminal or fecal microbes, but also in the surface-associated populations which are depleted and allow visualization (Figure 3) of spiral microbes emerging from deep within the crypts.[145] Studies of disturbances of surface-associated populations are sparse. However, in one report the luminal microbiota of rodents was unaffected while the surface-associated populations were grossly affected by a dietary component.[53]

## C. MANIPULATION

The concept of manipulating the colonic ecosystem dates back to 1907 when Metchnikoff proposed that one should avoid rich meats and alcohol to avoid intestinal putrefaction and proposed that ingestion of fermented milk could lead to a prolongation of life. Since then, the concept of manipulation of the colonic microbiota has been explored for a range of hosts including humans and several animal species. The field has expanded because of the need to control the microbiota in sensitive hosts such as the developing neonate, as well as adults subjected to disturbances of the ecosystem, as can occur during antibiotic therapy. The approach has been to orally administer preparations of living microorganisms, most frequently lactobacilli or bifidobacteria (see Chapter 11). Over the years, there has been considerable skepticism regarding this concept, largely due to contradictory or nonconclusive studies.[166] A more scientific approach is now being used with close attention to strain selection,[167] to understanding the mechanisms involved,[168] and to the factors affecting the microecology of the gut.[34]

Manipulation of the gut ecosystem by oral dosage of microbial preparations seems feasible and has shown promise in treating a number of intestinal disturbances including *C. difficile* infections.[169] Johansson et al.[62] report successfully increasing numbers of mucosal-associated lactobacilli when sampling the rectum 11 d after cessation of the oral administration of an oatmeal soup containing these bacteria. In this study, the increase in lactobacilli was accompanied by a significant decrease in Gram-negative anaerobes and Enterobacteriacea decreased 1000-fold in the rectal biopsies. In addition to using oral administration, some workers have used enemas of fecal microbiota for curing relapsing *C. difficile* infections.[170-172]

## VI. SUMMARY

The difficulties of studying the human colonic microbiota have been discussed and the present level of knowledge of the microbiota established. The acquisition and development of the major bacterial groups is well documented and detailed compositional tables are available. The present level of knowledge is limited by the types of sample available, inadequacies with selective media, difficulties with identification, and information available in the data base on gene sequences. Interactions between microbes need to be discussed in terms of the ecosystem rather than in isolation. Furthermore, the microbiota is not an isolated component but rather an active constituent of the colonic ecosystem, and results need to be examined by applying ecological principles. The system can be studied in its entirety, and can be investigated under controlled conditions, either by using model systems or human subjects. Such studies are feasible but technically demanding. An improved understanding of the interactions between components in the colonic ecosystem will allow one to better predict how the system will be affected by various external stresses and how it may be manipulated to the benefit of the host. Ultimately, one should be able to identify specific processes in the colon which can be attributable to particular species or groups of microorganisms. This will require complimentary *in vitro* and *in vivo* studies in which the functional moiety is identified *in vitro* and then demonstrated to be effective *in vivo*. It is an understatement to say that although there is a considerable amount of information already available on the colonic microbiota, we have as yet merely scratched the tip of an ecological iceberg.

## REFERENCES

1. **Atlas, R. M. and Bartha, R.,** *Microbial Ecology: Fundamentals and Applications,* 3rd ed., Benjamin Cummings Publishing Company Inc., Redwood City, CA, 1993.
2. **Lynch, J. M. and Hobbie, J. E.,** Introduction, in *Micro-organisms in Action: Concepts and Applications in Microbial Ecology,* Lynch, J. M. and Hobbie, J. E., Eds., Blackwell Scientific Publications, Oxford, 1988, 1.
3. **Stark, P. L. and Lee, A.,** The bacterial colonization of the large bowel of pre-term low birth weight neo-nates, *J. Hyg. Camb.,* 89, 59, 1982.
4. **Stark, P. L. and Lee, A.,** The microbial ecology of the large bowel of breast-fed and formulae-fed infants during the first year of life, *J. Med. Microbiol.,* 15, 189, 1982.
5. **Stark, P. L., Lee, A., and Parsonage, B. D.,** Colonization of the large bowel by *Clostridium difficile* in healthy infants: quantitative study, *Infect. Immun.,* 35, 895, 1982.
6. **Gustafsson, B. E.,** The germ-free animal: its potential and its problems, in *The Germfree Animal in Medical Research,* Coates, M. E. and Gustafsson, B. E., Eds., Laboratory Animals, Ltd., London, 1984, 1.
7. **Dubos, R., Schaedler, R. W., Costello, R., and Hoet, P.,** Indigenous, normal and autochthonous flora of the gastrointestinal tract, *J. Exp. Med.,* 122, 67, 1965.
8. **Rowland, I. R. and Wise, A.,** The effect of diet on the mammalian gut flora and its metabolic activities, *CRC Crit. Rev. Toxicol.,* 16, 31, 1985.
9. **Alexander, M.,** *Microbial Ecology,* John Wiley & Sons, New York, 1971.
10. **Savage, D. C.,** Interactions between the host and its microbes, in *Microbial Ecology of the Gut,* Clarke, R. T. J. and Bauchop, T., Eds., Academic Press, London, 1977, 277.
11. **Schaedler, R. W., Dubos, R., and Costello, R.,** The development of the bacterial flora in the gastrointestinal tract of mice, *J. Exp. Med.,* 122, 59, 1965.
12. **Savage, D. C.,** Microbial ecology of the gastrointestinal tract, *Ann. Rev. Microbiol.,* 31, 107, 1977.
13. **Lee, A.,** Neglected niches: The microbial ecology of the gastrointestinal tract, in *Advances in Microbial Ecology,* Vol. 8, Marshall, K. C., Ed., Plenum Press, New York, 1985, 115.
14. **Macfarlane, G. T. and Cummings, J. H.,** The colonic flora, fermentation and large bowel digestive function, in *The Large Intestine: Physiology, Pathophysiology and Disease,* Phillips, S. F., Pemberton, J. H., and Shorter, R. G., Eds., Raven Press, New York, 1991, 51.
15. **Cummings, J. H., Banwell, J. G., Englyst, H. N., Coleman, N., Segal, I., and Bersohn, D.,** The amount and composition of large bowel contents, *Gastroenterology,* 98, A408, 1990.
16. **Moore, W. E. C. and Holdeman, L. V.,** Human fecal flora: the normal flora of 20 Japanese-Hawaiians, *Appl. Microbiol.,* 27, 961, 1974.
17. **Finegold, S. M., Flora, D. J., Attebury, H. R., and Sutter, L. V.,** Fecal bacteriology of colonic polyp patients and control patients, *Cancer Res.,* 35, 3407, 1975.
18. **Reddy, B. S., Weisburger, J. H., and Wynder, E. L.,** Effect of high risk and low risk diets for colon carcinogens on fecal microflora and steroids in man, *J. Nutr.,* 105, 878, 1975.
19. **Simon, G. L. and Gorbach, S. L.,** Bacteriology of the colon, in *The Colon: Structure and Function,* Bustos-Fernandez, L., Ed., Plenum Medical Book Company, London, 1982, 103.

20. **Cummings, J. H. and Macfarlane, G. T.,** A review: the control and consequences of bacterial fermentation in the human colon, *J. Appl. Bacteriol.*, 70, 443, 1991.
21. **Midtvedt, T.,** Monitoring of the functional state of the microflora, in *Recent Advances in Microbial Ecology*, Hattori, T., Ishida, Y., Maruyama, Y., Morita, R., and Uchida, A., Eds., Japan Scientific Societies Press, Tokyo, 1989, 515.
22. **Miller, R. S. and Hoskins, L. C.,** Mucin degradation in human colon ecosystems. Fecal population densities of mucin-degrading bacteria estimated by a "most probable number" method, *Gastroenterology*, 81, 759, 1981.
23. **Crowther, J. S.,** Transport and storage of faeces for bacteriological examination, *J. Appl. Bacteriol.*, 34, 477, 1971.
24. **Rumney, C. J. and Rowland, I. R.,** *In vivo* and *in vitro* models of human colonic flora, *Crit. Rev. Food Sci. Nutr.*, 31, 299, 1992.
25. **Adams, R. F.,** Some effects of nutrients, non-nutrients and food poisoning organisms on the intestinal microbial system, *Food Technol. Aust.*, 30, 92, 1980.
26. **Miller, T. L. and Wolin, M. J.,** Fermentation by human large intestinal microbial community in an *in vitro* semicontinuous culture system, *Appl. Environ. Microbiol.*, 42, 400, 1981.
27. **Macfarlane, G. T., Hay, S., and Gibson, G. R.,** Influence of mucin on glycosidase, protease and arylamidase activities of human gut bacteria grown in a 3-stage continuous culture system, *J. Appl. Bacteriol.*, 66, 407, 1989.
28. **Hentges, D. J. and Freter, R.,** In vivo and in vitro antagonism of intestinal bacteria against *Shigella flexneri*. I. Correlation between different tests, *J. Infect. Dis.*, 110, 30, 1962.
29. **Freter, R., Stauffer, E., Cleven, D., Holdeman, L. V., and Moore, W. E. C.,** Continuous flow cultures as in vitro models of the ecology of large intestinal flora, *Infect. Immun.*, 39, 666, 1983.
30. **Veilleux, B. G. and Rowland, I.,** Simulation of the rat intestinal ecosystem using a two-stage continuous culture system, *J. Gen. Microbiol.*, 123, 103, 1981.
31. **Edwards, C. A., Duerden, B. I., and Read, N. W.,** Metabolism of mixed human colonic bacteria in a continuous culture mimicking the human cecal contents, *Gastroenterology*, 88, 1903, 1985.
32. **Wilson, K. H. and Freter, R.,** Interactions of *Clostridium difficile* and *E. coli* with microfloras in continuous flow cultures and gnotobiotic mice, *Infect. Immun.*, 54, 354, 1986.
33. **Bernhardt, H., Knoke, M., and Bootz, T.,** Stimulation of the intestinal microflora in a continuous flow culture, in *The Regulatory and Protective Role of the Normal Microflora*, Grubbe, R., Midtvedt, T., and Norin, E., Eds., Macmillan Press, Basingstoke, 1989, 345.
34. **Freter, R.,** Factors affecting the microecology of the gut, in *Probiotics: The Scientific Basis*, Fuller, R., Ed., Chapman and Hall, London, 1992, 111.
35. **Freter, R., Brickner, H., and Botney, M.,** Mechanisms that control bacterial populations in continuous-flow culture models of mouse large intestinal flora, *Infect. Immun.*, 39, 676, 1983.
36. **Freter, R., Brickner, H., Fekete, J., Vickerman, M. M., and Carey, K. E.,** Survival and implantation of *Escherichia coli* in the intestinal tract, *Infect. Immun.*, 39, 686, 1983.
37. **Klemm, P.,** Fimbrial adhesins of *Escherichia coli*, *Rev. Infect. Dis.*, 7, 321, 1985.
38. **Conway, P. L., Welin, A., and Cohen, P. S.,** Presence of a K88-specific receptor in porcine ileal mucus is age dependant, *Infect. Immun.*, 58, 3178, 1990.
39. **Fujiwara, S., Hirota, T., Nakazato, H., Muzutani, T., and Mitsuoka, T.,** Effect of Konjac mannan on intestinal microbial metabolism in mice bearing human flora and in conventional F344 rats, *Food Chem. Toxicol.*, 29, 601, 1991.
40. **Shedlofsky, S. and Freter, R.,** Synergism between ecologic and immunologic control of mechanisms of intestinal flora, *J. Infect. Dis.*, 129, 296, 1974.
41. **Rowland, I. R., Mallett, A. K., Bearne, C. A., and Farthing, M. J. G.,** Enzyme activities of the hindgut microflora of laboratory animals and man, *Xenobiotica*, 16, 519, 1986.
42. **Moughan, P. J., Birtles, M. J., Cranwell, P. D., Smith, W. C., and Pedraza, M.,** The piglet as a model animal for studying aspects of digestion and absorption in milk-fed human infants, in *Nutritional Triggers in Health and in Disease*, Simopoulos, A. P., Ed., *World Rev. Nutr. Diet*, Vol. 67, Karger, Basel, 1992, 40.
43. **Conway, P. L., Rang, C., Kennan, R., and Midtvedt, T.,** The frequency of transfer of RP1 plasmid *in vivo* and in intestinal extracts *in vitro*, unpublished data, 1994.
44. **Borriello, S. P.,** Clostridial flora of the gastrointestinal tract in health and disease, Ph.D. thesis, University of London, 1981.
45. **Itoh, K. and Freter, R.,** Control of *Escherichia coli* populations by a combination of indigenous clostridia and lactobacilli in gnotobiotic mice and continuous flow cultures, *Infect. Immun.*, 57, 559, 1989.
46. **Drasar, B. S. and Hill, M. J.,** *Human Intestinal Flora*, Academic Press, London, 1974.
47. **Rosenblatt, J. E., Fallon, A., and Finegold, S. M.,** Comparison of methods for the isolation of anaerobic bacteria from clinical specimens, *Appl. Microbiol.*, 25, 77, 1973.
48. **van Winkelhoff, A. J. and De Graff, J.,** Vancomycin as a selective agent for isolation of *Bacteroides* species, *J. Clin. Microbiol.*, 18, 1282, 1983.
49. **Finegold, S. M., Sutter, V. L., and Mathisen, G. E.,** Normal indigenous intestinal flora, in *Human Intestinal Microflora in Health and Disease*, Hentges, D. J., Ed., Academic Press, London, 1983, 3.
50. **Drasar, B. S. and Borrow P. A.,** *Intestinal Microbiology*, American Society for Microbiology, Washington, D.C., 1985.

51. **Holdeman, L. V., Good, I. J., and Moore, W. E. C.,** Human fecal flora: variation in bacterial composition within individuals and a possible effect of emotional stress, *Appl. Environ. Microbiol.*, 31, 359, 1976.

52. **Arnon, S. S.,** The clinical spectrum of infant botulism, *Rev. Infect. Dis.*, 1, 614, 1979.

53. **Adams, R. F. and Conway, P. L.,** The effect of erythrosine on the surface-associated bacteria of the rat stomach and caecum, *J. Appl. Bacteriol.*, 51, 171, 1981.

54. **Tannock, G. W.,** Molecular genetics: a new tool for investigating the microbial ecology of the gastrointestinal tract?, *Microbiol. Ecol.*, 15, 239, 1988.

55. **Stahl, D. A. and Amann, R. I.,** Development and application of nucleic acid probes, in *Nucleic Acid Techniques in Bacterial Systematics*, Stackebrandt, E. and Goodfellow, M., Eds., John Wiley & Sons, New York, 1991, 205.

56. **Olsen, G. J., Maidak, B. L., McCaughey, M. J., Overbeek, R., Macke, T. J., Marsh, T. L., and Woese, C. R.,** The ribosomal databases project, *Nucleic Acid Res.*, 21, 3021, 1993.

57. **DeLong, E. F., Wickham, G. S., and Pace, N. R.,** Phylogenetic strains: ribosomal RNA-based probes for identification of single cells, *Science*, 243, 1360, 1989.

58. **Amann, R. I., Krumholz, L., and Stahl, D. A.,** Fluorescent-oligonucleotide probing of whole cells for determinative, phylogenetic, and environmental studies in microbiology, *J. Bacteriol.*, 172, 762, 1990.

59. **Neidhardt, F. C., Ingraham, J. L., Low, K. B., Magasanik, B., Schaechter, M., and Umbarger, H. E., Eds.,** *Escherichia coli and Salmonella typhimurium: Cellular and Molecular Biology*, American Society for Microbiology, Washington, D.C., 1987.

60. **Poulsen, L. K., Ballard, G., and Stahl, D. A.,** Use of rRNA fluorescence in situ hybridization for measuring the activity of single cells in young and established biofilms, *Appl. Environ. Microbiol.*, 59, 1354, 1993.

61. **Eudy, W. W. and Burrous, S. E.,** Generation times of *Proteus mirabilis* and *Escherichia coli* in experimental infections, *Chemotherapy*, 19, 161, 1973.

62. **Johansson, M.-L., Molin, G., Jeppsson, B., Nobaek, S., Ahrné, S., and Bergmark, S.,** Administration of different *Lactobacillus* strains in fermented oatmeal soup: in vivo colonization of human intestinal mucosa and effect on the indigenous flora, *Appl. Environ. Microbiol.*, 59, 15, 1993.

63. **Henriksson, A., André, L., and Conway, P. L.,** Distribution of *Lactobacillus* spp. in the porcine gastrointestinal tract, *FEMS Microbiol. Ecol.*, 1994, (in press).

64. **Imberechts, H., De Greve, H., and Schlicker, C.,** Characterization of F107 fimbriae of *Escherichia coli* 107/86, which causes edema disease in pigs, and nucleotide sequence of the F107 major subunit gene, fedA, *Infect. Immun.*, 60, 1963, 1992.

65. **Cummings, J. H.,** Diet and short-chain fatty acids in the gut, in *Food and the Gut*, Hunter, J. O. and Jones, A., Eds., Balliere Tindall, London, 1985, 78.

66. **Midtvedt, T., Carlstedt-Duke, B., Haverstad, T., Midtvedt, A.-C., Norin, K. E., and Saxerholt, H.,** Establishment of a biochemically active intestinal ecosystem in ex-germfree rats, *Appl. Environ. Microbiol.*, 53, 2866, 1987.

67. **Rowland, I. R.,** Metabolic profiles of intestinal floras, in *Recent Advances in Microbial Ecology*, Hattori, T., Ishida, Y., Maruyama, Y., Morita, R. Y., and Uchida, A., Eds., Japan Scientific Societies Press, Tokyo, 1989, 510.

68. **Midtvedt, T., Bjørneklett, A., Carlstedt-Duke, B., Gustafsson, B. E., Hoverstad, T., Lingaas, E., Norin, K. E., Saxerholt, H., and Steinbakk, M.,** The influence of antibiotics upon microflora-associated characteristics in man and mammals, in *Germfree Research: Microflora Control and its Application to the Biomedical Sciences*, Wostmann, B. E., Ed., Alan R. Liss Inc., New York, 1985, 241.

69. **Norin, K. E., Gustafsson, B. E., Lindblad, B. S., and Midtvedt, T.,** The establishment of some microflora associated biochemical characteristics in feces from children during the first years of life, *Acta Paediatr. Scand.*, 74, 207, 1985.

70. **Midtvedt, A.-C., Carlstedt-Duke, B., Norin, K. E., Saxerholt, H., and Midtvedt, T.,** Development of five metabolic activities associated with the intestinal microflora of healthy infants, *J. Pediatr. Gastroenterol. Nutr.*, 7, 559, 1988.

71. **Tannock, G. W., Crichton, C., Welling, G. W., Koopman, J. P., and Midtvedt, T.,** Reconstitution of the gastrointestinal microflora of *Lactobacillus*-free mice, *Appl. Environ. Microbiol.*, 54, 2971, 1988.

72. **Allen, A.,** Structure and function of gastrointestinal mucus, in *Physiology of the Gastrointestinal Tract*, Johnson, L. R., Ed., Raven Press, New York, 1981, 617.

73. **Neutra, M. R., Phillips, T. L., and Phillips, T. E.,** Regulation of intestinal goblet cells in situ, in mucosal explants and in the isolated epithelium, in *Mucus and Mucosa*, Ciba Foundation Symposium 109, Nugent, J., Ed., Pittman, London, 1984, 20.

74. **Vercellotti, J. R., Salyers, A. A., Bullard, W. S., and Wilkins, T. D.,** Breakdown of mucins and plant polysaccharides in the human colon, *Can. J. Biochem.*, 55, 1190, 1977.

75. **Kim, Y. S., Morita, A., Miura, S., and Siddiqui, B.,** Structure of glycoconjugates of intestinal mucosal membranes. Role of bacterial adherence, in *Attachment of Organisms to the Gut Mucosa*, Boedecker, E. C., Ed., CRC Press, Boca Raton, FL, 1984, 99.

76. **Roediger, W. E. W.,** Interrelationship between bacteria and mucosa of the gastrointestinal tract, in *Microbial Metabolism in the Digestive Tract*, Hill, M. J., Ed., CRC Press, Boca Raton, FL, 1986, 201.

77. **Clamp, J. R. and Creeth, J. M.,** Some non-mucin components of mucus and their possible biological roles, in *Mucus and Mucosa*, Nugent, J. and O'Connor, M., Eds., Pitman Publishing Ltd., London, 1984, 121.

78. **Bienenstock, J. and Befus, A. D.,** Mucosal immunology, *Immunology*, 41, 249, 1980.

79. **Fubara, E. S. and Freter, R.,** Protection against enteric bacterial infections by secretory IgA antibodies, *J. Immunol.,* 111, 395, 1973.

80. **Edebo, L., Hed, J., Kihlström, E., Magnusson, K. E., and Stendahl, O.,** The adhesion of enterobacteria and the effect of antibodies of different immunoglobulin classes, *Scand. J. Infect. Dis. (Suppl.),* 24, 93, 1980.

81. **Stephens, S., Dolby, J. M., Montreuil, J., and Sprik, G.,** Differences in inhibition of the growth of commensal and enteropathogenic strains of *Escherichia coli* by lactoferrin and secretory immunoglobulin A isolated from human milk, *Immunology,* 41, 597, 1980.

82. **Arai, K., Lee, F., Miyajima, A., Miyatake, S., Arani, N., and Yokota, T.,** Cytokines: coordinators of immune and inflammatory responses, *Ann. Rev. Biochem.,* 59, 783, 1990.

83. **Macfarlane, G.T., Cummings, J. H., Macfarlane, S., and Gibson, G. R.,** Influence of retention time on degradation of pancreatic enzymes by human colonic bacteria grown in a 3-stage continuous culture system, *J. Appl. Bacteriol.,* 67, 521, 1989.

84. **Drasar, B. S. and Jenkins, D. J. A.,** Bacteria, diet and large bowel cancer, *Am. J. Clin. Nutr.,* 29, 1410, 1976.

85. **Finegold, S. M. and Sutter, V. L.,** Fecal flora in different populations, with special reference to diet, *Am. J. Clin. Nutr.,* 31, S116, 1978.

86. **Winitz, M., Adams, R. F., Seedman, D. A., Davis, P. N., Jakyo, L. G., and Hamilton, J. A.,** Studies in metabolic nutrition employing chemically defined diets. II. Effects on gut microflora populations, *Am. J. Clin. Nutr.,* 23, 546, 1970.

87. **Crowther, J. S., Drasar, B. S., Goddard, P., Hill, M. J., and Johnson, K.,** The effect of chemically defined diet on the faecal flora and faecal steroid concentration, *Gut,* 14, 790, 1973.

88. **Walker, A. R. P.,** Colon cancer and diet, with special reference to intakes of fat and fiber, *Am. J. Clin. Nutr.,* 29, 1417, 1976.

89. **Simon, G. L. and Gorbach, S. L.,** Intestinal flora in health and disease, in *Physiology of the Gastrointestinal Tract,* Johnson, L. R., Ed., Raven Press, New York, 1981, 1361.

90. **Borriello, S. P.,** Microbial flora of the gastrointestinal tract, in *Microbial Metabolism in the Digestive Tract,* Hill, M. J., Ed., CRC Press, Boca Raton, FL, 1986, 2.

91. **Tannock, G. W.,** Effect of dietary and environmental stress on the gastrointestinal microbiota, in *Human Intestinal Microflora in Health and Disease,* Hentges, D. J., Ed., Academic Press, London, 1983, 517.

92. **Bond, J. H., Currier, B. E., Buchwald, H., and Levitt, M. D.,** Colonic conservation of malabsorbed carbohydrates, *Gastroenterology,* 78, 444, 1980.

93. **Macfarlane, G. T., Gibson, G. R., and Cummings, J. H.,** Comparison of fermentation reactions in different regions of the colon, *J. Appl. Bacteriol.,* 72, 57, 1992.

94. **Gorbach, S. L., Nahas, L., Plaut, A. G., Weinstein, L., Patterson, J. F., and Levitan, R.,** Studies of intestinal microflora. V. Fecal microbial ecology in ulcerative colitis and regional enteritis: relationship to severity of disease and chemotherapy, *Gastroenterology,* 54, 575, 1968.

95. **Gorbach, S. L., Banwell, J. G., Jacobs, B., Chatterjee, B. D., Mitra, R., Brigham, K. L., and Neogy, K. N.,** Intestinal microflora in Asiatic cholera. I. "Rice-water" stool, *J. Infect. Dis.,* 121, 32, 1970.

96. **van der Merwe, J. P., Stegeman, J. H., and Hazenberg, M. P.,** The resident faecal flora is determined by genetic characteristics of the host. Implications for Crohn's disease, *Antonie van Leeuwenhoek,* 49, 119, 1983.

97. **Grove, D.,** Amoebic dysentery, intestinal protozoa and helminths, in *Microbes and Infections of the Gut,* Goodwin, C. S., Ed., Blackwell Scientific Publications, Oxford, 1984, 209.

98. **Hoskins, L. C. and Boulding, E. T.,** Mucin degradation in human colon ecosystems. Evidence for the existence and role of bacterial subpopulations producing glycosidases as extracellular enzymes, *J. Clin. Invest.,* 67, 163, 1981.

99. **Prins, R. A.,** Biochemical activities of gut micro-organisms, in *Microbial Ecology of the Gut,* Clarke, R. T. J. and Bauchop, T., Eds., Academic Press, London, 1977, 74.

100. **Hungate, R. E.,** Gut microbiology, in *Microbial Ecology,* Loutit, M. W. and Miles, J. A. R., Eds., Springer-Verlag, Berlin, 1978, 258.

101. **Freter, R.,** Mechanisms that control the microflora in the large intestine, in *Human Intestinal Microflora in Health and Disease,* Hentges, D. J., Ed., Academic Press, London, 1983, 33.

102. **Conway, P. L., Blomberg, L., Welin, A., and Cohen, P. S.,** The role of piglet intestinal mucus in pathogenicity of *Escherichia coli* K88, in *FEMS Symposium No. 58: Molecular Pathogenesis of Gastrointestinal Infections,* Wadström, T., Mäkelä, P. H., Svennerholm, A.-M., and Wolf-Watz, H., Eds., Plenum Press, New York, 1991, 335.

103. **Freter, R.,** Prospects for preventing the association of harmful bacteria with host mucosal surfaces, in *Bacterial Adherence,* Beachey, E. H., Eds., Chapman and Hall, London, 1980, 441.

104. **Roediger, W. E. W.,** Role of anaerobic bacteria in the metabolic welfare of the colonic mucosa in man, *Gut,* 21, 793, 1980.

105. **Abrams, G. D.,** Impact of intestinal microflora on intestinal structure and function, in *Human Intestinal Microflora in Health and Disease,* Hentges, D. J., Ed., Academic Press, London, 1983, 292.

106. **Tannock, G. W.,** Microbial interference in the gastrointestinal tract, *Asian J. Clin. Sci.,* 2, 2, 1981.

107. **Foo, M. C., Lee, A., and Cooper, G. N.,** Natural antibodies and the intestinal flora of rodents, *Aust. J. Exp. Biol. Med. Sci.,* 52, 321, 1974.

108. **Foo, M. and Lee, A.,** Immunological response of mice to members of the autochthonous intestinal flora, *Infect. Immun.,* 6, 525, 1972.

109. **Van Saene, H. K. F. and Van der Waaij, D.,** A novel technique for detecting IgA coated potentially pathogenic microorganisms in the human intestine, *J. Immunol. Meth.,* 30, 37, 1979.

110. **Van der Waaij, D.,** The immunoregulation of the intestinal flora: Experimental investigations on the development and the composition of the microflora in normal and thymus less mice, *Microecol. Ther.,* 14, 63, 1984.

111. **Berg, R. D.,** Host immune response to antigens of the indigenous intestinal flora, in *Human Intestinal Microflora in Health and Disease,* Hentges, D. J., Ed., Academic Press, London, 1983, 101.

112. **Wold, A. E., Dahlgren, U. I. H., and Hanson, L. A.,** Difference between bacterial and food antigens in mucosal immunogenicity, *Infect. Immun.,* 57, 2666, 1989.

113. **Kamin, R. A., Henry, C., and Fudenberg, H.,** Suppressor cells in the rabbit appendix, *J. Immunol.,* 113, 1151, 1974.

114. **Ament, M. E., Ochs, H. D., and Davis, S. D.,** Structure and function of the gastrointestinal tract in primary immunodeficiency syndromes. A study of 39 patients, *Medicine,* 52, 227, 1973.

115. **Severijnen, A. J., van Kleef, R., Hazenberg, M. P., and van de Merwe, J. P.,** Cell wall fragments from major residents of the human intestinal flora induce chronic arthritis in rats, *J. Rheumatol.,* 16, 1061, 1989.

116. **Severijnen, A. J., van Kleef, R., Grandia, A. A., van der Kwast, T. H., and Hazenberg, M. P.,** Histology of joint inflammation induced in rats by cell wall fragments of the anaerobic intestinal bacterium *Eubacterium aerofaciens, Rheumatol. Int.,* 11, 203, 1991.

117. **Kool, J., Ruseler-van Embden, J. G., van Lieshout, L. M., de Visser, H., Gerrits-Boeye, M. Y., van den Berg, W. B., and Hazenberg, M. P.,** Induction of arthritis in rats by soluble peptidoglycan-polysaccharide complexes produced by human intestinal flora, *Arthr. Rheumatol.,* 34, 1611, 1991.

118. **Hentges, D. J.,** Gut flora and disease resistance, in *Probiotics: The Scientific Basis,* Fuller, R., Ed., Chapman and Hall, London, 1992, 87.

119. **Hobbs, J. R.,** Primary immune paresis, in *Immunology and Development,* Clinics in Developmental Medicine No. 34, Adinolfi, M., Ed., Heinemann, London, 114, 1969.

120. **Murrell, W. G., Ouvrier, R. A., Stewart, B. J., and Dorman, D. C.,** Infant botulism in a breast-fed infant from rural New South Wales, *Med. J. Aust.,* 1, 583, 1981.

121. **Conway, P. L. and Adams, R. F.,** unpublished data, 1994.

122. **Wada, N., Nishida, N., and Iwaki, S.,** Neutralizing activity against *Clostridium difficile* toxin in the supernatants of cultured colostral cells, *Infect. Immun.,* 29, 545, 1980.

123. **Sonnabend, O. A., Sonnabend, W. F. F., Krech, U., Molz, G., and Sigrist, T.,** Continuous microbiological and pathological study of 70 sudden and unexpected infant deaths: toxigenic intestinal *Clostridium botulinum* infection in 9 cases of sudden infant death syndrome, *Lancet,* 1, 237, 1985.

124. **Wilkins, T. D., Krivan, H., Stiles, B., Carman, R., and Lyerly, D.,** Clostridial toxins active locally in the gastrointestinal tract, in *Microbial Toxins and Diarrhoeal Disease,* CIBA Foundation Symposium 112, Pittman, London, 1985, 230.

125. **Mims, L. C., Medawar, M. S., Perkins, J. R., and Grubb, W. R.,** Predicting neonatal infections by evaluation of the gastric aspirate: a study in two hundred and seven patients, *Am. J. Obstet. Gynecol.,* 114, 232, 1972.

126. **Bettelheim, K. A., Breardon, A., Faiers, M. C., and O'Farrell, S. M.,** The origin of O serotypes of *Escherichia coli* in babies after normal delivery, *J. Hyg.,* 72, 67, 1974.

127. **Mata, L. J. and Urrutia, J. J.,** Intestinal colonization of breast-fed children in a rural area of low socioeconomic level, *Ann. N.Y. Acad. Sci.,* 176, 93, 1971.

128. **Benno, Y., Sawada, K., and Mitsuoka, T.,** The intestinal microflora of infants: composition of fecal flora in breast-fed and bottle-fed infants, *Microbiol. Immunol.,* 28, 975, 1984.

129. **Bennet, R. and Nord, C. E.,** The intestinal microflora during the first weeks of life: Normal development and changes induced by caesarean section, pre-term birth and antimicrobial treatment, in *The Regulatory and Protective Role of the Normal Microflora,* Grubb, R., Midtvedt, T., and Norin, E., Eds., Macmillan Press, Basingstoke, England, 1989, 19.

130. **Bezirtzoglou, E. and Romond, C.,** Apparition of *Clostridium* sp. and *Bacteroides* in the intestine of the newborn delivered by cesarean section, *Comp. Immunol. Microbiol. Infect. Dis.,* 13, 217, 1990.

131. **Bezirtzoglou, E., Romond, M. B., and Romond, C.,** Regulation of the bacterial intestinal implantation in infants born by Caesarean section, *Comp. Immunol. Microbiol. Infect. Dis.,* 15, 71, 1992.

132. **Gorbach, S. L., Nahas, L., and Lerner, P. I.,** Studies of intestinal microflora. I. Effects of diet, age and periodic sampling on numbers of faecal microorganisms in man, *Gastroenterology,* 53, 845, 1967.

133. **Mata, L. J., Carrillo, C., and Villatoro, E.,** Fecal microflora in healthy persons in a preindustrial region, *Appl. Microbiol.,* 17, 596, 1969.

134. **Finegold, S. M., Attebery, H. R., and Sutter, V. L.,** Effect of diet on human fecal flora: comparison of Japanese and American diets, *Am. J. Clin. Nutr.,* 27, 1456, 1974.

135. **Finegold, S. M., Sutter, V. L., Sugihara, P. T., Elder, H. A., Lehmann, S. M., and Phillips, R. L.,** Fecal microbial flora in Seventh Day Adventist populations and control subjects, *Am. J. Clin. Nutr.,* 30, 1781, 1977.

136. **Duerden, B. I.,** The isolation and identification of *Bacteroides* spp. from normal human faecal flora, *J. Med. Microbiol.,* 13, 69, 1980.

137. **Akama, K. and Otani, S.,** *Clostridium perfringens* as the flora in the intestine of healthy persons, *Jpn. J. Med. Sci. Biol.*, 23, 161, 1970.
138. **Drasar, B. S., Goddard, P., Heaton, S., Peach, S., and West, B.,** Clostridia isolated from faeces, *J. Med. Microbiol.*, 9, 63, 1976.
139. **Bentley, D. W., Nichols, R. L., Condon, R. E., and Gorbach, S. L.,** The microflora of the human ileum and intraabdominal colon, *J. Lab. Clin. Med.*, 79, 421, 1972.
140. **Lee, A.,** Normal flora of animal intestinal surfaces, in *Adsorption of Microorganisms to Surfaces*, Bitton, G. and Marshall, M., Eds., John Wiley & Sons, New York, 1980, 145.
141. **Savage, D. C. and Blumershine, R. V. H.,** Surface-surface association in microbial communities populating epithelial habitats in the murine gastrointestinal ecosystem: scanning electron microscopy, *Infect. Immun.*, 10, 240, 1974.
142. **Savage, D. C.,** Adherence of the normal flora to mucosal surfaces, in *Bacterial Adherence*, Beachey, E. H., Ed., Chapman and Hall, London, 1980, 31.
143. **Jones, G. W. and Isaacson, R. E.,** Proteinaceous bacterial adhesins and their receptors, *CRC Crit. Rev. Microbiol.*, 10, 229, 1983.
144. **Leach, W. D., Lee, A., and Stubbs, R. P.,** Localization of bacteria in the gastrointestinal tract: a possible explanation of intestinal spirochaetosis, *Infect. Immun.*, 7, 961, 1973.
145. **Conway, P. L., Maki, J., Mitchell, R., and Kjelleberg, S.,** Starvation of marine flounder, squid and laboratory mice and its effect on the intestinal microbiota, *FEMS Microbiol. Ecol.*, 38, 187, 1986.
146. **Lee, F. D., Kraszewski, A., Gordon, J., Howie, J. G. R., McSeveney, R., and Harland, W. A.,** Intestinal spirochaetosis, *Gut*, 12, 126, 1971.
147. **Takeuchi, A., Jervis, H. R., Nakagawa, H., and Robinson, D. M.,** Spiral-shaped organisms on the surface of colonic epithelium of the monkey and man, *Am. J. Clin. Nutr.*, 27, 1287, 1974.
148. **Nelson, D. P. and Mata, L. J.,** Bacterial flora associated with the human gastrointestinal mucosa, *Gastroenterology*, 58, 56, 1970.
149. **Croucher, R., Houston, A. P., Bayliss, C. E., and Turner, R. J.,** Populations associated with different regions of the human colon wall, *Appl. Environ. Microbiol.*, 45, 1025, 1983.
150. **Peach, S., Lock, M. R., Katz, D., Todd, I. P., and Tabaqchali, S.,** Mucosal-associated bacterial flora of the intestine in patients with Crohn's disease and in a control group, *Gut*, 19, 1034, 1978.
151. **Hartley, C. L., Neumann, C. S., and Richmond, M. H.,** Adhesion of commensal bacteria to the large intestine wall in humans, *Infect. Immun.*, 23, 128, 1979.
152. **Edmiston, C. E., Jr., Avant, G. R., and Wilson, F. A.,** Anaerobic bacterial populations on normal and diseased human biopsy tissue, *Appl. Environ. Microbiol.*, 43, 1173, 1982.
153. **Smith, M. B., Goradia, V. K., Holmes, J. W., McCluggage, S. G., Smith, J. W., and Nichols, R. L.,** Suppression of the human mucosal-related colonic microflora with prophylactic parenteral and/or oral antibiotics, *World J. Surg.*, 14, 636, 1990.
154. **Miller, T. L., Wolin, M. J., de Macario, E. C., and Macario, A. J. L.,** Isolation of *Methanobrevibacter smithii* from human feces, *Appl. Environ. Microbiol.*, 43, 227, 1982.
155. **Miller, T. L. and Wolin, M. J.,** Methanogens in human and animal intestinal tract systems, *Appl. Environ. Microbiol.*, 7, 223, 1986.
156. **Tompkins, D. S., Waugh, M. A., and Cooke, E. M.,** Isolation of intestinal spirochaetes from homosexuals, *J. Clin. Pathol.*, 34, 1385, 1981.
157. **Hovind-Hougen, K., Birch-Andersen, A., Henrik-Nielson, R., Orholm, M., Pederson, J. O., Tegibjaerg, P. S., and Thaysen, E. H.,** Intestinal spirochetosis: morphological characterization and cultivation of the spirochete *Brachysira aalborgi* gen. nov., sp. nov., *J. Clin. Microbiol.*, 16, 1127, 1982.
158. **Malnick, H., Thomas, M. E. M., Lotay, H., and Robbins, M.,** Anaerobiospirillum species isolated from humans with diarrhoea, *J. Clin. Pathol.*, 36, 1097, 1983.
159. **Drasar, B. S. and Roberts, A. K.,** Control of the large bowel microflora., in *Human Microbial Ecology*, Hill, M. J. and Marsh, P. D., Eds., CRC Press, Boca Raton, FL, 1989, 87.
160. **Harder, W., Kuenen, J. G., and Martin, A.,** A review: microbial selection in continuous culture, *J. Appl. Bacteriol.*, 43, 1, 1977.
161. **Smith, D. T.,** The disturbance of the normal bacterial ecology by the administration of antibiotics with the development of new clinical syndromes, *Ann. Intern. Med.*, 37, 1135, 1952.
162. **Lipson, A.,** Infecting dose of *Salmonella*, *Lancet*, 1, 969, 1976.
163. **Nord, C. E.,** Studies on the ecological impact of antibiotics, *Eur. J. Clin. Microbiol.*, 9, 517, 1990.
164. **Brismar, B., Edlund, C., Malmborg, A. S., and Nord, C. E.,** Ecological impact of antimicrobial prophylaxis on intestinal microflora in patients undergoing colorectal surgery, *Scand. J. Infect. Dis. Suppl.*, 70, 25, 1990.
165. **Midtvedt, T.,** The influence of quinolones on the faecal flora, *Scand. J. Infect. Dis. Suppl.*, 68, 14, 1990.
166. **Conway, P. L.,** Lactobacilli: Fact and fiction, in *The Regulatory and Protective Role of the Normal Microflora*, Grubb, R., Midtvedt, T., and Norin, E., Eds., The Macmillan Press, Basingstoke, Great Britain, 1989, 263.
167. **Havenaar, R., Ten Brink, B., and Huis In't Veld, J. H. J.,** Selection of strains for probiotic use, in *Probiotics: The Scientific Basis*, Fuller, R., Ed., Chapman and Hall, London, 1992, 209.

168. **Kitazawa, H., Matsumura, K., Itoh, T., and Yamaguchi, T.,** Interferon induction in murine peritoneal macrophages by stimulation with *Lactobacillus acidophilus, Microbiol. Immun.,* 36, 311, 1992.

169. **Goldin, B. R. and Gorbach, S. L.,** Probiotics for humans, in *Probiotics: The Scientific Basis,* Fuller, R., Ed., Chapman and Hall, London, 1992, 355.

170. **Bowden, T. A., Mansberger, A. R. J., and Lykins, L. E.,** Pseudomembranous enterocolitis: mechanisms of restoring floral homeostasis, *Am. J. Surg.,* 47, 178, 1981.

171. **Schwan, A., Sjolin, S., Trottestam, U., and Aronsson, B.,** Relapsing *Clostridium difficile* enterocolitis cured by rectal infusion of normal faeces, *Scand. J. Infect. Dis.,* 16, 211, 1984.

172. **Tvede, M. and Rask-Madsen, J.,** Bacteriotherapy for chronic relapsing *Clostridium difficile* diarrhoea in six patients, *Lancet,* I, 1156, 1989.

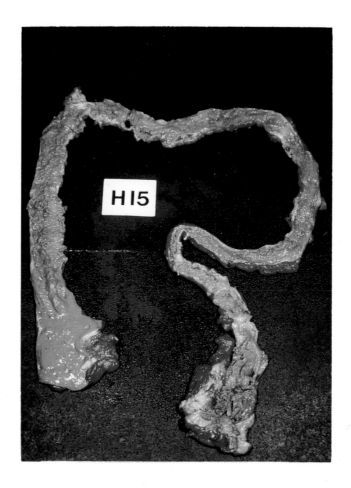

**Plate 1**   Human large intestine from an adult male subject removed at autopsy. Gut contents in the proximal colon (i.e., cecum, ascending colon) have a liquid consistency, whereas those in distal regions (i.e., descending colon, sigmoid rectum) are more solid. This is due to the absorptive nature of the colon. (Photograph courtesy of J. H. Cummings.)

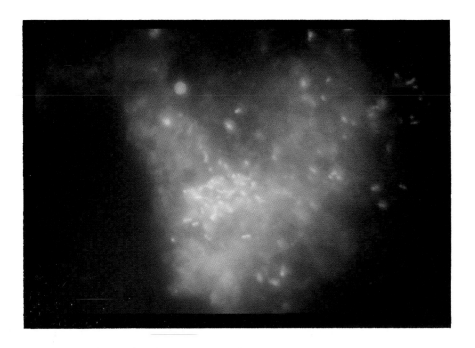

**Plate 2** Fluorescent appearance of methanogenic bacteria in a fecal slurry (10%w/v) after microscopic examination under ultraviolet light at 420 nm.

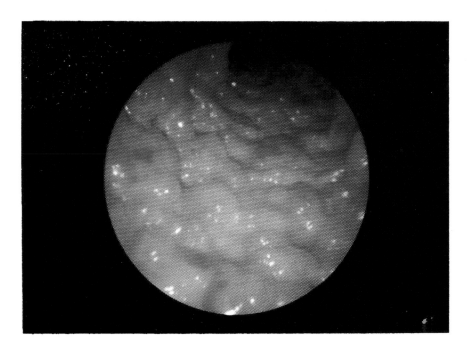

**Plate 3** Gas-filled cysts typical of pneumatosis cystoides intestinalis. Endoscopic view of the descending colon of a 56-year-old male patient.

Chapter 2

# Molecular Genetics of Intestinal Anaerobes

*Jeremy M. Wells\* and Clive Allison*

## CONTENTS

I. Introduction ........................................................................................................................ 25
II. Genetic Diversity of Intestinal Anaerobes ...................................................................... 26
   A. Introduction ................................................................................................................. 26
   B. Genetic Methods of Identification and Classification ................................................ 26
      1. Whole Chromosome Analysis ............................................................................... 27
      2. Specific Genes ........................................................................................................ 27
   C. Phylogenetic Analysis of Intestinal Anaerobes ......................................................... 27
      1. Classification of Clostridia ..................................................................................... 28
      2. Classification of Bifidobacteria ............................................................................. 28
      3. Classification of Sulfate-Reducing Bacteria ......................................................... 28
      4. Classification of *Bacteroides* Species ................................................................... 28
   D. Identification and Enumeration of Intestinal Microorganisms ................................... 29
III. Indigenous Plasmids and Mobile Genetic Elements ....................................................... 29
   A. Introduction ................................................................................................................. 29
   B. Plasmids of Lactobacilli and Bifidobacteria .............................................................. 30
   C. Plasmids of Clostridia ................................................................................................. 30
   D. Plasmids of Bacteroides .............................................................................................. 32
   E. Plasmids of Methanogens and Sulfate-Reducing Bacteria ......................................... 32
   F. Transposons and Insertion Sequence Elements of Lactobacilli .................................. 32
   G. Transposons and Insertion Sequences of Bacteroides ................................................ 33
   H. Transposons of Clostridia ........................................................................................... 34
IV. Cloning and Analysis of Genetic Determinants .............................................................. 34
   A. Cloning Strategies ....................................................................................................... 34
   B. Virulence Determinants ............................................................................................... 35
   C. Transposons and Targeted Insertional Mutagenesis ................................................... 36
   D. Physical Mapping ........................................................................................................ 37
V. Genetic Manipulation of Intestinal Bacteria ................................................................... 37
   A. Introduction ................................................................................................................. 37
   B. Mobilization of Genetic Material by Plasmid Conjugation ........................................ 38
   C. Transduction ................................................................................................................ 39
   D. Protoplast Transformation ........................................................................................... 41
   E. Electroporation ............................................................................................................ 41
   F. Cloning Vectors ........................................................................................................... 42
   G. Gene Expression .......................................................................................................... 47
   H. Release of Genetically Engineered Bacteria .............................................................. 48
VI. Conclusions ...................................................................................................................... 51
References ............................................................................................................................... 51

## I. INTRODUCTION

The molecular genetics of anaerobic bacteria is of increasing biological importance as many anaerobes are now accepted as having central roles to play in industry and in the pathogenesis of some diseases. Moreover, recognition that the indigenous colonic microflora of humans is predominantly anaerobic has generated an interest in the role of these bacteria in nutrition, health, and disease. This chapter gives an appraisal of the current knowledge of the genetics of intestinal anaerobes to provide a basis for future

\* Jeremy M. Wells, as an Advanced Fellow, is supported by the Biotechnology and Biological Sciences Research Council (BBSRC).

studies of mechanisms central to the colonic fermentation, especially at the molecular level. Our emphasis will be on discussing aspects perceived as relevant to future work. Because the field is too large for all contributions to be mentioned, general references are included throughout to provide access to the wide literature. Generally, bacteria which exist in a transient manner are included only where genetic studies have implications for the normal ecological component organisms of the large intestine.

Molecular approaches offer several advantages for studying bacteria that exist in complex ecosystems such as the large gut.[1] The previously difficult task of enumerating, identifying, and classifying anaerobic bacterial species using conventional phenotypic analyses is undergoing a revolution as a result of new genetic techniques. These techniques include direct polymerase chain reaction (PCR) sequencing to determine the nucleotide sequence of the rRNA gene. Moreover, there have been important advances in the areas of inter- and intra-species gene transfer. In particular, novel antibiotic resistance determinants have been identified in some gut anaerobes and their mechanisms of transfer are now more clearly understood.[2,3] Such studies have led to the development of essential molecular genetic tools such as cloning vectors and transposons, as well as the development of heterologous gene expression systems in intestinal anaerobes. Methods for the isolation and characterization of genetic determinants, which have been developed in aerobic species, together with new genetic tools specific for anaerobe gene transfer and expression in *Escherichia coli*, have provided direct genetic approaches for the study of anaerobic gut microbiology at the molecular level.

Genetic information on gut anaerobes is available only for a limited number of species, while DNA sequence data are still relatively sparse. Undoubtedly, studies of intestinal anaerobic bacteria at the molecular level will benefit greatly from the now large base of knowledge of genetic systems and molecular techniques that have been developed in less fastidious bacterial species. Many aerobes are now accessible to genetic study. Much of the information on central molecular mechanisms and the regulation of bacterial gene expression, has been elucidated in *E. coli*, its bacteriophages, and species belonging to the genera *Bacillus*, *Pseudomonas*, and *Salmonella*.[4] While genetic tools are available for some anaerobic bacteria, this is not the case for the majority of colonic species in this regard. Of particular importance is the need for a thorough knowledge of plasmid biology and gene transfer systems. This will allow development of efficient cloning strategies and vectors for genetic studies of intestinal anaerobes, especially in genera only distantly related phylogenetically to *E. coli*. The availability of these tools should facilitate the use of *in vivo* mutagenesis and complementation methods to identify and exploit the genetic determinants of anaerobic hosts.

## II. GENETIC DIVERSITY OF INTESTINAL ANAEROBES

### A. INTRODUCTION

Taxonomically, the human large intestine represents one of the most diverse and heavily colonized microbial ecosystems.[1,5] Diversity at the genetic level reflects immense physiological differences in the colon. Bacteria that inhabit the human large intestine are predominantly anaerobic but range from the strictly anaerobic methanogens to the relatively oxygen tolerant species of *Bacteroides*, *Lactobacillus*, and *Bifidobacterium*, and some facultative anaerobes such as *E. coli*. The importance of taxonomy and classification of bacteria in this community lies in the necessity to differentiate between pathogenic and commensal organisms, and to identify species involved in the metabolism of substrates and metabolic interactions that may affect the host.

Metabolic and morphological characteristics are the basis of traditional taxonomic studies. However, molecular genetic methods of phylogenetic analysis now provide a means by which to determine species not dependent on particular phenotypic tests. While knowledge of the overall DNA composition of the cell (mol% G+C) is a valuable preliminary tool for determining relatedness, it can only be used to exclude a particular bacterium from a taxon. A more reliable method for the differentiation of bacterial species is therefore the determination of DNA sequence homologies. The latter have been measured by competition experiments, 16S rRNA cataloging[6] and following direct PCR sequencing of rRNA genes.[7,8]

### B. GENETIC METHODS OF IDENTIFICATION AND CLASSIFICATION

Numerous methods of identification and classification exploit the uniqueness of a nucleotide sequence, or composition of cellular DNA. These methods may be divided into those which include the whole chromosome in the analysis, and those which focus on specific gene sequences.

## 1. Whole Chromosome Analysis

Perhaps the simplest genetic criterion by which to compare species is the mol% G+C of total cellular DNA, which is determined either by melting temperature or chemical analysis. Although strains within a species generally should not differ by more than 4%, unrelated strains may have the same % G+C, making this method only useful for the exclusion of new strains from particular taxa. Knowledge of the relative G+C content of given bacterial DNA can also be valuable for the design of experiments where detailed genetic information is lacking for a given species. For example, for the choice of appropriate restriction enzymes and for the design of DNA hybridization probes based on known sequences in other organisms.

A more reliable method of phylogenetic analysis of total cellular DNA involves determination of homology between two bacterial genomes by hybridization competition experiments. Here, membrane-bound, denatured cellular DNA from a reference strain is hybridized with a mixture of radio-labeled homologous DNA and unlabeled DNA from either the same or an unknown (competitor) strain. The hybridization signal obtained with the mixture of reference and competitor DNAs is expressed as a percentage of the homologous competitor-DNA. Signals with 100% indicate full genotypic identity. This method has been used successfully for the phylogenetic analysis of anaerobic genera such as *Bifidobacterium*[9] and, using both rRNA-DNA and DNA-DNA hybridization, for *Clostridium*.[6,10] Although these methods are relatively expensive and time consuming, they have provided a first means by which members of highly heterogeneous genera can be characterized when the previously accepted phenotypic methods have proved inadequate for rigorous systematization.[11]

The primary structure of the bacterial chromosome is highly strain specific and several techniques exist that exploit the uniqueness of restriction endonuclease cleavage sites in bacterial genomic DNA for identification and classification. One technique that has been used is analysis of restriction fragment length polymorphisms (RFLP) using specific endonucleases to cleave total cellular DNA, followed by electrophoresis to size fractionate the fragments. Identification of pattern differences is carried out either by ethidium bromide staining of gels, or by Southern hybridization with labeled probes homologous with specific genes of interest. This method has been given a further dimension by the development of pulsed field gel electrophoresis (PFGE) which, in conjunction with restriction endonucleases that recognize rarely occurring sequences in the chromosome (up to about 30 sites), can be used to identify differences between arrays of DNA restriction fragments representing the entire bacterial genome.

## 2. Specific Genes

Ribosomal RNA is the preferred target molecule for modern phylogenetic analyses. Because the ribosome is of ancient origin and is found in all self replicating cells, it is ideal for phylogenetic comparisons. Moreover, the 16S rRNA primary structure is highly conserved, but contains regions of confined sequence divergence. Thus, rRNA based probes may be designed which, according to their respective target sequences, recognize either single species or a number of different taxa. This allows assessment of relatedness between both distant and closely related species. Early studies involved cataloging in which the 16S rRNA was isolated and digested with ribonuclease T1. The resulting oligonucleotides, which were separated by two-dimensional electrophoresis, were sequenced to provide data for different organisms.[8] This approach highlighted for the first time the surprisingly high degree of phylogenetic heterogeneity among clostridia and anaerobic Gram-positive cocci.[12,13] The logical extension of the cataloging approach is the determination and comparison of partial or complete rRNA gene sequences. These objectives are no longer impractical because of the development of direct PCR sequencing techniques and economical oligonucleotide primer synthesis. Indeed, rRNA gene sequence data are accumulating rapidly in computer information banks. Such sequence comparisons allow analysis of relatedness between species across a broad range of taxa, and have already begun to improve our understanding of phylogenetic relationships between the species of several anaerobic genera including intestinal bacteria such as *Bacteroides*, *Clostridium*, *Lactobacillus*, and sulfate-reducing bacteria (SRB).[14,15] As a result, genetic analyses have provided information to complement established phenotypic tests, and have allowed misclassified species to be placed into appropriate genera.

## C. PHYLOGENETIC ANALYSIS OF INTESTINAL ANAEROBES

A detailed description of the current taxonomy of all intestinal bacteria is beyond the scope of this chapter. However, some recent developments will be discussed, particularly where they illustrate how molecular genetics has been used to resolve phenotypically complex taxa into more coherent phylogenetic groups.

## 1. Classification of Clostridia

Within the genus *Clostridium*, species exhibit a wide diversity of metabolic activities, nutritional requirements, and mol%G+C ratios.[3,16] Approved standards for classification of the whole genus still do not exist and the inherent genotypic heterogeneity is illustrated by a range of mol%G+C of between 21 and 54 mol%. The majority of intestinal species have G+C contents in the lower half of this range.[17] Clostridia have been grouped into three homology groups, based on rRNA hybridization experiments and mol%G+C.[6] However, these methods were unable to resolve this phenotypically diverse genus into a phylogenetically coherent group. It was later confirmed, by rRNA cataloging, that clostridial species occupy six independent and deep branching phylogenetic sublines. These also embrace nonclostridial species such as *Sarcina*, *Desulfotomaculum*, *Eubacterium*, *Peptostreptococcus*, *Selenomonas*, and *Megasphaera*. The use of PCR techniques is likely to enhance the detailed classification of clostridia. Gurtler et al.[18] have analyzed eight clostridia including the gut species, *C. bifermentans*, *C. perfringens*, *C. sporogenes*, and *C. difficile* by restriction analysis of PCR-amplified 16S rRNA genes. A hierarchical structure for the lower G+C clostridia has been proposed based on 16S and cellular rRNA homology data.[19,20] This tree splits the clostridia from their nonspore forming relatives into six taxons, which are still somewhat heterogeneous. Their acceptance as separate genera awaits the discovery of suitable phenotypic traits for individual differentiation.

## 2. Classification of Bifidobacteria

The genus *Bifidobacterium* is composed of irregularly shaped, obligately anaerobic Gram-positive rods that produce large quantities of acetate and lactate from their saccharolytic fermentation. Their diverse cell morphology coupled with a relative lack of accurate typing methods leads to a large number of bacteria being ascribed to species status within the genus. DNA hybridization analysis has provided a first means by which to confirm the uniqueness of several species. This method, together with immunological analysis of their transketolase enzymes, demonstrated that the existence of many other separate species was not justified.[10,21] The exact position of the genus *Bifidobacterium* in bacterial taxonomic structure had been controversial for some time with the primary uncertainty relating to whether the genus belonged within the lactobacilli, catenabacteria, propionibacteria, or corynebacteria. The analysis of mol%G+C demonstrated unambiguous differences between these genera, with 28 strains of bifidobacteria giving a value of $60.1 \pm 0.33$ mol%G+C compared with 67.6% (propionibacteria), 54.7% (corynebacteria), 35.5% (catenabacteria), and 33–49% (lactobacilli). However, the definitive placing of bifidobacteria in a specific taxonomic family was not possible until Fox et al.[12,22] applied an rRNA cataloging approach to these bacteria. These methods placed bifidobacteria in the family *Actinomycetaceae*, adjacent to clusters including the propionibacteria. A more detailed analysis of the phylogeny of bifidobacteria is now possible following the determination of nucleotide sequence data for bifidobacterial 16S rRNA genes, many of which have been deposited in the European Molecular Biology Laboratory (EMBL) nucleotide sequence library.

## 3. Classification of Sulfate-Reducing Bacteria

Sulfate-reducing bacteria have been identified, from diverse environments, that have a wide range of metabolic activities. These include representatives from the Gram-negative and Gram-positive eubacteria, the thermophilic eubacteria, and the archaebacteria.[23,24] Devereux et al.[25] constructed a phylogenetic tree based on 16S rRNA sequence homologies that illustrates this diversity, and classified SRB into seven Gram-negative groups and one Gram-positive group. The intestinal species identified so far have been found within the eubacterial group. Hence, the power of the 16S rRNA primary structure for determination of relatedness over a wide taxonomic range is well illustrated by phylogenetic analysis of SRB.

## 4. Classification of *Bacteroides* Species

Determination of the phenotype and species mol%G+C content of many bacterial groups previously included in the genus *Bacteroides* has indicated that they are unrelated. The mol%G+C composition has now been used to define a revised genus of *Bacteroides* as those species with a DNA-base composition in the range 40–48 mol%G+C.[15] This reclassification of the bacteroides has also relied greatly on the determination of rRNA homology and, more recently, on 16S rRNA cataloging and sequencing. Examples include reclassification of asaccharolytic pigmented species as *Porphyromonas* (formerly *Bacteroides asaccharolyticus*, *Bacteroides endodontalis*, and *Bacteroides gingivalis*).[26] The most frequently isolated colonic species now belong to the *Bacteroides fragilis* group (40–48 mol%G+C), and the nonbacteroides that are yet to be reclassified: *Bacteroides capillosus* (60 mol%G+C), *Bacteroides coagulans, Bacteroides putredinis*. The reclassified groups include *Anaerorhabdus furcosus* (*Bacteroides furcosus,* 34 mol%G+C), *Mitsuokella multiacidus* (*Bacteroides multiacidus*, 56–58 mol%G+C), *Tissierella praeacuta* (*Bacteroides praeacutus*, 28 mol%G+C).

## D. IDENTIFICATION AND ENUMERATION OF INTESTINAL MICROORGANISMS

Major limitations to studies of the microbial ecology of the human large gut are the time consuming and laborious methods of identification and enumeration of the resident bacteria. In particular, there is no simple, rapid, and reproducible assay for anaerobic species that are frequently fastidious and require specialist handling and techniques for culture. The use of nucleic acid probes can potentially circumvent many of these problems. Salyers and coworkers[27] first used DNA hybridization to detect *Bacteroides thetaiotaomicron* by a method which involved screening of randomly cloned fragments of chromosomal DNA for specific hybridization to denatured bacteria, compared with other *Bacteroides* species. Two such probes were identified and, when radiolabeled, allowed about 50 ng of *B. thetaiotaomicron* DNA to be detected. A similar technique was employed to enumerate and identify *Bacteroides vulgatus* in human feces, using colony hybridization with a 700 bp *B. vulgatus* specific probe,[28] and subsequently to enumerate fecal polysaccharide degrading *Bacteroides*.[29] The method was refined by using a 16S rRNA gene probe to detect a broad range of clinically important species to a detection limit of ca. 10 ng DNA.[30] Morotomi et al.[31] used a dot blot hybridization method to develop a method for identification of *Bacteroides* species in fecal samples. The method, which used $^{32}$P-labeled chromosomal DNA probes, correctly identified 95% of *Bacteroides* spp. from the fecal samples of 19 individuals.

Rumen microbiologists also need to rapidly detect and identify individual species within a complex microbiota. The advantages of the rRNA molecule as a target for phylogenetic classification also have implications for using rRNA gene sequences to detect intestinal bacteria by hydridization. Stahl et al.[32] used 16S rRNA targeted oligonucleotide probes, with species-specific and group-specific homologies, to enumerate rumen species by hybridization with bacteria denatured on filters. Furthermore, 16S rRNA probes linked to fluorescent molecules have been used for the visualization of SRB in biofilms from anaerobic bioreactors.[33] Nucleic acid hybridization offers the advantages of high specificity of probe for target sequence and rapid, cheap probe generation, particularly when using synthetic oligonucleotides. Furthermore, these may be designed with the required degree of specificity, e.g., when constructed from sequenced rRNA loci, that provide an additional evolutionary marker. The disadvantages of using DNA probes lie in the need for sequence information for the target DNA and in a relatively poor sensitivity (approximately $10^5$–$10^6$ bacteria) even when probes are radiolabeled. Moreover, many Gram-positive species are refractory to rapid lysis which makes conventional filter hybridization difficult. The requirement for sequence data can be avoided if randomly cloned gene fragments are used. The use of non-radioactive methods has encountered a number of problems including interference from autofluorescing sample components and interaction of the label with assay constituents.[34,35] Therefore, the use of nucleic acid probes to study the large intestinal microflora is currently limited to studies of numerically predominant species. There is hope, however, that the accumulation of DNA sequences in appropriate data bases and the rapid development of improved probe labeling, detection, and hybridization strategies will lead to a more widespread use of DNA probes in the future.

## III. INDIGENOUS PLASMIDS AND MOBILE GENETIC ELEMENTS

### A. INTRODUCTION

Plasmids and mobile genetic elements are widespread among many anaerobic bacterial species and have been studied in intestinal bacteroides, clostridia, and lactic acid bacteria. They are of enormous importance to genetics studies because they may encode genes required for virulence, various metabolic properties, and confer resistance to bacteriophages and antibiotics.

Extrachromosomal elements can be transferred by the natural processes of conjugation, transduction, and transformation. Thus, they may serve as carriers of genetic information which can be disseminated throughout the bacterial population. In this way, plasmids and transposons have contributed greatly to the genetic evolution of microorganisms.

Many of the smaller plasmids in lactobacilli, clostridia, and other Gram-positive bacteria are members of a widespread family of plasmids (the ssDNA plasmids) which replicate by the rolling-circle mechanism (reviewed by Gruss and Ehrlich[36]). The single-stranded DNA intermediate which is generated during replication of ssDNA plasmids has an increased chance of undergoing homologous and illegitimate recombination. This may explain why they are highly interrelated and widely disseminated among several genera of Gram-positive bacteria. The ssDNA plasmids are often maintained at 20 to 100 copies per chromosome, whereas larger plasmids (usually conjugative) are maintained at only one or two copies per chromosome. Replication of these large plasmids is usually by the bidirectional or unidirectional elongation of a replication fork formed on the double-stranded plasmid. It is important to note that common methods of plasmid isolation generally do not detect large (>15–20 kbp), low copy number plasmids. Therefore, the particular isolation technique used should be carefully considered before any bacterium is considered as plasmid free. Recently, the development of PFGE has increased the chance of detection of these plasmids by allowing the separation and visualization of large DNA molecules in agarose gels.

Conjugative plasmids have mobilization genes which enable them to transfer themselves from one bacterium to another by a process known as conjugation. These plasmids have been identified in many species of gut anaerobes and are probably one of the most important means of natural gene transfer *in vivo*.

Two types of mobile genetic elements, transposons and insertion sequences (IS), that are capable of transposing themselves from one replicon to another have been identified. Both types of elements are generally flanked by inverted repeat sequences and encode a transposase enzyme involved in transfer of the element. Transposons differ from IS elements in that they carry a phenotypic marker, usually an antibiotic resistance gene. These elements are particularly prevalent among species of the genera *Bacteroides* and *Clostridium*.

## B. PLASMIDS OF LACTOBACILLI AND BIFIDOBACTERIA

Lactobacilli commonly contain several plasmids and the plasmid content or profile of a bacterial isolate can be used to distinguish between different strains.[37-39] Normally, no physiological role is ascribed to the plasmids, but some have been shown to encode genes involved in metabolic processes, bacteriocin production, and antibiotic resistance.[40-43] One plasmid has also been reported to confer a growth advantage in studies of isogenic strains of *Lactobacillus reuteri* in the cecum of gnotobiotic mice.[44] The isolation and sequence analysis of several small lactobacilli plasmids has revealed the presence of features typical of the widespread family of plasmids, which replicate via a rolling circle mechanism.[45-48] In addition, a few conjugative plasmids have been identified in lactobacilli but little is known of their mechanism of conjugation.[41,49]

Sgorbati et al.[50] have surveyed 1461 isolates, representing 24 different species of the genus *Bifidobacterium* for the presence and distribution of plasmids. Seventy percent of *Bif. longum* isolates harbored plasmids. In contrast, none of the 446 strains of other human bifidobacterial species, i.e., *dentium*, *breve*, *bifidum*, *pseudocatenulatum*, and *catenulatum* were found to contain plasmids. Several different plasmid profiles were evident among the 123 different isolates of *Bif. longum* examined. A few of these plasmids were large enough to be conjugative but no phenotypic traits were ascribed to those identified. In a separate study, Iwata and Morishita[51] surveyed 45 human isolates for the presence of plasmids. In contrast to the results of the previous survey, they found that 40% of these strains contained one or more plasmids. The identification and further characterization of these cryptic plasmids should enable shuttle vectors and gene transfer systems to be developed for *Bifidobacterium* spp.

## C. PLASMIDS OF CLOSTRIDIA

The clostridial plasmid pIP404 from *C. perfringens* (10.2 kb; Table 1) has been extensively studied. The complete nucleotide sequence encodes ten open reading frames (ORFs), of which four still have an unknown function. Three of the genes (*bcn*, *uviA*, and *uviB*) are involved with bacteriocin production and immunity (reviewed by Rood and Cole[52]). The *bcn* gene encodes a 96.5 kDa secreted bacteriocin which is produced upon UV induction. Although the exact mechanism of secretion is unclear, the presence of a highly lipophilic region at the C-terminus of the protein had led workers to suggest that it acts as an ionophore.[53]

In order to identify the replication loci of pIP404, overlapping restriction fragments of the plasmid DNA were first cloned into a vector carrying selectable markers but lacking a replicon. Plasmids containing a replicon were selected by transformation of *Bacillus subtilis*. The minimal origin of replication was defined as containing two ORFs encoding a 48.7 kDa Rep protein and a 23.3 kDa Cop

**Table 1**  Indigenous Plasmids

| Plasmid | Source | Size (kb) | Features and genetic determinants | Reference |
|---|---|---|---|---|
| pIP404 | *C. perfringens* | 10.2 | Bacteriocin production and immunity genes *bcn*, *uviA*, and *uviB*, Copy number control and maintenance genes *cop* and *res;* theta replication mechanism, *rep* gene identified | 226 |
| pIP401 | *C. perfringens* | 53 | Conjugative; CamP and TetP resistance; identical to pCW3 except for insertion of 6.2 kb transposon Tn*4451* carrying CatP | 227 |
| pCW3 | *C. perfringens* | 47 | Conjugative TetP; see above | 60 |
| pBF4 | *B. fragilis* | 41 | Conjugative; aerobic Tc$^r$ and MLS (*ermF*) resistance | 65 |
| pBFTM10 | *B. fragilis* | 14.9 | Conjugative; aerobic Tc$^r$ and transposable MLS determinant homologous to that of pBF4 and pBI136; Similar to pCP1 from *B. thetaiotaomicron* | 66 |
| pBFKW1 | *B. fragilis* | 40 | Conjugative; β-lactam resistance | 70 |
| pIP417 | *B. vulgatus* | 7.7 | Conjugative; Resistance to 5-nitroimidazoles (NiR). Rep, Mob, and NiR determinants different to pIP419 | 71 |
| pIP419 | *B. thetaiotaomicron* | 10 | Resistance to 5-nitroimidazoles; see above | 71 |
| pCP1 | *B. thetaiotaomicron* | 15 | Highly related to pBFTM10, and carrying the same determinants | 67 |
| pBI136 | *B. ovatus* | 80.6 | Conjugative; transposable MLS determinant homologous to that of pBF4 and pBFMT10; Transfer genes *btgA* and *btgB* and transfer origin *oriT* identified | 68 |
| pRR14 | *B. ruminicola* | 19.5 | Conjugative; Tc$^r$ determinant | 228 |
| pB8-51 | *Bacteroides* sp. | 4.2 | Nonconjugative; containing mobilization region recognized by the transfer functions of pBF4 and the Tc$^r$ conjugal elements | 94 |
| p353-2 | *L. pentosus* | 2.4 | Sequencing has revealed features typical of RCR plasmids; 81.5% similarity to p8014-2 isolated from *L. plantarum* | 171 |
| pPM68 & pPM52 | *L. acidophilus 88* | 68 MDa and 52 MDa | Plasmids associated with conjugal transfer of bacteriocin (Lactacin F) production and immunity. | 41 |
| p1 & p3 | *L. acidophilus* | 1.6 and 4.2 | Related plasmids; sequencing and functional analysis of a divergent promoter in plasmid p1 | 47 |
| pLAR33 | *Lactobacillus* sp. | 18 | MLS$^r$ determinant | 43 |

protein which was membrane associated, and shown to influence plasmid copy number.[54] The *ori* of replication comprised two related families of direct repeats, also found at the replication origins of numerous Gram-negative and Gram-positive plasmids, e.g., F, RK2, RK6, pSC101 and pRAT11 (reviewed by Scott[55]). A strong promoter situated in the *ori* transcribes a 150 nucleotide RNA which is complementary to the terminal 125 nt of the *rep* mRNA, suggesting that it may regulate copy number by sequestering the Rep protein mRNA. A third mechanism of replication control which serves to ensure distribution of the plasmid to daughter cells during cell division is provided by the *res* gene product. The predicted amino acid sequence of the 6.0 kDa Res protein is homologous to the site-specific resolvases of various transposons and may be involved in resolving plasmid multimers into monomers at specific resolution binding sites on the plasmid.

There are numerous reports of cryptic clostridial plasmids and the replicons of several small plasmids of *C. butyricum* have been analyzed. Sequencing and deletion analysis of the replicons has provided evidence for replication by the rolling circle mechanism which is supported by homology with Rep proteins encoded on other Gram-positive plasmids.[3,56]

Several closely related conjugative plasmids have been found in *C. perfringens* that confer resistance to tetracycline (*Tet P* determinant) and, in some cases, a chloramphenicol resistance determinant (*Cat P*).[3,52] Bacteriocin production in *C. perfringens* and resistance to mercury in *C. cochlearium* have also been transferred by conjugation (reviewed by Odelson et al.[57]). Furthermore, there is evidence that some clostridial toxins are plasmid encoded although conjugative transfer of these determinants has not yet been demonstrated.[52,58,59]

Two conjugative *C. perfringens* plasmids (pIP401 and pCW3; Table 1) have been extensively studied and shown to be identical, except for the presence of a 6.2 kb transposon (Tn*4451*) carrying a chloramphenicol resistance determinant in pIP404.[52,60] Conjugative plasmids from *C. perfringens* strains obtained from extremely diverse environmental and geographical origin have been shown to be identical, or closely related to, pCW3 indicating that they have all evolved from one major progenitor plasmid.[61,62]

## D. PLASMIDS OF BACTEROIDES

Numerous small cryptic plasmids have been identified in intestinal species of *Bacteroides*. Some workers have grouped these into three different classes based on DNA homology and restriction endonuclease digestion patterns.[63] Reports of all three classes maintained in a single isolate indicated that they may be compatible. One cryptic plasmid pB8-51 (homology class 2; Table 1) has been found in several colonic *Bacteroides* species and carries a region which enables it to be mobilized by the conjugal chromosomal tetracycline resistance (Tc^r) elements and the conjugative plasmid pBF4.[64]

Several large conjugative plasmids have also been identified in *Bacteroides* species (Table 1). The first to be studied (pBF4) was obtained from a clinical isolate of *B. fragilis* and was shown to transfer resistance to macrolide-lincosamine-streptogramin B (MLS^r).[65] Subsequently, three other conjugative plasmids (pBFTM10, pCP1, and pBI136) which confer resistance to MLS were reported,[66-68] all of which share homology with the MLS^r determinant of plasmid pBFTM10, suggesting that this determinant is widely distributed in nature.[67] A conjugation transferable plasmid-linked resistance to chloramphenicol has also been described for *B. uniformis*.[69] Recently, R-plasmid mediated transfer of β-lactam resistance and resistance to 5-nitroimidazoles has been reported in *B. fragilis*.[70,71] The resistance determinants of some *Bacteroides* plasmids have been cloned and sequenced, but information on their replication, maintenance, mobilization, and transfer functions is limited. Recently, a transfer deficient derivative of pBFTM10 contained in the *E. coli* shuttle vector pGAT400, and a similar shuttle vector pGAT550, were used to determine regions of pBFTM10 that were essential for mobilization by the broad host range *E. coli* plasmid R751. Transposon insertional mutagenesis in this same region was used to map and sequence the determinant. Two ORF encoding proteins of 23.2. and 33.8 kD (genes *btgA* and *btgb*) were necessary for transfer and an *oriT* gene with DNA sequence homology to that of the IncP group of *E. coli* plasmids was also identified.[72]

## E. PLASMIDS OF METHANOGENS AND SULFATE-REDUCING BACTERIA

Several cryptic plasmids have been isolated from methanogenic bacteria but none from species that are commonly found in the gut.[73] The best characterized of these is the circular multicopy plasmid pME2001 isolated from *Methanobacterium thermautotrophicum* which has been completely sequenced.[74] Plasmid pME2001 may be used to construct future cloning vectors for methanogens by incorporating selective resistance markers into nonessential regions of the plasmid. If they are found to replicate in a wide range of species, such derivatives will greatly facilitate the development of efficient transformation procedures in methanogens.

The *Desulfovibrio* species of SRB have been shown to contain indigenous plasmids but these have not been extensively characterized.[75,76] One 130 MDa plasmid found in at least three different strains of *D. vulgaris* has been shown, however, to carry nitrogenase genes by hybridization studies with the *nif* genes of *Klebsiella pneumoniae*.[76] A small cryptic plasmid isolated from *D. desulfuricans* is currently being evaluated as a potential cloning vector in a number of *Desulfovibrio* strains.[77]

## F. TRANSPOSONS AND INSERTION SEQUENCE ELEMENTS OF LACTOBACILLI

Only one insertion sequence of *Lactobacillus* origin has been identified,[78] even though they are wide-spread among the lactococcal members of the lactic acid bacteria. This insertion sequence (ISL1) was identified by a detailed comparison of the homology between a prophage and virulent bacteriophage for the host strain *L. casei* S1. The presence of a 1256 bp IS in the virulent bacteriophage prevented maintenance of the lysogenic state by causing it to constitutively express genes required for lytic growth. This determinant contains two ORFs encoding proteins with predicted molecular weights of 10.7 and 31.7 kDa which are flanked by inverted repeat sequences at the termini. Transposons and IS from the numerically predominant enteric species of lactobacilli have yet to be characterized but work with ISL1 from *L. casei* has exemplified how the DNA rearrangements, brought about by transfer of an insertion sequence can influence gene expression on DNA of bacteriophage, plasmid, or chromosomal origin.[78]

**Table 2** Mobile Genetic Elements

| Element | Source | Size (kb) | Features | References |
|---|---|---|---|---|
| Tn*4351* | *B. fragilis* (pBF4) | 5.5 | MLS[r], Tc[r] | 81 |
| Tn*4400* | *B. fragilis* (pBFTM10) | | MLS[r], Tc[r] | 82 |
| Tn*5030* | *B. fragilis* | 45 | MLS[r], Tc[r] | 101 |
| Tc[r] V479 | *B. fragilis* | 70–80 | Tc[r], Tra[+] conjugal element | 64, 100 |
| Tc[r] Em[r] CEST | *B. fragilis* | 70–80 | Tc[r], Em[r], Tra[+] conjugal element | 64, 100 |
| Tc[r] ERL | *B. fragilis* | 70–80 | Tc[r], Tra[+] conjugal element | 64, 100 |
| 12256 | *B. fragilis* | 150–200 | Tc[r], Em[r] Tra[+] conjugal element | 64, 100 |
| Tc[r] Em[r] ERL | *B. fragilis* | 70–80 | Tc[r], Em[r], Tra[+] conjugal element | 64, 100 |
| Tn*4399* | *B. fragilis* | 9.6 | Mobilizes nonconjugal plasmids in *cis*; no antibiotic marker | 89 |
| IS*942* | *B. fragilis* | 1.598 | Inserted upstream of *CcrA* gene | 88 |
| IS*4400* | *B. fragilis* (pBFTM10) | 1.15 | | 82 |
| IS*4351* | *B. fragilis* (pBF4) | 1.15 | | 86 |
| Tn*4551* | *B. ovatus* (pBI136) | | MLS[r] | 83 |
| Tc[r] Em[r] DOT | *B. thetaiotaomicron* | 70–80 | Tc[r], Em[r], Tra[+] conjugal element | 64, 100 |
| XBU4422 | *B. uniformis* | 65 | Chromosomal; conjugal | 99 |
| Tn*4451* | *C. perfringens* | | Cm[r] (*catP*) Tra[+] | 102 |
| Tn*4452* | *C. perfringens* | | Cm[r] (*catP*) Tra[+] | 102 |
| ISM1 | *Methanobrevibacter smithii* | 1.38 | | 229 |

## G. TRANSPOSONS AND INSERTION SEQUENCES OF BACTEROIDES

In the course of studying the spontaneously occurring MLS sensitive variants of *Bacteroides*-containing plasmids pBF4, pBFTM10, and pBI136, it was observed that the MLS resistance determinant was bound by directly repeated (DR) sequences characteristic of transposable elements (Table 2).[57,67,68,79] Later Shoemaker et al.[80] showed that the 5.5 kb MLS determinant of pBF4 flanked at both ends by DR could transpose from a *Bacteroides–E. coli* shuttle vector into the *E. coli* conjugative plasmid R751 and also into the chromosome of *E. coli*. This transposable element, designated Tn*4351*, was also shown to be transposable in *Bacteroides* spp.[81] Similarly, the MLS[r] resistance determinants of *Bacteroides* plasmids pBFTM10 and pBI136 were shown to undergo transposition in *Bacteroides* and *E. coli* and have been designated Tn*4400* and Tn*4551*, respectively.[82,83]

An unusual feature of transposons Tn*4400* and Tn*4351* is that in addition to the MLS resistance determinant, they carry a tetracycline resistance gene (*tetX*) that only confers resistance in aerobically grown *E. coli*. Genetic and biochemical analysis of this determinant has shown that *tetX* encodes an nicotinamide adenine dinveleotide phosphate (NADP)-requiring oxidoreductase that acts by chemically modifying tetracycline in the presence of oxygen.[84] Transposons Tn*4351* and Tn*4400* are closely related, sharing more than 90% sequence homology.[85] In contrast, Tn*4551* is nearly 3kb longer and only shares sequence homology to the others in DRs and clindamycin resistance determinants.[85] Transposition of the DR of *Bacteroides* transposons to form IS lacking intervening sequences, has also been demonstrated (Table 2).[82,86,87]

Recently, a novel 1.6 kb *B. fragilis* insertion sequence was isolated and sequenced.[88] One copy of this IS was integrated 19 bp upstream of the metallo-β-lactamase gene (*ccrA*) and probably provides transcription start sites.

In order to investigate the presence of chromosomal conjugal transposons, Hecht and Malamy[89] carried out a series of *B. fragilis–E. coli* filter mating experiments with a Tra[−] derivative of the conjugal shuttle plasmid pGAT400. During mating experiments, a 9.6 kb *Bacteroides* chromosomal transposon, Tn*4399* was found to transpose randomly into the Tra[−] derivatives of pGAT400 and mediate plasmid transfer between *Bacteroides* and *E. coli* and from one strain of *B. fragilis* to another. This element may therefore be carrying an *ori*T-like sequence recognized by the chromosomal conjugal transfer apparatus (see discussion below), or may encode its own transfer functions.

A prevalent class of conjugative elements that transfer combined resistance to clindamycin and tetracycline, or to tetracycline alone, have also been described in *Bacteroides* species (Table 2).[90-95] These elements are capable of transferring themselves as well as co-resident plasmids from one *Bacteroides* strain to another or from *Bacteroides* to *E. coli* recipients.[64] In many cases, the frequency of self transfer and of mobilization of co-resident plasmids was enhanced by pregrowth of the strains in low concentrations of tetracycline. This induction with tetracycline has been shown to lead to the appearance of plasmid-like forms, designated nonreplicating *Bacteroides* units (NBUs) which originate from the chromosome.[96] The region of the conjugal Tc[r] Em[r] DOT element responsible for production of the NBUs lies within a 10 kb region adjacent to the Tc[r] gene. Insertional mutations in this region also disrupt transfer functions of the element, suggesting that this region might also encode genes involved in the transfer.[97] DNA sequence analysis of the region has revealed the presence of the tetracycline resistance gene (*tetQ*) and two other genes (*rteA* and *rteB*) encoding proteins with amino acid similarities to known two component regulatory systems.[97]

The Tc[r] Em[r] DOT conjugal elements from *B. thetaiotaomicron* DOT and the *B. uniformis* Tc[r] Em[r] DOT transconjugant strain have been cloned and physically mapped in *E. coli* using a shuttle cosmid vector. Southern hybridization experiments indicated that the Tc[r] elements from different strains of *Bacteroides* are closely related but not identical.[98] As expected, some of the cosmid clones which hybridized to the chromosomal DNA of the recipient (*B. uniformis* strain 0061), contained DNA from the junction between the element and the host chromosome. However, some clones containing internal regions of the conjugal element also hybridized to this strain which suggests that a related cryptic element is present in the chromosome lacking detectable resistance genes.[99] This cryptic element, designated XBU4422, was later identified on *Bacteroides–E. coli* shuttle vectors, when one of the Tc[r] elements was being used to mobilize these vectors to *Bacteroides* recipients. XBU4422 is self transmissible and shares DNA sequence homology with other *Bacteroides* conjugal Tc[r] elements, particularly at its ends. The ends of this element were also used as probes to identify its insertion sites in NotI fragments of chromosomal DNA separated in agarose gels by transverse alternating field electrophoresis. It appears that each Tc[r] element has three to seven specific insertion sites in the chromosome of *B. thetaiotaomicron*. By comparing the sizes of NotI enzyme digested DNA fragments derived from *B. thetaiotaomicron* strain 4001, lacking any Tc[r] elements, and transconjugants of the same strain containing a Tc[r] element, it was shown that most of the elements were 70 to 80 kb in size, except Tc[r] Em[r] 12256 which was 150 to 200 kb and may be a hybrid element.[100] The Tc[r] elements sometimes also transfer unlinked chromosomal segments or NBUs and these also integrate specifically at alternative sites in the host chromosome. A comparison of the junction sequences and target sequences of different insertions showed that XBU4422 carried 4–5 kb of adjacent chromosomal DNA when excised from the chromosome and integrated into a plasmid. A short region of sequence similarity between one of the ends of XBU4422 and its target site has been identified which may be important for integration.

Halula and Macrina[101] used the MLS[r] determinant as a probe to screen a cosmid library prepared from a *B. uniformis* transconjugant following conjugal mating with a strain of *B. fragilis* that transfers MLS[r] in the absence of detectable plasmid DNA. Analysis of overlapping cosmid and lambda clones revealed that the chromosomal conjugal element was approximately 45 kb in size. This transposon, designated Tn*5030*, lacked the characteristic terminal repeat sequences typically associated with other transposons and preferentially integrated into the chromosome of *B. uniformis* recipients.

## H. TRANSPOSONS OF CLOSTRIDIA

Two 6.2 kb chloramphenicol resistance transposons, Tn*4451* and Tn*4452*, have been characterized from *C. perfringens*. These transposons were first identified due to spontaneous loss of chloramphenicol resistance (*catP* gene) from recombinant plasmids pIP401 and pJIR27 in *recA* strains of *E. coli*. Transposition of these Cm[r] elements from temperature sensitive plasmids into the chromosome of *E. coli* was also observed.[102] Heteroduplex analysis of the transposons has shown that they are closely related and only differ in a 0.4 kb region at the right end of the transposon. Sequence analysis of the ends of the transposons has revealed features common to many transposable elements.[103]

Similarly, the erythromycin resistance determinant of *C. difficile* (*ermBP*, formerly known as *ermP*) shows instability in *recA* strains of *E. coli* with one copy of the DR sequences which flank the *ermBP* gene being lost at high frequency.[104] Evidence for the ability of this element to transpose itself is currently lacking, however.

## IV. CLONING AND ANALYSIS OF GENETIC DETERMINANTS

### A. CLONING STRATEGIES

Randomly generated fragments of chromosomal DNA are usually cloned into plasmid, cosmid, or bacteriophage vectors to generate gene banks for screening with antibodies or hybridization probes. A frequently cutting enzyme is usually employed to generate partially digested fragments of genomic DNA for insertion into the vector. However, care must be taken with the choice of enzyme, because those with a G + C rich recognition sequence for example would not be expected to cut the A + T rich lactobacilli and clostridial DNA randomly. The enzyme Sau3A which is often used to generate partial digests of *E. coli* and mammalian DNA may not cut DNA from *C. perfringens* due to the action of a restriction modification system which methylates a cytosine in the recognition site of the enzyme.[53] Depending on whether the gene libraries are constructed in plasmid, cosmid, or bacteriophage vectors, about 1000–5000 clones must be screened in order to have a 99% probability of isolating an average sized gene from an organism with a genome size equivalent to *E. coli* (approximately 4.5 Mb). A simple visual screening procedure is particularly valuable when the cloned genes are expressed and have an activity which is apparent on solid media containing a suitable substrate. For example, clones of *E. coli* expressing *C. perfringens* sialidase have been detected by screening for hydrolysis of MU-α-*N*-acetylneuraminic acid in this way.[105]

Some genes may be cloned by complementation of mutant strains of the same bacterium or other species. Typically, such an approach might involve construction of a gene bank by transformation of ligated DNA directly into the mutant strain or by pooling all the clones of the genomic library, extracting the recombinant plasmid DNAs and then transforming these into the appropriate mutant strain. Complementation relies on expression of the foreign gene in the host (e.g., *E. coli*), therefore it is often advantageous to construct the genomic library in a pUC type vector. The basal level of transcription arising from the *E. coli lac* promoter present in vectors of the pUC series generates transcripts of the cloned gene which may then be translated. The glutamine synthetase gene of *B. fragilis* and *C. acetobutylicum* are examples of genes which have been cloned by complementation.[106,107]

## B. VIRULENCE DETERMINANTS

Bacterial pathogenesis involves a complex progression of events that damage the host following infection, usually as a result of the action of several virulence factors. A virulence factor may be defined as a characteristic whose loss decreases the pathogenic potential of the organism, excluding those changes which also affect its ability to grow in the laboratory. Because molecular genetics offers the opportunity to mutate genes for the study of putative virulence determinants in isolation, it is a powerful tool by which to investigate bacterial pathogenicity. Mobile genetic elements provide the means to insertionally inactivate chromosomal virulence genes and further to rapidly isolate (clone) the genes themselves prior to more rigorous genetic characterization and nucleotide sequencing.[108] Ironically, mobile DNA sequences frequently encode virulence traits. In particular, the transmissible resistance to antibiotics has been extensively studied in clostridia. This resistance has been studied in bacteroides and is discussed elsewhere in this chapter. Oligonucleotide probes, based on the determined N-terminal amino acid of a purified protein, may be used to identify and isolate virulence genes by screening genomic DNA libraries derived from the pathogen. This technique, termed reverse genetics, and is readily applicable to secreted virulence proteins that may be easily purified.[109]

Several species of *Clostridium* are pathogenic, hence there has been a broad interest in the cloning of their virulence determinants. Many virulence factors have been identified in *C. perfringens* including 12 toxins, enterotoxin, non α-δ-θ haemolysin, and neuraminidase. The various strategies employed in the cloning of these determinants provide useful examples of genetic methods available for the anaerobes. The α toxin (or phospholipase C) is produced by all five types (A to E) of *C. perfringens*. The gene has been cloned and mapped between two rRNA operons in the most conserved segment of the genome, which also contains many housekeeping genes.[52] The α toxin is expressed from a constitutive promoter, whereas the expression of θ, κ, and λ toxins and neuraminidase is coordinately regulated by the *pfoR* gene product.[110] The enterotoxin of *C. perfringens* (CPE) accumulates in large quantities intracellularly as an insoluble inclusion body. This allows purification, and fragments of the gene have been isolated by reverse genetics using synthetic oligonucleotide probes based on the protein sequence of the enterotoxin. Neither of these fragments encoded a full length product, although they overlapped sufficiently to enable elucidation of the complete nucleotide sequence.[111] At about the same time, the C-terminal half of the *cpe* gene was cloned by Hanna et al.[112] These workers screened a lambda gt11 expression library, with an anti-CPE monoclonal antibody, to isolate a clone which expressed a β-galactosidase-CPE fusion protein. In addition to its toxigenicity, the rapid growth of *C. perfringens in vivo* contributes to its virulence. In this respect, it is significant that the *C. perfringens* chromosome carries 10 rRNA operons interspersed with many tRNA genes which provide a good supply of translational components to the cell.

The cloning and expression of virulence genes in *E. coli* has greatly facilitated the purification and biochemical analysis of many pathogenic factors. A large body of literature exists which deals with the structure–function relationships of the virulence determinants of *C. perfringens*,[52] but it is beyond the scope of this review to summarize all of the available knowledge.

## C. TRANSPOSONS AND TARGETED INSERTIONAL MUTAGENESIS

Transposons are valuable tools for *in vivo* mutagenesis by virtue of their ability to transpose randomly into plasmid and chromosomal DNA. The antibiotic resistance genes that they carry can be used to select for bacteria that have acquired a transposon. Transposons can be introduced into cells by conjugation using suicide vectors which are unable to replicate in the new host. If the transposon integrates randomly into the genome, it can be used for a general mutational analysis as demonstrated for *B. thetaiotaomicron*, by the isolation of mutants which are unable to utilize chondroitin sulfate and starch.[113,114] In contrast to chemical methods of mutagenesis, transposons are more precise in that only one region of the chromosome is usually altered by their insertion. Since the mutations are generally stable, the mutant phenotype is usually able to be investigated *in vivo*. However, the polar nature of some transposon insertions may also affect the transcription of genes downstream from the insertion.

Once a mutation has been generated, a restriction fragment containing the transposon and flanking chromosomal DNA can be cloned into a suitable plasmid vector in *E. coli* by selecting for the transposon encoded resistance. The cloned chromosomal DNA may then be recovered in sufficient quantity for it to be analyzed by restriction mapping and sequencing. The Tn*916*-like transposable elements originally found in *Enterococcus faecalis* and *Streptococcus sanguis* are especially valuable for mutagenesis, because precise excision of the integrated element from the cloned DNA can be achieved. The mutated genes may be isolated by cloning restriction fragments of genomic DNA, which have been digested with an enzyme not present in the transposon, into an appropriate vector in *E. coli*. Again, the clones can be selected by virtue of the tetracycline determinant encoded within the transposon. In the absence of tetracycline selection the transposon excises from the plasmid thus restoring the structural integrity of the cloned gene (reviewed by Clewell and Gawron-Burke[115]).

Tn*916* has been successfully introduced into *C. perfringens* by conjugal filter mating with *Ent. faecalis* at a frequency of $10^{-7}$ to $10^{-8}$ per viable recipient and also by transformation with a suicide vector carrying the transposon.[52] Southern hybridization showed that the transposon had integrated randomly into the *C. perfringens* chromosome and multiple insertion events occurred at high frequency. In *C. acetobutylicum*, Tn*916* has been used to isolate mutants with defects in fermentation pathways[116] but has yet to be exploited for the isolation of genes and analysis of gene function in intestinal clostridia.

Tn*919*, which is structurally similar to Tn*916* and shares similar properties, has been transferred into *Lactobacillus plantarum*, *Leuconostoc mesenteroides*, and *Lactobacillus curvatus* Lc 2-c by agar surface matings.[117,118] The introduction of Tn*917* into indigenous plasmids of *Lactobacillus plantarum* following electroporation of plasmid pTV1 carrying the transposon has also been reported.[119,120] For these species, however, the frequency of conjugal transfer is too low for mutational screening procedures. Future applicability of these transposons for the cloning of chromosomal and plasmid genes in *Lactobacillus* species will depend on the development of improved transfer systems and the demonstration of random Tn insertion in the DNA of the hosts. Some progress has been made in this area for the lactococcal species of lactic acid bacteria.[121]

Gene function may be further studied by targeted insertional activation of specific gene sequences. This process is accomplished by cloning a fragment of the target gene of interest, or a mutated gene, into a suicide vector. Following plasmid transfer to the host, homologous recombination events can then be selected for using the antibiotic resistance marker present on the suicide vector. This approach has the advantage that it does not introduce polar mutations and has been used to study the function of genes involved in chondroitin sulfate and starch degradation by colonic species of bacteroides.

*Bacteroides thetaiotaomicron* produces two very similar chondroitin lyase enzymes (I and II) which cleave chondroitin sulfate, a mucopolysaccharide found in large amounts in the lumin of the gut. In order to determine whether chondroitin lyase II was necessary to cleave chondroitin sulfate, a fragment of the cloned gene was cloned into the suicide vector pE-3. When this plasmid was conjugally transferred to *B. thetaiotaomicron* from *E. coli*, the chromosomal chondroitin lyase II gene was inactivated by homologous recombination and insertion of the plasmid into the chromosome.[122] Mutants lacking activity of this particular lyase were still able to grow on chondroitin sulfate as a sole carbon source. Subsequently, strains in which part of the chondroitin lyase II gene had been deleted grew in the intestinal tracts of germ-free mice and without any measurable disadvantage over wild type organisms.[123] The reason for the presence of two separate chondroitin lyase enzymes in *B. thetaiotaomicron* remains unresolved.

Insertional mutagenesis has also been used to determine whether the cloned pullulanase gene of *B. thetaiotaomicron* was essential for the degradation of pullulan, which is a starch-like polysaccharide found in the colon.[124] The pullulanase activity of the mutant was only reduced by 55% suggesting the presence of a second enzyme in *B. thetaiotaomicron*. Later work did identify a second pullulanase gene in this organism.[125] Transposon insertional mutagenesis has also been used to map the determinants of pBFTM10 required for its transfer from *B. fragilis* and for plasmid mobilization by the IncP transfer system in *E. coli*.[72]

## D. PHYSICAL MAPPING

The Hfr-like and transduction-based systems which have been used to map the chromosomes of several aerobic bacteria are not generally applicable to intestinal anaerobes. However, the development of PFGE has made it possible to determine the genome sizes and generate physical and genetic maps for many bacteria.[126-130] The physical mapping of a bacterial chromosome by PFGE requires the initial screening of several different restriction endonucleases to identify "rare cutters" which recognize rare DNA sequences and yield fragments greater than 50 kb in size. The choice of enzymes is governed by the base pair composition of the DNA and by the length of recognition sequence for the enzyme. Enzymes with a G + C rich recognition sequence such as ApaI FspI MluI, NruI, SacII and SmaI have proved to be especially valuable for PFGE analysis of A + T rich clostridial chromosomes. The genome size can be estimated from the sum of the lengths of genomic fragments generated by digestion with individual restriction endonucleases. The relative order of restriction sites can also be determined by performing digests with different combinations of enzymes. It is often difficult, however, to unambiguously locate all fragments by this means. Thus, other techniques such as partial digestion and detection of the fragments by hybridization with a probe complementary to one end of the fragment may help determine the order of restriction sites.[131] Such end derived probes can be obtained by cloning fragments of chromosomal DNA which have been cut with a rare enzyme and a second, more frequently cut, enzyme into plasmid vectors cut with restriction endonucleases which generate compatible ends. End probes can also be used to screen genomic libraries for missing end clones and linking clones that contain DNA fragments which span the region containing a site for a rare enzyme. These linking clones can then be used as hybridization probes to unambiguously demonstrate the contiguity of two restriction fragments.[128]

Once a physical map of the chromosome has been established, genes of interest may be localized by means of DNA hybridization using probes prepared from the host organism or from heterologous genes preferably with a similar G + C content. Using the techniques described, a genome map has been established for *C. perfringens* strain CPN50 (serotype A) on which more than 100 restriction sites and 24 gene loci have been positioned.[130] This study has recently been extended to other strains of *C. perfringens* including members of the serotypes B, D, and E.[52,132] Moreover, type A, B, D, and E strains were shown to be closely related. It has not yet been possible, however, to directly correlate differences in pathogenesis or virulence to variant regions of the chromosome which may contain small insertions and deletions.

# V. GENETIC MANIPULATION OF INTESTINAL BACTERIA

## A. INTRODUCTION

Genetic transfer systems are essential tools with which to study and manipulate the genetic traits of bacteria. Natural conjugation systems have been especially useful for the study of plasmid structure and function and also for the manipulation of bacteria by gene transfer and the mobilization of co-resident plasmids. For example, the *E. coli* IncP and IncQ conjugative plasmids have played a major role in the development of natural gene transfer in *Bacteroides* spp. and have also enabled similar systems to be established for *Desulfovibrio* spp. Transposons and chromosomal conjugal elements have also been used to target genes for mutagenesis, transfer genetic elements, and mobilize additional genetic material. Many bacteria are naturally competent to take up DNA at low frequency which represents a further transformation mechanism. For many organisms more rapid and efficient methods have also been developed to introduce DNA directly into cells by protoplast transformation and electroporation.

In addition to a transformation system, recombinant cloning vehicles are required for genetic manipulation. Plasmids are usually the starting point for the development of cloning vectors, because they are easily purified and manipulated *in vitro*. Moreover, the necessary origins of replication and specific transformation markers are commonly associated with plasmids. However, mobile elements such as transposons, as well as transducing bacteriophages, may also be used as vectors. Many of the plasmid vectors are shuttle vectors that have been designed to replicate in one or more host, since genetic manipulations are often performed more easily in *E. coli* or *Bacillus subtilis*. In most cases, these shuttle vectors have been developed by combining part or all of a cryptic plasmid from the host organism with a replicon derived from an *E. coli* plasmid. There may also be fundamental differences between the expression signals of the various hosts which make it necessary to incorporate different resistance markers for selection in *E. coli* and the anaerobic host. A major obstacle for the development of cloning vectors for the methanogens has been the lack of suitable markers because these organisms are resistant to antibiotics normally used for the selection of recombinant plasmids.

The structural or segregational stability of newly constructed vectors may be poor. For example, many of the small cryptic plasmids of lactobacilli are unstable when combined with *E. coli* DNA sequences. These plasmids are members of a large family of Gram-positive plasmids which generate a single-stranded intermediate during a rolling circle replication mechanism. This mechanism of replication is easily perturbed by the recombination of sequences in the single-stranded replication intermediate.[36]

The presence of host restriction and modification systems, as well as plasmid incompatibility, are also important considerations when developing genetic systems for new species. Some plasmids are unable to co-exist in one bacterium, a property used to classify incompatibility groups. Although several different compatible plasmids are able to exist in one cell, cured strains lacking any extrachromosomal DNA are preferred as hosts for genetic experiments. Evidence for the presence or absence of a restriction modification system in a particular host can be determined by comparing the transformation frequency obtained with homologous and heterologous plasmid DNA. For example, a 1000-fold decrease in the numbers of bacteroides transformants has been observed using plasmid DNA isolated from *E. coli* rather than *Bacteroides* spp.[133,134] Similarly, fewer *C. perfringens* transformants were obtained when plasmid DNA was isolated from *E. coli* rather than *C. perfringens*, which suggests that restriction barriers may exist between these two organisms.[135] Smith and co-workers[134] also found that *B. fragilis* was 8 to 12-fold less efficiently transformed with plasmid DNA from heterologous *Bacteroides* species, presumably indicating species specific differences in the restriction and modification systems. In contrast, however, no differences in the transformation efficiencies were found with *C. perfringens* strain L-13, using plasmid DNA isolated from *E. coli* or *C. perfringens*.[136,137]

Type II restriction endonucleases have been identified in *D. desulfuricans* (strain Norway)[138] and in three different species of *Bifidobacterium* which indicates that they might also be present in strains chosen for genetic manipulation.[139-141]

## B. MOBILIZATION OF GENETIC MATERIAL BY PLASMID CONJUGATION

Gene transfer by conjugation has only been extensively exploited for genetic studies in the genus *Bacteroides*, although conjugative plasmids have been discovered in both lactobacilli and clostridia. In lactobacilli, the ability to ferment lactose and the production of bacteriocin have been associated with natural transfer mechanisms. However, little is known about the plasmid biology of these systems (Table 1).[42,142,143] Significant potential exists to exploit the broad host range streptococcal plasmid pAMβ1, and its derivatives, for genetic manipulation of lactobacilli. This plasmid has been successfully transferred into *L. plantarum* and *L. acidophilus* by conjugation.[144-146]

Native conjugative plasmids have not been used to introduce additional genetic material into clostridia, but pAMβ1 has proved especially useful for this purpose. Although use of this plasmid has been limited to industrially important species of clostridia, pAMβ1 and its derivatives have clear potential for gene transfer and genetic manipulation of intestinal species. Plasmid pAMβ1 has been transferred to *C. acetobutylicum* from *Lactococcus lactis*, *Ent. faecalis*, *Bacillus subtilis*, and other strains of *C. acetobutylicum*.[147-149] The plasmid has been successfully used as a cloning vehicle for inter-species complementation in clostridia, and also to express a pseudomonad *xylE* gene in *C. acetobutylicum*.[150,151] Both of these strategies exploited the ability of pAMβ1 to form co-integrate molecules with Rep⁻ plasmids transformed into *Bac. subtilis*. Although the plasmids transferred to *Bac. subtilis* were unable to replicate, they encoded a selectable marker and a region of homology with pAMβ1 which promoted co-integrate formation. The plasmids containing the pAMβ1 transfer determinants could then be conjugally transferred to *C. acetobutylicum*. Derivatives of pAMβ1, lacking a replication origin, have also been used to mobilize other Gram-positive plasmids into *C. acetobutylicum*.[152]

The conjugal transfer system described by Trieu-Cuot et al.[153] has also been used to transfer the prototype vector pAT187 containing a pAMβ1 replicon to *C. acetobutylicum* from *E. coli*. In this system, the transfer functions of the IncP plasmid pRK212.2 efficiently mobilize vectors such as pAT187, containing the RK2 origin of transfer (*oriT*) from *E. coli* to many Gram-positive bacteria, by the filter mating procedure.

In contrast to other gut anaerobes, *Bacteroides* species have been extensively manipulated by conjugation. Conjugal transfer of MLS resistance among bacteroides was first described by three independent groups of workers.[65,154,155] Since then, a number of vectors have been developed for conjugal transfer from *E. coli* to bacteroides which contain replicons and marker genes derived from both hosts. These plasmids may be transferred by helper IncP plasmids of *E. coli* such as R751 or RK231, which cannot replicate in bacteroides.[80,156,157]

In *Bacteroides* spp., the chromosomal Tc$^r$ elements have been used to conjugally transfer co-resident plasmids between strains or from *Bacteroides* to *E. coli* recipients.[64,94] This process has allowed those strains which do not receive shuttle vectors from *E. coli* at satisfactory efficiencies to be manipulated. One cryptic plasmid (pB8-51), which is common to several species, has been used to construct vectors which can be transferred by conjugation from one strain of bacteroides to another, and to *E. coli* by the chromosomal Tc$^r$ elements and the conjugative plasmid pBF4 (Table 1).[64,94]

An important advance in the development of genetic transfer systems for the SRB has been the successful IncP dependent transfer of the *E. coli* broad-host range IncQ plasmid R300B to two strains of *Desulfovibrio*.[158] This required the development of a medium which would support the co-culture of the obligately anaerobic sulfate reducer and the facultative anaerobe *E. coli*. In crosses between *Desulfovibrio* and *E. coli*, plasmid R300B, carrying resistance markers for streptomycin (Sm), and sulfadiazine (Su) were transferred at very high frequencies ($5 \times 10^{-2}$ to 1 per recipient). However, the sulfamide resistance marker was only expressed in one *Desulfovibrio* strain, indicating possible taxonomic divergence between species. Matings between *Desulfovibrio* sp. 8031 and *Alcaligenes eutrophus* harboring the IncP1 plasmid pMOL4, also resulted in high frequency transfer, but the plasmid appeared to be unstable, as the resistance marker was rapidly lost during growth. Similarly, van den Berg et al.[159] have demonstrated IncQ and IncP plasmid transfer to *D. vulgaris* at similar frequencies ($1 \times 10^{-2}$ per recipient; Table 3). These workers also showed that the IncP1 plasmid was stable in *D. vulgaris*.

## C. TRANSDUCTION

Transduction is a method of gene transfer mediated by bacteriophage. Zinder and Lederberg[160] were the first to show that genetic markers of *Salmonella* spp. were transferred by a temperate bacteriophage. The phage, named P22, was able to incorporate any genetic marker into a phage particle with approximately equal efficiency. This type of gene transfer is known as generalized transduction and differs from specialized transduction mediated by the *E. coli* lambda bacteriophage. During specialized transduction, only those markers adjacent to the prophage attachment site have the potential to be transferred.[161]

Transduction has not been described in *Clostridium* or *Bacteroides* and there are only a few reports of transduction in *Lactobacillus*, despite much work on the temperate and virulent bacteriophages of the latter genus. One early study described the transfer of auxotrophic markers of lysine, proline, serine, and lactose metabolism by a temperate bacteriophage PLS-1 in *L. salivarius*.[162] More recently, the temperate bacteriophage øadh was shown to mediate transfer of several different plasmids in *L. acidophilus* in a process typical of generalized transducing phage.[163] As many key functions of the lactic acid bacteria are plasmid encoded, transduction should prove extremely useful for genetic studies in these organisms. Transduction systems such as these warrant further investigation as they may prove to be effective means by which to integrate and stabilize plasmid encoded traits in lactobacilli.

The first system of genetic exchange described for SRB was generalized transduction mediated by a defective bacteriophage.[164] In this study, transfer of antibiotic resistance between different derivatives of *D. desulfricans* strain ATCC 27774 was mediated by a defective bacteriophage (Dd1) at frequencies of transfer in the range of $10^{-5}$ to $10^{-6}$ per (recipient) colony forming unit. Evidence for bacteriophage-mediated transduction was provided by showing that the removal of bacteriophage-like particles from

**Table 3** Gene Transfer Systems

| Organism and strain | Method | Efficiency | References |
|---|---|---|---|
| **Conjugation systems** | | | |
| E. coli to Bacteroides spp. | IncP dependent mobilization | $10^{-6}$ to $10^{-3}$ [a] | 94, 156 |
| Bacteroides spp. to other Bacteroides spp. or E. coli | Mobilization by Bacteroides Chromosomal Tc$^r$ elements | $10^{-3}$ to $10^{-5}$ | 64, 94 |
| Streptococcus to L. plantarum | Conjugal transfer, pAMβ1 and pIP501 | $3.9 \times 10^{-6}$ to $1.1 \times 10^{-5}$ [a] | 144 |
| Streptococcus faecalis to L. plantarum | Conjugal transfer of pAMβ1 | $4.5 \times 10^{-8}$ to $1.1 \times 10^{-7}$ [a] | 145 |
| Lactococcus lactis to Lact. reuteri L. acidophilus, and L. salivarius | Conjugal transfer of pAMβ1 | $4 \times 10^{-9}$ to $1.9 \times 10^{-5}$ [a] | 146 |
| E. coli to Desulfovibrio spp. | IncP dependent transfer | $10^{-2}$ to 1 [a] | 158 |
| E. coli to D. vulgaris | InQ and IncP plasmid transfer | $1 \times 10^{-2}$ [a] | 159 |
| **Transduction** | | | |
| L. acidophilus | Transduction of plasmids by øadh | $3.6 \times 10^{-8}$ to $8.3 \times 10^{-10}$ [b] | 163 |
| L. salivarius | Generalized transduction of auxotrophic markers by PLS-1 | $10^{-7}$ to $10^{-8}$ [b] | 162 |
| **Protoplast transformation** | | | |
| C. perfringens strain 11268 | PEG treatment of protoplasts or autoplasts | $1.4 \times 10^{-6}$ [c] | 173 |
| C. perfringens strain 11268 | PEG treatment of protoplasts | $2 \times 10^{-7}$ [c] | 135 |
| C. perfringens strain L-13 | PEG treatment of protoplasts | $5 \times 10^{-7}$ [c] | 137 |
| C. perfringens strain L-13 | PEG treatment of protoplasts | $3.9 \times 10^{-5}$ [c] | 136 |
| L. acidophilus strain 100-33 | PEG treatment of protoplasts | $2 \times 10^{-7}$ to $8 \times 10^{-8}$ [c] | 167 |
| L. plantarum NC1 and NC4 | PEG assisted transfection of bacteriophage DNA | 25 to 230 PFU/mg DNA | 170 |
| L. fermentum strains 604 and 605 | Intragenic protoplast fusion | $1 \times 10^{-8}$ to $3 \times 10^{-8}$ [d] | 168 |
| L. acidophilus and Lact. reuteri | PEG treatment of protoplasts | 5 to 1000/μg DNA | 169 |
| **Chemical methods** | | | |
| B. fragilis 638 | PEG facilitated transformation | $4.2 \times 10^3$ [e] | 230 |
| **Electroporation** | | | |
| C. perfringens strain 3624A | Electroporation of intact cells | $1.2 \times 10^3$ [e] | 175 |
| C. perfringens strain 3624A | Electroporation of intact cells | $1.0 \times 10^4$ [e] | 176 |
| C. perfringens strain 3624A | Electroporation of late log phase cells | $9.2 \times 10^4$ [e] | 177 |
| C. perfringens strain 13 | Electroporation of late log phase cells | $9.8 \times 10^6$ [e] | 177 |
| C. perfringens strain P90.2.2 | Electroporation of late log phase cells | $1.2 \times 10^4$ [e] | 178 |
| C. perfringens strain 13 | Electroporation of lysostaphin-treated cells | $3 \times 10^5$ [e] | 179 |
| B. uniformis 1100 | Anaerobic electroporation | $1 \times 10^6$ [e] | 133 |
| B. fragilis strain 638 | Anaerobic electroporation | $8.1 \times 10^5$ [e] | 134 |
| B. ovatus strain V211 | Anaerobic electroporation | $1.1 \times 10^4$ [e] | 134 |
| B. uniformis strain V528 | Anaerobic electroporation | $1.9 \times 10^4$ [e] | 134 |
| Lactobacillus spp. | Electroporation of intact cells | $1.7 \times 10^2$ to $9.7 \times 10^5$ [e] | 180 |
| L. plantarum ATCC 8014 | Cells grown in 1% glycine | $2 \times 10^2$ [e] | 182 |
| Lactobacillus spp. | Cells grown in 1% glycine or 0.2 mol/l D-L threonine | $1 \times 10^2$ to $1 \times 10^7$ [e] | 183 |
| Propionibacterium jensenii | Electroporation of intact cells | $3.2 \times 10^1$ [e] | 180 |
| Desulfovibrio fructosovorans | Anaerobic electroporation | Chromosomal DNA used | 188 |
| Methanococcus voltae | Anaerobic electroporation | Chromosomal DNA used | 186 |

*Note:* Further information on how the transformation efficiency obtained by a particular method varies with different conditions, strains, and plasmids may often be found in the reference cited.

[a] Per recipient. [b] Per PFU. [c] Per viable cell per μg. [d] Of the total cell input. [e] Per μg DNA.

cell-free filtrates prevented gene transfer. No induction or plaque formation has been observed with this bacteriophage suggesting that it may be defective in its ability to preferentially package phage DNA.

The transduction of genetic markers has been reported in two different nonintestinal species of methanogens but no data are available for the large gut species.[73,165]

## D. PROTOPLAST TRANSFORMATION

The cell walls of Gram-positive bacteria present a natural barrier to the uptake of exogenous DNA. However, methods developed in the late 1970s for the preparation of osmotically stabilized cells, which had been enzymatically stripped of their cell walls (protoplasts) has enabled workers to induce organisms to assimilate DNA in the presence of polyethylene glycol (PEG). The protoplast transformation procedures described so far (Table 3) have often been tailored to optimize protoplast induction and the regeneration of particular strains. Consequently, these procedures have gained a reputation for being difficult to reproduce, especially with strains of bacteria other than those used in the original methodology.

Procedures for the preparation and regeneration of protoplasts among species of the genus *Lactobacillus* were first developed for *L. casei*.[166] Similar conditions were also adopted for plasmid transformation of *L. acidophilus*[167] and protoplast fusion of different strains of *L. fermentum*.[168] In addition, PEG-assisted protoplast transformation and transfection has been reported by several groups.[169-172] As with many protoplast transformation procedures, however, the frequency is low and restricted to selected strains only.

The transformation of intestinal clostridia using these methods has been less successful. Although procedures have been established for the transformation of protoplasts (L-forms), they are of limited use as it has not been possible to regenerate the cell wall afterwards.[136,137] A procedure for regenerating bacillary walled bacteria from autoplasts (protoplasts which form as a result of autolytic activity in osmotically stabilizing buffer) has been described.[173] However, the method is tedious and possibly strain specific, since it could not be reproduced by other workers.[136]

## E. ELECTROPORATION

The most simple and rapid method for transformation of anaerobic bacteria is electroporation which is increasingly becoming the method of choice, especially for new species. The procedure involves discharging a high voltage electric field (2–12.5 kV/cm) through a suspension of cells in a pulse lasting between 2.5 and 10 ms. The electric field creates pores in the cell membrane which allow DNA molecules to enter the cell.[174]

The first attempts to transform *C. perfringens* by electroporation of intact cells gave only low numbers of transformants ($10^2$ to $10^3$ transformants per $\mu$g of DNA).[175] Subsequent improvements to this method yielded approximately tenfold higher efficiencies of transformation with the same strain.[176] While this method was applicable to A and C type strains, no transformants were obtained with three different type B strains of *C. perfringens*.[177] The improvement in transformation efficiency may have been due to the use of cells in the late stationary phase of growth, which may have undergone partial autolysis resulting in weakening of the cell wall and improved accessibility of DNA to the cell membrane. This idea is supported by similar studies on the effects of culture age on the frequency of transformation with different strains of *C. perfringens*.[178] Other workers have treated cells with lysostaphin to cleave pentaglycine bridges in the cell wall before electroporation (Table 3).[179] The variable results obtained with this procedure are probably not due to strain variation or quality of the DNA, but could arise from the use of different batches of culture medium which may cause alterations in either the thickness or composition of the cell wall.[52]

In a detailed study of the effects of various parameters on the transformation of *L. acidophilus* by electroporation, a protocol was developed which could be used to transform several different strains of enteric lactobacilli with efficiencies varying from $3.8 \times 10^2$ to $9.7 \times 10^5$.[180] It is significant that the same procedure was used to successfully transform several other Gram-positive bacteria, even though conditions were not optimized for each strain. Such results are in marked contrast to protoplast methods for transformation and demonstrate its applicability for a broad range of bacteria. Improvements in the efficiency of transformation of lactobacilli by electroporation have also been achieved by strategies which increase the permeability of the cell wall. Growth of strains in the presence of amino acids such as threonine and glycine have been shown to affect the sensitivity of strains to lysozyme.[181] These procedures were used by other workers to develop methods for transformation of enteric species of lactobacilli by electroporation.[182,183] For some strains, the efficiency of transformation was too low to enable direct cloning of genes, or the preparation of reasonably sized gene libraries. However, it should be possible to define more precisely the degree of glycine-induced cell wall weakening required to improve the accessibility of DNA to the cell membrane without increasing the osmotic fragility of the cells.[184,185]

Electroporation of cells under anaerobic conditions has also been used to transform the strict anaerobe *B. uniformis* with frequencies greater than $10^5$ transformants per mg of DNA.[133] Subsequently, an electroporation procedure for several species of bacteroides has been developed which does not rely on the use of anaerobic conditions.[134] However, incorporation of 1 mmol/l $MgCl_2$ to the electroporation buffer (10% glycerol) improved the recovery of transformants 100-fold. Moreover, the cells could be stored frozen in the electroporation buffer with only a twofold reduction in the recovery of transformants.

Electroporation methods for transformation have also recently been applied to *Methanococcus* and *Desulfovibrio* species. Rescue of an auxotrophic mutant of *M. voltae* by electroporation with chromosomal DNA, was 50- to 80-fold more efficient than natural transformation.[186] Similar frequencies were observed with an integration vector (pMip2) carrying the puramycin acetyltransferase resistance marker from *Streptomyces alboniger*, and the cloned *hisA* gene of *M. voltae* to promote recombination. The low efficiency of transformation with this vector may be due to restriction of plasmid DNA isolated from *E. coli* in *M. voltae*.[187] However, the successful use of a selectable marker for genetic transformation of a methanogen is an important step towards developing systems for genetic manipulating of these organisms.

Transformation of intestinal species of SRB has not as yet been reported. However, electroporation has been used to construct a *hydN* gene mutant of *Desulfovibrio fructosovorans* by marker exchange using the kanamycin resistance gene as a reporter.[188] The demonstration that transformation can be achieved by electroporation in these organisms should allow a more rapid development of the vectors required for genetic manipulation.

## F. CLONING VECTORS

Some of the most valuable cloning vectors currently available for use in gut anaerobes are shown in Table 4. The *C. perfringens* cloning vectors are shuttle vectors capable of replication and selection in both *C. perfringens* and *E. coli*. These contain a replicon derived from a *C. perfringens* plasmid in addition to an *E. coli*-derived replicon. The clostridial antibiotic resistance determinants *tetP*, *catP*, and *ermBP* have been employed as transformation markers and have the advantage that they are all expressed in *E. coli* and *C. perfringens*. As pointed out by Rood and Cole[52], however, many of these vectors are unsuitable for studies of pathogenicity in *C. perfringens*, since they carry the β-lactamase (*bla*) gene which confers resistance to ampicillin. This gene has not been found in *C. perfringens* and, for safety reasons, it would be prudent not to introduce it. Vector pAK201 does not carry an intact (*bla*) gene but might also be considered unsuitable as a vector for *C. perfringens* because it does harbor a complete copy of the λ toxin gene. The toxin gene is apparently not expressed, however, the reasons for this have not been investigated. The most recent generation of shuttle vectors for *C. perfringens* including pJIR418,[189] pJIR750, and pJIR751[190] do not have an intact copy of the *bla* gene and are based on the well-characterized *C. perfringens* plasmid pIP404. These vectors have the advantage that they are likely to be more structurally stable than other *C. perfringens* shuttle vectors, which are based on plasmids for which the mechanism of replication is uncharacterized. Many small plasmids found in Gram-positive organisms replicate by the rolling circle mechanism of replication (RCR) which involves generation of a single stranded intermediate. It is well recognized that the insertion of DNA into plasmids which replicate by this mechanism can lead to structural instability, probably as a result of sequence recombination and excision in the single-stranded intermediate.[36] Plasmid pJIR 418 (Figure 1) for which the entire sequence is known is perhaps the most advanced cloning vector available for *C. perfringens*. It contains the pIP1404 replicon, *catP* and *ermBP* (which function in both hosts) and the multiple cloning site and *lacZ'* gene of pUC18 which enables the screening of recombinants on 5-bromo-4-chloro-3-indolyl β-galactopyranoside (X-Gal) medium in *E. coli*.[189] Two derivatives of pJIR 418 have recently been described which carry only one of the two antibiotic resistance determinants present on the original vector.[190] The number of selectable markers available for *C. perfringens* are limited, therefore these new vectors (pJIR750 and pJIR751) considerably increase the potential options for genetic manipulation of this organism.

**Table 4**  Representative Cloning Vectors

| Plasmid | Size (kb) | Source of replicons | | Phenotypes[a], comments | Reference |
| | | Host | E. coli | | |
| --- | --- | --- | --- | --- | --- |
| **Vectors for Clostridium perfringens** | | | | | |
| pJU12 | 11.6 | pJU121 (*C. perfringens*) | pBR332 | Tc[r], [Ap[r], Tc[r]], rudimentary vector | 135 |
| pJU13 | 12.2 | pJU122 (*C. perfringens*) | pBR332 | Tc[r], [Ap[r], Tc[r]], rudimentary vector | 135 |
| pJU16 | 12.2 | pJU122 (*C. perfringens*) | pBR332 | Tc[r], [Ap[r], Tc[r]], rudimentary vector | 135 |
| pHR106 | 7.9 | pJU122 (*C. perfringens*) | pSL100 | Cm[r], [Ap[r], Cm[r]] | 137 |
| pSB92A2 | 7.9 | pCP1 (*C. perfringens*) | pHG165 | Cm[r], [Ap[r], Cm[r]] | 178 |
| pAK201 | 8.0 | pHB101 (*C. perfringens*) | pBR332 | Cm[r], [Cm[r]] | 176 |
| pTG67 | 6.6 | pIP404 (*C. perfringens*) | pUC18 | Cm[r], [Ap[r], Cm[r]] | 54 |
| pJIR418 | 7.4 | pIP404 (*C. perfringens*) | pUC18 | Cm[r], Em[r], [Cm[r], Em[r] XG screening] | 189 |
| pJIR750 | 6.6 | pIP404 (*C. perfringens*) | pUC18 | Cm[r], [Cm[r], XG screening] | 190 |
| pJIR751 | 5.9 | pIP404 (*C. perfringens*) | pUC18 | Em[r], [Em[r] XG screening] | 190 |
| **Vectors for Lactobacillus** | | | | | |
| pLPE323 | 3.6 | p353-2 (*L. pentosus*) | None | Em[r], vector for *Lactobacillus* spp. | 171 |
| pLPE350 | 3.9 | p353-2 (*L. pentosus*) | None | Em[r], vector for *Lactobacillus* spp. | 171 |
| pLPC37 | 3.7 | p8014-2 (*L. plantarum*) | None | Em[r], vector for *Lactobacillus* spp. | 171 |
| pLE16 | 7.6 | pLB10 (*Lactobacillus* sp.) | pBR328 | Cm[r], [Cm[r]] Further studies on structural stability and use as a cloning vector required | 193 |
| pTRK159 | 10.3 | pPM4 (*Lactobacillus* sp.) | pACYC184 | Em[r], Tc[r], Cm[r] [Tc[r], Cm[r]] Based on a Rep⁻ derivative of pSA3 | 231 |
| **Vectors for Bacteroides** | | | | | |
| pFD176 | 7.3 | pBI143 (*B. ovatus*) | pUC19 | MLS[r], [Ap[r]] | 202 |
| pBI191 | 5.3 | pBI143 (*B. ovatus*) | None | MLS[r], only replicates in *Bacteroides* | 202 |
| pDP1 | 19 | pCP1 (*B. fragilis*) | pBR322 | MLS[r], [Ap[r], Tc[r]] | 67 |
| pE5-2 | 17 | pB8-51 (*B. eggerthii*) | pRSF1010 | MLS[r], [Su[r], Tc[r]] | 80 |
| pSS-2 | 46 | None | pRSF1010 | MLS[r], [Su[r], Tc[r]] suicide vector | 80 |
| pDK3 | 8.5 | pCP1 (*B. fragilis*) | pBR322 | MLS[r], [Ap[r]] | 156 |
| pDK4.1/4.2 | 8.5 | pCP1 (*B. fragilis*) | pBR322 | MLS[r], [Ap[r]] pDK3 derivatives | 156 |
| pVAL-1 | 11 | pB8-51 (*B. eggerthii*) | pBR328 | MLS[r], [Ap[r], Tc[r]] | 64 |
| pVAL-7 | 9.5 | None | pBR328 | MLS[r], [Ap[r]] suicide vector | 232 |
| pE3-1 | 13 | None | pBR328 | MLS[r], [Ap[r], Tc[r]] suicide vector | 122 |
| pFD325TT | 10.2 | pBI143 (*B. ovatus*) | pUC19 | MLS[r], [Ap[r]] | 204 |
| pFD288 | 8.8 | pBI143 (*B. ovatus*) | pUC19 | Cc[r], [Sp[r]]; *oriT*; pUC19/pBI143 chimera | 233 |
| pFD290 | 7.8 | pBI143 (*B. ovatus*) | pUC19 | Cc[r], [Ap[r]], expression vector | 204 |
| pFD340 | 8.6 | pBI143 (*B. ovatus*) | pUC19 | Cc[r], [Ap[r]], *oriT*, expression vector | 204 |
| pFD214 | 6.3 | pBI143 (*B. ovatus*) | pUC19 | (MLS[r])[b] [Ap[r]], promoter probe vector | 205 |
| pKBF367-12 | 10.5 | pBF367 (*B. fragilis*) | pBR322 | Cc[r], [Km[r]] | 157 |
| pJST61 | 12.3 | pBFTM2006 (*B. fragilis*) | pBR322 | MLS[r], [Ap[r], Tc[r]] *oriT* of RK2; l pR-*endRI* fusion product lethal to *E. coli* in the absence of cI repressor; cloning into a unique BglII site inactivates the *endRI* gene | 201 |
| **Vectors for methanogens** | | | | | |
| pMip1 | 7.4 | None | pUC18 | Pm[r], [Ap[r]] integration vector for *M. voltae* | 187 |
| pMip2 | 7.4 | None | pUC18 | Pm[r], [Ap[r]] integration vector for *M. voltae* | 187 |
| **Broad host-range vectors for Gram-positive organisms** | | | | | |
| pGK12 | 4.4 | pWVO1 (*Lact. lactis*)[c] | (None) | Em[r], Cm[r], [Cm[r]] | 195 |
| pMIG1 | 4.8 | pSH71 (*Lact. lactis*)[c] | (None) | Cm[r], Km[r], [Cm[r], Km[r]], MCS | 185 |

**Table 4** Continued.

| Plasmid | Size (kb) | Source of replicons | | Phenotypes[a], comments | Reference |
|---|---|---|---|---|---|
| | | Host | E. coli | | |
| pMIG4 | 4.6 | pSH71 (*Lact. lactis*)[c] | (None) | Em[r], MCS [Em[r]] (unpublished[d]) | |
| pIL252 | 4.7 | pAMβ1 (*Ent. faecalis*) | (None) | Em[r], MCS; low-copy-number version | 198 |
| pIL253 | 5.0 | pAMβ1 (*Ent. faecalis*) | (None) | Em[r], MCS; high-copy-number version | 198 |
| pMTL500E | 6.4 | pAMβ1 (*Ent. faecalis*) | pMTL20 | Em[r] [Ap[r]], MCS | 199 |
| pMTL502E | 7.5 | pAMβ1 (*Ent. faecalis*) | pMTL20 | Low-copy-number version of pMTL500E | |

*Note:* All of the vectors are potentially capable of being transferred by transformation procedures and some are also capable of being conjugally mobilized. For example, the indigenous plasmids of *Bacteroides* which have been used to construct shuttle vectors (i.e., pBI143, pB8-51, pCP1, and pBFTM10) can be mobilized from *E. coli* to *Bacteroides* by IncP plasmids and from one species of *Bacteroides* to another by *Bacteroides* conjugative Tc[r] elements.

[a] The phenotypic resistance markers of the plasmids are abbreviated as follows: r, resistance; Em, erythromycin; Cc, clindamycin; Cm, chloramphenicol; Km, kanamycin; Pm, puromycin; Tc, tetracycline; MLS, macrolide-lincosamide-streptogramin B (e.g., clindamyin and erythromycin); Sp, spectinomycin. The selection for the vector in *E. coli* is bracketed. MCS, multiple cloning sites; XG, X-Gal screening; *oriT*, *E. coli* RK2 transfer origin to promote high frequency conjugal transfer from *E. coli* donors. [b]The MLS resistance marker in this promoter probe vector can only be activated by insertion of a promoter which functions in *Bacteroides*. [c]These replicons will function in *E. coli* as well as a broad host-range of Gram-positive organisms. [d]Unpublished data (J. M. Wells, 1994).

The enteric lactobacilli often harbor one or more cryptic plasmids, which could potentially serve as cloning vehicles. The sequence analysis of plasmids from different species of lactobacilli has revealed that they belong to the large family of Gram-positive vectors which replicate via a single-stranded intermediate.[45-47,171] Early attempts to construct *Lactobacillus*–*E. coli* shuttle vectors by ligating cryptic lactobacilli plasmids to pBR322, pUC8, pACY184, or an Em[r] derivative of pAMβ1, were unsuccessful as none of the constructions were able to replicate in *E. coli* or *Streptococcus sanguis*.[191] A potential shuttle vector has been described for *L. plantarum* which was obtained by ligating a small high copy cryptic *Lactobacillus* plasmid into the EcoRI site of pUC19.[192] However, this plasmid lacks a marker for selection in lactobacilli and requires further evaluation of its structural and segregational stability in both hosts. Recently, Posno et al.[183] reported on the general instability of vectors derived from lactobacilli. However, one vector, pLPE323, derived from a 2.4 kb cryptic plasmid from *L. pentosus* (p353-2), was structurally and segregationally stable in three out of four lactobacilli. It was later shown that the presence of *E. coli* or λ bacteriophage DNA in the vectors derived from p353-2, or another structurally related plasmid p8014-2, resulted in segregational instability but not structural instability of the vectors in lactobacilli.[171] As with the other small *Lactobacillus* plasmids characterized so far, p353-2 and p8014-2 carried sequence elements typical of plasmids which replicated via a RCR mechanism.

A *Lactobacillus*–*E. coli* shuttle vector was recently constructed by ligating pBR328 and a cryptic *Lactobacillus* plasmid, pLB10.[193] When this shuttle vector (designated pLE16) was transferred to *Lactobacillus* spp., it was able to grow in the presence of chloramphenicol. However, the growth of transformants was very slow and only low amounts of plasmid were recovered from the bacteria. This result may have been because of the high level of chloramphenicol (30 μg/ml) used for selection and the fact that the *cat* gene was under the control of an *E. coli* promoter. The plasmid was segregationally unstable in the absence of selection and evidence for its structural stability in *Lactobacillus* spp., was not shown.

A number of vectors have been constructed which function in a broad range of Gram-positive hosts. The replicons of the *Lactococcus lactis* cryptic plasmids pWV01 and pSH71 are highly homologous and have been used to construct broad host range shuttle vectors.[194-196] Plasmid pGK12, which incorporates the replicon of pWV01 and the *cat* and *erm* resistance determinants from pC194, has been successfully introduced into several *Lactobacillus* spp. as well as species of *Bacillus*, *Enterococcus*, *Pediococcus*, *Propionibacterium*, *Listeria*, *Staphylococcus*, and *E. coli*.[180] Small, high copy derivatives of these plasmids containing multiple cloning sites and *E. coli* replicons, as well as specialized promoter and terminator cloning vectors are also available (Figure 1).[185,197] These derivatives have considerable potential for the manipulation of lactobacilli because pGK12, which contains the same replicon, has been

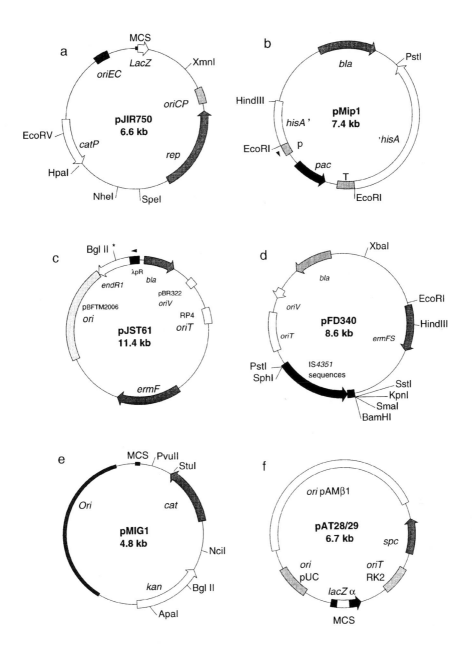

**Figure 1** Physical and genetic maps of representative cloning vectors for the intestinal anaerobes. (a) Shuttle vector pJIR750 carrying replicons which function in *Escherichia coli* (*oriEC*) and *Clostridium perfringens* (*oriCP*). A derivative is available (pJIR751) in which the *catP* gene is replaced with the *C. perfringens* erythromycin resistance gene (*ermBP*); (b) pMip1 integration vector based on *E. coli* pUC18 containing the *hisA* gene sequences of *Methanococcus voltae* interrupted by the puromycin resistance (*pac*) gene; (c) pJST61 *Bacteroides*–*E. coli* shuttle vector. The *endR1* gene product is lethal to *E. coli* in the absence of the lambda CI repressor but can be inactivated by cloning into the unique Bgl II site indicated. EndR1 is not expressed in *Bacteroides*; (d) *Bacteroides*–*E. coli* shuttle vector pFD340 for gene expression in *Bacteroides*. Genes cloned into the unique SstI, KpnI, SmaI, or BamHI sites will be transcribed by the IS4351 promoter; (e) pMIG1, one of the pMIG series shuttle vectors carrying a Gram-positive broad host-range replicon derived from *Lactococcus lactis*. A derivative pMIG4 carries an erythromycin resistance gene marker in place of the *cat* gene; (f) The broad host-range Gram-positive plasmids pAT28 and pAT29 differ with respect to the orientation of the multiple cloning site (MCS). Other abbreviations are as follows: ori, origin of replication; p, promoter; t, terminator; *OriT*, the transfer origin of *E. coli* RK2 which enables high frequency conjugal transfer from *E. coli* donors by IncP helper plasmids; bla, β-lactamase gene from *E. coli*; *ermF* and *ermFS*, macrolide-lincosamide-streptogramin B resistance (MLS$^r$) genes; *spc*, spectinomycin resistance gene.

shown to be very stable in *L. plantarum* (0.2% loss of plasmid per generation in the absence of antibiotic selection).

A number of broad-host range Gram-positive shuttle vectors based on the streptococcal plasmid pAMβ1 have also been described.[198,199] These vectors should have an extremely broad host range but there have been no reports of their use in intestinal species of clostridia or lactobacilli.

Bacteroides shuttle vectors have been developed for both transformation and conjugative gene transfer. One of the first to be developed was pDP, a Tc[r] sensitive deletion derivative of pBR322 containing the *ori*T from RK2 fused to the 15 kb *Bacteroides* plasmid pCP1. This plasmid was mobilized from *E. coli* by the IncP plasmid pRK231 to *B. fragilis* (frequency, $10^{-6}$ per recipient) but the mobilizing plasmid could not be maintained.[200] The pDP1 vectors were subsequently improved by incorporating multiple cloning sites and reducing their overall size (see pDK4.1 and pDK4.2 in Table 4).[156]

A second series of shuttle vectors (the pE5-2 series) which could be mobilized by the RK2 transfer system were constructed by Shoemaker et al.[80] The IncP-dependent origin of transfer and sulfanilamide resistance marker were provided by RSF1010, which was fused to a 4.4-kb cryptic plasmid pB8-51 and the 3.8 kb fragment from pBF4 containing the MLS[r] determinant. The frequency of transfer from *E. coli* to *B. uniformis* was 1000-fold higher than that observed with pDP1 when matings were performed under anaerobic conditions.[80] These vectors could also be mobilized to other species of *Bacteroides* or *E. coli* by the chromosomal conjugative Tc[r] elements found in many bacteroides. The IncP plasmids are not maintained in *Bacteroides*, despite their ability to replicate in many other Gram-negative bacteria. More recently, a shuttle vector (pVAL-1) was constructed by fusing the bacteroides cryptic plasmid pB8-51 with plasmid pTB1, carrying the bacteroides MLS[r] determinant, cloned into the EcoRI site of pBR328.[64] The bacteroides conjugal Tc[r] elements and conjugative plasmid pBF4 were able to recognize a mobilization region provided by the pB8-51 sequences and mobilize the shuttle vector pVAL-1 from one *Bacteroides* strain to another and from *Bacteroides* spp. to *E. coli*. Derivatives of the pVAL-1 and pE5-2 lacking the *Bacteroides* replicon have been constructed as suicide delivery vectors for *Bacteroides* (see pVAL-7, pSS-2, and pE3-1 in Table 4).[81,122]

Pheulpin et al.[157] have described the construction of small IncP mobilizable *E. coli*–*Bacteroides* shuttle vectors based on a 4.6 kb cryptic plasmid from *B. fragilis* and pKC7, a mobilizable derivative of pBR322. The clindamycin resistance determinant from *B. fragilis* plasmid pBFTM10 was incorporated into the plasmid to provide a transformation marker for bacteroides.

Recently, Thompson and Malamy[201] have constructed a *Bacteroides*–*E. coli* shuttle vector pJST61, which permits efficient library construction in *E. coli* as well as IncP dependent transfer of the library to *Bacteroides* spp. (Figure 1). The special feature of this vector is the inclusion of the λ *endRI* gene, which is lethal to *E. coli* in the absence of the cI repressor. The *endRI* gene product can be inactivated by cloning DNA fragments into the unique BglII site within the vector. Use of this site therefore ensures that all of the recombinant clones isolated in *E. coli* contain inserts. An additional important feature of this vector is that the cloned genes may be expressed from the strongly regulated promoter which is located upstream of *endRI*.

One other cryptic plasmid pBI143 (2.7 kb), which is widely disseminated among all the intestinal tract *Bacteroides* species, has been used to construct shuttle vectors.[202,203] For example, the shuttle vector pFD176 was constructed by ligating pBI143 to the pUC19 cloning vector and a 1.9 kb fragment of the *Bacteroides* plasmid pBF4 carrying the selectable MLS[r] determinant. Promoter probe vectors and expression vectors have also been constructed for bacteroides based on pBI143 and pUC19 fusions (see Table 4).[204,205] As for the pB8-51 based shuttle vectors, derivatives based on pBI143 can be mobilized by bacteroides chromosomal Tc[r] elements and IncP plasmids of *E. coli*.

Although transformation procedures exist for methanogens, and several plasmids have been isolated from certain species, derivatives of them have not been reintroduced.[73,206-208] However, a major advance in the development of transformation systems for these bacteria has been the development of an integration vector (Mip1) and the demonstration that a eubacterial puromycin resistance marker could be expressed in *Methanococcus voltae*.[187] This vector is based on the *E. coli* vector pUC18, and incorporates the cloned *hisA* gene of *M. voltae* to provide sequences for homologous recombination. The *hisA* gene is interrupted by a fragment of cloned DNA from *Streptomyces alboniger* encoding the puromycin acetyltransferase (*pac* gene) resistance marker which has been placed under control of the *M. voltae* methyl coenzyme M reductase promoter. Following electroporation of *M. voltae* with pMip1 DNA, puromycin resistant clones were recovered in which the vector had stably integrated into the chromosome via recombination between the *hisA* gene sequences. The *pac* gene therefore provides a potentially valuable marker for the development of cloning vectors for methanogenic bacteria.

## G. GENE EXPRESSION

There are numerous examples of heterologous genes not expressed in *Bacteroides*, particularly antibiotic resistance determinants found on the various shuttle vectors. One exception is the *tetM* gene of streptococci present on pDK3, for which a small increase in tetracycline resistance was detected in *Bacteroides*.[156] Similarly, some *Bacteroides* resistance determinants are not expressed in *E. coli*, even when transcribed from the *E. coli lac* promoter. The reasons for the failure to express heterologous genes in different organisms are complex and require systematic analysis of the activity of different transcription and translation control signals in the various hosts. Differences between hosts in respect to codon usage, proteolytic degradation, and secretion signals may also play an important role in the efficiency of expression or activity of the protein produced. Genetic tools are now available to study these factors in some species of intestinal anaerobes.

Recently, Smith and co-workers[204] developed two expression vectors for *Bacteroides* spp. which use the outwardly directing promoters of a *Bacteroides* insertion element, IS*4351*, to express heterologous genes. These vectors are based on the general purpose *Bacteroides*–*E. coli* shuttle vector constructed by Smith[203] (Table 4) in which a 1.6 kb fragment containing one copy of IS431 less 27 bp from one end had been inserted into the multiple cloning site. This construct (pFD290) has the IS4351 promoters upstream of the unique cloning sites for BamHI, SmaI, KpnI, and SstI. A translation stop codon is present at the end of the IS fragment in the same frame as the putative transposase reading frame. A mobilizable derivative of pFD290 was constructed by incorporating the *ori*T of plasmid RK2 to generate pFD340 (Figure 1). The *cat* gene of Tn9, *tetM* gene of *Streptococcus agalactiae*, and *udiA* β-glucuronidase gene of *E. coli* were expressed in *Bacteroides* using these vectors. The *E. coli tet*C gene was also tested, but this did not confer resistance to tetracycline in *Bacteroides*.[204] This might be due to the poor match between the *E. coli* ribosome binding site of the *tetC* gene (3/6 bp match) and the 3′ end of the 16S rRNA. Only low or undetectable levels of activity were detected in *Bacteroides* when these antibiotic resistance genes were placed under control of their native promoters. This supports the notion that there are fundamental differences between the expression signals in *E. coli* and *Bacteroides*. Further evidence for this idea was provided by analyzing the activity of strong *E. coli* promoters in *Bacteroides* and *E. coli*. For this purpose, Smith et al.[204] constructed a promoter probe vector by cloning a promoterless *cat* gene into the shuttle vector pFD288. DNA fragments encoding translation stop codons in all three reading frames were placed 35 bp upstream of the *cat* gene initiation codon to prevent translational fusions. Efficient transcription terminators were also placed downstream of the *cat* gene to prevent aberrant transcription of plasmid DNA sequences. When several strong *E. coli* promoters were cloned into pFD325TT, they displayed no detectable activity in *Bacteroides* while they proved to be highly active in *E. coli*. The important conclusion of this work was that indigenous *Bacteroides* promoters are needed to obtain gene expression in these organisms.

Apart from antibiotic resistance genes, there are no examples of heterologous gene expression in the enteric species of *Clostridium*, although numerous clostridial genes have been expressed in *E. coli*.[3] A problem with the expression of some genes of Gram-negative or eukaryotic origin in clostridia may arise as a consequence of the considerably high dA + dT content of some species (e.g., 75% for *C. perfringens*). Codons present in genes of Gram-negative bacteria or eukaryotic origin, which are rarely used in clostridia, may therefore limit the level of heterologous translation. Conversely, the failure to express the *C. perfringens bcn* gene in *E. coli* has been attributed to the strong codon bias of mesophilic clostridia, since there was no detectable expression even when the *bcn* gene was placed under control of the the *E. coli lacZ* transcriptional and translational signals.[53] Several *C. perfringens* genes have been analyzed for their codon usage, and degree of expression in *E. coli* using the T7 expression system.[52] These findings suggested that there was no correlation between the level of expression and the number of rare codons but that gene length was the decisive factor. These workers suggested that the longer the mRNA, the greater the probability that a ribosome will stall at rare codons and allow the mRNA to be degraded by RNases. However, before firm conclusions can be drawn regarding the influence of codon usage on gene expression levels, other factors such as toxicity of the product in a heterologous host, proteolytic degradation, mRNA stability, and the influence of the secondary structure of the mRNA on translation initiation require investigation.

A significant impetus exists to develop improved strains of industrially important lactobacilli by genetic manipulation. One reason is that they have a major role in the fermentation and preservation of silage, meat, vegetables, and dairy products.[49] Progress has already been made with strains of *L. plantarum* that are used as commercial inocula for grass silage. The preservation of crops by lactic acid bacteria is brought about by the fermentation of carbohydrate to lactic acid. This process lowers pH and prevents the growth of harmful organisms. However, the limiting amount of fermentable sugar present in grass silage makes it necessary to add additional carbohydrate or commercial preparations of enzymes that degrade plant polysaccharides. As an alternative approach, some workers have engineered strains of *L. plantarum* to produce an endoglucanase from *Clostridium thermocellum*.[209,210] In one case, a DNA fragment encoding *celE*, the *C. thermocellum* endoglucanase gene, and its native promoter was cloned into the broad host range shuttle vector pSA3 to generate pM25.[209] Endonuclease activity was detected in the culture medium and in much lower amounts within cell extracts suggesting that the clostridial signal secretion signal is efficiently recognized and processed by lactobacilli. This work provides evidence for the ability of clostridial transcription and translation signals to function in *Lactobacillus* spp., although it is likely that more efficient expression of *celE* would result from using homologous expression signals. The recombinant plasmid was lost at a frequency of 5% per generation in the absence of selection, indicating a requirement for stabilization of these genes by integration. Similarly, other workers have found segregational instability of plasmids carrying an amylase gene from *Bacillus stearothermophilus* and the *celA* endoglucanase gene from *C. thermocellum* in *L. plantarum*.[210] An improvement in the stability of the selective marker was obtained by insertion of the transforming genes into the chromosome of *L. plantarum* by means of single homologous recombination. Both genes were expressed in lactobacilli but the expression of α amylase was reduced when inserted into the chromosome, probably due to the lower gene dosage. Interestingly, the integrated DNA could be amplified in some of the transformants by selection on increasing concentrations of antibiotic (as shown for integrated genes in *B. subtilis*).[211]

A strain of *L. curvatus* transformed with a staphylococcal vector containing the lipase gene from *Staphylococcus hyicus* under control of a constitutive *Staphylococcus carnosus* promoter expressed and secreted the lipase into the growth medium.[212]

## H. RELEASE OF GENETICALLY ENGINEERED BACTERIA

Much of this chapter deals with methods of manipulating DNA of anaerobic bacteria either by mutation in the host organism or by modification and cloning in a new host species, such as *E. coli* (see Table 5). The rapid advances in technology which enable such manipulation have been mirrored by increasing anxiety about the consequences of releasing genetically engineered microorganisms (GEMs) into the environment. One concern is that the pathogenic potential of these bacteria may be increased by the introduction of recombinant DNA or by genetic modification in the laboratory. However, the multifactorial nature of virulence in the majority of bacterial pathogens would seem to mitigate against the accidental production of a new pathogen and its effective maintenance in the environment.[213-215] Another perceived risk is to the environment itself, particularly if the engineered strains should become pests, either themselves or by transmission of their new DNA to other species.[216] In this context, it is the survival and persistence of the organism in the environment that will decide its likely impact on the ecosystem, and it is this issue which requires serious consideration when contemplating the release of GEMs. The increasing interest in the modification of intestinal bacteria for oral administration to humans is particularly relevant to workers involved in the genetic manipulation of gut anaerobes.

The possibility of using live recombinant bacteria as antigen delivery vehicles is now a major area of vaccine research. Studies with attenuated pathogenic bacteria are furthest advanced with the *aro* mutants of *Salmonella* spp.[217] and with BCG.[218] However, there is now a growing interest in the use of non-pathogenic Gram-positive lactic acid bacteria and oral streptococci for this purpose, as these food-grade organisms can be repeatedly administered without any pathogenic effect.[219-222]

There is a body of genetic information on anaerobic rumen bacteria, primarily focused on plant polysaccharide degrading enzymes of *Fibrobacter*, *Ruminococcus*, *Butyrivibrio*, and *Bacteroides* and those involved in lactose metabolism by *Streptococcus bovis*.[223-225] One long-term objective of these studies may be to alter the overall rumen fermentation by the introduction of genetically modified organisms with enhanced degradative abilities and hence increase energy released from animal feeds. Clearly, such studies have potential parallels in the human large intestine, where bacteria play a key role in production of both toxic and nontoxic metabolites. Genetic manipulation of the fermentation in the

**Table 5**  Cloned Genes of Intestinal Anaerobes

| Species | Gene | Size (kbp) | Sequence | Expression | Reference |
|---|---|---|---|---|---|
| *Bacteroides fragilis* | Glutamine synthetase (*glnA*) | 2.19 | c | EC | 107 |
| | β-lactamase (*ccrA*) | 0.75 | c | + | 234 |
| | Carbapenemase | — | | + | 235 |
| | Conjugal transfer (*btgA, btgB*) | 2.16 | c | + | 72 |
| | Methyltransferase, CLNr (*ermFU*) | 0.80 | c | − | 236 |
| | Iron superoxide dismutase (*sod*) | 0.58 | c | − | NP |
| | Imipenem-cefoxitin resistance (*cfiA*) | 0.79 | c | + | 201 |
| | Sialidase (*nanH*) | 0.80 | p | EC | 237 |
| | Levanase (*scrL*) | 1.25 | c | − | NP |
| | Sucrase | — | | EC | 238 |
| | Polysaccharide utilization (*chuR*) | 1.20 | − | − | 239 |
| | Two component signal transducer (*rprX, rprY*) | 1.56 0.71 | c | EC | 240 |
| | RecA-like gene | 1.0 | c | EC | 241 |
| | Metronidazole resistance | 0.20 | c | + | 242 |
| *B. thetaiotaomicron* | Chondroitin-4-sulfatase | − | n | − | 243 |
| | Chondroitinase II | − | n | − | 244 |
| | Plasmid transfer operon (*rteAB*) | 2.30 1.32 | c | + | 245 |
| | Pullulanase | − | n | − | 232 |
| | Regulatory protein (*chuR*) | 1.20 | c | + | 239 |
| *B. uniformis* | β-lactamase (ORF2, *cblA*) | 1.39 | c | − | NP |
| *B. ovatus* | α-galactosidase | − | n | EC | 246 |
| | Arabinosidase | − | n | EC | 247 |
| | Xylanase | − | n | EC | 248 |
| | Xylosidase | − | n | EC | 249 |
| *B. ruminicola* | Xylanase | — | | + | 248 |
| | β(1-4) Endoglucanase | 1.09 | c | EC | 249 |
| *Clostridium bifermentans* | Acid soluble spore proteins | − | n | − | 250 |
| *C. difficile* | Chloramphenicol acetyl transferase (*catD*) | 0.63 | c | − | 251 |
| | Toxin A (*toxA*) | 8.13 | c | + | 252 |
| | Toxin B (*toxB*) | 7.10 | c | + | 253 |
| | Enterotoxin A (*cdtA*) | − | p | − | NP |
| | Glutamate dehydrogenase | 1.27 | c | − | 254 |
| *C. perfringens* | Sialidase (*nanH*) | 1.14 | c | EC | 105 |
| | Acid-soluble spore protein C1 | 0.61 | c | − | 255 |
| | Acid-soluble spore protein C2 | 0.54 | c | − | 255 |
| | Histidine decarboxylase | 0.96 | c | EC | 256 |
| | Enterotoxin (*cpe*) | 1.36 | c | EC | 111 |
| | α-toxin (phospholipase C) (*cpa*) | 1.20 | c | EC | 257 |
| | ε-toxin | 0.99 | c | EC | 258 |
| | Chloramphenicol acetyl transferase (*catQ*) | 0.66 | c | EC | 259 |
| | Chloramphenicol acetyl transferase (*catP*) | 0.62 | c | EC | 260 |
| | β-galactosidase α-peptide (*lacZ*) | 2.13 | c | EC | 189 |
| | Hyaluronidase (*nagH*) | 6.5 | c | EC | NP |
| | Perfringolysin (theta toxin) (*pfoA*) | 1.7 | c | EC | 111 |
| | Perfringolysin regulatory protein (*pfoR*) | 1.02 | | EC | 261 |
| | Transposase (*tnp*) | 1.43 | c | EC | 262 |
| | MLS resistance (*ermBP*) | 1.0 | n | + | 104 |
| | Tetracycline resistance (*tetP*) | − | p | + | 62 |

**Table 5** Continued.

| Species | Gene | Size (kbp) | Sequence | Expression | Reference |
|---------|------|-----------|----------|------------|-----------|
| *C. septicum* | Sialidase | 3.04 | c | EC | 263 |
| *Eubacterium* strain VPI12708 | Bile acid inducible operon (*baiBCDEA2F*) | 6.2 | c | + | 264 |
| *Lactobacillus fermentum* | Glutamate racemase | 0.81 | c | – | NP |
| *L. plantarum* | Rep protein | 0.95 | c | – | 46 |
| | D-lactate dehydrogenase | 1.00 | c | EC | 265 |
| *Megasphaera elsdenii* | Short chain acyl-CoA dehydrogenase | 1.6 | c | + | NP |
| *Ruminococcus albus* | β-1,4-D-glucanase (*celA*) | 1.22 | c | + | 266 |
| | β-glucosidase (*ralbglT*) | 2.80 | c | + | 267 |
| *Desulfovibrio desulfuricans* | Restriction endonuclease DdeI methylase (*hsdM, R*) | 0.72 1.25 | c | + | 268 |
| | Flavodoxin (*fla*) | 0.45 | c | + | 269 |
| | Prismane protein | 1.64 | c | + | 270 |
| *D. vulgaris* | [Fe] hydrogenase (*hydA, B*) | 1.26 | c | + | 271 |
| | Cytochrome c (*hmc*) | 1.64 | c | EC | 272 |
| | Cytochrome c3 (*cyc*) | 0.39 | c | EC | 273 |
| | Cytochrome c553 (*cyf*) | – | – | – | 274 |
| | Chemoreceptor protein A (*dcrA*) | 2.00 | c | EC | 275 |
| | Rubredoxin oxidoreductase (*rbo*) | 0.38 | c | + | ??? |
| | OMPase (*pyrF*) | | n | EC | 276 |
| | Desulfoviridin (*dsvC*) | 0.32 | c | EC | NP |
| | Flavodoxin | 0.45 | c | EC | 277 |

*Note:* NP, not published; c, complete sequence published or deposited in a database; p, partial sequence published or deposited in a database; n, sequence data not currently available; +, gene expression reported; –, not reported; EC, gene expressed in *Escherichia coli*.

human colon has not yet been attempted *in vivo*. However, with increasing knowledge of the role of specific metabolites such as butyrate in the health of the large intestine, and improved methods for enumeration and identification of the resident bacteria, recombinant DNA technology might be expected to play an important role in future studies of gut microbiology.

There are a number of factors that must be addressed before an application can be made to government authorities regarding the use of mutant or engineered bacterial strains in human trials. The following are the major aspects of a proposed program of research which should be addressed with detailed planning in research proposals involving the eventual release of GEMs.

1. Selection of the recipient strain, (i.e., a disabled host).
2. Genetic transfer, i.e., stability of the modified DNA and its ability to transfer to other species.
3. Release of GEMs into the environment, i.e., human oral consumption.
4. Establishment of GEMs in the environment.
5. Population changes of GEMs, i.e., pattern of population growth and spread. Interaction of the construct with the normal flora.
6. Procedures for monitoring the progress of GEM and its presence in the environment.
7. Termination and cleanup, i.e., procedures for removal of the GEM in the event of an early termination of the experiment.

## VI. CONCLUSIONS

In many anaerobic species, a genetic approach to understanding bacterial pathogenesis is now possible. Studies of plasmids and mobile genetic elements in species belonging to genera such as *Bacteroides* and *Clostridium* have paved the way for relatively straightforward experiments, such as *in vivo* transposon mutagenesis, to test the relative importance of factors associated with virulence. Furthermore, the development of efficient cloning vectors and methods of genetic exchange have important implications for future studies of heterologous genes and the production of recombinant proteins in anaerobic intestinal hosts of potential biomedical or biotechnological importance. It is significant, however, that many intestinal species remain virtually ignored by molecular genetics, including almost all intestinal species belonging to genera such as *Bifidobacterium*, *Eubacterium*, and *Fusobacterium* for which only very preliminary genetic data are available. Future work on the molecular biology of these important gut anaerobes may provide particularly exciting insights into their role in the large intestine. In addition, recent studies suggest that currently recognized intestinal species may represent a minor portion of the diversity of species actually present in the colon. Genetic methods could find a role in the identification and characterization of the nonculturable bacteria in the human large intestine. The continued application of genetic methods for enumeration, identification, and classification of gut anaerobes, coupled with studies of genes central to virulence and growth of important species in the colon, promises a rapid acceleration in our understanding of the ecology, biochemistry, and pathology of the large intestine.

## REFERENCES

1. **Finegold, S. M., Sutter, V. L., and Mathisen, G. E.,** Normal indigenous intestinal flora, in *Human Intestinal Microflora in Health and Disease*, Hentges, D. J., Ed., Academic Press, New York, 1983, chap. 1.
2. **Béchet, M., Pheulpin, P., Joncquiert, J.-C., Tierny, Y., and Guillaume, J.-B.,** A survey of recent advances in genetic engineering in *Bacteroides*, in *Microbiology and Biochemistry of Strict Anaerobes Involved in Interspecies Hydrogen Transfer*, Bélaich, J.-P., Bruschi, M., and Garcia, J.-L., Eds., Plenum Press, New York, 1990, 313.
3. **Young, M., Staudenbauer, W. L., and Minton, N. P.,** Genetics of *Clostridium*, in *Clostridia. Biotechnology Handbooks 3*, Minton, N. P. and Clarke, D. J., Eds., Plenum Publishing, New York, 1991, chap. 3.
4. **Ingraham, J. L., Low, K. B., Magasanik, B., Schaechter, M., and Umbarger, H. E.,** *Escherichia coli and Salmonella typhimurium Cellular and Molecular Biology*, American Society for Microbiology, Washington, D.C., 1987.
5. **Moore, W. E. C. and Holdeman, L. V.,** Human fecal flora: the normal flora of 20 Japanese-Hawaiians, *Appl. Microbiol.*, 27, 961, 1974.
6. **Johnson, J. L.,** DNA reassociation and RNA hybridization of bacterial nucleic acids, in *Methods in Microbiology*, Gottschalk, G., Ed., Academic Press, London, 1985, 34.
7. **Lane, D. J., Pace, B., Olsen, G. J., Stahl, D. A., Sogin, M. L., and Pace, N. R.,** Rapid determination of 16S ribosomal RNA sequences for phylogenetic analysis, *Proc. Natl. Acad. Sci. USA*, 82, 6995, 1985.
8. **Woese, C. R.,** Bacterial evolution, *Microbiol. Rev.*, 51, 221, 1987.
9. **Scardovi, V., Trovatelli, L. D., Zani, G., Crociani, F., and Mateuzzi, D.,** Deoxyribonucleic acid homology relationships among species of the genus *Bifidobacterium*, *Int. J. Syst. Bacteriol.*, 21, 276, 1971.
10. **Schleifer, K. H. and Stackebrandt, E.,** Molecular systematics of prokaryotes, *Ann. Rev. Microbiol.*, 37, 143, 1983.
11. **Bezkorovainy, A.,** Classification of bifidobacteria, in *Biochemistry and Physiology of Bifidobacteria*, Bezkorovainy, A. and Miller-Catchpole, R., Eds., CRC Press, Boca Raton, FL, 1989, chap. 1.
12. **Fox, G. E., Stackebrandt, E., Hespell, R. B., Gibson, J., Maniloff, J., Dyer, T. A., Wolfe, R. S., Balch, W. E., Tanner, R. S., Magrum, L. J., Chen, K. N., and Woese, C. R.,** The phylogeny of prokaryotes, *Science*, 209, 457, 1980.
13. **Ludwig, W., Weizenegger, M., Kilpper-Bälz, R., and Schleifer, K. H.,** Phylogenetic relationships of anaerobic streptococci, *Int. J. Syst. Bacteriol.*, 38, 15, 1988.
14. **Deveraux, R., Delaney, M., Widdel, F., and Stahl, D. A.,** Natural relationships among sulfate-reducing eubacteria, *J. Bacteriol.*, 171, 6689, 1989.
15. **Shah, H. N.,** The genus *Bacteroides* and related taxa, in *The Prokaryotes: A Handbook on the Biology of Bacteria Ecophysiology, Isolation, Identification, Application 2nd ed.*, Balows, A., Trüper, H. G., Dworkin, M., Harder, W., and Schleifer, K. H., Eds., Springer-Verlag, New York, 1992, chap. 196.
16. **Cato, E. P., George, W. L., and Finegold, S. M.,** Genus *Clostridium*, in *Bergey's Manual of Systematic Bacteriology*, Sneath, P. H. A., Mair, N. S., Sharpe, M. E., and Holt, J. G., Eds., Williams and Wilkins, Baltimore, 1986, 1141.

17. **Hippe, H., Andreesen, J. R., and Gottschalk, G.,** The genus *Clostridium* — nonmedical aspects, in *The Prokaryotes: A Handbook on the Biology of Bacteria Ecophysiology, Isolation, Identification, Application 2nd ed.,* Balows, A., Trüper, H. G., Dworkin, M., Harder, W., and Schleifer, K. H., Eds., Springer-Verlag, New York, 1992, chap. 81.

18. **Gurtler, V., Wilson, V. A., and Mayall, B. C.,** Classification of medically important clostridia using restriction endonuclease site differences of PCR-amplified 16S rDNA, *J. Gen. Microbiol.,* 137, 2673, 1991.

19. **Fox, G. E. and Stackebrandt, E.,** The application of 16S rRNA cataloguing and 5S rRNA sequencing in bacterial systematics, in *Methods in Microbiology,* Colwell, R. R. and Grigorova, R., Eds., Academic Press, Orlando, 1987, 405.

20. **Rainey, F. A. and Stackebrandt, E.,** 16S rDNA analysis reveals phylogenetic diversity among poly-saccharolytic clostridia, *FEMS Microbiol. Lett.,* 113, 125, 1993.

21. **Scardovi, V.,** Genus *Bifidobacterium,* in *Bergey's Manual of Systematic Bacteriology,* Sneath, P. H. A., Mair, N. S., Sharpe, M. E., and Holt, J. G., Eds., Williams and Wilkins, Baltimore, 1986, 1418.

22. **Fox, G. E., Pechman, K. R., and Woese, C. R.,** Comparative cataloging of 16S rRNA: molecular approach to prokaryotic systematics, *Int. J. Syst. Bacteriol.,* 27, 44, 1977.

23. **Pfennig, N., Widdel, F., and Trüper, H. G.,** The dissimilatory sulfate-reducing bacteria, in *The Prokaryotes: A Handbook on the Biology of Bacteria Ecophysiology, Isolation, Identification, Application 2nd ed.,* Balows, A., Trüper, H. G., Dworkin, M., Harder, W., and Schleifer, K. H., Eds., Springer-Verlag, New York, 1992, 926.

24. **Widdel, F.,** Microbiology and ecology of sulfate- and sulfur-reducing bacteria, in *Biology of Anaerobic Microorganisms,* Zehnder, A. J. B., Ed., Wiley, New York, 1988, 469.

25. **Devereux, R., Delaney, M., Widdel, F., and Stahl, D. A.,** Natural relationships among sulfate-reducing eubacteria, *J. Bacteriol.,* 171, 6689, 1989.

26. **Shah, H. N.,** The genus *Porphyromonas,* in *The Prokaryotes: A Handbook on the Biology of Bacteria Ecophysiology, Isolation, Identification, Application 2nd ed.,* Balows, A., Trüper, H. G., Dworkin, M., Harder, W., and Schleifer, K. H., Eds., Springer-Verlag, New York, 1992, chap. 197.

27. **Salyers, A. A., Lynn, S. P., and Gardner, J. F.,** Use of randomly cloned DNA fragments for identification of *Bacteroides thetaiotaomicron, J. Bacteriol.,* 154, 287, 1983.

28. **Kuritza, A. P. and Salyers, A. A.,** Use of a species-specific DNA hybridization probe for enumerating *Bacteroides vulgatus* in human feces, *Appl. Environ. Microbiol.,* 50, 958, 1985.

29. **Kuritza, A. P., Shaughnessy, P., and Salyers, A. A.,** Enumeration of polysaccharide degrading *Bacteroides* species in human feces by using species-specific DNA probes, *Appl. Environ. Microbiol.,* 51, 385, 1986.

30. **Kuritza, A. P., Getty, C. E., Shaughnessy, P., Hesse, R., and Salyers, A. A.,** DNA probes for identification of clinically important *Bacteroides* species, *J. Clin. Microbiol.,* 23, 343, 1986.

31. **Morotomi, M., Ohno, T., and Mutai, M.,** Rapid and correct identification of intestinal *Bacteroides* spp. with chromosomal DNA probes by whole-cell dot blot hybridization, *Appl. Environ. Microbiol.,* 54, 1158, 1988.

32. **Stahl, D. A., Flesher, B., Mansfield, H. R., and Montgomery, L.,** Use of phylogenetically based hybridization probes for studies of ruminal microbial ecology, *Appl. Environ. Microbiol.,* 54, 1079, 1988.

33. **Amann, R. I., Stromley, J., Devereux, R., Key, R., and Stahl, D. A.,** Molecular and microscopic identification of sulfate-reducing bacteria in multispecies biofilms, *Appl. Environ. Microbiol.,* 58, 614, 1992.

34. **Jablonski, E. G.,** Detection systems for hybridization reactions, in *DNA Probes for Infectious Diseases,* Tenover, F. C., Ed., CRC Press, Boca Raton, FL, 1989, 15.

35. **Smith, G. L. F., Socransky, S. S., and Smith, C. M.,** Non-isotopic DNA probes for the identification of subgingival microorganisms, *Oral Microbiol. Immunol.,* 4, 41, 1989.

36. **Gruss, A. and Ehrlich, S. D.,** The family of highly interrelated single-stranded deoxyribonucleic acid plasmids, *Microbiol. Rev.,* 53, 231, 1989.

37. **Tannock, G. W.,** Biotin-labelled plasmid DNA probes for detection of epithelium-associated strains of lactobacilli, *Appl. Environ. Microbiol.,* 55, 461, 1989.

38. **Tannock, G. W., Fuller, R., and Pedersen, K.,** *Lactobacillus* succession in the piglet digestive tract demonstrated by plasmid profiling, *Appl. Environ. Microbiol.,* 56, 1310, 1990.

39. **Tannock, G. W., Fuller, R., Smith, S. L., and Hall, M. A.,** Plasmid profiling of members of the family Enterobacteriaceae, lactobacilli, and bifidobacteria to study the transmission of bacteria from mother to infant, *J. Clin. Microbiol.,* 28, 1225, 1990.

40. **Vescovo, M., Morelli, L., and Bottazzi, V.,** Drug resistance plasmids in *Lactobacillus acidophilus* and *Lactobacillus reuteri, Appl. Environ. Microbiol.,* 43, 50, 1982.

41. **Muriana, P. M. and Klaenhammer, T. R.,** Conjugal transfer of plasmid-encoded determinants for bacteriocin production and immunity in *Lactobacillus acidophilus* 88, *Appl. Environ. Microbiol.,* 53, 553, 1987.

42. **Chassy, B. M. and Rokaw, E.,** Conjugal transfer of lactose plasmids in *Lactobacillus casei,* in *Molecular Biology, Pathogenicity and Ecology of Bacterial Plasmids,* Levy, S. B. and Clowes, R. C., Eds., Plenum Press, New York, 1981.

43. **Rinckel, L. A. and Savage, D. C.,** Characterization of plasmids and plasmid-borne macrolide resistance from *Lactobacillus* sp. strain 100-33, *Plasmid,* 23, 119, 1990.

44. **Sarra, P. G., Vescovo, M., and Bottazzi, V.,** Antagonism and adhesion among isogenic strains of *Lactobacillus reuteri* in the caecum of gnotobiotic mice, *Microbiologica,* 12, 69, 1989.

45. **Bates, E. E. M. and Gilbert, H. J.,** Characterization of a cryptic plasmid from *Lactobacillus plantarum, Gene,* 85, 253, 1989.

46. **Bouia, A., Bringel, F., Frey, L., Kammerer, B., Belarbi, A., Guyonvarch, A., and Hubert, J.-C.,** Structural organization of pLP1, a cryptic plasmid from *Lactobacillus plantarum* CCM 1904, *Plasmid,* 22, 185, 1989.

47. **Damiani, G., Romagnoli, S., Ferretti, L., Morelli, L., Bottazzi, V., and Sgaramella, V.,** Sequence and functional analysis of a divergent promoter from a cryptic plasmid of *Lactobacillus acidophilus* 168 S, *Plasmid,* 17, 69, 1987.

48. **Skaugen, M.,** The complete nucleotide sequence of a small cryptic plasmid from *Lactobacillus plantarum, Plasmid,* 22, 175, 1989.

49. **Chassy, B. M.,** Prospects for the genetic manipulation of lactobacilli, *FEMS Microbiol. Rev.,* 46, 297, 1987.

50. **Sgorbati, B., Scardovi, V., and Leblanc, D. J.,** Plasmids in the genus *Bifidobacterium, J. Gen. Microbiol.,* 128, 2121, 1982.

51. **Iwata, M. and Morishita, T.,** The presence of plasmids in *Bifidobacterium breve, Lett. Appl. Microbiol.,* 9, 165, 1989.

52. **Rood, J. I. and Cole, S. T.,** Molecular genetics and pathogenesis of *Clostridium perfringens, Microbiol. Rev.,* 55, 621, 1991.

53. **Garnier, T. and Cole, S. T.,** Characterization of a bacteriocinogenic plasmid from *Clostridium perfringens* and molecular genetic analysis of the bacteriocin-encoding gene, *J. Bacteriol.,* 168, 1189, 1986.

54. **Garnier, T. and Cole, S. T.,** Identification and molecular genetic analysis of replication functions of the bacteriocinogenic plasmid pIP404 from *Clostridium perfringens, Plasmid,* 19, 151, 1988.

55. **Scott, J. R.,** Regulation of plasmid replication, *Microbiol. Rev.,* 48, 1, 1984.

56. **Brehm, J. K., Pennock, A., Bullman, H. M., Young, M., Oultram, J. D., and Minton, M. P.,** Physical characterization of the replication origin of cryptic plasmid pCB101 isolated from *Clostridium butyricum* NCIB 7423, *Plasmid,* 28, 1, 1992.

57. **Odelson, D. A., Rasmussen, J. L., Smith, C. J., and Macrina, F. L.,** Extrachromosomal systems and gene transmission in anaerobic bacteria, *Plasmid,* 17, 87, 1987.

58. **Finn, C. W., Jr., Silver, R. P., Habig, W. H., Hardegree, M. C., Zon, G., and Garon, C. F.,** The structural gene for tetanus neurotoxin is on a plasmid, *Science,* 224, 881, 1984.

59. **Duncan, C. L., Rokos, E. A., Christensen, C. M., and Rood, J. I.,** Multiple plasmids in different toxigenic types of *Clostridium perfringens*: possible control of beta-toxin production, in *Microbiology 1978,* Schlessinger, D., Ed., American Society for Microbiology, Washington D.C., 1978, 246.

60. **Abraham, L. J. and Rood, J. I.,** Cloning and analysis of the *Clostridium perfringens* tetracycline resistance plasmid pCW3, *Plasmid,* 13, 155, 1985.

61. **Abraham, L. J. and Rood, J. I.,** Identification of Tn*4451* and Tn*4452,* chloramphenicol resistance transposons from *Clostridium perfringens, J. Bacteriol.,* 169, 1579, 1987.

62. **Abraham, L. J., Wales, A. J., and Rood, J. I.,** World-wide distribution of the conjugative *Clostridium perfringens* tetracycline resistance plasmid pCW3, *Plasmid,* 14, 37, 1985.

63. **Callihan, D., Young, F., and Clark, V.,** Identification of three homology classes of small cryptic plasmids in intestinal *Bacteroides* species, *Plasmid,* 9, 17, 1983.

64. **Valentine, P. J., Shoemaker, N. B., and Salyers, A. A.,** Mobilization of *Bacteroides* plasmids by *Bacteroides* conjugal elements, *J. Bacteriol.,* 170, 1319, 1988.

65. **Welch, R., Jones, K., and Macrina, F.,** Transferable lincosamide-macrolide resistance in *Bacteroides, Plasmid,* 2, 261, 1979.

66. **Tally, F., Snydman, D., Shimell, M., and Malamy, M.,** Characterization of pBFTM10, a clindamycin-erythromycin resistance transfer factor from *Bacteroides fragilis, J. Bacteriol.,* 151, 686, 1982.

67. **Guiney, D. G., Jr., Hasegawa, P., and Davies, C. E.,** Homology between clindamycin resistance plasmids in *Bacteroides, Plasmid,* 11, 268, 1984.

68. **Smith, C. J. and Macrina, F.,** Large transmissible clindamycin resistance plasmid in *Bacteroides ovatus, J. Bacteriol.,* 158, 739, 1984.

69. **Martinez-Suarez, J., Baquero, F., Reig, M., and Perez-Diaz, J.,** Transferable plasmid-linked chloramphenicol acetyltransferase conferring high-level resistance in *Bacteroides uniformis, Antimicrob. Agents Chemother.,* 28, 113, 1985.

70. **Yamaoka, K., Wantanabe, K., Muto, Y., Katoh, N., Ueno, K., and Tally, F. P.,** R-plasmid mediated transfer of beta-lactam resistance in *Bacteroides fragilis, J. Antibiot.* (Tokyo), 43, 1302, 1990.

71. **Reysset, G., Haggoud, A., Su, W. J., and Sebald, M.,** Genetic and molecular analysis of pIP417 and pIP419: *Bacteroides* plasmids encoding 5-nitroimidazole resistance, *Plasmid,* 27, 181, 1992.

72. **Hecht, D. W., Jagielo, T, J., and Malamy, M. H.,** Conjugal transfer of antibiotic resistance factors in *Bacteroides fragilis:* the *btgA* and *btgB* genes of plasmid pBFM10 are required for its transfer from *Bacteroides fragilis* and for its mobilization by IncPb plasmid R751 in *Escherichia coli, J. Bacteriol.,* 173, 7471, 1991.

73. **Leisinger, T. and Meile, L.,** Approaches to gene transfer in methanogenic bacteria, in *Microbiology and Biochemistry of Strict Anaerobes Involved in Interspecies Hydrogen Transfer,* Bélaich, J.-P., Bruschi, M., and Garcia, J.-L., Eds., Plenum Press, New York, 1990, 11.

74. **Bokranz, M., Klein, A., and Meile, L.,** Complete nucleotide sequence of plasmid pME2001 of *Methanobacterium thermoautotrophicum* (Marburg), *Nucleic Acids Res.*, 18, 363, 1990.

75. **Postgate, J. R., Kent, H. M., Robson, R. L., and Chesshyre, J. A.,** The genomes of *Desulfovibrio gigas* and *D. vulgaris*, *J. Gen. Microbiol.*, 130, 1597, 1984.

76. **Postgate, J. R., Kent, H. M., and Robson, R. L.,** DNA from diazotrophic *Desulfovibrio* strains is homologous to *Klebsiella pneumoniae* structural *nif* DNA and can be chromosomal or plasmid-borne, *FEMS Microbiol. Lett.*, 33, 159, 1986.

77. **Wall, J. D., Rapp-Giles, B. J., and Concannon, S. P.,** Identification of a small cryptic plasmid in *Desulfovibrio desulfuricans*, in *Abstracts, 90th Annual Meeting of the American Society for Microbiology*, American Society for Microbiology, Washington, D.C., 1990, H-158.

78. **Shimizu-Kadota, M., Kiraki, M., Hirokawa, H., and Tsuchida, N.,** ISL1: a new transposable element in *Lactobacillus casei*, *Mol. Gen. Genet.*, 200, 193, 1985.

79. **Shimell, M., Smith, C. J., Tally, F., Macrina, F., and Malamy, M.,** Hybridization studies reveal homologies between pBF4 and pBFTM10, two clindamycin-erythromycin resistance transfer plasmids of *Bacteroides fragilis*, *J. Bacteriol.*, 152, 950, 1982.

80. **Shoemaker, N. B., Guthrie, E. P., Salyers, A. A., and Gardener, J. F.,** Evidence that the clindamycin-erythromycin resistance gene of *Bacteroides* plasmid pBF4 is on a transposable element, *J. Bacteriol.*, 162, 626, 1985.

81. **Shoemaker, N. B., Getty, C., Gardener, J. F., and Salyers, A. A.,** Tn*4351* transposes in *Bacteroides* and mediates integration of R751 into the *Bacteroides* chromosome, *J. Bacteriol.*, 165, 929, 1986.

82. **Robillard, N. J., Tally, F. P., and Malamy, M. H.,** Tn*4400*, a compound transposon isolated from *Bacteroides fragilis* functions in *Escherichia coli*, *J. Bacteriol.*, 164, 1248, 1985.

83. **Smith, C. J. and Spiegel, H.,** Transposition of Tn*4551* in *Bacteroides fragilis:* identification and properties of a new transposon from *Bacteroides* spp., *J. Bacteriol.*, 169, 3450, 1987.

84. **Speer, B. S., Bedzyk, L., and Salyers, A. A.,** Evidence that a novel tetracycline resistance gene found on two *Bacteroides* transposons encodes a NADP-requiring oxidoreductase, *J. Bacteriol.*, 173, 176, 1991.

85. **Smith, C. J. and Gonda, M.,** Comparison of the transposon-like structures encoding clindamycin resistance in *Bacteroides* R plasmids, *Plasmid*, 13, 182, 1985.

86. **Rasmussen, J. L., Odelson, D. A., and Macrina, F. L.,** Complete nucleotide sequence of insertion element IS*4351* from *Bacteroides fragilis*, *J. Bacteriol.*, 169, 3573, 1987.

87. **Hwa, V., Shoemaker, N. B., and Salyers, A. A.,** Direct repeats flanking the *Bacteroides* transposon Tn*4351* are insertion sequence elements, *J. Bacteriol.*, 170, 449, 1988.

88. **Rasmussen, J. L. and Kovacs, E.,** Identification and DNA sequence of a new *Bacteroides fragilis* insertion sequence-like element, *Plasmid*, 25, 141, 1991.

89. **Hecht, D. W. and Malamy, M. H.,** Tn*4399*, a conjugal mobilizing transposon of *Bacteroides fragilis*, *J. Bacteriol.*, 171, 3603, 1989.

90. **Tally, F. P., Shimell, M. J., Carson, G. R., and Malamy, M. H.,** Chromosomal and plasmid mediated transfer of clindamycin resistance in *Bacteroides fragilis*, in *Molecular Biology, Pathogenicity and Ecology of Bacterial Plasmids*, Levy, S. B. and Clowes, R. C., Eds., Plenum Press, New York, 1981, 51.

91. **Macrina, F., Mays, T., Smith, C., and Welch, R.,** Non-plasmid associated transfer of antibiotic resistance in *Bacteroides*, *J. Antimicrob. Chemother.*, D8, 77, 1981.

92. **Mays, T., Smith, C., Welch, R., Delfini, C., and Macrina, F.,** Novel antibiotic resistance transfer in *Bacteroides*, *Antimicrob. Agents Chemother.*, 21, 110, 1982.

93. **Rashtchian, A., Dubes, G., and Booth, S.,** Tetracycline-inducible transfer of tetracycline resistance in *Bacteroides fragilis* in the absence of detectable plasmid DNA, *J. Bacteriol.*, 150, 141, 1982.

94. **Shoemaker, N. B., Getty, C., Guthrie, E. P., and Salyers, A. A.,** Regions in *Bacteroides* plasmids pBFTM10 and pB8-51 that allow *Escherichia coli-Bacteroides* shuttle vectors to be mobilized by IncP plasmids and by a conjugative *Bacteroides* tetracycline resistance element, *J. Bacteriol.*, 166, 959, 1986.

95. **Smith, C., Welch, R., and Macrina, F.,** Two independent conjugal transfer systems operate in *Bacteroides fragilis* V479-1, *J. Bacteriol.*, 151, 281, 1982.

96. **Shoemaker, N. B. and Salyers, A. A.,** Tetracycline-dependent appearance of plasmid-like forms in *Bacteroides uniformis* 0061 mediated by conjugal *Bacteroides* tetracycline resistance elements, *J. Bacteriol.*, 170, 1651, 1988.

97. **Stevens, A. M., Shoemaker, N. B., and Salyers, A. A.,** The region of a *Bacteroides* conjugal chromosomal tetracyline resistance element which is responsible for the production of plasmid-like forms from unlinked chromosomal DNA might also be involved in transfer of the element, *J. Bacteriol.*, 172, 4271, 1990.

98. **Shoemaker, N. B., Barber, R. D., and Salyers, A. A.,** Cloning and characterization of a *Bacteroides* conjugal tetracycline-erythromycin resistance element by using a shuttle cosmid vector, *J. Bacteriol.*, 171, 1294, 1989.

99. **Shoemaker, N. B. and Salyers, A. A.,** A cryptic 65-kilobase-pair transposon like element isolated from *Bacteroides uniformis* has homology with *Bacteroides* conjugal tetracycline resistance elements, *J. Bacteriol.*, 172, 1694, 1990.

100. **Bedzyk, L. A., Shoemaker, N. B., Young, K. E., and Salyers, A. A.,** Insertion and excision of *Bacteroides* conjugative chromosomal elements, *J. Bacteriol.*, 174, 166, 1992.

101. **Halula, M. and Macrina, F. L.,** Tn*5030*: a conjugative transposon conferring clindamycin resistance in *Bacteroides* species, *Rev. Infect. Dis.*, 12, S235, 1990.

102. **Abraham, L. J. and Rood, J. I.,** Identification of Tn*4451* and Tn*4452* chloramphenicol resistance transposons from *Clostridium perfringens, J. Bacteriol.,* 169, 1579, 1987.

103. **Abraham, L. J. and Rood, J. I.,** The *Clostridium perfringens* chloramphenicol resistance transposon Tn4451 excises precisely in *Escherichia coli, Plasmid,* 19, 164, 1988.

104. **Berryman, D. I. and Rood, J. I.,** Cloning and hybridization analysis of ermP, a macrolide-lincosamide-streptogramin B resistance determinant from *Clostridium perfringens, Antimicrob. Agents Chemother.,* 33, 1346, 1989.

105. **Roggentin, P., Rothe, B., Lottspeich, F., and Schauer, R.,** Cloning and sequencing of a *Clostridium perfringens* sialidase gene, *FEBS Lett.,* 238, 31, 1988.

106. **Janssen, P. J., Jones, W. A., Jones, D. T., and Woods, D. R.,** Molecular analysis and regulation of the *glnA* gene of the Gram-positive anaerobe *Clostridium acetobutylicum, J. Bacteriol.,* 170, 400, 1988.

107. **Southern, J. A., Parker, J. R., and Woods, D. R.,** Expression and purification of glutamine synthetase cloned from *Bacteroides fragilis, J. Gen. Microbiol.,* 132, 2827, 1986.

108. **Macrina, F. L.,** Molecular cloning of bacterial antigens and virulence determinants, *Ann. Rev. Microbiol.,* 38, 193, 1984.

109. **Bolin, I. and Wolf-Watz, H.,** Molecular cloning of the temperature-inducible outer membrane protein 1 of *Yersinia pseudotuberculosis, Infect. Immun.,* 43, 72, 1984.

110. **Shimizu, T., Okake, A., Minami, J., and Hagashi, H.,** An upstream regulatory sequence stimulates the expression of the perfringolysin O gene of *Clostridium perfringens, Infect. Immun.,* 59, 137, 1991.

111. **Iwanejko, L. A., Routledge, M. N., and Stewart, G. S. A. B.,** Cloning in *E. coli* of the enterotoxin gene from *Clostridium perfringens* type A, *J. Gen. Microbiol.,* 135, 90, 1989.

112. **Hanna, P. C., Wnek, A. P., and McClane, B. A.,** Molecular cloning of the 3' half of the *Clostridium perfringens* enterotoxin gene and demonstration that this region encodes receptor-binding activity, *J. Bacteriol.,* 171, 6815, 1989.

113. **Salyers, A. A., Pajeau, M., and McCarthy, R. E.,** Importance of mucopolysaccharides as substrates for *Bacteroides thetaiotaomicron* growing in the intestinal tracts of exgermfree mice, *Appl. Environ. Microbiol.,* 54, 1970, 1988.

114. **Anderson, K. and Salyers, A. A.,** Genetic evidence that outer membrane binding of starch is required for starch utilization by *Bacteroides thetaiotaomicron, J. Bacteriol.,* 171, 3199, 1989.

115. **Clewell, D. B. and Gawron-Burke, C.,** Conjugative transposons and the dissemination of antibiotic resistance in streptococci, *Ann. Rev. Microbiol.,* 170, 3046, 1988.

116. **Bertam, J., Kuhn, A., and Dürre, P.,** Tn916-induced mutants of *Clostridium acetobutylicum* defective in regulation of solvent formation, *Arch. Microbiol.,* 153, 373, 1990.

117. **Knauf, H. J., Vogel, R. F., and Hammes, W. P.,** Introduction of the transposon Tn919 into *Lactobacillus cuvatus,* Lc2-c, *FEMS Microbiol. Lett.,* 65, 101, 1989.

118. **Hill, C., Daly, C., and Fitzgerald, G. F.,** Conjugative transfer of the transposon Tn919 to lactic acid bacteria, *FEMS Microbiol. Lett.,* 30, 115, 1985.

119. **Aukrust, T. and Nes, I. F.,** Transformation of *Lactobacillus plantarum* with the plasmid pTV1 by electroporation, *FEMS Microbiol. Lett.,* 52, 127, 1988.

120. **Cosby, W. M., Axelsson, L. T., and Dobrogosz, W. J.,** Tn*917* transposition in *Lactobacillus plantarum* using the highly temperature sensitive plasmid pTV1Ts as a vector, *Plasmid,* 22, 236, 1989.

121. **Fitzgerald, G. F. and Gasson, M. J.,** *In vivo* gene transfer systems and transposons, *Biochimie,* 70, 489, 1988.

122. **Guthrie, E. P. and Salyers, A. A.,** Use of targetted insertional mutagenesis to determine whether chondroitin lyase II is essential for chondroitin sulfate utilization by *Bacteroides thetaiotaomicron, J. Bacteriol.,* 166, 966, 1986.

123. **Salyers, A. A. and Guthrie, E. P.,** A deletion in the chromosome of *Bacteroides thetaiotaomicron* that abolishes production of chondroitinase II does not affect the survival of this organism in the gastrointestinal tracts of exgermfree mice, *Appl. Environ. Microbiol.,* 54, 1964, 1988.

124. **Smith, K. A. and Salyers, A. A.,** A cell-associated pullulanase from *Bacteroides thetaiotaomicron*: cloning, characterization and insertional mutagenesis to determine its role in pullulan utilization, *J. Bacteriol.,* 171, 2116, 1989.

125. **Smith, K. and Salyers, A. A.,** Characterization of a neopullulanase and an α-glucosidase from *Bacteroides thetaiotaomicron* 95-1, *J. Bacteriol.,* 173, 2962, 1991.

126. **Ely, B. and Gerardot, C. J.,** Use of pulse-field gradient gel electrophoresis to construct a physical map of the *Caulobacter crescentus* genome, *Gene,* 68, 323, 1988.

127. **Kauc, L., Mitchell, M., and Goodgal, S. H.,** Size and physical map of the chromosome of *Haemophilus influenza, J. Bacteriol.,* 171, 2474, 1989.

128. **Smith, C. L., Econome, J. G., Schutt, A., Klco, S., and Cantor, C. R.,** A physical map of the *Escherichia coli* K-12 genome, *Science,* 236, 1448, 1987.

129. **Chang, N. and Taylor, D. E.,** Use of pulse-field agarose gel electrophoresis to size genomes of *Campylobacter* species and to construct a *Sal*I Map of *Campylobacter jejuni* UA580, *J. Bacteriol.,* 172, 5211, 1990.

130. **Canard, B. and Cole, S. T.,** Genome organization of the anaerobe pathogen *Clostridium perfringens, Proc. Natl. Acad. Sci. USA,* 96, 6676, 1989.

131. **Graham, M. Y., Otani, T., Boime, I., Olson, M. V., Carle, G. F., and Chaplin, D. D.,** Cosmid mapping of the human chorionic gonadotropin β subunit genes by field-inversion gel electrophoresis, *Nucleic Acids Res.,* 15, 4437, 1987.

132. **Cole, S. T. and Canard, B.,** Genome mapping of *Clostridium perfringens,* in *Genetics and Molecular Biology of Anaerobic Bacteria,* Sebald, M., Ed., Springer-Verlag, New York, 1993, chap. 18.

133. **Thompson, A. M. and Flint, H. J.,** Electroporation induced transformation of *Bacteroides ruminicola* and *Bacteroides uniformis*, by plasmid DNA, *FEMS Microbiol. Lett.*, 61, 101, 1989.

134. **Smith, C. J., Parker, C. J., and Rogers, M.,** Plasmid transformation of *Bacteroides* spp. by electroporation, *Plasmid*, 24, 100, 1990.

135. **Squires, C. H., Heefner, D. L., Evans, R. J., Kopp, B. J., and Yarus, M. J.,** Shuttle plasmids for *Escherichia coli* and *Clostridium perfringens*, *J. Bacteriol.*, 159, 465, 1984.

136. **Mahoney, D. E., Mader, J. A., and Dubel, J. R.,** Transformation of *Clostridium perfringens* L-forms with shuttle plasmid DNA, *Appl. Environ. Microbiol.*, 54, 264, 1988.

137. **Roberts, I., Holmes, W. M., and Hylemon, P. B.,** Development of a new shuttle plasmid system for *Escherichia coli* and *Clostridium perfringens*, *Appl. Environ. Microbiol.*, 54, 268, 1988.

138. **Sznyter, L. A., Slakto, B., Moran, L., O'Donnell, K. H., and Brooks, J. E.,** Nucleotide sequence of the DdeI restriction-modification system and characterization of the methylase protein, *Nucleic Acids Res.*, 15, 8249, 1987.

139. **Khosaka, T., Kiwaki, M., and Rak, B.,** Two site-specific endonucleases *Bin*SI and *Bin*SII from *Bifidobacterium infantis*, *FEBS Lett.*, 163, 170, 1983.

140. **Khosaka, T. and Kiwaki, M.,** Restriction endonucleases from *Bifidobacterium bifidium*, *FEBS Lett.*, 177, 57, 1984.

141. **Khosaka, T., Sakurai, T., Takahashi, H., and Saito, H.,** A new site-specific endonuclease *Bbe*I from *Bifidobacterium breve*, *Gene*, 17, 117, 1982.

142. **Hofer, F.,** Transfer of lactose-fermenting ability in *Lactobacillus lactis*, *N.Z. J. Dairy Sci. Technol.*, 20, 179, 1985.

143. **Muriana, P. M. and Klaenhammer, T. R.,** Conjugal transfer of plasmid-encoded determinants for bacteriocin production and immunity in *Lactobacillus acidophilus* 88, *Appl. Environ. Microbiol.*, 53, 553, 1987.

144. **West, C. A. and Warner, P. J.,** Plasmid profiles and transfer of plasmid-encoded antibiotic resistance in *Lactobacillus plantarum*, *Appl. Environ. Microbiol.*, 50, 1319, 1985.

145. **Shrago, A. W., Chassy, B. M., and Dobrogosz, W. J.,** Conjugal plasmid transfer (pAMβ1) in *Lactobacillus plantarum*, *Appl. Environ. Microbiol.*, 52, 574, 1986.

146. **Vescovo, M., Morelli, L., Bottazzi, V., and Gasson, M. J.,** Conjugal transfer of broad-host-range plasmid pAMβ1 into enteric species of lactic acid bacteria, *Appl. Environ. Microbiol.*, 46, 753, 1983.

147. **Oultram, J. D. and Young, M.,** Conjugal transfer of plasmid pAMβ1 from *Streptococcus lactis* and *Bacillus subtilis* to *Clostridium acetobutylicum*, *FEMS Microbiol. Lett.*, 27, 129, 1985.

148. **Reyset, G. and Sebald, M.,** Conjugal transfer of plasmid-mediated antibiotic resistance from streptococci to *Clostridium acetobutylicum*, *Ann. Microbiol. Inst. Pasteur.*, 136, 275, 1985.

149. **Yu, P.-L. and Pearce, L. E.,** Conjugal transfer of streptococcal antibiotic resistance plasmids into *Clostridium acetobutylicum*, *Biotechnol. Lett.*, 8, 469, 1986,

150. **Oultram, J. D., Peck, H., Brehm, J. K., Thompson, D. E., Swinfield, T.-J., and Minton, N. P.,** Introduction of genes for leucine biosynthesis from *Clostridium pasteurianum* into *Clostridium acetobutylicum*, *Mol. Gen. Genet.*, 214, 177, 1988.

151. **Minton, N. P., Brehm, J. K., Oultram, J. D., Swinfield, T. J., and Thompson, D. E.,** Construction of plasmid vector systems for gene transfer in *Clostridium acetobutylicum*, in *Anaerobes Today*, Hardy, J. M. and Borriello, S. P., Eds., Wiley & Sons, Chichester, England, 1988, 125.

152. **Davies, A., Oultram, J. D., Pennock, A., Wiliams, D. R., Richards, D. F., Minton, N. P., and Young, M.,** Conjugal gene transfer in *Clostridium acetobutylicum*, in *Genetics and Biotechnology of Bacilli*, Vol. 2, Ganesan, A. T. and Hock, J. A., Eds., Academic Press, London, 1988, 391.

153. **Trieu-Cuot, P., Carlier, C., Martin, P., and Courvalin, P.,** Plasmid transfer by conjugation from *E. coli* to gram-positive bacteria, *FEMS Microbiol. Lett.*, 48, 289, 1987.

154. **Privitera, G., Dublanchet, A., and Sebald, M.,** Transfer of multiple antibiotic resistance between subspecies of *Bacteroides fragilis*, *J. Infect. Dis.*, 139, 97, 1979.

155. **Tally, F., Snydman, D., Gorbach, S., and Malamy, M.,** Plasmid mediated, transferable resistance to clindamycin and erythromycin in *Bacteroides fragilis*, *J. Infect. Dis.*, 139, 83, 1979.

156. **Guiney, D. G., Bouic, K., Hasegawa, P., and Matthews, B.,** Construction of shuttle cloning vectors for *Bacteroides fragilis* and use in assaying foreign tetracycline resistance gene expression, *Plasmid*, 20, 17, 1988.

157. **Pheulpin, P., Tierny, Y., Béchet, M., and Guillaume, J.-B.,** Construction of new plasmid vectors for *Escherichia coli-Bacteroides* transgenetic cloning, *FEMS Microbiol. Lett.*, 55, 15, 1988.

158. **Powell, B., Mergeay, M., and Christofi, N.,** Transfer of broad host-range plasmids to sulfate-reducing bacteria, *FEMS Microbiol. Lett.*, 59, 269, 1989.

159. **van den Berg, W. A. M., Stokkermans, J. P. W. G., and van Dongen, W. M. A. M.,** Development of a plasmid transfer system for the anaerobic sulphate reducer *Desulfovibrio vulgaris*, *J. Biotechnol.*, 12, 173, 1989.

160. **Zinder, N. and Lederberg, J.,** Transduction in bacteria, *J. Bacteriol.*, 64, 679, 1952.

161. **Morse, M. L., Lederberg, E. M., and Lederberg, J.,** Transduction in *E. coli* K12, *Genetics*, 41, 142, 1956.

162. **Tohyama, K., Sakurai, T., and Arai, H.,** Transduction by temperate phage PLS-1 in *Lactobacillus salivarius*, *Jpn. J. Bacteriol.*, 26, 482, 1971.

163. **Raya, R. R., Kleeman, E. G., Luchansky, J. B., and Klaenhammer, T. R.,** Characterization of the temperate bacteriophage fadh and plasmid transduction in *Lactobacillus acidophilus* ADH, *Appl. Environ. Microbiol.*, 55, 2206, 1989.

164. **Rapp, B. J. and Wall, J.,** Genetic transfer in *Desulfovibrio desulfuricans*, *Proc. Natl. Acad. Sci. U.S.A.*, 84, 9128, 1987.

165. **Bertani, G. and Baresi, L.,** Genetic transformation in the methanogen *Methanococcus voltae* PS, *J. Bacteriol.*, 169, 2730, 1989.

166. **Lee-Wickner, L. J. and Chassy, B. M.,** The production and regeneration of protoplasts of *Lactobacillus caseii*, *Appl. Environ. Microbiol.*, 48, 994, 1984.

167. **Lin, J. H.-C. and Savage, D.,** Genetic transformation of rifampicin resistance in *Lactobacillus acidophilus*, *J. Gen. Microbiol.*, 132, 2107, 1986.

168. **Iwata, M., Mada, M., and Ishiwa, H.,** Protoplast fusion of *Lactobacillus fermentum*, *Appl. Environ. Microbiol.*, 52, 392, 1986.

169. **Morelli, L., Cocconcelli, P. S., Bottazzi, V., Damiani, G., Ferretti, L., and Sgaramella, V.,** *Lactobacillus* protoplast transformation, *Plasmid*, 17, 73, 1987.

170. **Cosby, W. M., Casas. I. A., and Dobrogosz, W. J.,** Formation, regeneration, and transfection of *Lactobacillus plantarum* protoplasts, *Appl. Environ. Microbiol.*, 54, 2599, 1988.

171. **Leer, R. J., van Luijk, N., Posno, M., and Pouwels, P. H.,** Structural and functional analysis of two cryptic plasmids from *Lactobacillus pentosus* MD353 and *Lactobacillus plantarum* ATCC 8014, *Mol. Gen. Genet.*, 234, 265, 1992.

172. **McCarthy, D. M., Lin, J. H.-C., Rinckel, L. A., and Savage, D. C.,** Genetic transformation in *Lactobacillus* sp. strain 100-33 of the capacity to colonize the nonsecreting gastric epithelium in mice, *Appl. Environ. Microbiol.*, 54, 416, 1988.

173. **Heffner, D. L., Squires, C. H., Evans, R. J., Kopp, B. J., and Yarus, M. J.,** Transformation of *Clostridium perfringens*, *J. Bacteriol.*, 159, 460, 1984.

174. **Calvin, N. M. and Hanawalt, P. C.,** High-efficiency transformation of bacterial cells by electroporation, *J. Bacteriol.*, 170, 2796, 1988.

175. **Allen, S. P. and Blaschek, H. P.,** Electroporation-induced transformation of intact cells of *Clostridium perfringens*, *Appl. Environ. Microbiol.*, 54, 2322, 1988.

176. **Kim, A. Y. and Blaschek, H. P.,** Construction of an *E. coli-Clostridium perfringens* shuttle vector and plasmid transformation of *Clostridium perfringens*, *Appl. Environ. Microbiol.*, 55, 360, 1989.

177. **Allen, S. P. and Blaschek, H. P.,** Factors involved in the electroporation-induced transformation of *Clostridium perfringens*, *FEMS Microbiol. Lett.*, 70, 217, 1990.

178. **Phillips-Jones, M. K.,** Plasmid transformation of *Clostridium perfringens* by electroporation methods, *FEMS Microbiol. Lett.*, 66, 221, 1990.

179. **Scott, P. T. and Rood, J. I.,** Electroporation mediated transformation of lysostaphin-treated *Clostridium perfringens*, *Gene*, 82, 327, 1989.

180. **Luchansky, J. B., Muriana, P. M., and Klaenhammer, T. R.,** Application of electroporation for transfer of plasmid DNA to *Lactobacillus, Lactococcus, Leuconostoc, Listeria, Pediococcus, Bacillus, Staphylococcus, Enterococcus* and *Propionibacterium*, *Mol. Microbiol.*, 2, 637, 1988.

181. **Chassy, B. M.,** A gentle method for the lysis of oral streptococci, *Biochem. Biophys. Res. Comm.*, 68, 603, 1976.

182. **Aukrust, T. and Nes, I. F.,** Transformation of *Lactobacillus plantarum* with the plasmid pTV1 by electroporation, *FEMS Microbiol. Lett.*, 52,127, 1988.

183. **Posno, M., Leer, R. J., van Luijk, N., van Giezen, M. J. F., Heuvelmans, P. T. M. H., Lokman, B. C., and Pouwels, P. H.,** Incompatibility of *Lactobacillus* vectors with replicons derived from small cryptic *Lactobacillus* plasmids and segregational instability of the introduced vectors, *Appl. Environ. Microbiol.*, 57, 1822, 1991.

184. **Holo, H. and Ness, I. F.,** High-frequency transformation of *Lactococcus lactis* subsp. *cremoris* grown with glycine in osmotically stabilized media, *Appl. Environ. Microbiol.*, 55, 3119, 1989.

185. **Wells, J. M., Wilson, P. W., and Le Page, R. W. F.,** Improved cloning vectors and transformation procedure for *Lactococcus lactis*, *J. Appl. Bacteriol.*, 74, 629, 1993.

186. **Micheletti, P. A., Sment, K. A., and Konisky, J.,** Isolation of a coenzyme M-auxotrophic mutant and transformation by electroporation in *Methanococcus voltae*, *J. Bacteriol.*, 173, 3414, 1991.

187. **Gernhardt, P., Possot, O., Foglino, M., Sibold, L., and Klein, A.,** Construction of an integration vector for use in the archaebacterium *Methanococcus voltae* and expression of a eubacterial resistance gene, *Mol. Gen. Genet.*, 221, 273, 1990.

188. **Rousset, M., Dermoun, Z., Chippaux, M., and Belaich, J. P.,** Marker exchange mutagenesis of the *hydN* genes in *Desulfovibrio fructosovorans*, *Mol. Microbiol.*, 5, 1735, 1991.

189. **Sloan, J., Warner, T. A., Scott, P. T., Bannam, T. L., Berryman, D. I., and Rood, J. I.,** Construction of a sequenced *Clostridium perfringens-E. coli* shuttle plasmid, *Plasmid*, 27, 207, 1992.

190. **Bannam, T. L. and Rood, J. I.,** *Clostridium perfringens-Escherichia coli* shuttle vectors that carry single antibiotic resistance determinants, *Plasmid*, 29, 233, 1993.

191. **Lee-Wickner, L.-J. and Chassy, B. M.,** Molecular cloning and characterization of cryptic plasmids isolated from *Lactobacillus casei*, *Appl. Environ. Microbiol.*, 49, 1154, 1985.

192. **Mayo, B., Hardon, C., and Brana, A. F.,** Selected characteristics of several strains of *Lactobacillus plantarum*, *Microbiologia SEM*, 5, 105, 1989.

193. **Chagnaud, P., Chan Kwo Chion, C. K., Duran, R., Naouri, P., Arnaud, A., and Galzy, P.,** Construction of a new shuttle vector for *Lactobacillus*, *Can. J. Microbiol.*, 38, 69, 1992.

194. **de Vos, W. M.,** Gene cloning in lactic streptococci, *Neth. Milk Dairy J.*, 40, 141, 1986.

195. **Kok, J., van der Vossen, J. M. B. M., and Venema, G.,** Construction of plasmid cloning vectors for lactic streptococci which also replicate in *Bacillus subtilis* and *Escherichia coli*, *Appl. Environ. Microbiol.*, 48, 726, 1984.

196. **Gasson, M. J. and Anderson, P. H.,** High copy number plasmid vectors for use in lactic streptococci, *FEMS Microbiol. Lett.*, 30, 193, 1985.

197. **Kok, J.,** Special purpose vectors for lactococci, in *Genetics and Molecular Biology of Streptoccoci, Lactococci, and Enterococci*, Dunny, G. M., Cleary, P. P., and McKay, L. L., Eds., American Society for Microbiology, Washington, D.C., 1991, 97.

198. **Simon, D. and Chopin, A.,** Construction of a vector plasmid family and its use for molecular cloning in *Streptococcus lactis*, *Biochimie*, 70, 559, 1988.

199. **Oultram, J. D., Loughlin, M., Swinfield, T.-J., Brehm, J. K., Thompson, D. E., and Minton, N. P.,** Introduction of plasmids into whole cells of *Clostridium acetobutylicum* by electroporation, *FEMS Microbiol. Lett.*, 56, 83, 1988.

200. **Guiney, D. G., Hasegawa, P., and Davis, C.,** Plasmid transfer from *Escherichia coli* to *Bacteroides fragilis*: differential expression of antibiotic resistance phenotypes, *Proc. Natl. Acad. Sci. U.S.A.*, 83, 7204, 1984.

201. **Thompson, J. S. and Malamy, M. H.,** Sequencing the gene for imipenem-cefoxitin-hydrolyzing enzyme (CfiA) from *Bacteroides fragilis* TAL2480 reveals strong similarity between CfiA and *Bacillus cereus* β-lactamase II, *J. Bacteriol.*, 172, 2584, 1990.

202. **Smith, C. J.,** Development and use of cloning systems for *Bacteroides fragilis*: cloning of a plasmid-encoded clindamycin resistance determinant, *J. Bacteriol.*, 164, 294, 1985.

203. **Smith, C. J.,** Clindamycin resistance and the development of genetic systems in the *Bacteroides*, *Dev. Ind. Microbiol.*, 30, 23, 1989.

204. **Smith, C. J., Rogers, M. B., and McKee, M. L.,** Heterologous gene expression in *Bacteroides fragilis*, *Plasmid*, 27, 141, 1992.

205. **Smith, C. J.,** Nucleotide sequence analysis of Tn*4551*: use of ermFS operon fusions to detect promoter activity in *Bacteroides fragilis*, *J. Bacteriol.*, 169, 4589, 1987.

206. **Brown, J. W., Daniels, C. J., and Reeve, J. N.,** Gene structure, organization, and expression in archaebacteria, *Crit. Rev. Microbiol.*, 16, 287, 1989.

207. **Thomm, M., Buchner, J., and Stetter, K. O.,** Evidence for a plasmid in a methanogenic bacterium, *J. Bacteriol.*, 153, 1060, 1983.

208. **Meile, L., Kiener, A., and Leisinger, T.,** A plasmid in the archaebacterium *Thermoautotrophicum*, *Mol. Gen. Genet.*, 191, 480, 1983.

209. **Bates, E. M., Gilbert, H. J., Hazlewood, G. P., Huckle, J., Laurie, J. I., and Mann, S. P.,** Expression of a *Clostridium thermocellum* endoglucanase gene in *Lactobacillus plantarum*, *Appl. Environ. Microbiol.*, 55, 2095, 1989.

210. **Scheirlinck, T. Mahillon, J., Joos, H., Dhaese, P., and Michiels, F.,** Integration and expression of α-amylase and endoglucanase genes in the *Lactobacillus plantarum* chromosome, *Appl. Environ. Microbiol.*, 55, 2130, 1989.

211. **Janniere, L., Niaudet, B., Pierre, E., and Hairlick, S. D.,** Stable gene amplification in the chromosome of *Bacillus subtilis*, *Gene*, 40, 47, 1985.

212. **Vogel, R. F., Gaier, W., and Hammes, W. P.,** Expression of the lipase gene from *Staphylococcus hyicus* in *Lactobacillus cuvatus* Lc2-c, *Microbiol. Lett.*, 69, 289, 1990.

213. **Smith, H.,** The mounting interest in bacterial and viral pathogenicity, *Ann. Rev. Microbiol.*, 43, 1, 1989.

214. **Brill, W. J.,** Why engineered microorganisms are safe, *Issues Sci. Technol.*, 4, 44, 1988.

215. **Mekalanos, J. J.,** Environmental signals controlling expression of virulence determinants in bacteria, *J. Bacteriol.*, 174, 1, 1992.

216. **Stotsky, G. and Babich, H.,** Survival of, and genetic transfer by, genetically engineered bacteria in natural environments, *Adv. Appl. Microbiol.*, 31, 93, 1986.

217. **Chatfield, S., Li, J. L., Sydenham, M., Douce, G., and Dougan, G.,** Salmonella genetics and vaccine development, in *Molecular Biology of Bacterial Infection*, Hormaeche, C. E., Penn, C. W., and Smyth, C. J., Eds., Cambridge University Press, Cambridge, 1992.

218. **Stover, C. K., de la Cruz, V. F., Fuerst, T. R., Burlein, J. E., Benson, L. A., Bennett, L. T., Bamsal, G. P., Young, J. F., Lee, M. H., and Hatfull, G. F.,** New use of BCG for recombinant vaccines, *Nature*, 351, 456, 1991.

219. **Wells, J. M., Wilson, P. W., Norton, P., Gasson, M., and Le Page, R. W. F.,** *Lactococcus lactis* high level expression of tetanus toxin fragment C and protection from lethal challenge, *Mol. Microbiol.*, 8, 1155, 1993.

220. **Gerritse, K., Posno, M., Schellekens, M. M., Boersma, W. J. A., and Claasen, E.,** Oral administration of TNP-*Lactobacillus* conjugates in mice: a model for evaluation of mucosal and systemic immune responses and memory formation elicited by transformed lactobacilli, *Res. Microbiol.*, 141, 955, 1990.

221. **Pozzi, G., Contorni, M., Oggioni, M. R., Manganelli, R., Tommasino, M., Cavalieri, F., and Fischetti, V. A.,** Delivery and expression of a heterologous antigen on the surface of streptococci, *Infect. Immun.*, 60, 1902, 1992.

222. **Iwaki, M., Okahashi, N., Takahashi, I., Kanamoto, T., Sugita-Konishi, Y., Aibara K., and Koga, T.,** Oral immunization with recombinant *Streptococcus lactis* carrying the *Streptococcus mutans* surface protein antigen gene, *Infect. Immun.*, 58, 2929, 1990.

223. **Hazlewood, G. P. and Teather, R. M.,** The genetics of rumen bacteria, in *The Rumen Microbial Ecosystem,* Hobson, P. N., Ed., Elsevier Press, London, 1988, 323.

224. **Teather, R. M.,** Application of gene manipulation to rumen microflora, *Can. J. Anim. Sci.,* 65, 563, 1985.

225. **Gilbert, H. J. and Hall, J.,** Molecular cloning of *Streptococcus bovis* lactose catabolic genes, *J. Gen. Microbiol.,* 133, 2285, 1987.

226. **Garnier, T. and Cole, S. T.,** Complete nucleotide sequence and genetic organization of the bacteriocinogenic plasmid pIP404 from *Clostridium perfringens, Plasmid,* 1, 134, 1988.

227. **Magot, M.,** Physical characterization of the *Clostridium perfringens* tetracycline-chloramphenicol resistance plasmid pIP401, *Ann. Inst. Pasteur Microbiol.,* 135B, 269, 1984.

228. **Flint, H. J., Thomson, A. M., and Bisset, J.,** Plasmid-associated transfer of tetracycline resistance in *Bacteroides ruminicola, Appl. Environ. Microbiol.,* 54, 855, 1988.

229. **Hamilton, P. T. and Reeve, J. N.,** Structure of genes and an insertion element in the methane producing archaebacterium *Methanobrevibacter smithii, Mol. Gen. Genet.,* 200, 47, 1985.

230. **Smith, C. J.,** Polyethylene glycol facilitated transformation of *Bacteroides fragilis* with plasmid DNA, *J. Bacteriol.,* 164, 466, 1985.

231. **Muriana, P. M. and Klaenhammer, T. R.,** Cloning, phenotypic expression, and DNA sequence of the gene for lactacin F, an antimicrobial peptide produced by *Lactobacillus* spp., *J. Bacteriol.,* 173, 1779, 1991.

232. **Smith, K. A. and Salyers, A. A.,** A cell-associated pullulanase from *Bacteroides thetaioatomicron:* cloning characterization and insertional mutagenesis to determine its role in pullulan utilization, *J. Bacteriol.,* 171, 2116, 1989.

233. **Smith, C. J.,** Clindamycin resistance and the development of genetic systems in the *Bacteroides, Dev. Ind. Microbiol.,* 30, 23, 1989.

234. **Rasmussen, B. A., Gluzman, Y., and Tally, F. P.,** Cloning and sequencing of the class B beta-lactamse gene (ccrA) from *Bacteroides fragilis* TAL366, *Antimicrob. Agents Chemother.,* 34, 1590, 1990.

235. **Podglajen, I., Brenil, J., Bordon, F., Gutman, L., and Collatz, E.,** A silent carbapenemase gene in strains of *Bacteroides fragilis* can be expressed after a one-step mutation, *FEMS Microbiol Lett.,* 70, 21, 1992.

236. **Halula, M., Manning, S., and Macrina, F. L.,** Nucleotide sequence of ermFU, macrolide-lincosamide-streptogramin B (MLS) resistance gene encoding an RNA methylase from the conjugal element of *Bacteroides fragilis* V503, *Nucleic Acids Res.,* 19, 3453, 1991.

237. **Russo, T. A., Thompson, J. S., Codoy, V. G., and Malamy, M. H.,** Cloning and expression of the *Bacteroides fragilis* TAL2480 neuraminidase gene *nan*H in *E. coli, J. Bacteriol.,* 172, 2594, 1990.

238. **Scholle, R., Steffan, H. E., Goodman, H. J. K., and Woods, D. R.,** Expression and regulation of a *Bacteroides fragilis* sucrose utilization system cloned in *Escherichia coli, Appl. Environ. Microbiol.,* 56, 1944, 1990.

239. **Cheng, Q., Hwa, V., and Salyers, A. A.,** A locus that contributes to colonization of the intestinal tract by *Bacteroides thetaiotaomicron* contains a single regulatory gene (*chu*R) that links two polysaccharide utilization pathways, *J. Bacteriol.,* 174, 7185, 1992.

240. **Rasmussen, B. A. and Kovacs, E.,** Cloning and identification of a two component signal-transducing regulatory system from *Bacteroides fragilis, Mol. Microbiol.,* 7, 765, 1993.

241. **Goodman, H. J. K. and Woods, D. R.,** Molecular analysis of the *Bacteroides recA* gene, *Gene,* 94, 77, 1990.

242. **Wehnert, G. U., Abratt, V. R., and Woods, D. R.,** Molecular analysis of a gene from *Bacteroides fragilis* involved in metronidazole resistance in *Escherichia coli, Plasmid,* 27, 242, 1992.

243. **Guthrie, E. P. and Salyers, A. A.,** Evidence that the chondroitin lyase gene of *Bacteroides thetaiotaomicron* is adjacent to the gene for a chondro-4-sulfatase, *J. Bacteriol.,* 169, 1192, 1987.

244. **Hwa, V. and Salyers, A. A.,** Evidence for differential regulation of genes in the chondroitin sulfate utilization pathway of *Bacteroides thetaiotaomicron, J. Bacteriol.,* 174, 342, 1992.

245. **Stevens, A. M., Sanders, J. M., and Salyers, A. A.,** Genes involved in production of plasmid-like forms by a *Bacteroides* conjugal chromosomal element share significant amino acid homology with two component regulatory systems, *J. Bacteriol.,* 174, 2935, 1992.

246. **Valentine, P. J., Gherardini, F. C., and Salyers, A. A.,** A *Bacteroides ovatus* chromosomal locus which contains an α-galactosidase gene may be important for the colonization of the gastrointestinal tract, *Appl. Environ. Microbiol.,* 57, 1615, 1991.

247. **Whitehead, T. R. and Hespell, R. B.,** Heterologous expression of the *Bacteroides ruminicola* xylanase gene in *Bacteroides fragilis* and *Bacteroides uniformis, Appl. Environ. Microbiol.,* 55, 893, 1989.

248. **Whitehead, T. R. and Hespell, R. B.,** Cloning and expression in *E. coli* of a xylanase gene from *Bacteroides ruminicola, FEMS Microbiol. Lett.,* 54, 61, 1990.

249. **Matsushita, O., Russell, J. B., and Wilson, D. B.,** A *Bacteroides ruminicola* 1,4-beta-endoglucanase is encoded in two reading frames, *J. Bacteriol.,* 173, 6919, 1991.

250. **Cabrera-Martinez, R. M., Masson, J. M., Setlow, B., Waites, W. M., and Setlow, P.,** Purification and amino acid sequence of two small, acid-soluble proteins from *Clostridium bifermentans* spores, *FEMS Microbiol. Lett.,* 52, 139, 1989.

251. **Wren, B. W., Mullany, P., Clayton, C., and Tabaqchali, S.,** Nucleotide sequence of a chloramphenicol acetyl transferase gene from *Clostridium difficile, Nucleic Acids Res.,* 17, 4877, 1989.

252. **Sauerborn, M. and von Eichel-Streiber, C.,** Nucleotide sequence of *Clostridium difficile* Toxin A, *Nucleic Acids Res.*, 18, 1629, 1990.

253. **Barroso, L. A., Wang, S. Z., Phelps, C. J., Johnson, J. L., and Wilkins, T. D.,** Nucleotide sequence of *Clostridium difficile* toxin B gene, *Nucleic Acids Res.*, 18, 4004, 1990.

254. **Lyerly, D. M., Barroso, L. A., and Wilkins, T. D.,** Identification of the latex reactive protein of *Clostridium difficile* as glutamate dehydrogenase, *J. Clin. Microbiol.*, 29, 2639, 1991.

255. **Holck, A., Blom, H., and Granum, P.,** Cloning and sequencing of the genes encoding acid soluble spore proteins from *Clostridium perfringens*, *Gene*, 91, 107, 1990.

256. **van Poelje, D. D. and Snell, E. E.,** Cloning, sequencing, expression and site-directed mutagenesis of the gene from *Clostridium perfringens* encoding pyruvoyl-dependent histidine decarboxylase, *Biochemistry*, 29, 132, 1990.

257. **Leslie, D., Fairweather, N., Pickard, D., Dougan, G., and Kehoe, M.,** Phospholipase C and haemolytic activities of *Clostridium perfringens* alpha-toxin cloned in *E. coli*: sequence and homology with *Bacillus cereus* phospholipase C, *Mol. Microbiol.*, 3, 383, 1989.

258. **Harvard, H. L., Hunter, S. E., and Titball, R. W.,** Comparison of a nucleotide sequence and development of a PCR test for the epsilon toxin gene of *Clostridium perfringens* type B and type D, *FEMS Microbiol. Lett.*, 97, 77, 1992.

259. **Bannam, T. L. and Rood, J. I.,** The relationship between the *Clostridium perfringens catQ* gene product and chloramphenicol acetyl transferases from other bacteria, *Antimicrob. Agents Chemother.*, 35, 471, 1991.

260. **Steffen, C. and Matzutra, H.,** Nucleotide sequence analysis and expression studies of a chloramphenicol-acetyl-transferase-encoding gene from *Clostridium perfringens*, *Gene*, 75, 349, 1991.

261. **Tweten, R. K.,** Cloning and expression in *E. coli* of the perfringolysin O (theta toxin) gene from *Clostridium perfringens* and characterization of the gene product, *Infect. Immun.*, 56, 3228, 1988.

262. **Daube, G., Simon, P., and Kaeckenbeeck, A.,** "IS*1151*" and IS-like element of *C. perfringens*, *Nucleic Acids Res.*, 21, 352, 1993.

263. **Rothe, B., Roggentin, P., and Schauer, R.,** The sialidase gene from *Clostridium septicum*: cloning, sequencing, expression in *E. coli* and identification of conserved sequences in sialidases and other proteins, *Mol. Gen. Genet.*, 226, 190, 1991.

264. **Mallonee, D. H., White, W. B., and Hylemon, P. B.,** Cloning and sequencing of a bile acid-inducible operon from *Eubacterium* sp. strain VPI 12708, *J. Bacteriol.*, 172, 7011, 1990.

265. **Taguchi, H. and Ohta, T.,** D-lactate dehydrogenase is a member of the D-isomer-specific 2-hydroxyacid dehydrogenase family. Cloning, sequencing and expression in *E. coli* of the D-lactate dehydrogenase gene of *Lactobacillus plantarum*, *J. Biol. Chem.*, 266, 1258, 1991.

266. **Ohmiya, K., Kajino, T., Kato, A., and Shimizu, S.,** Structure of a *Ruminococcus albus* endo-1,4-beta-glucanase gene, *J. Bacteriol.*, 171, 6771, 1989.

267. **Ohmiya, K., Takano, M., and Shimizu, S.,** DNA sequence of a beta-glucosidase from *Ruminococcus albus*, *Nucleic Acids Res.*, 18, 671, 1990.

268. **Sznyter, L. A., Slatko, B., Moran, L., O'Donnell, K. H., and Brooks, J. E.,** Nucleotide sequence of the DdeI restriction-modification system and characterization of the methylase protein, *Nucleic Acids Res.*, 15, 8249, 1987.

269. **Helms, L. R. and Swenson, R. P.,** Cloning and characterization of the flavodoxin gene from *Desulfovibrio desulfuricans*, *Biochim. Biophys. Acta*, 1089, 417, 1991.

270. **Stokkermans, J. P., den Berg, W. A., van Dongen, W. M., and Veeger, C.,** The primary structure of a protein containing a putative [6Fe-6S] prismane cluster from *Desulfovibrio desulfuricans*, *Biochim. Biophys. Acta*, 1132, 83, 1992.

271. **Voordouw, G. and Brenner, S.,** Nucleotide sequence of the gene encoding the hydrogenase from *Desulfovibrio vulgaris* (Hildenborough), *Eur. J. Biochem.*, 148, 515, 1985.

272. **Pollock, W. B. R., Loutfi, M., Bruschi, M., Rapp-Giles, B. J., Wall, J. D., and Voordouw, G.,** Cloning, sequencing and expression of the gene encoding the high-molecular weight cytochrome c from *Desulfovibrio vulgaris* Hildenborough, *J. Bacteriol.*, 173, 220, 1991.

273. **Voordouw, G. and Brenner, S.,** Cloning and sequencing of the gene encoding cytochrome c3 from *Desulfovibrio vulgaris* (Hildenborough), *Eur. J. Biochem.*, 159, 347, 1986.

274. **van Rooijen, G. J. H., Bruschi, M., and Voordouw, G.,** Cloning and sequencing of the gene encoding cytochrome c553 from *Desulfovibrio vulgaris* Hildenborough, *J. Bacteriol.*, 171, 3573, 1989.

275. **Dolla, A., Fu, R., Brumlik, M. J., and Voordouw, G.,** Nucleotide sequence of *dcrA*, a *Desulfovibrio vulgaris* Hildenborough chemoreceptor gene, and its expression in *E. coli*, *J. Bacteriol.*, 174, 1726, 1992.

276. **Li, C., Peck, H. D., and Przybyla, A. E.,** Complementation of an *E. coli pyrF* mutant with DNA from *Desulfovibrio vulgaris*, *J. Bacteriol.*, 165, 644, 1986.

277. **Krey, G. D., Vanin, E. F., and Swenson, R. P.,** Cloning, nucleotide sequence and expression of the flavodoxin gene from *Desulfovibrio vulgaris* (Hildenborough), *J. Biol. Chem.*, 236, 15436, 1988.

*Chapter 3*

# Carbohydrate Metabolism in the Colon

*Michael J. Hudson and Philip D. Marsh*

## CONTENTS

   I. Introduction ................................................................................................................ 61
     A. The Nature of the Colonic Microflora ...................................................................... 61
     B. Utilization of Carbohydrate by the Colonic Microflora ........................................... 62
  II. Sources of Carbohydrate in the Colon ....................................................................... 63
 III. Metabolism of Endogenous Carbohydrates ................................................................ 64
     A. Metabolism of Mucin ............................................................................................... 64
     B. Metabolism of Other Glycoproteins ......................................................................... 65
 IV. Metabolism of Dietary Carbohydrate ......................................................................... 66
     A. Nonstarch Polysaccharides ....................................................................................... 67
     B. Resistant Starch ........................................................................................................ 67
     C. Low Molecular Weight Carbohydrates ..................................................................... 68
  V. Bifidogenic Potential of Malabsorbed Carbohydrates ................................................ 68
     A. Raffinose ................................................................................................................... 69
     B. Fructo-Oligosaccharides and Inulin ......................................................................... 69
     C. Isomalto-Oligosaccharides ....................................................................................... 70
     D. *trans*-Galactosylated-Oligosaccharides .................................................................. 70
 VI. Summary and Concluding Remarks ............................................................................. 71
References ............................................................................................................................ 72

## I. INTRODUCTION

### A. THE NATURE OF THE COLONIC MICROFLORA

The human colon is colonized by a relatively stable, dense, and complex microflora comprising diverse species and metabolic activities (see Chapter 1).[1-3] There is evidence for at least some degree of compartmentalization of the flora through physical associations with particulate matter and mucus shreds, and by the bacteria proliferating as microcolonies. In addition, there is a distinct mucosal flora attached to, or closely associated with, the mucosal surface within the crypts and in the overlying mucus layer.[4-6] The relative contribution made by the mucosal flora to carbohydrate metabolism in the large intestine, and particularly the use of endogenous carbohydrates in host secretions, is not known and nor is this known for other aspects of metabolism. For the purposes of this review, the mucosal flora will not be considered in detail except to observe that it colonizes a unique and nutritionally privileged site and must surely contribute significantly to the initial modification and breakdown of the oligosaccharide side-chains of mucin glycoprotein through saccharolytic activity (see Chapter 8).

The overall density of the colonic microflora is approximately $10^{11}$ microorganisms per gram wet weight, with a slight increasing gradient from the proximal to distal colon, which is in part attributable to the decreasing water content of the digesta.[7] The predominant flora is comprised of obligately anaerobic bacteria. Facultative organisms comprise less than 1% of the cultivable flora, filamentous molds, yeasts, obligate aerobes, and protozoa are rarely found as anything but very minor components.[1-3] The predominant flora is represented by only about 20 species, which together constitute 99% of the cultivable flora,[3] whereas perhaps 120 to 130 other species comprise less than 1% of the flora in any one individual. This microbiota remains relatively constant as regards time and variables such as diet, but does differ qualitatively and quantitatively between individual subjects.[1-3] Minor components are a complex mixture of resident species and a variable number and proportion of transients which are unable to compete with or displace resident strains in the synergistic associations and complex food chains which, in part, determine the composition and stability of the flora.

The stability of the resident flora is of pivotal importance in maintaining a colonic flora that is able to exclude pathogens. The balance of the flora is relatively easily disrupted by the presence of low

**Table 1**  Major Saccharolytic Bacterial Genera
in the Human Large Intestine[a]

| Genus | Mean numbers in feces ($\log_{10}$/g dry wt) | Major fermentation end products |
|---|---|---|
| *Bacteroides* | 11.3 | Acetate, propionate, succinate |
| *Bifidobacterium* | 10.2 | Acetate, lactate, ethanol, formate |
| *Eubacterium* | 10.7 | Acetate, butyrate, lactate |
| *Ruminococcus* | 10.2 | Acetate |
| *Peptostreptococcus* | 10.1 | Acetate, lactate |
| *Lactobacillus* | 9.6 | Lactate |
| *Clostridium* | 9.8 | Acetate, propionate, butyrate, lactate |
| *Streptococcus* | 8.3 | Lactate, acetate |

[a]  Table taken from References 1 to 3.

concentrations of antimicrobial agents, as evidenced from colonization by abnormal flora consisting of Gram-negative species, candida yeast and *Clostridium difficile*, but there are no similar instances of changes in the diet which cause a marked disruption. Interest in the influence of diet on the colonic or fecal flora stems from observations on differences in populations studied with variable diets and in the incidence of colonic disease, including cancer. It has proved particularly difficult, however, to induce significant change in the fecal flora by dietary manipulation alone. It is possible that the relative insensitivity of the bacteriological techniques used in such studies would have missed small variations occurring in the balance of particular species, biotypes, or metabolic groups during short-term dietary modification.[1] It is also possible that important components of the microbiota are not cultivable by current methods. More serious are the limitations imparted by the study of fecal bacteria rather than cecal and colonic floras, because it is changes in the colonic population that are of interest during short-term diet manipulations; any changes in the flora may be negated or diminished by the time of maturation.

## B. UTILIZATION OF CARBOHYDRATE BY THE COLONIC MICROFLORA

The overall catabolism of the array of substrates entering the colon involves metabolic interactions among the predominant and minor species present. The degradation of some complex molecules may require the concerted action of several species, while the products of metabolism of one organism may be used as nutrients by other species (the secondary feeders) such that food webs and food chains develop. Indeed, the stable persistence of such a diverse microflora is indicative that no single nutrient is limiting the growth of all the resident organisms.

The vast majority of bacteria in the human large intestine are saccharolytic *in vitro* and derive carbon and energy from the breakdown of carbohydrates (Table 1).[1-3] Most bacteria are not reliant on the availability of simple sugars, but are able to derive carbon and energy from the breakdown of complex carbohydrates, either alone or as part of a microbial consortium or food chain. Significant quantities of carbohydrate enter the colon daily. They are derived from undigested and unabsorbed food residues and from endogenous sources such as mucin and other host secretions. Quantity and type of dietary carbohydrates are largely determined by the extent to which they escape digestion and absorption from the upper gastrointestinal tract. These carbohydrates are potentially available to the microflora for hydrolytic breakdown, a process referred to generally as fermentation; the colonic environment is, of course, extremely anaerobic with only trace amounts of oxygen present and with a very low redox potential resulting from the considerable metabolic activity of the normal flora. Complex polysaccharides are broken down in a series of hydrolytic steps to simple sugars that are then metabolized by the flora to a range of fermentation end products which include short chain fatty acids (SCFA) and other organic acids, alcohols, $H_2$, and $CO_2$ (Figure 1). The final products of bacterial metabolism are SCFA and gases. The ultimate fate of these end products depends on the amount of absorption of metabolites, including the gaseous products, from the colon and rectum and on the balance of activity of the three $H_2$ disposal mechanisms which are thought to operate in the colon: methanogenesis, dissimilatory sulfate reduction, and acetogenesis.

The extent to which a particular carbohydrate is available for fermentation clearly depends on the complexity of the carbohydrate polymer, the residence time in the colon as well as the nature and metabolic activity of the colonic flora, and possible specific adaptation to utilization of a particular substrate. The

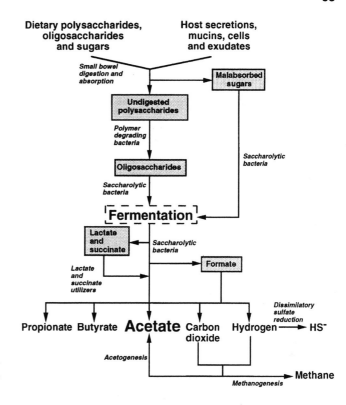

**Figure 1** Simplified scheme of the major pathways of carbohydrate metabolism by bacteria in the large intestine.

benefits of carboxylic acids to host nutrition and the relative importance of different methods of $H_2$ disposal are discussed elsewhere in this volume (see Chapters 5 and 6), but the striking importance, and practical benefit of colonic fermentation is shown in studies of patients with short bowel syndrome, i.e., resection with consequent loss of absorptive capacity for nutrients, electrolytes, and water. The very short residence time of carbohydrate in the bowel of such patients severely limits the degree to which dietary carbohydrate is digested and absorbed, and high-fat and elemental diets are commonly recommended to improve energy intake. Recent studies have shown that patients with short bowel anastomosed to colon are better able to salvage carbon and energy from malabsorbed carbohydrate (by as much as 2 MJ/d) than are patients with jejunostomies and no colon in continuity.[8,9] The energy is recovered from SCFA produced by fermentation of malabsorbed carbohydrate in the colon; up to one-half of the carbohydrate intake is malabsorbed in such patients.[9] In the absence of a colon, patients with short bowel syndrome must use other strategies, such as high-fat diets, in order to maintain an equivalent energy intake.

Our understanding of the metabolism of carbohydrates in the large intestine has been derived from a variety of sources, including carbohydrate balance studies in volunteers, microbiological analyses of human material such as feces, and laboratory-based studies. The latter range from investigation of the metabolism of simple substrates by pure cultures of enteric bacteria in batch culture to the utilization of complex molecules by diverse mixed cultures in multistage chemostats,[10] the latter being designed to model aspects of the complex ecology of the colon. By combining the results from all of these types of study, a comprehensive view of the role of intestinal microorganisms in the metabolism of carbohydrates in the colon can be obtained. The aim of this chapter, therefore, is to review the literature on the metabolism of endogenous and exogenous carbohydrates by the normal colonic microflora.

## II. SOURCES OF CARBOHYDRATE IN THE COLON

The relative contributions of dietary and endogenous sources of carbohydrate to the totality of fermentable substrate entering the cecum, that is potentially available to the colonic microflora, can only be estimated from indirect measurements and by the study of ileal effluent of ileostomists. The ability of endogenous sources of carbohydrate alone to support the colonic flora is evident from persistence of the colonic or fecal flora during periods of fasting and during ingestion of elemental, completely absorbable diets, and also in patients undergoing parenteral (intravenous) feeding. In all of these instances, the bulk of intestinal contents is markedly reduced, but the intestinal flora persists in both its original complexity and metabolic activity.[10-14] This situation suggests that endogenous nutrients provided by the host are able to sustain the growth of the resident microflora of the gut in the same way as has been observed elsewhere in the digestive tract.[15] Similarly, studies have shown the persistence of a colonic or fecal-type microflora in defunctioned segments of colon and rectum in patients who have undergone colonic resection. In these cases, the only sources of nutrients are from the blood supply to the bowel and from endogenous secretions such as goblet cell mucins.[16]

The large intestine contains about 220 g wet weight of contents (range *circa* 60–900 g), 35 g dry weight, giving a dry matter content of approximately 16%.[17] Stephen and Cummings[18] developed a gravimetric procedure for the estimation of the solid fractions in feces and found that bacteria comprised up to 55% of the dry weight. While daily fecal losses are not necessarily constant, this estimate puts the average loss of bacterial biomass from the large intestine at up to 20 g (dry weight). The large intestinal contents and the flora are anatomically on the "outside" of the host, and form an open fermentor system which is fed nutrients semi-continuously through the ileocecal valve. Some metabolites are lost by absorption and incorporation into bacterial biomass, which together with undigested dietary residues and gastrointestinal tract epithelial cells are lost by defecation. Salyers[19] found that growth yields for saccharolytic bacteroides were between 0.2 and 0.3 g dry weight of bacteria per gram of carbohydrate fermented; if these values are typical of other saccharolytic genera and for different carbohydrates, then some 40–60 g of fermentable carbohydrate are required daily to balance these losses. The major source of this carbohydrate is undigested starch, and recent estimates suggest that up to 40 g pass into the colon; dietary fiber intake in western diets, calculated as nonstarch polysaccharides (NSP), usually provides less than 20 g and a normal mixed diet will provide a similar amount again of low molecular weight oligosaccharides, sugars, and sugar alcohols. In addition to utilization of dietary and endogenous carbohydrates, intestinal bacteria can obtain carbon and energy from the fermentation of amino acids obtained from the undigested proteinaceous fractions of diet and endogenous secretions (see Chapter 4), from urea hydrolysis, and also from digestion and recycling of dead bacteria.

## III. METABOLISM OF ENDOGENOUS CARBOHYDRATES

### A. METABOLISM OF MUCIN

Mucus is found throughout the digestive tract, and acts as a protective barrier over mucosal surfaces. It is secreted by goblet cells that line the gastrointestinal tract,[20] and is believed to play an essential role in maintaining the health of the colon. Chapter 8 in this volume reviews the structure and metabolism of mucin. However, because this glycoprotein plays an important role in carbohydrate fermentation in the colon, a brief overview will also be given here.

Mucins consist of a peptide core with oligosaccharide side chains; these side chains can be linear or branched structures of varying length, linked via α-N-acetyl-galactosamine to serine and threonine residues in the core polypeptide (Figure 2). The oligosaccharide side chains help protect the peptide core from proteolytic digestion. The outer, nonreducing end of completed chains terminates in α-linked glycosides, whereas the inner residues include β-linked sequences of D-galactose and N-acetyl-D-glucosamine. The terminal residues include sialic acids and the ABH blood group antigens whose antigenic specificity is conferred by α-(1-3)-N-acetyl-D-galactosamine, α-(1-3)-D-galactose, and α-(1,2)-L-fucose, respectively, linked to β-D-galactose (Figure 2).[20] Since the structure of these glycoproteins is complex, complete degradation of the side chains requires the action of many different glycosidases, each with the appropriate anomeric and sugar specificity. Thus, enzymes such as sialidase, the blood group ABH-degrading α-glycosidases, β-D-galactosidase, and β-N-acetyl-glucosaminidase are required to catalyze the cleavage of each glycosidic linkage sequentially down the chain, starting from the terminal end.

Evidence for mucin degradation by enteric bacteria has come from three main sources. First, the levels of nondialyzable carbohydrates and levels of A, B, and H blood group immunoreactions were reduced in feces of conventional mice compared with germ-free animals.[21] Second, biochemical analyses of mucin recovered from different regions of the gut lumin have shown that the degree of degradation correlates directly with the extent of bacterial colonization.[22] Third, fecal extracts and cultures of certain gut bacteria have been shown in the laboratory to produce the relevant enzymes, and to be able to degrade, in part, model glycoprotein substrates that are closely related in structure to colonic mucin.[22]

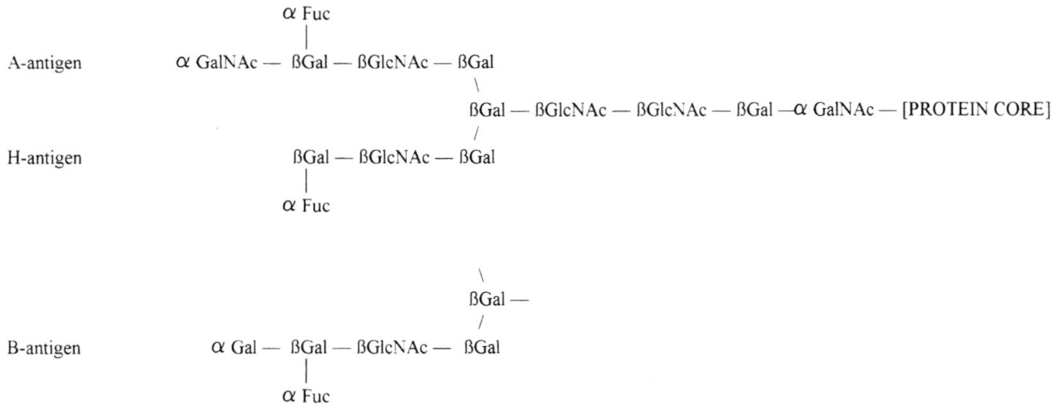

```
                        α Fuc
                         |
A-antigen      α GalNAc — ßGal — ßGlcNAc — ßGal
                                              \
                                               ßGal — ßGlcNAc — ßGlcNAc — ßGal —α GalNAc — [PROTEIN CORE]
                                              /
H-antigen              ßGal — ßGlcNAc — ßGal
                         |
                        α Fuc


                                              \
                                               ßGal —
                                              /
B-antigen      α Gal — ßGal — ßGlcNAc — ßGal
                                 |
                                α Fuc
```

**Figure 2**  Schematic representation of typical oligosaccharide side chains of a glycoprotein showing blood group A, B, and H antigens. Fuc, L-fucose; Gal, D-galactose; GalNAc, N-acetyl-D-galactosamine; GlcNAc, N-acetyl-D-glucosamine.

Relatively few gut bacteria appear to be capable of totally degrading the oligosaccharide side-chains of mucin, and they represent only about 1% of the total fecal microflora.[23] These bacteria are *Ruminococcus* species, including *R. torques*, and *Bifidobacterium* spp.[24] Similarly, only 11 strains out of 342 fecal isolates (representing 32 species and 8 genera) were able to degrade model mucin substrates;[25,26] the strains again belonged to the genera *Ruminococcus* and *Bifidobacterium*, together with some *Bacteroides* species. Enzyme activities included blood group A- and H-degrading α-glycosidases, sialidase, and β-glycosidases; oligosaccharide breakdown was associated most strongly with extracellular glycosidase production rather than cell-bound activities.[27] Most enzymes appear to be produced constitutively. Significantly perhaps, when the activities of blood group A-, B-, and H-degrading glycosidases were measured in fecal extracts from ABH secretors, the enzyme activity corresponding to the subject's blood type antigen was greatest.[22]

Many of the studies described above have assayed for gross utilization of a mucin substrate by organisms in pure culture. Studies of a number of species of gut bacteria, including representatives of some of the predominant species in the colon have shown them to possess some, but not all, of the glycosidases required to degrade the glycoproteins.[24,28-30] Pure cultures of these bacteria would be expected to break down mucins to a limited extent only so that, in the host, mucin degradation is more likely to be a cooperative process with several bacteria sequentially, and competitively, cleaving glyco-sidic bonds.[24,26,28] The substantial degradation of such complex molecules may require, therefore, the concerted action of several species. For example, when an α-galactosidase-producing strain of *Ruminococcus* spp. was co-cultured with a strain of *Bifidobacterium bifidum* that possessed several β-N-acetyl-hexosaminidases that the former strain lacked, the release of hexoses from a blood group B-containing glycoprotein increased from <50 to >90% in the mixed culture, while bacterial growth was several-fold greater than was obtained with either strain alone (Table 2).[24] Evidence has also emerged recently that sugars released during the microbial breakdown of glycoproteins can support the growth of wider (nonmucin-degrading) populations of fecal bacteria.[31] The data also suggest that although most of the enzymes involved are exo-glycosidases, some endo-N-acetylhexosaminidases are also produced.[31]

## B. METABOLISM OF OTHER GLYCOPROTEINS

The extracellular glycosidases produced by subpopulations of the colonic microflora can diffuse into the general luminal environment and therein degrade other glycoconjugates that are structurally related to mucin oligosaccharides. These include substances found on bacterial cell walls and glycosphingolipids on gut epithelial cells.

The outer surfaces of Gram-negative bacteria contain repeated oligosaccharide units that confer the antigenic specificity of somatic O-antigens that are used, for example, in the serotyping of Enterobacteriaceae. The somatic O-antigen of *Escherichia coli* O86 includes a terminal trisaccharide that appears to be identical to the human blood group B antigen. Incubation of either fecal extracts or supernatants of human anaerobic fecal bacteria caused a loss of group B-like immunoreactivity from the surface of cells of *E. coli* O86.[32] Likewise, structurally similar glycoproteins on the membrane of

**Table 2**  Concerted Action Between Two Species of Enteric Bacteria in the Degradation of Oligosaccharide Side Chain of Blood Group B Salivary Mucin Glycoprotein

| | Enzyme activity | | | |
| --- | --- | --- | --- | --- |
| | α-galactosidase | β-*N*-acetyl hexosaminidase | Percentage loss of mucin hexoses | Bacterial growth[a] |
| *Ruminococcus* sp. alone | + | – | 43 | 32 |
| *Bifidobacterium* sp. alone | – | + | 46 | 22 |
| *Ruminococcus* sp. and *Bifidobacterium* sp. | + | + | 96 | 139 |

[a]  Bacterial growth is expressed as a change in optical density ($\times 10^3$) at 600 nm.

From Hoskins, L. C., et al., *J. Clin. Invest.*, 75, 944, 1985.

mammalian cells can be degraded in an analogous manner by bacterial glycosidases.[24,33] The bacterial enzymes demonstrated a high degree of specificity towards the main lactoseries type 1 and 2 oligosaccharide chains produced by the human gastrointestinal epithelium; other glycosphingolipids with different core chains were unaffected.[33,34]

Glycosphingolipids are analogous to glycoproteins since they have oligosaccharide side-chains of up to 60 sugar molecules attached to a lipid ceramide base group, which usually consists of sphingosine with an amide-linked fatty acid group. The oligosaccharide chain comprises the same sugars that are found in glycoprotein side-chains and are arranged in straight and branched arrays that may also express ABH and Lewis blood group antigens. The ceramide group acts as an anchor in the epithelial cell membrane. The normal function of cell-surface glycosphingolipids is obscure, but they are an important group of molecules acting as receptors for some bacterial toxins, including cholera toxin, and as attachment groups for a range of intestinal bacteria. Glycosphingolipids are lost from the epithelium, probably as a result of normal epithelial cell turnover, and in the lumin milieu, the oligosaccharide side-chains are progressively degraded by the normal flora in much the same way as glycoproteins are degraded.[35] It is not known whether the normal mucosal flora is able to modify the oligosaccharide chain, and thus the receptor specificity, when the glycosphingolipids are *in situ* on the mucosal epithelium cell surface.

The action of bacterial glycosidases in the human colon will modify, therefore, the antigenic specificity, structure, and function of host macromolecules. The consequences of this process will be to generate nutrients directly for the enzyme-producing bacteria, and to liberate mono- and disaccharides that could be used by other species. For example, it has been proposed that disaccharides similar in structure to the "Bifidus factor" can be released during mucin breakdown which may promote and maintain the growth of enteric bifidobacteria in the human colon.[31] The degradation of glycoconjugates on the surface of epithelial cells could also lead to the uncovering of binding sites for members of the resident microflora, and could also alter putative receptors for exotoxins on the mucosal epithelium.[34] These studies emphasize the importance of the metabolism of endogenous carbohydrates in the ecology of the human colon.

## IV. METABOLISM OF DIETARY CARBOHYDRATE

A broad range of carbohydrates are consumed in the normal mixed diet; moreover, the types and balance of different carbohydrates eaten differ between subjects on a seasonal, daily, and meal-to-meal basis. In addition, there are large differences in the intakes of various ethnic groups. The bulk of research on carbohydrate metabolism has emanated from studies on first world man and his diet. Western man is atypical in several respects from the global norm; in particular, diets consumed in the industrialized and developed world tend not to rely on any one particular carbohydrate-rich staple food, and lactase deficiency is uncommon. The opposite is true of most societies in the developing world.

A mixed vegetable, cereal, and legume diet will contain a considerable range of different carbohydrates that escape digestion and absorption, and which differ in molecular weight and in the complexity of the polymers. Once in the colon the extent to which these substances are degraded by the flora depends on its species composition, its adaptation to the carbohydrate substrate, and the residence or contact time between the substrate and the bacteria in the large intestine.

**Table 3** The Proportion of Dietary Carbohydrate Escaping Digestion and Recoverable in Ileostomy Fluid

| | Nonstarch Polysaccharides | Starch |
|---|---|---|
| White bread | 100 | 2 |
| Cornflakes | 100 | 5 |
| Oats | 95 | 2 |
| Banana | 95 | 90 |
| Cooked potato (hot) | 100 | 4 |
| Cooked potato (cold) | 100 | 12 |

*Note:* Data are percentage recovery in ileostomy effluent of starch and nonstarch polysaccharide components of test meals.

From Englyst, H. N. and Kingman, S. M., *Dietary Fiber*, Kritchevsky, D. and Anderson, J. A., Eds., Plenum Press, New York, 1990, 49.

## A. NONSTARCH POLYSACCHARIDES

Analyses of dietary fiber in foods are based on the fraction of plant cell walls that are not broken down in the upper digestive tract. Since starch was considered, at the time, to be almost completely digested by mammalian enzymes in the small bowel, NSP were used to estimate the dietary fiber that enters the colon and was, therefore, potentially available for fermentation. In fact, a considerable, but variable, amount of starch escapes digestion and enters the colon (Table 3), where it is rapidly fermented.[36] The NSP fraction contains the major components of plant cell walls (i.e., cellulose, hemicellulose, xylan, pectin) together with vegetable and seed-storage polysaccharides (i.e., inulin, β-glucan, guar gum, psyllium). It is conveniently estimated by hydrolysis and gas chromatographic analysis of the monosaccharide residues as alditol acetates as described by Englyst et al.[37] Lignin has been considered to form part of dietary fiber, although it is not a carbohydrate but a random cross-linked polymer of coniferyl alcohol that is not degraded by colonic bacteria.

Digestion of NSP is carried out by various components of the colonic flora, but principally by members of the genus *Bacteroides*, and mainly the numerically predominant *Bacteroides fragilis* group, together with bifidobacteria, ruminococci, and eubacteria.[38] Hemicelluloses such as arabinogalactan, the noncellulosic β-glucans such as laminarin, and pectin are all rapidly degraded in the colon, and in fecal incubations *in vitro*. In contrast to rumen fermentations, cellulose digestion in the human colon is slow but very variable, and depends on the source of cellulose and its physical structure.[38] Modified celluloses such as methylcellulose and carboxymethylcellulose, which are manufactured for use as food additives and texture modifiers, are more rapidly fermented *in vitro* although the fate of these polymers in the colon is not clear.

## B. RESISTANT STARCH

It is clear from the data shown in Table 3 that a great deal of starch passes from the small bowel undigested or partially digested to become available for bacterial fermentation in the colon. Many common starchy foods, including staple diets of most societies, contain a form of starch that is resistant to digestion in the small bowel, and this fraction is termed "resistant starch."[39] Starchy foods become or are inherently resistant to α-amylase and α-amyloglucosidase digestion in the small bowel for several reasons:

1. A physical inaccessibility to enzymatic digestion, as in unmilled grains and legumes.
2. The specific and highly compact microcrystalline structure (type B X-ray diffraction pattern) of ungelatinized starch granules, such as those found in uncooked potato and banana.
3. The formation of retrograded starch polymers; these types of starch are present in many staples such as potato, maize, legumes, and rice which have been subjected to cooking or other processing, and particularly cycles of heating and cooling. This causes changes in the microcrystalline structure that confer a resistance to digestion which resembles that of natural ungelatinized starch granules. This property is ascribed not to the amylopectin fraction which makes up the bulk of starch, but to the amylose fraction which usually

constitutes only about 20% of the starch polymer. Recovery experiments with standard maize and high amylose maize meals have shown the latter to be far less degradable in the small bowel.[40]

4. Inhibition of enzyme action by complexing with other diet components such as phytates and lignin.

The recognition that fractions of starchy foods could escape digestion and absorption from the small bowel has led to a reevaluation of the importance of the contribution of resistant starch in physiological and therapeutic studies of malabsorbed and undigested carbohydrate, and the suggestion that resistant starch be included in analyses of total dietary fiber. An inherent difficulty in this aim is to accurately predict the amount of resistant starch in foodstuffs subjected to cycles of heating, cooling, and freezing during normal cooking and in the food processing industry. For example, Scheppach et al.[41] monitored breath $H_2$ output after ingestion of test meals of potato which had been cooked by boiling and then subjected to different treatments. Hydrogen production was about threefold higher after ingestion of the cooked potatoes which had been frozen and thawed to 20°C, compared to potatoes eaten fresh at 60°C. Reheating the frozen and thawed potatoes to 90°C and eating at 60°C resulted in a less pronounced 1.5-fold increase in breath $H_2$ levels. These data show not only that the formation of resistant starch by retrogradation during freezing is partially reversible but also the difficult task of assigning a resistant starch estimate to foods that may be cooked and preserved by freezing and consumed at different temperatures.

## C. LOW MOLECULAR WEIGHT CARBOHYDRATES

The digestive and absorptive processes of the upper digestive tract are highly efficient but even in normal healthy individuals a small but significant proportion of disaccharides such as sucrose and lactose escape hydrolysis and pass into the colon, where they are rapidly fermented by bacteria. Bond et al.[42] demonstrated that up to 4% of the sucrose in a meal can enter the hindgut in this way. In a simple test meal, these losses may be due, in part, to rapid transit of the food through the bowel, thus escaping digestion, but in normal mixed meal low molecular weight carbohydrate might be partially protected from digestion in complexes with high molecular weight material. Of course, in lactase deficient individuals, almost all lactose is malabsorbed and is very rapidly fermented in the colon, with characteristic side effects of bloating, abdominal cramps, and osmotic diarrhea.

In addition to malabsorbed dietary sugars, there are other low-molecular weight dietary carbohydrates not hydrolyzed by small bowel disaccharides that pass into the colon. These include natural oligosaccharides such as stachyose and raffinose which are present at low levels in onions, garlic, and artichokes.

Naturally-occurring and chemically-modified sugars and sugar alcohols have been manufactured for use as noncariogenic sweeteners for use in diabetic and "low-calorie" food products and also as mild laxatives. These sugars include lactulose, palatinose, xylitol, sorbitol, maltitol, and lactitol. Although lactulose is not a normal constituent of milk, it is formed during heating and particularly during heat-sterilization procedures; infant formulae, for example, may contain as much as 5% lactulose as a result of heat treatment.

Lactulose and lactitol are both used clinically to treat portal systemic encephalopathy, and act by inducing rapid fermentation in the colon with concomitant fall in pH and SCFA production. Experimental studies suggest that any malabsorbed and rapidly fermented carbohydrate would possess similar efficacy.[43] The ammonia produced by the colonic flora, whose absorption from the colon is the cause of the encephalopathy, is thought not to be trapped by the low pH *per se*, but is reduced as a result of diminished urea and amino acid deamination rates, combined with an increase in conversion into bacterial protein nitrogen and biomass.

One interesting oddity is erythritol, which is a 4-carbon polyol that occurs in lower plants and mushrooms. In feeding experiments, erythritol appears to be well absorbed from the small bowel and excreted unaltered in the urine.[44] However, mushrooms are particularly fibrous and poorly digested, and it is likely that naturally complexed erythritol might well escape the small bowel and pass into the colon. Once there, the fate of this substance is uncertain, because in batch fermentations at least, this polyol is not degraded at all (as judged from $H_2$ evolution).

## V. BIFIDOGENIC POTENTIAL OF MALABSORBED CARBOHYDRATES

Among the great variety of metabolic processes that occur through bacterial activity in the human intestine, there are some that are considered to be potentially harmful because of the production of cytotoxic and mutagenic compounds.[1,45,46] These include substances produced by the hydrolysis of glucuronide and glycoside conjugates (see Chapter 7), the reduction of azo-linked and nitro compounds, the formation of phenols and indoles from aromatic amino acids (see Chapter 4), the reduction of nitrate to nitrite, and the transformation of bile acids. The components of the bacterial flora responsible for these transformations are not generally known, but measurements of the respective enzyme activities as markers have shown that they are reduced in subjects and experimental animals when supplements of saccharolytic bacteria such as lactobacilli and bifidobacteria are supplied, together with a nonabsorbed fermentable carbohydrate, often as a yogurt or fermented milk preparation.

Naturally-occurring populations of bifidobacteria and lactobacilli can themselves be stimulated by the feeding of a nonabsorbed fermentable carbohydrate. This ability to stimulate the protective and beneficial flora is often loosely termed the "bifidogenic" potential of a nonabsorbed fermentable carbohydrate, and there is currently much commercial interest in developing a product which invokes a marked stimulation of the intestinal populations of bifidobacteria and which is safe to use as a food supplement. However, the ingestion of most nonabsorbed carbohydrate with strong bifidogenic potential suffers from the common drawback of a tendency to cause bloating, abdominal discomfort, and pain due to the gas produced during colonic fermentation. To assess the bifidogenic potential of a carbohydrate intended for use as a food additive or dietary supplement requires time-consuming and costly microbiological analysis of feces from properly controlled volunteer diet studies. Although many fecal flora studies have been reported in the literature, disappointingly few have included, for example, placebo feeding. Examples of different low-molecular weight sugars with bifidogenic properties are presented to illustrate these points.

## A. RAFFINOSE

Goldin et al.[47] described marked changes in the fecal flora of volunteers consuming daily supplements of 15 g of raffinose (galactosyl-sucrose) for four weeks. During this time, bifidobacterial numbers rose threefold with an accompanying reduction in the numbers of putrefactive organisms, including clostridia and bacteroides. During the period of raffinose ingestion, bifidobacteria increased as a proportion of the total flora from less than 15% to 58–80%, displacing bacteroides as the predominant microorganisms in the gut.

## B. FRUCTO-OLIGOSACCHARIDES AND INULIN

Fructo-oligosaccharides exist naturally in many plants, including onion, garlic, chicory, Jerusalem artichoke, and wheat. They consist of up to 60 fructose units linked together by $\beta(1\text{-}2)$ linkages with a terminal D-glucose moiety, and are essentially a sucrose molecule with a fructose polymer chain. Inulin has 20 or more fructose molecules while smaller polymers are termed fructo-oligosaccharides. Mixtures of short polymers with up to four fructose groups are synthesized from sucrose by the action of $\beta$-fructofuranosidase from *Aspergillus niger*, and this method is used to prepare novel sweetening agents for use as mild laxative agents and in diabetic and bifidogenic health foods.

*In vitro* studies with fecal bacteria have shown that fructo-oligosaccharides support the growth of many species in the colon, but show a much more pronounced stimulatory effect on the bifidobacteria (Table 4). In contrast, other carbohydrates including starch, polydextrose, and pectin produced a more general effect on the different colonic genera tested.[48] A similar and rather striking bifidogenic response was reported in studies of the fecal flora of a group of elderly adults given a diet supplement of 8 g/d of low molecular weight fructo-oligosaccharides for two weeks.[49] During the administration period, the mean bifidobacterial count increased by almost 10-fold, with an even greater response of $10^2$- to $10^4$-fold in individuals with low initial counts. There was a concomitant increase in the total bacterial count during fructo-oligosaccharide ingestion, but little change in the numbers of bacteroides. Interestingly, lactobacilli also increased up to tenfold during the study but other bacterial groups such as enterobacteria and clostridia did not change markedly. The lactate-utilizing veillonellae increased gradually during the study, perhaps as an adaptation to increased levels or availability of lactic acid being produced by elevated populations of bifidobacteria and lactobacilli. An interesting inverse relationship was noted between the presence of high counts of bifidobacteria and the carriage rate of *Clostridium perfringens* in the patients. These workers and others have recorded several general benefits of fructo-oligosaccharides, including mild laxative effects with relief from constipation, and a significant lowering of serum lipids and blood pressure, particularly in elderly and diabetic subjects.[50] However, intestinal gas production with bloating and abdominal discomfort are common complaints following ingestion of inulin and fructo-oligosaccharides and at least one study suggests that prolonged ingestion of this carbohydrate does not invoke an adaptive response.[51]

**Table 4** Increase in Bacterial Counts in Fecal Slurries
Incubated Anaerobically for 12 h After the Addition of Sugars

|  | FOS | Inulin | Fructose | Starch | Polydextrose | Pectin |
|---|---|---|---|---|---|---|
| Total anaerobes | 1.0 | 1.7 | 1.2 | 1.3 | 1.1 | 1.4 |
| Bacteroides | 0.3 | 0.3 | 0.7 | 0.9 | 1.4 | 0.7 |
| Bifidobacteria | 1.4 | 1.3 | 0.5 | 0.4 | 0.8 | 0.8 |
| Clostridia | 0.1 | 0.2 | 0.5 | 0.0 | 2.0 | 0.1 |
| Total aerobes | 0.7 | 0.0 | 1.3 | 0.7 | 1.0 | 1.1 |
| Enterobacteria | 0.3 | 0.6 | 1.6 | 1.9 | 2.2 | 0.2 |
| Lactobacilli | 0.4 | 0.0 | 0.4 | 1.0 | 0.5 | 0.4 |

*Note:* Values are differences in $\log_{10}$ counts of mean values from triplicate determinations
from 3 fecal samples. FOS, fructo-oligosaccharides.

From Wang, X. and Gibson, G. R., *J. Appl. Bacteriol.*, 75, 373, 1993.

## C. ISOMALTO-OLIGOSACCHARIDES

Kohmoto and colleagues[52] have described changes in the intestinal flora during ingestion of oligosaccharides based on glucose units linked $\alpha(1\text{-}6)$, as in isomaltose, isomaltotriose, and isomaltotetrose, and with both $\alpha(1\text{-}6)$ and $\alpha(1\text{-}4)$ bonds as in panose. Such oligosaccharides are found naturally in honey and in fermented soya products such as miso and soy sauce, but are synthesized commercially from corn starch by enzymatic digestion with $\alpha$-amylase, pullulanase, and $\alpha$-glucosidase. Cultures of bifidobacteria and bacteroides were able to utilize the isomalto-oligosaccharides for growth, however clostridia, enterobacteria, and lactobacilli were unable to do so. A pronounced and significant increase in numbers of bifidobacteria was found in separate studies of two groups of adult volunteers and elderly patients given 13.5 g isomalto-oligosaccharides daily for 10 to 14 d. In the adult volunteers, this increase was about 4-fold, but in elderly patients there was a 10-fold increase, and in patients with low initial counts of bifidobacteria, the increase was as much as 100-fold.

## D. *trans*-GALACTOSYLATED-OLIGOSACCHARIDES

Similar studies to those discussed above for fructo- and isomalto-oligosaccharides have shown that ingestion of *trans*-galactosylated-oligosaccharides, a complex mixture of up to four galactose units linked $\beta(1\text{-}6)$, $\beta(1\text{-}4)$, or $\beta(1\text{-}3)$ to a lactose molecule, causes a significant increase in fecal bifidobacteria in volunteers.[53,54] A similar bifidogenic effect is observed if this malabsorbed oligosaccharide is fed to rats which have been implanted with a human gut flora,[55] with a significant fourfold increase in cecal bifidobacteria, an increase in the total and lactobacilli counts, and a marked decrease in enterobacteria. These investigators also showed that concomitant with these changes in species composition, cecal nitrate reductase and $\beta$-glucuronidase activities were each reduced to approximately 60% of control values (Table 5), but $\beta$-glucosidase activity more than doubled. Bifidobacteria possess $\beta$-glucosidase activity but neither nitrate reductase nor $\beta$-glucuronidase[56] and the difference in $\beta$-glucosidase levels in the cecal contents was not found when enzyme activities were corrected for bacterial numbers rather than protein.

Of particular interest was the observation that prior feeding with *trans*-galactosyl-oligosaccharides reduced the rate at which cecal suspensions converted 2-amino-3-methyl-3H-imidazo[4,5-*f*]quinoline (IQ), a mutagen which is found widely in grilled, fried, barbecued, and other cooked foods, to the highly reactive and genotoxic 7-hydroxy metabolite (7-OHIQ). The rates of transformation of IQ by cecal suspensions prepared from oligosaccharide-fed animals were about 20, 60, and 80% of the control values after 2, 4, and 24 h incubation, respectively. What is unclear is whether this effect is the result of a change in the balance of the flora *per se*, or the enzymatic activity of the bacteria, or alterations in the kinetics of the transformation due, perhaps, to the lower cecal pH values in oligosaccharide-fed animals. Clearly, further experiments are necessary to answer this question, but it is tempting to ascribe the reduced rate of metabolism in terms of a change in the balance of the colonic microflora as Van Tassell et al.[57] have shown that the IQ to 7OH-IQ transformation is catalyzed by clostridia and eubacteria, bacterial genera that are generally suppressed in absolute numbers, or as a proportion of the total flora, during ingestion of *trans*-galactosyl-oligosaccharide and other similar bifidogenic, nonabsorbed carbohydrates.

**Table 5**  Effect of Feeding *trans*-Galactosyl-
Oligosaccharides (TOS) on the Bacterial Flora
and Enzyme Activities of Cecal Contents of Rats
Implanted with a Human Flora

|  | Control diet | TOS diet |
|---|---|---|
| Direct microscopic count | 10.7 | 11.3 |
| Total anaerobes | 10.4 | 11.1 |
| Total aerobes | 9.6 | 9.6 |
|   Bifidobacteria | 10.0 | 10.6 |
|   Bacteroides | 9.6 | 9.7 |
|   Lactobacilli | 9.5 | 9.9 |
|   Enterobacteria | 8.3 | 6.5 |
| pH | 6.85 | 6.16 |
| Ammonia (mmol/g) | 35.9 | 32.5 |
| β-Glucosidase |  |  |
|   (mmol/h/mg protein) | 0.89 | 2.05 |
|   (mmol/h/$10^{10}$cells) | 3.06 | 3.43 |
| β-Glucuronidase |  |  |
|   (mmol/h/mg protein) | 4.08 | 2.53 |
|   (mmol/h/$10^{10}$cells) | 13.03 | 3.20 |
| Nitrate reductase |  |  |
|   (mmol/h/mg protein) | 0.86 | 0.50 |
|   (mmol/h/$10^{10}$cells) | 2.99 | 0.62 |
| IQ to 7-OHIQ (% conversion) |  |  |
|   2 h | 20 | 4 |
|   4 h | 36 | 22 |
|   24 h | 93 | 77 |

*Note:*  Bacterial counts are mean values of 6 rats expressed as
$\log_{10}$ bacteria/g gut contents. Units of other measurements
are as stated, and are mean values for 4 to 6 rats.

From Rowland, I. R. and Tanaka, R., *J. Appl. Bacteriol.*, 74,
667, 1993.

## VI. SUMMARY AND CONCLUDING REMARKS

In this chapter we have discussed some aspects of the important role played by carbohydrate in supporting the ecology and metabolism of the colonic microflora. The carbohydrate used by the flora comes from both diet and endogenous sources. Dietary sources include undigested complex carbohydrates ("dietary fiber"), malabsorbed polymers and low-molecular weight carbohydrates (starch and lactose), and natural or manufactured simple carbohydrates, that are malabsorbed, for use as dietary supplements. The quantity of dietary carbohydrate from different foods which reaches the colon varies greatly, as is evident from data that are emerging on the effects of cooking and other processing on the proportion of resistant starch present in foods. It is also evident, however, that the normal colonic microflora can persist with little or no diet-derived carbohydrate entering through the ileo-cecal valve, albeit at reduced overall cell population densities. The indigenous bacteria of the large intestine have evolved specific enzymes for the degradation and utilization of the oligosaccharide side-chains of mucin and epithelial glycoproteins and glycolipids. To do this, they make use of complex interactions which result in food chains and webs that underpin the stability of the colonic ecosystem.

In recent years we have seen many significant advances in our understanding of the amount and types of carbohydrate utilizing bacteria in the large bowel, of the nature of carbohydrate metabolism, and of the fate and importance of the fermentation end-products. We also now recognize that the intestinal flora contributes to health, for example, through resistance to gut infection, but also to disease, through production of potentially toxic or mutagenic and carcinogenic metabolites. The current interest in bifidogenic or similar probiotic substances and organisms that can alter the indigenous flora or its metabolism is aimed at addressing this relationship. We can expect further developments in the interactions between dietary carbohydrate and the improved health of humans (and animals) through the manipulation and control of the intestinal flora and its metabolism.

# REFERENCES

1. **Drasar, B. S. and Hill, M. J.,** *Human Intestinal Flora*, Academic Press, London, 1974.
2. **Finegold, S. M., Sutter, V. L., and Mathisen, G. E.,** Normal indigenous intestinal flora, in *Human Intestinal Microflora in Health and Disease*, Hentges, D. J., Ed., Academic Press, New York, 1983, 3.
3. **Moore, W. E. C. and Holdeman, L. V.,** Human fecal flora: the normal flora of 20 Japanese-Hawaiians, *Appl. Microbiol.*, 27, 961, 1974.
4. **Savage, D. C.,** Associations of indigenous microorganisms with gastrointestinal epithelial surfaces, in *Human Intestinal Microflora in Health and Disease*, Hentges, D. J., Ed., Academic Press, New York, 1983, 55.
5. **Croucher, S. C., Houston, A. P., Bayliss, C. E., and Turner, R. J.,** Bacterial populations associated with different regions of the human colon wall, *Appl. Environ. Microbiol.*, 45, 1025, 1983.
6. **Hartley, M. G., Hudson, M. J., Swarbrick, E. T., Hill, M. J., Gent, A. E., Hellier, M. D., and Grace, R. H.,** The rectal mucosa-associated microflora in patients with ulcerative colitis, *J. Med. Microbiol.*, 36, 96, 1992.
7. **Macfarlane, G. T., Gibson, G. R., and Cummings, J. H.,** Comparison of fermentation reactions in different regions of the human colon, *J. Appl. Bacteriol.*, 72, 57, 1992.
8. **Royall, D., Wolever, T. M. S., and Jeejeebhoy, K. N.,** Evidence for colonic conservation of malabsorbed carbohydrate in short bowel syndrome, *Am. J. Gastroenterol.*, 87, 751, 1992.
9. **Nordgaard, I., Hansen, B. S., and Mortensen, P. B.,** Colon as a digestive organ in patients with short bowel, *Lancet*, 343, 373, 1994.
10. **Macfarlane, G. T., Hay, S., and Gibson, G. R.,** Influence of mucin on glycosidase, protease and arylamidase activities of human gut bacteria grown in a 3-stage continuous culture system, *J. Appl. Bact.*, 66, 407, 1989.
11. **Winetz, M., Adams, R., Seedman, D., Peyton, D. N., Jayko, L. G., and Hamilton, J. A.,** Studies in metabolic nutrition employing chemically defined diets. II. Effects on gut microflora populations, *Am. J. Clin. Nutr.*, 23, 546, 1970.
12. **Attebury, H. R., Sutter, V. I., and Finegold, S. M.,** Effect of a partially chemically defined diet on normal fecal flora, *Am. J. Clin. Nutr.*, 25, 1391, 1972.
13. **Bounous, G. and Devroede, G. J.,** Effects of an elemental diet on human fecal flora, *Gastroenterology*, 66, 210, 1974.
14. **Hudson, M. J., Borriello, S. P., and Hill, M. J.,** Elemental diets and the bacterial flora of the gastrointestinal tract, in *Elemental Diets*, Russell, R. I., Ed., CRC Press, Boca Raton, FL, 1981, 105.
15. **Littleton, N. W., McCabe, R. M., and Carter, C. H.,** Studies of oral health in persons nourished by stomach tube. II. Acidogenic properties and selected bacterial components of plaque material, *Arch. Oral Biol.*, 12, 601, 1967.
16. **Miller, T. L., Weaver, G. A., and Wolin, M. J.,** Methanogens and anaerobes in a colon segment isolated from the normal fecal stream, *Appl. Environ. Microbiol.*, 48, 449, 1984.
17. **Cummings, J. H., Banwell, J. G., Englyst, H. N., Coleman, N., Segal, I., and Berson, D.,** The amount and composition of human large bowel contents, *Gastroenterology*, 98, A408, 1990.
18. **Stephen, A. M. and Cummings, J. H.,** The microbial contribution to human faecal mass, *J. Med. Microbiol.*, 13, 45, 1980.
19. **Salyers, A. A.,** Energy sources of major intestinal fermentative anaerobes, *Ann. Rev. Microbiol.*, 32, 158, 1979.
20. **Horowitz, M. I.,** Gastrointestinal glycoproteins, in *The Glycoconjugates, Vol. I*, Horowitz, M. I. and Pigman, W., Eds., Academic Press, New York, 1977, 189.
21. **Hoskins, L. C. and Zamcheck, N.,** Bacterial degradation of gastrointestinal mucins. I. Composition of mucus constituents in the stools of germ-free and conventional rats, *Gastroenterology*, 54, 210, 1968.
22. **Hoskins, L. C.,** Mucin degradation by enteric bacteria: Ecological aspects and implications for bacterial attachment to gut mucosa, in *Attachment of Organisms to the Gut Mucosa, Vol. II*, Boedeker, E.C., Ed., CRC Press, Boca Raton, FL, 1984, 51.
23. **Miller, R. S. and Hoskins, L. C.,** Mucin degradation in human colon ecosystems. Fecal population densities of mucin-degrading bacteria estimated by a 'most probable number' method, *Gastroenterology*, 81, 759, 1981.
24. **Hoskins, L. C., Augustines, M., McKee, W. B., Boulding, E. T., Kriaris, M., and Niedermeyer, G.,** Mucin degradation in human colon ecosystems. Isolation and properties of fecal strains that degrade ABH blood group antigens and oligosaccharides from mucin glycoproteins, *J. Clin. Invest.*, 75, 944, 1985.
25. **Salyers, A. A., West, S. E. H., Vercellotti, J. R., and Wilkins, T. D.,** Fermentation of mucins and plant polysaccharides by anaerobic bacteria from the human colon, *Appl. Environ. Microbiol.*, 34, 529, 1977.
26. **Salyers, A. A., Vercellotti, J. R., West, S. E. H., and Wilkins, T. D.,** Fermentation of mucin and plant polysaccharides by strains of *Bacteroides* from the human colon, *Appl. Environ. Microbiol.*, 33, 319, 1977.
27. **Hoskins, L. C. and Boulding, E. T.,** Mucin degradation in human colon ecosystems. Evidence for the existence and role of bacterial subpopulations producing glycosidases as extracellular enzymes, *J. Clin. Invest.*, 67, 163, 1981.
28. **Roberton, A. M. and Stanley, R. A.,** In vitro utilization of mucin by *Bacteroides fragilis*, *Appl. Environ. Microbiol.*, 43, 325, 1982.
29. **Bayliss, C. E. and Houston, A. R.,** Characterization of plant polysaccharide- and mucin-fermenting anaerobic bacteria from human feces, *Appl. Environ. Microbiol.*, 48, 626, 1984.

30. **Macfarlane, G. T. and Gibson, G. R.,** Formation of glycoprotein degrading enzymes by *Bacteroides fragilis, FEMS Microbiol. Lett.*, 77, 289, 1991.
31. **Hoskins, L. C., Boulding, E. T., Gerken, T. A., Harouny, V. R., and Kriaris, M. S.,** Mucin glycoprotein degradation by mucin oligosaccharide-degrading strains of human fecal bacteria. Characterisation of saccharide cleavage products and their potential role in the nutritional support of larger faecal bacterial populations, *Microb. Ecol. Health Dis.*, 5, 193, 1992.
32. **Cromwell, C. L. and Hoskins, L. C.,** Antigen degradation in human colon ecosystems: Host's ABO blood type influences enteric bacterial degradation of a cell surface antigen on *Escherichia coli* 086, *Gastroenterology*, 73, 37, 1977.
33. **Falk, P., Hoskins, L. C., and Larson, G.,** Bacteria of the human intestinal microbiota produce glycosidases specific for lacto-series glycosphingolipids, *J. Biochem. (Japan)*, 108, 466, 1990.
34. **Hoskins, L. C.,** Bacterial glycosidases and degradation of glycoconjugates in the human gut, in *Molecular Pathogenesis of Gastrointestinal Infections*, Wädstrom, T., Makela, P. H., Svennerholm, A. M., and Wolf-Watz, H., Eds., Plenum Press, New York, 1991, 37.
35. **Larson, G.,** The normal microflora and glycosphingolipids, in *The Regulatory and Protective Role of the Normal Microflora*, Grubb, P., Midvedt, T., and Norin, E., Eds., Macmillan, London, 1989, 129.
36. **Englyst, H. N. and Kingman, S. M.,** Dietary fibre and resistant starch: a nutritional classification of plant polysaccharides, in *Dietary fiber*, Kritchevsky, D., and Anderson, J. A., Eds., Plenum Press, New York, 1990, 49.
37. **Englyst, H. N., Wiggins, H. S., and Cummings, J. H.,** Determination of non-starch polysaccharides in plant foods by gas-liquid chromatography of constituent sugars as alditol acetates, *Analyst*, 107, 307, 1982.
38. **Salyers, A. A. and Leedle, J. A.,** Carbohydrate metabolism in the human colon, in *Human Intestinal Microflora in Health and Disease*, Hentges, D. J., Ed., Academic Press, New York, 1983, 129.
39. **Englyst, H. N. and Cummings, J. H.,** Resistant starch, a 'new' food component: a classification of starch for nutritional purposes, in *Cereals in a European Context*, Morton, I. D., Ed., Ellis Horwood, Chichester, 1987, 221.
40. **Wolf, M. J., Khoo, U., and Inglett, G. E.,** Partial digestibility of cooked amylomaize in humans and mice. *Die Stärke*, 29, 401, 1977.
41. **Scheppach, W., Bach, M., Bartram, P., Christl, S., Bergthaller, W., and Kasper, H.,** Colonic fermentation of potato starch after a freeze-thaw cycle, *Dig. Dis. Sci.*, 36, 1601, 1991.
42. **Bond, J. H., Currier, B. E., Buchwald, H., and Levitt, M. D.,** Colonic conservation of malabsorbed carbohydrate, *Gastroenterology*, 78, 444, 1980.
43. **Vince, A., McNeil, N. I., Wager, J. D., and Wrong, O. M.,** The effect of lactulose, pectin, arabinogalactan and cellulose on the production of organic acids and metabolism of ammonia by intestinal bacteria in a faecal incubation system, *Br. J. Nutr.*, 63, 17, 1990.
44. **Hiele, M., Ghoos, Y., Rutgeerts, P., and Vantrappen, G.,** Metabolism of erythritol in humans: comparison with glucose and lactitol, *Br. J. Nutr.*, 69, 169, 1993.
45. **Goldin, B. R.,** In situ bacterial metabolism and colon mutagens, *Ann. Rev. Microbiol.*, 40, 367, 1986.
46. **Rowland, I. R.,** Ed., *Role of the Gut Flora in Toxicity and Cancer*, Academic Press, London, 1988.
47. **Benno, Y., Endo, K., Shiragami, N., Sayama, K., and Mitsuoka, T.,** Effects of raffinose intake on human fecal flora, *Bifid. Microflora*, 6, 59, 1987.
48. **Wang, X. and Gibson, G. R.,** Effects of the *in vitro* fermentation of oligofructose and inulin by bacteria growing in the human large intestine, *J. Appl. Bacteriol.*, 75, 373, 1993.
49. **Hidaka, H., Eida, T., Takizawa, T., Tokunaga, T., and Tashiro, Y.,** Effects of fructooligosaccharides on intestinal flora and human health, *Bifid. Microflora*, 5, 37, 1986.
50. **Yamashita, K., Kawai, K., and Itakura, M.,** Effects of fructooligosaccharides on blood glucose and serum lipids in diabetic subjects, *Nutr. Res.*, 4, 961, 1984.
51. **Stone-Dorshow, T. and Levitt, M. D.,** Gaseous response to ingestion of a poorly absorbed fructooligosaccharide sweetener, *Am. J. Clin. Nutr.*, 46, 61, 1987.
52. **Kohmoto, T., Fukui, F., Takaku, H., Machida, Y., Arai, M., and Mitsuoka, T.,** Effect of isomalto-oligosaccharides on human fecal flora, *Bifid. Microflora*, 7, 61, 1988.
53. **Tanaka, R., Takayama, H., Morotomi, M., Kuroshima, T., Ueyama, S., Matsumoto, K., Kuroda, A., and Mutai, M.,** Effects of administration of TOS and *Bifidobacterium breve* 4006 on the human fecal flora, *Bifid. Microflora*, 2, 17, 1983.
54. **Ito, M., Deguchi, Y., Miyamori, K., Matsumoto, H., Kikuchi, H., Matsumoto, K., Kobayashi, Y., Yajima, T., and Kan, T.,** Effects of administration of galactooligosaccharides on the human faecal microflora, stool weight and abdominal sensation, *Microb. Ecol. Health Dis.*, 3, 285, 1990.
55. **Rowland, I. R. and Tanaka, R.,** The effects of transgalactosylated oligosaccharides on gut flora metabolism in rats associated with a human faecal microflora, *J. Appl. Bacteriol.*, 74, 667, 1993.
56. **Saito, Y., Tanako, T., and Rowland, I. R.,** Effects of soybean oligosaccharides on the human gut microflora in *in vitro* culture, *Microb. Ecol. Health Dis.*, 5, 105, 1992.
57. **Van Tassell, R. L., Carman, R. J., Kingston, D. G. I., and Wilkins, T. D.,** Bacterial metabolism in humans of the carcinogen IQ to the direct acting mutagen hydroxy-IQ, *Microb. Ecol. Health Dis.*, 2, 123, 1989.

Chapter 4

# Proteolysis and Amino Acid Fermentation

*Sandra Macfarlane and George T. Macfarlane*

## CONTENTS

I. Introduction ........................................................................................................................ 75
II. Proteolytic Activity in the Large Intestine ................................................................... 77
    A. Comparison of Proteolytic Activities in the Small Intestine and Large Bowel ................. 78
    B. Characterization of Proteolysis in the Large Intestine ........................................................ 79
III. Proteolytic Bacteria ......................................................................................................... 83
    A. Isolation and Enumeration ................................................................................................... 83
    B. Proteolytic Bacteroides ........................................................................................................ 84
        1. *Bacteroides fragilis* Group ............................................................................................ 84
        2. *Bacteroides splanchnicus* .............................................................................................. 85
    C. Proteolytic Clostridia .......................................................................................................... 86
        1. *Clostridium perfringens* ................................................................................................. 86
        2. *Clostridium bifermentans* .............................................................................................. 86
    D. Other Bacteria ...................................................................................................................... 86
IV. Amino Acid Fermentation ............................................................................................... 87
    A. Fermentation of Organic N-Containing Compounds ......................................................... 87
    B. Short Chain Fatty Acids ....................................................................................................... 88
    C. Ammonia ............................................................................................................................... 89
    D. Amines .................................................................................................................................. 90
    E. Phenols and Indoles ............................................................................................................. 91
    F. Gas Metabolism ................................................................................................................... 94
References .............................................................................................................................. 94

## I. INTRODUCTION

The last 20 years or so have witnessed a prodigious increase in the research efforts of microbiologists, nutritionists, and clinicians investigating the breakdown and fermentation of complex carbohydrates by anaerobic bacteria growing in the human large intestine. The main focus of this work has been to characterize the apparent beneficial implications of these processes with respect to health (see Chapters 3 and 5). In comparison, the digestion and metabolism of proteins and peptides by colonic microorganisms, as well as the ecological and physiological significance of these substances as carbon, nitrogen, and energy sources has almost been ignored. This is somewhat surprising in view of the fact that in the early years of this century, the baleful effects of putrefaction in the large bowel were widely recognized as a significant cause of morbidity through, among others, the writings of the Russian microbiologist Metchnikoff[1,2] and the British surgeon Arbuthnot Lane.[3] They realized that the absorption of toxic substances or "poisons" from the bowel was responsible, in many cases, for autointoxication, and that fetid breath, increased body temperature, depression, sleeplessness, vomiting, severe headaches, attacks of skin diseases, such as acne, as well as other physiological manifestations could be attributed to the activities of intestinal bacteria. Indeed, Lane advocated surgical removal of the colon, in extreme cases, to relieve symptoms of malaise. Metchnikoff[2] believed that the onset of senility and shortening of lifespan resulted from putrefaction in the bowel, and although there was no real firm scientific evidence to support this contention, he did appreciate that ammonia, phenol, and $H_2S$ were among the toxic agents produced during the degradation of proteins in the large gut. Today, we can also add amines, mercaptans, and indoles to this list. An outline of the major pathways involved in the catabolism of organic N-containing compounds is shown in Figure 1.

    Unlike the situation with carbohydrate availability in the colon, there is no shortage of organic N-containing compounds available for fermentation by bacteria throughout the length of the bowel,[4,5] at least not in individuals living on western diets. Studies with ileostomists show that small intestinal

0-8493-4524-3/95/$0.00+$.50

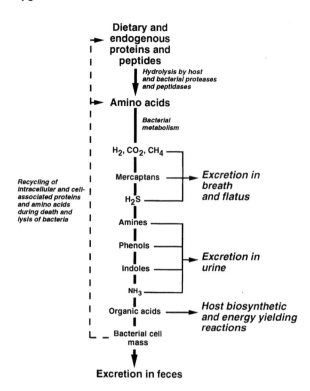

**Figure 1** Overview of the breakdown of complex organic N-sources and amino acid fermentation in the large gut.

contents contain very low levels of urea and free amino acids, and that the majority of nitrogen in ileal effluent is present in the form of proteins and peptides.[6] Estimates suggest that somewhere between 3 and 25 g of these substances enter the colon each day from the small bowel.[6-8] They occur partly in the form of dietary residues (for example, sarcoplasmic and myofibrillar muscle proteins), but a large proportion comes from the host itself, in oral, gastric, pancreatic, and small intestinal secretions such as enzymes, mucins, and other glycoproteins. The large intestine also provides protein in bacterial secretions and lysis products, colonic mucins, and desquamated mucosal cells, which can provide an important N-source, because the turnover of the lining of the colon is rapid, with columnar epithelial cells and goblet cells having a half-life of about 6 d.[9] Mucins are believed to be important carbon and energy sources for some groups of saccharolytic bacteria in the colon (see Chapter 8), and there is a body of evidence which suggests that the breakdown of this mucopolysaccharide in the large gut is facilitated by bacterial proteases.[10] Further support for this notion is afforded by observations that enzymes produced by the stomach pathogen *Helicobacter pylori* may also be directly involved in the breakdown of gastric mucin.[11] From the foregoing, it becomes apparent that there are many different types and classes of proteins potentially available for use by bacteria growing in the large intestine. Among them are nonstructural proteins such as serum albumins, recirculating antigen-antibody complexes, the connective tissue proteins collagen, elastin, as well as substances derived from various plant and bacterial sources.

Quantitatively, the most important and interesting sources of protein in the large intestine are the hydrolytic enzymes elaborated by exocrine cells of the pancreas, which include proteases (trypsin, chymotrypsin, elastase), lipases, amylase, and nucleic acid hydrolases. Although good quantitative data on human exocrine pancreas activity are few, there is evidence that between 1 to 3 g each of both trypsin and chymotrypsin are produced daily, with a further 0.5 g of elastase.[12]

To conclude this introduction, it is probably reasonable to say that since the publication of Drasar and Hill's 1974 monograph[13] on the metabolism of intestinal bacteria, our knowledge of nitrogen metabolism in human colonic anaerobes has remained fundamentally unchanged. Concerning proteolysis, most authors have drawn inferences from the work of the rumen microbiologists, which may in some circumstances be useful in highlighting certain general biological principles. However, for a variety of nutritional and physiological reasons, as well as the profound differences in species composition of the rumen and colonic microbiotas, this approach is not particularly satisfactory when considering the human large bowel. In this chapter, we have focused on human proteolytic bacteria and proteolysis in the colon,

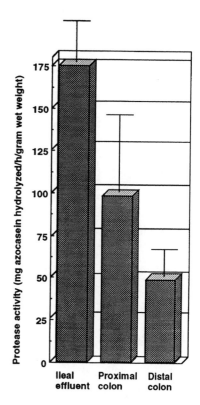

**Figure 2** Comparison of proteolytic activities in human ileal effluent and in different regions of the large bowel. Results are from samples obtained from four persons. Error bars show SEM. From *Macfarlane, G. T., Cummings, J. H., Macfarlane, S., and Gibson, G. R., J. Appl. Bacteriol., 67, 521, 1989.*

which is an unusually neglected area of intestinal microbiology. Few studies have been made in this field, which is probably due to the widespread and incorrect belief that human gut microorganisms are not particularly proteolytic. An overview of amino acid metabolism is also given that reviews the various toxic metabolites arising from amino acid fermentation.

## II. PROTEOLYTIC ACTIVITY IN THE LARGE INTESTINE

The human large intestine is one of the most proteolytic natural environments known. Studies aimed at determining the contribution made by bacteria to proteolysis and protein degradation in this organ are complicated by an unusual situation in which substantial, though variable, amounts of endopeptidases produced by the pancreas together with small intestinal peptidases enter the large bowel mixed with digesta from the terminal ileum. Pancreatic trypsin and chymotrypsin have been shown to bind to the intestinal mucosa and to adsorb onto particulate materials in the gut without loss of activity,[14] although significant amounts of chymotrypsin are known to be present in fecal water.[15] Measurements of protease activity in small intestinal contents and in material taken from the proximal and distal colons show that protease activities progressively decline as contents move through the bowel (Figure 2). This condition is due to number of reasons. First, pancreatic proteins are broken down by bacteria in the large bowel. Second, host produced antiproteases may affect the activities of pancreatic endopeptidases in the gut.[16,17] The role of bacteria in degrading these enzymes is shown in animal studies, where pancreatic protease activities are considerably higher in the feces of germ-free rats compared with their conventional counterparts.[18] Other work has shown that fecal trypsin levels increased 100-fold in antibiotic treated patients, although chymotrypsin and elastase activities were only two to three times higher.[17] Pancreatic endopeptidases probably undergo autodigestion in the colon, and there may be synergistic effects with bacterial proteases.[19,20]

The transit time of digestive material through the gut is likely to have a major effect on the type and amount of proteolytic activity occurring in the colon, since it is likely that with short transits, higher levels of pancreatic protease activity would survive passage through the gut. Moreover, under these circumstances, there would be less intracellular proteases released into the external milieu from dead and lysed bacteria. Although few *in vivo* data exist to directly support this hypothesis, studies have been made on

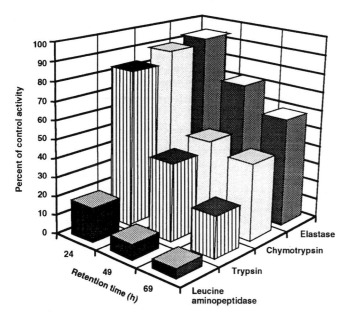

**Figure 3** Effect of retention time on the breakdown of leucine aminopeptidase and pancreatic proteases by mixed populations of human colonic bacteria growing in a three-stage continuous culture system. The figure shows data obtained from the third stage vessel. From *Macfarlane, G. T., Cummings, J. H., Macfarlane, S., and Gibson, G. R., J. Appl. Bacteriol., 67, 521, 1989.*

the effect of system retention time (loosely analogous to transit time) on the ability of mixed populations of colonic bacteria to degrade pancreatic proteases and leucine aminopeptidase, using a three-stage continuous culture system.[20] The results showed that as retention time was increased, recoveries of peptidolytic activity diminished, and that leucine aminopeptidase and trypsin in particular, were extensively degraded by the bacteria (Figure 3).

## A. COMPARISON OF PROTEOLYTIC ACTIVITIES IN THE SMALL INTESTINE AND LARGE BOWEL

Unlike the small gut where proteolytic activities are almost entirely due to host-produced enzymes, proteolysis in the colon consists of a complex mixture of host and bacterial proteases. An indication of the contribution made by bacteria to large intestinal proteolysis is shown in Table 1. In this study,[21] fecal proteolysis was determined in 10 healthy volunteers, and a patient who did not have a pancreas. The results indicated that fecal proteolysis varied considerably between different individuals, and demonstrated that protease activities in the pancreatectomy patient were within the range found in healthy people, albeit at the lower end. The data also show that fecal proteolysis is associated with bacteria and particulate materials, as well as occurring extracellularly.

Proteolysis differs both quantitately and qualitatively in the small and large intestines. This difference is immediately apparent from the study outlined in Table 2, where proteases in ileal effluent and different fractions of feces were tested for their ability to degrade a range of native and diazotized protein substrates.[22] These results contrast the massive difference in proteolytic activities that are recoverable in small intestinal contents and feces. It is also clear that pancreatic and bacterial proteases break down different proteins in different ways. For example, BSA and ovalbumin were poorly degraded, or completely recalcitrant to proteases in ileal effluent and feces, compared to casein. Highly globular proteins have also been shown to be resistant to hydrolysis by rumen microorganisms.[23,24] From these and subsequent studies,[25] it was concluded that structure and configuration were important determinants of a protein's susceptibility to enzymic hydrolysis. In the colon, however, other factors may affect protein breakdown, including pH (see discussion later), and their occurrence in complexes with other polymers that restrict the accessibility of the substrate to hydrolytic enzymes.

Table 2 also shows that, with the exception of azoalbumin, native proteins were degraded to a greater extent than their diazotized derivatives by both small intestinal and fecal proteases. This situation may be due to steric effects resulting from the presence of the sulfanilamide adduct on the protein molecule, at or near the sites of proteolytic cleavage, or to conformational changes in the proteins caused by diazotization, that restrict access to the proteases. In the case of azoalbumin, the increase in susceptibility to proteolytic digestion probably results from diazotization altering protein tertiary structure and making potential cleavage sites more available to proteolytic enzymes.

**Table 1** Distribution of Proteolytic Activity
in Different Fractions of Feces

| | Protease activity[a] | |
|---|---|---|
| | **Healthy volunteers[b]** | **Pancreatectomy patient[c]** |
| Mean total activity | 15.8 | 6.8 |
| Range | 3.3–49.5 | 4.5–8.8 |

| **Fecal fraction** | **Mean % activity recovered** | |
|---|---|---|
| 10% (w/v) slurry | 100 | 100 |
| Extracellular fraction | 32 | 26 |
| Particulate fraction | 48 | 36 |
| Bacterial fraction | 17 | 36 |

[a] Milligrams azocasein hydrolyzed $h^{-1}$ (gram wet weight feces)$^{-1}$.
[b] Ten persons.
[c] Results from five consecutive daily samples from one individual.

From Macfarlane, G. T., Allison, C., Gibson, S. A. W., and Cummings, J. H.,
*Appl. Environ. Microbiol.*, 55, 679, 1989.

## B. CHARACTERIZATION OF PROTEOLYSIS IN THE LARGE INTESTINE

Proteolytic enzymes can be broadly classified as belonging to one of four main groups: serine proteases, aspartic or acid proteases, thiol or cysteine proteases, and the metalloproteases.[26] The enzymes are then further categorized on the basis of their substrate side-chain specificities. Serine proteases are generally active at neutral to alkaline pH whereas thiol proteases are most active at neutral pH, and are inhibited by sulfhydryl agents and activated by reducing agents. Aspartic proteases have acid pH optima, whereas metalloproteases are sensitive to metal chelators and are active at neutral and alkaline pH.

The major groups of proteases present in the human gastrointestinal tract can be determined using protease inhibitors, although these methods are relatively indiscriminate when dealing with such a complex system as the colon which contains hundreds of different peptide hydrolases. In general, however, the greater the degree of inhibition of proteolysis exerted by a protease inhibitor, the greater the contribution made by the class of protease on which it is active to overall proteolysis in the samples being tested. Table 3 lists some of the more common inhibitors and their properties.

Studies using protease inhibitors to compare proteolysis in small bowel contents and feces[22] have shown that ileal protease activities are almost completely inhibited by addition of the serine protease inhibitor PMSF, and to a lesser degree, by chymostatin and soybean trypsin inhibitor (STI), which respectively inhibit chymotrypsin and trypsin (Figure 4). These data are therefore consistent with pancreatic endopeptidases being serine enzymes. In feces, however, the effects of these inhibitors decline markedly, whereas inhibition by thiol (thimerosal, iodoacetate) and metalloprotease (EDTA, cysteine) inhibitors increases. These results show that human colonic contents contain a mixture of serine, thiol, and metalloproteases, and that aspartic proteases make only a minor contribution to protein breakdown in this organ. Consequently, not only do absolute amounts of proteolytic activity decline in the colon, but the character of the enzymes involved changes markedly.

Superficially, proteolysis in the human large bowel might be considered to share certain similarities with the rumen, where the majority of proteases (about 75%) are associated with particulate materials, and are of the serine, metalloprotease, and thiol type.[27] Apart from bacteriological considerations, however, there is another major difference in these two ecosystems, in that unlike the rumen, the healthy colon does not harbor large populations of protozoa. In the rumen, many different species of these organisms have been shown to produce a variety of proteases of low and high molecular mass that are sensitive to thiol protease inhibitors.[28]

Table 4 shows the results of more detailed experiments designed to determine the types of proteases present in different fecal fractions. They indicate that there are significant qualitative differences in proteolytic activity associated with particulate materials, bacterial cells, and fecal water. For example, the particulate fraction in particular shows high levels of inhibition by PMSF, STI, and thimerosal, whereas the extracellular fraction has comparatively higher levels of inhibition by chymostatin, the metalloprotease inhibitor EDTA, and the thiol protease inhibitor iodoacetate. These data are in many ways comparable with a similar series of studies that investigated the distribution of polysaccharidases and glycosidases in

**Table 2  Hydrolysis of Proteins by Small Intestinal Contents and Different Fractions of Feces**

| Type of sample | Test substrates | | | | | | | |
|---|---|---|---|---|---|---|---|---|
| | Casein | BSA | Collagen | Ovalbumin | Azocasein | Azoalbumin | Azocoll | Azosoybean flour |
| Ileal effluent | 1214 ± 546 | 0 ± 0 | 281 ± 35 | 131 ± 48 | 319 ± 45 | 208 ± 34 | 150 ± 21 | 333 ± 55 |
| Feces | 20 ± 2 | 2 ± 1 | 6 ± 2 | 2 ± 0.5 | 11 ± 6 | 9 ± 7 | 5 ± 4 | 6 ± 3 |
| Bacterial fraction | 5.20 ± 2.70 | 0.09 ± 0.13 | 0.50 ± 0.26 | 0.23 ± 0.16 | 0.41 ± 0.15 | 0.03 ± 0.02 | 0.19 ± 0.14 | 0.29 ± 0.04 |
| Extracellular fraction | 3.00 ± 2.00 | 0.09 ± 0.13 | 0.67 ± 0.26 | 0.47 ± 0.27 | 0.17 ± 0.24 | 0.01 ± 0.01 | 0.31 ± 0.15 | 0.34 ± 0.23 |
| Particulate fraction | 7.10 ± 3.30 | 0.08 ± 0.09 | 0.38 ± 0.36 | 0.18 ± 0.17 | 0.89 ± 0.47 | 0.07 ± 0.03 | 0.29 ± 0.14 | 0.54 ± 0.23 |

*Note:* Values are milligrams of protein hydrolyzed $h^{-1}$ (gram wet weight sample)$^{-1}$. Data are mean results obtained from feces of five healthy individuals or small gut contents of five ileostomy subjects ± SD.

From Gibson, S. A. W., McFarlan, C., Hay, S., and Macfarlane, G. T., *Appl. Environ. Microbiol.*, 55, 679, 1989.

**Table 3  Some Inhibitors Used in Characterizing Microbial Proteases**

| Protease inhibitor | Concentration in assay | Description | Reference |
|---|---|---|---|
| EDTA | 1–5 mmol/l | Metal chelator; inhibits metalloproteases; action reversible. | 165 |
| Cysteine | 2 mmol/l | Reduces disulfide bonds; inhibits metalloproteases. | 166 |
| PMSF[a] | 1–3 mmol/l | Irreversibly inhibits serine proteases, including trypsin and chymotrypsin; inhibition of some thiol proteases reversed by reducing agents such as dithiothreitol. | 167 |
| Iodoacetate | 5 mmol/l | Nonspecific inhibition of thiol proteases. | 167 |
| Thimerosal | 5 mmol/l | Inhibits thiol and some serine proteases. | 168 |
| Elastinal | 5 μmol/l | Reversible inhibitor of elastase-like serine proteases. | 169 |
| STI[b] | 100 μg/ml | Reversible inhibitor of trypsin-like enzymes. | 170 |
| Chymostatin | 100 μg/ml | Specific and reversible inhibitor of chymotrypsin-like proteases. | 169 |
| Pepstatin A | 50 μg/ml | Reversibly inhibits aspartic proteases. | 169 |
| TLCK[c] | 30–50 μg/ml | Irreversible inhibition of trypsin-like serine proteases, and some thiol proteases. | 171 |
| TLPK[d] | 50–100 μg/ml | Irreversibly inhibits chymotrypsin-like serine proteases; also affects some thiol proteases. | 171 |

[a] Phenylmethylsulfonyl fluoride. [b] Soybean trypsin inhibitor. [c] L-1-Chloro-3-[4-tosylamido]-7-amino-2-heptanone. [d] L-1-Chloro-3-[4-tosylamido]-4-phenyl 2-butanone.

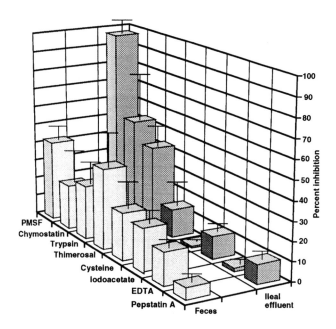

**Figure 4** Characterization of proteolytic activities using human ileal effluent (N=5) and feces (N=10) using protease inhibitors. Error bars show SD. From *Macfarlane, G. T., Allison, C., Gibson, S. A. W., and Cummings, J. H., Appl. Bacteriol., 64, 37, 1988.*

**Table 4** Inhibition of Proteolytic Activity Associated with Different Fractions of Large Intestinal Contents

| | **Percent inhibition** | | |
|---|---|---|---|
| Protease inhibitor | Extracellular fraction | Particulate fraction | Bacterial fraction |
| PMSF | 32.0 ± 16.7 | 59.8 ± 13.6 | 47.8 ± 7.9 |
| STI | 37.2 ± 10.0 | 49.6 ± 18.3 | 24.2 ± 14.8 |
| Chymostatin | 40.9 ± 22.5 | 38.8 ± 17.9 | 20.1 ± 7.8 |
| Pepstatin A | 5.0 ± 1.2 | 5.8 ± 1.0 | 6.8 ± 4.5 |
| Thimerosal | 28.0 ± 12.2 | 46.9 ± 20.9 | 37.0 ± 5.9 |
| Iodoacetate | 28.0 ± 12.2 | 13.7 ± 8.8 | 7.8 ± 1.5 |
| EDTA | 19.1 ± 5.4 | 2.0 ± 1.8 | 4.6 ± 5.6 |
| Cysteine | 22.0 ± 7.3 | 18.2 ± 10.6 | 32.0 ± 7.3 |

*Note:* Data are mean results from 10 healthy individuals ± SD.

human intestinal contents.[29] The explanation given by the authors was that either different populations of bacteria existed on the surface of particulate material in the gut, compared to those that were "free living," or that bacteria growing on surfaces were simply expressing different enzymes.

Studies with washed fecal bacteria demonstrate that high levels of protease activity are present inside these organisms, the majority of which is soluble (Table 5). This situation suggests that large amounts of proteases could potentially be released during lysis of the cells, making a significant contribution towards extracellular proteolysis in the large intestine. In addition, the inhibitor data in Table 5 indicate that although there is a multiplicity of proteases associated with the cells, serine enzymes predominate. Interestingly, they also provide some evidence for the synthesis of trypsin and chymotrypsin-like proteases by colonic bacteria.

Many different nutritional and environmental factors are likely to affect the synthesis and activities of proteolytic enzymes in the large bowel. One of the most important is pH.[15,30-33] The physiological

**Table 5** Characterization of Cell-Associated Proteolytic Activity of Intestinal Bacteria

| Bacterial fraction | Mean protease activity[a] | Percent inhibition of proteolysis | | | | | | | |
|---|---|---|---|---|---|---|---|---|---|
| | | PMSF | Chymostatin | Thimerosal | Iodoacetate | Pepstatin A | Cysteine | STI | EDTA |
| Whole bacteria | 74 ± 15 | 56 ± 8 | 10 ± 5 | 27 ± 9 | 25 ± 5 | 8 ± 3 | 29 ± 16 | 14 ± 1 | 24 ± 6 |
| Crude cell homogenate | 460 ± 58 | 48 ± 8 | 20 ± 23 | 26 ± 22 | 12 ± 4 | 12 ± 6 | 20 ± 23 | 20 ± 22 | 28 ± 14 |
| Cell wall material | 58 ± 35 | 46 ± 7 | 9 ± 12 | 22 ± 20 | 4 ± 3 | 5 ± 4 | 16 ± 14 | 19 ± 9 | 7 ± 6 |
| Soluble fraction | 342 ± 72 | 52 ± 6 | 15 ± 12 | 44 ± 7 | 11 ± 2 | 7 ± 4 | 15 ± 13 | 21 ± 14 | 18 ± 11 |

*Note:* Results are mean values of measurements made on bacteria from five different fecal samples ± SD.

[a] μg Azocasein hydrolyzed h$^{-1}$(ml sample)$^{-1}$.

From Gibson, S. A. W., McFarlan, C., Hay, S., and Macfarlane, G. T., *Appl. Environ. Microbiol.*, 55, 679, 1989.

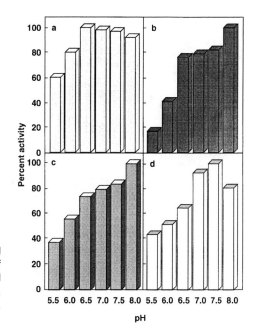

**Figure 5** Effect of pH on proteolyic activities associated with (a) the extracellular fraction of feces; (b) cell extracts of fecal bacteria; (c) washed particulate material in feces; and (d) whole fecal bacteria. From *Gibson, S. A. W., McFarlan, C., Hay, S., and Macfarlane, G. T., Appl. Environ. Microbiol., 55, 679, 1989.*

significance of this parameter lies in the fact that the pH of intestinal contents can vary by as much as two units in different parts of the colon. Thus, in healthy individuals, the proximal colon typically has an overall pH of around 5, whereas in the distal gut, pH begins to approach neutrality.[34,35] Of course, this range does not take into account the more extreme pH values that are likely to occur in the center of microcolonies, or in other microenvironments within the gut. Figure 5 shows the influence of pH on general protease activities in different fractions of feces. From this study, it can be seen that proteolysis was active over a wide pH range in all samples tested, but had marked optima at alkaline pH in the particulate fraction, and in extracts from fecal bacteria. In all cases, protease activity was particularly sensitive to low pH which, as discussed earlier, correlates with the presence of serine, thiol, and metalloproteases in the large intestine. Consequently, factors that influence pH in the colon such as carbohydrate availability[36] and transit time[37] may also indirectly affect protein digestion.

## III. PROTEOLYTIC BACTERIA

### A. ISOLATION AND ENUMERATION

Enumeration and characterization of proteolytic bacteria in human feces has shown that a large number of intestinal species are proteolytic, but that the nature and types of proteases produced by different organisms varies considerably.[5] Possibly the most simple way of screening for proteolytic bacteria in the gut is to plate serial dilutions of feces onto agar plates containing a protein, and then looking for precipitation or clearing of the protein around bacterial colonies after growth. Results of such a study are shown in Figure 6. They show that where casein was used as the protein substrate, carriage rates of some proteolytic species varied considerably in different individuals, and members of the *Bacteroides fragilis* group were the numerically predominant proteolytic bacteria in the large gut. A major problem with this type of experiment, however, is that whereas it is relatively straightforward to identify extracellular protease production by clostridia, for example, it is generally much more difficult to visualize weakly proteolytic bacteria, or species that produce cell-bound proteases. In the case of colonic bacteroides, which primarily form these types of enzymes, it was necessary to use long incubation periods to detect the small zones of clearing that formed around the colonies. Therefore, this method of screening for protease forming bacteria lacks sensitivity and can potentially give a false indication of the numbers and types of organisms that are present.

With the exception of *Clostridium perfringens*, where research interest has focused on areas other than its role as a member of the normal gut flora, protease formation has been studied in very few large

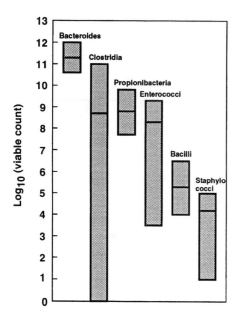

**Figure 6** Enumeration of proteolytic bacteria in feces using traditional plate counting procedures. Data show the range and mean values (bars) in feces obtained from five healthy individuals. From *Macfarlane, G. T., Cummings, J. H., and Allison, C., J. Gen. Microbiol., 132, 1647, 1986.*

intestinal bacteria. Recently, however, some progress has been made with respect to members of the *B. fragilis* group, *B. splanchnicus*, and *Clostridium bifermentans*, which will shortly be reviewed.

## B. PROTEOLYTIC BACTEROIDES
### 1. *Bacteroides fragilis* Group

These bacteria constitute the single most numerous assemblage of microorganisms in the human colon, accounting for approximately 30% of the total cultivable flora.[38] Much of the research on these organisms has concentrated on their ability to break down complex carbohydrates.[39] However, they are nutritionally versatile with proteolytic abilities variable throughout the group. In general, species identified as *B. fragilis* tend to be the most proteolytic, whereas *B. thetaiotaomicron* and *B. ovatus* are either weakly proteolytic or nonproteolytic.[40] Members of the *B. fragilis* group are able to utilize ammonia and peptides as N-sources, but do not grow well on mixtures of individual amino acids.[41] In view of their ubiquity in the gut, the bacteroides are likely to contribute significantly towards the turnover of proteins in that organ.

Interestingly, a recent study has indicated that trypsin and chymotrypsin are the preferred N-sources for some members of the *B. fragilis* group,[19] and that because of autodigestion of the pancreatic endopeptidases, nonproteolytic species are also able to grow on these substrates, although less efficiently. Some members of the *B. fragilis* group have been reported to be able to degrade colonic mucins[42,43] through their ability to synthesize a range of enzymes that hydrolyze glycosidic bonds on the carbohydrate side-chains, but the role of bacteroides proteases in mucin degradation is not clear. Like proteolytic enzymes produced by many other bacteria, peptide hydrolases formed by members of the *B. fragilis* group including *B. vulgatus*, *B. distasonis*, and somewhat surprisingly, *B. ovatus*, have been reported to have pathogenic effects, in that they are able to destroy brush border enzymes in bacterial overgrowth syndromes in the gut.[44] On the basis of inhibition data, the proteases were believed to be serine enzymes which had elastase-like characteristics, but subsequent studies suggested that the enzymes were unable to break down substrates derived from elastin.[45]

*Bacteroides fragilis* constitutively forms at least three proteases, although the process is strongly affected by cultural conditions.[45] These enzymes are sensitive to serine and thiol protease inhibitors such as PMSF and thimerosal, but cysteine and EDTA inhibit proteolysis to a lesser degree, suggesting the existence of metalloproteases.[45-47] Batch culture experiments demonstrate that the enzymes are firmly cell bound during exponential growth, and that high levels of proteolytic activity accumulate within the cells.[46] As the bacteria enter the stationary phase, there is a massive release of the intracellular proteases into the culture medium. This event does not appear to result from lysis of the cells, but the signaling mechanism that initiates the process is as yet unknown. Unlike other Gram-negative species, such as *Porphyromonas gingivalis*,[48] where a similar release of proteases occurs at the end of active growth, the majority (>90%) of proteolytic activity released from *B. fragilis* is soluble, and not associated with particulate materials or membrane vesicles.

Chemostat studies show that *B. fragilis* produces higher levels of proteases during growth under N-limiting conditions, at high dilution rates, which is consistent with the enzymes being constitutive in nature.[46] In these experiments, extracellular proteolytic activity, in trace amounts, was only found at very low dilution rates, which was probably due to lysis of the bacteria under severe nutritional stress. Moreover, the ratio of intracellular to whole cell protease activities increased concomitantly with dilution rates, which correlates with the high levels of intracellular enzymes seen during exponential growth in the batch cultures. The fact that protease activity was firmly associated with the bacteria in the chemostats further suggests that protease secretion requires a signal that is produced only by late log or stationary phase bacteria. Experiments with batch cultures indicated that when *B. fragilis* was grown with casein as the sole N-source, protease synthesis was about fourfold higher compared to ammonia or peptone-grown cells.[45]

Although they were carried out using pure cultures of bacteria, these physiological studies suggest that high levels of cell-bound protease would be synthesized in the large bowel by *B. fragilis*-type bacteria during rapid growth in the cecum, where carbohydrate availability is greatest. However, with progressive nutrient depletion in the distal bowel, these enzymes would be expected to be released into gut contents, thereby contributing to the extracellular protease pool.

*Bacteroides fragilis* proteases appear to have relatively limited substrate specificities in that they are not especially active against elastin, collagen, gelatin, or globular proteins such as BSA, although they do efficiently break down casein and azocasein, and as discussed earlier, trypsin and chymotrypsin.[45] Studies relating to the biochemical characteristics of these enzymes are facilitated by their arylamidase activities towards a range of β-naphthylamide and p-nitroanilide substrates. Proteases formed by *B. fragilis* group bacteria exhibit significant arylamidase activity towards leucine p-nitroanilide (LPNA), valylalanine p-nitroanilide (VAPNA), and glycylprolyl p-nitroanilide (GPRPNA).[46] No activity is seen when synthetic trypsin, chymotrypsin, or elastase substrates (benzoylarginine p-nitroanilide, glutarylphenylalanine p-nitroanilide, succinylalanylalanylalanine p-nitroanilide, respectively) are used. The degree of proteolytic activity expressed by these organisms appears to be particularly related to the levels of LPNA degrading activity in bacterial cultures,[40] which suggests that the thiol protease which is responsible is the major peptidase formed by these bacteria. As mentioned earlier, there is some evidence for the involvement of bacteroides proteases in tissue damage in the gut. Some correlation with these observations may be made with the GPRPNA degrading protease formed by *B. fragilis*, because peptidolytic enzymes that have arylamidase activity towards this substrate have been linked with tissue destruction in periodontal disease,[49] and the pathogenicity of *Porphyromonas gingivalis*.[50]

## 2. *Bacteroides splanchnicus*

Like members of the *B. fragilis* group, this bacterium also occurs in high numbers in the large gut. Although it is saccharolytic, *B. splanchnicus* differs from members of the *B. fragilis* group by virtue of the fact that it is unable to grow on sucrose and certain other disaccharides.[51] Together with other colonic bacteroides, *B. splanchnicus* is nutritionally versatile in that it is able to use ammonia, peptides, or proteins, including pancreatic proteases, as N-sources.[19] This ability to scavenge pancreatic and other host secretions in the colon probably provides a significant competitive advantage for *B. splanchnicus* and other bacteroides with similar metabolic capabilities.

Batch and continuous culture studies have shown that protease formation by *B. splanchnicus* is similar in many respects to *B. fragilis*.[52] Multiple enzymes are produced which are cell bound during active growth and released from the bacteria in the stationary phase. Overall, protease activity is generally growth rate associated in chemostats, expressed maximally under carbon-excess conditions, and sensitive to serine, thiol, and metalloprotease inhibitors. *Bacteroides splanchnicus* proteases do not manifest trypsin, chymotrypsin, or elastase-like characteristics, but LPNA, VAPNA, and GPRPNA are hydrolyzed. Valyl, alanyl, or glycyl p-nitroanilides are not degraded, indicating a requirement by some enzymes for C-terminus blocked dipeptidyl substrates. As with *B. fragilis*, the LPNA hydrolyzing enzyme is a thiol protease, whereas the proteases active against the dipeptidyl substrates are serine enzymes. *Bacteroides splanchnicus* proteases have neutral to alkaline pH optima, and are stimulated by $Ca^{2+}$ and inhibited by reducing agents such as dithiothreitol and dithioerythritol. Biochemical characterization of the GPRPNA hydrolyzing enzyme[53] suggests that it has a molecular mass of 160 kDa, consisting of two 80 kDa monomers. This enzyme is also active against azocasein and VAPNA.

## C. PROTEOLYTIC CLOSTRIDIA

### 1. *Clostridium perfringens*

This bacterium has a ubiquitous distribution in a wide range of natural environments, and is a pathogenic organism in humans and many other animal species.[54] *Clostridium perfringens* is also a member of the normal gut flora, typically being recovered in numbers ranging from $10^4$ to $10^{12}$/g dry weight of feces.[55] It produces a variety of extracellular toxins including collagenase ($\kappa$ toxin)[56] that hydrolyzes collagen or gelatin and a general protease ($\lambda$ toxin)[54] which has no collagenolytic activity. *Clostridium perfringens* proteases have been reported to be sensitive to thiol and metalloprotease inhibitors.[57]

Protease formation by *C. perfringens* has been extensively documented, mainly in relation to food quality evaluation studies[58,59] and tissue destruction.[60,61] These experiments have shown that bacteria grown in protein containing media secrete higher levels of protease than those grown on amino acids[58] but that some organic N-containing compounds, such as casein, repressed protease formation.[59]

Physiological studies with a human fecal isolate of *C. perfringens* demonstrated that proteases were produced during exponential growth in batch culture, and that this process ceased as the bacteria entered stationary phase.[31] These studies also indicated that glucose and ammonia availability had little effect on protease production, but that increasing peptone concentrations in the culture medium was stimulatory. Culture pH also influenced protease secretion, which was optimal at pH 6. Continuous-culture experiments in which this bacterium was grown in glucose or peptide-limited chemostats showed that protease formation increased concomitantly with growth rate, irrespective of the form of nutrient limitation. Nutrient availability only affected protease synthesis at low dilution rates, where higher levels were observed in peptide-limited cultures.

### 2. *Clostridium bifermentans*

Like *C. perfringens*, *C. bifermentans* is both saccharolytic and strongly proteolytic. It is a member of the normal human gut flora, and is usually isolated in significant numbers (approximately $10^9$/g dry weight) from feces.[55] Physiological studies aimed at characterizing protease formation by *C. bifermentans* have shown that synthesis occurs during active growth, and is stimulated by glucose and peptone.[32] This process is generally similar to that found with *C. perfringens*, but contrasts markedly with that of *C. sporogenes*, in which proteolytic activity is only secreted in stationary phase cultures, and is strongly repressed by carbohydrates and amino acids.[62] Unlike many other bacteria that form extracellular proteases, such as *Clostridium botulinum*,[63] *Vibrio alginolyticus*,[64] *Aeromonas hydrophila*,[65] and *Pseudomonas aeruginosa*[66] ammonia did not repress protease synthesis in *C. bifermentans*.

SDS-PAGE of cell-free supernatants from *C. bifermentans* chemostats (Figure 7) indicated that under certain growth conditions, the organism produced as many as 18 different proteolytic enzymes, with molecular masses ranging from 36 to 125 kDa, some of which appeared to be primarily formed at high dilution rates.[32] However, the information in Figure 7, when considered by itself, is somewhat misleading, because measurements of total protease activity in these chemostats indicated that taken as whole, protease production by *C. bifermentans* was repressed at high dilution rates, during both carbon and nitrogen-limited growth.[32] Why *C. bifermentans* produces such a large number of proteolytic enzymes is not clear, however, formation of multiple extracellular enzymes putatively designed to fulfill the same task, is not particularly unusual in microorganisms.[67-69]

Proteolysis in *C. bifermentans* cultures was optimal at neutral pH, and although individual proteases were not characterized, inhibitor data showed that the majority of the enzymes formed by this bacterium were of the metalloprotease type, with possibly a minor serine and thiol protease element. None of the proteases produced by *C. bifermentans* were collagenolytic, or had trypsin, chymotrypsin, or elastase-like activities. Moreover, of 21 different synthetic $\beta$-naphthylamide and p-nitroanilide substrates tested, none were hydrolyzed by any of the proteases produced by this bacterium. Despite this situation, which was due in part to the diversity of *C. bifermentans* peptide hydrolases, a wide range of native and diazotized proteins were digested by culture filtrates, including highly globular substrates such as BSA.

## D. OTHER BACTERIA

Table 6 lists some other proteolytic species that occur in the large intestine and, where known, summarizes certain biological properties of their proteases. The table shows that proteolysis has a wide taxonomic distribution, being found in many different genera and species, although in most cases, detailed information is limited.

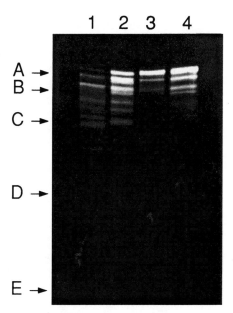

1  2  3  4

A →
B →

C →

D →

**Figure 7** SDS-PAGE of extracellular proteases produced by *Clostridium bifermentans* at different dilution rates in glucose-excess chemostats. Lane 1, D=0.09 h$^{-1}$; Lane 2, D=0.35 h$^{-1}$; Lane 3, D=0.50 h$^{-1}$; Lane 4, D=0.70 h$^{-1}$. Molecular weight markers are indicated by arrows: (A) β-galactosidase (116 kDa); (B) phosphorylase B (97.4 kDa); (C) bovine serum albumin (66 kDa); (D) egg albumin (45 kDa); (E) carbonic anhydrase (29 kDa). From *Macfarlane, G. T. and Macfarlane, S., Appl. Environ. Microbiol., 58, 1195, 1992.*

E →

**Table 6** Characteristics of Protease Activities of Some Bacteria that Occur in the Large Intestine

| Bacteria | Location and types of proteases | References |
|---|---|---|
| Fusobacteria (e.g., *F. necrophorum*) | Cell-associated proteolytic activity. Mainly produced at low dilution rates in continuous culture. Synthesis repressed by glucose and stimulated by peptides. | 172 |
| *Prevotella ruminicola* | Mainly cell-associated proteases produced during active growth. Protease activity stimulated by reducing agents and sensitive to serine, thiol, and aspartic protease inhibitors. | 170, 173 |
| *Prevotella melaninogenica* | Cell-associated collagenase released by autolysis. Maximal levels produced at the end of active growth. Proteolytic activity of this bacterium cleaves Fab/Fc of IgA$_1$. | 174–176  174–176 |
| Lactobacilli | Intracellular and cell wall-bound proteases. Sensitive to serine and thiol protease inhibitors. | 177, 178 |
| Enterococci (e.g., *E. faecium*) | Cell-associated and extracellular protease activities. | 5 |
| *Clostridium ramosum* | Protease reported to be active against IgA$_1$ and IgA$_2$. | 179 |
| *Propionibacterium acnes* | Multiple proteases produced. Extracellular enzymes sensitive to serine and thiol protease inhibitors. Secreted by bacteria during exponential growth in batch culture and at high dilution rates in chemostats. Synthesis may be repressed by glucose. | 180 |
| *Clostridium septicum* | Up to six cell-associated gelatinases produced. | S. Macfarlane, unpublished results |
| *Clostridium sporogenes* | Multiple extracellular proteases secreted at the end of active growth in batch culture, and at low dilution rates in chemostats. Synthesis repressed by glucose, amino acids, and ammonia. Sensitive to serine, thiol, and metalloprotease inhibitors. Proteases active against a wide range of substrates. | 181, 182 |

## IV. AMINO ACID FERMENTATION

### A. FERMENTATION OF ORGANIC N-CONTAINING COMPOUNDS

As with much of our knowledge concerning the proteolytic activities of anaerobic bacteria, the vast majority of studies on peptide and amino metabolism of these organisms has been carried out by rumen and oral microbiologists. In some cases, however, the organisms studied are metabolically similar to, or are also members of, the normal human gut flora, which allows for some relevant observations to be made.

Peptides and amino acids are used as carbon, nitrogen, and energy sources by both saccharolytic and asaccharolytic bacteria. Some saccharolytic species are able to derive small amounts of energy from the carbon skeletons of peptides and amino acids, although not enough to serve as energy sources by themselves.[70] In contrast, other bacteria are obligate amino acid fermenters, and are unable to obtain energy from the metabolism of carbohydrates. Some anaerobes, such as *Fusobacterium nucleatum*, are able to ferment amino acids and sugars concurrently,[71] with sugar transport being driven by amino acid metabolism.[72] Studies with rumen organisms indicate that different amino acids are fermented at different rates. For example, arginine and threonine are metabolized most rapidly, followed by lysine, phenylalanine, leucine, and isoleucine.[73]

Although large populations of amino acid fermenting bacteria occur in the human large intestine,[55,74] there is a considerable body of evidence which suggests that anaerobic bacteria, in general, prefer to assimilate amino nitrogen in the form of peptides, rather than as free amino acids.[75,76] Peptides have been shown to stimulate growth of many gut species.[70,77,78] Their transport is often accompanied by excretion of parts of the molecule, which are not required for fermentation or biosynthetic reactions by the organisms.[79] The release of large amounts of amino acids has been reported to occur in cultures of a variety of clostridia and bifidobacteria.[80]

Peptides and amino acids are transported into bacteria in different ways. Peptide uptake is relatively nonspecific, whereas amino acid transport systems are highly specific for groups of structurally related amino acids, and have high levels of physiological activity.[81] This situation is seen in protein fermentation experiments with mixed populations of colonic bacteria, where free amino acids only accumulate in small amounts during digestion of these polymers, and where the rate of peptide transport has been shown to be an important limiting factor in their utilization by the microorganisms.[82] This shows that while the initial events associated with protein depolymerization occur extracellularly, the destruction of a protein is completed inside the cells, where the component amino acids are made available for fermentation through the activities of intracellular peptidases. The mechanisms that control peptide transport in anaerobic bacteria are unknown, but experiments with pure and mixed cultures of rumen bacteria have indicated that the chemical composition of peptides affects their rate of breakdown. Thus, hydrophilic peptides appear to be more rapidly degraded than those which contain high levels of aliphatic, proline, or bulky aromatic amino acid residues.[83,84]

A range of transamination, oxidation, and reduction reactions are involved in the fermentation of amino acids by human intestinal anaerobes. Many different types of electron donors/acceptors are potentially available to participate in these reactions, including other amino acids, keto acids, molecular $H_2$ and unsaturated fatty acids.[85] In Stickland reactions, which are carried out by many clostridia, pairs of amino acids are simultaneously fermented, with one acting as the electron donor and the other as the electron acceptor. Isoleucine, leucine, phenylalanine, alanine, and serine are common electron donors, with tryptophan, proline, leucine, glycine, and proline, respectively, acting as their electron acceptors.[85-87]

Substrate availability is not an important factor that affects amino acid fermentation in the large bowel. However, other nutritional, physiological, host and bacteriological factors undoubtedly influence the outcome of fermentation. For example, the amino acid composition of different protein substrates will affect the amounts and types of fermentation products that can be formed as demonstrated in *in vitro* studies on collagen, BSA, and casein breakdown.[82] Carbohydrate affects amino acid fermentation reactions in some anaerobic bacteria, including *Megasphaera elsdenii*,[88] and some clostridia.[89,90] This result may be of significance in different anatomical regions of the colon, since carbohydrate availability is markedly reduced in the distal gut. The time needed to pass material through the large intestine is also important because extended gut transits reduce the amounts of carbohydrate available for fermentation, and allow more time for proteins and peptides to be digested.

## B. SHORT CHAIN FATTY ACIDS

Short chain fatty acids (SCFA) are major end products of protein breakdown and amino acid fermentation in the large intestine. Protein fermentation studies with mixed populations of human intestinal bacteria show that acetate is the major SCFA produced, followed by butyrate, propionate, and a variety of branched chain fatty acids (BCFA).[91] This section will deal with certain bacteriological aspects of SCFA formation from amino acids, while the nutritional and physiological significance of these metabolites to the host are discussed in more detail in Chapter 5.

Table 7 lists some of the bacteria that ferment amino acids, their relative numbers in the colon, and their substrates and fermentation products. It is clear from this table that clostridia are particularly

**Table 7**  Substrates and Acidic End Products of Some Amino Acid Fermenting Anaerobes that Occur in the Colon

| Bacteria | Amino acids fermented | Organic acids produced |
|---|---|---|
| Clostridia (9.8) | Glycine, threonine, valine, leucine, glutamate, lysine, isoleucine, tyrosine, phenylalanine, tryptophan | Acetate, propionate, isobutyrate, butyrate, 2-methylbutyrate, isovalerate, phenylacetate, phenylpropionate, phenyllactate, OH-phenylacetate, OH-phenylpropionate, indoleacetate, indolepropionate |
| *Eubacterium lentum* (9.3) | Arginine, citrulline | Acetate, lactate, succinate |
| Fusobacteria (8.4) | Glutamate, lysine | Acetate, butyrate |
| Acidaminococci (8.5) | Glutamate | Acetate, butyrate |
| Peptostreptococci (10.0) | Glutamate | Acetate, butyrate |
| *Prevotella melaninogenica* (9.6) | Glutamate | Acetate, butyrate |

*Note:* Values in parentheses show approximate log bacterial numbers per gram dry weight of feces.

From Finegold, S. M., Sutter, V. L., and Mathisen, G. E., *Human Intestinal Microflora in Health and Disease*, Hentges, D. J., Ed., Academic Press, London, 1983, 3.

versatile in their ability to utilize a wide range of amino acids. Correspondingly, many different types of acidic substances are produced, including phenol-substituted fatty acids (see discussion later). Clostridia produce propionate from alanine via the acrylate pathway,[92] and from threonine in reactions involving threonine dehydrase and keto acid dehydrogenases.[93] Acetate is formed from glycine in Stickland reactions, but also from serine, glutamate, lysine, γ-aminobutyrate and ornithine.[86]

The BCFA isobutyrate, 2-methylbutyrate, and isovalerate are the respective reduced carbon skeletons of the branched chain amino acids valine, isoleucine, and leucine. BCFA are formed by many bacteria; they are usually one carbon shorter than their amino acid precursors, and are characteristically produced in large amounts in Stickland reactions by clostridia.[93] Isocaproate is a relatively minor BCFA in the large intestine,[34,91] which is formed by the oxidation of leucine.

Because BCFA are mainly produced as a result of amino acid fermentation, their formation by colonic bacteria has been used as a marker of protein digestion in gut contents.[91] In these studies, intestinal material obtained from sudden death victims was incubated *in vitro* in the absence of exogenously added substrates. Measurements of fermentation acids showed that BCFA production was four to five times greater in material taken from the distal bowel, compared to that removed from the cecum and proximal colon. These results suggest that the catabolism of proteins by intestinal bacteria increases in importance as digestive materials pass through the large bowel.

Review of the other amino acid fermenting bacteria in Table 7 shows that glutamate is a common amino acid substrate, and acetate and butyrate are its principal fermentation acids. Two main pathways of glutamate metabolism occur in anaerobic bacteria. They are the methylaspartate pathway, which is used by the majority of clostridia, and the hydroxyglutamate pathway found in fusobacteria, peptostreptococci, and acidaminococcus.[85] *Prevotella melaninogenica* prefers peptides,[94] but aspartate is fermented as the free amino acid. However, growth of this bacterium is inhibited by a variety of other amino acids, including valine and serine.[95] Aspartate fermentation in *P. melaninogenica* yields succinate and acetate. It has been shown that 96% of the aspartate was deaminated to form ammonia,[96] with the remainder presumably being used for biosynthetic purposes.

## C. AMMONIA

The majority of ammonia in the large gut probably derives from the deamination of amino acids. Although urea is considered by many researchers to be the major source of this metabolite,[97-99] there is a sizable body of evidence that conflicts with this view. For example, very low levels of urea are present in ileal effluent[6,7] and the colon itself appears to be impermeable to urea.[100,101] Moreover, intravenous infusion studies with [15]N-labeled urea in human volunteers showed that only a small proportion of the label (about 8%) reached the large bowel.[102] Fecal ammonia concentrations are on average about 14 mmol/l, although somewhat higher levels are found in the gut.[103,104]

Ammonia formation in the large intestine has a number of health implications for the host. Concentrations as low as 5 mmol/l are known to have several cytopathic effects on colonic epithelial cells.

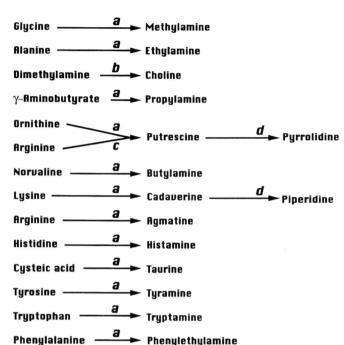

**Figure 8** Amines produced by intestinal bacteria and their precursor molecules. (a) Decarboxylation reactions; (b) N-dealkylation; (c) decarboxylation and hydrolysis; and (d) oxidative decarboxylation followed by ring closure.

Ammonia has been shown to affect their intermediary metabolism and DNA synthesis, and reduces the lifespan of mucosal cells.[105,106] Ammonia is generally more toxic towards healthy than transformed cells in the colon, hence it may potentially select for neoplastic growth. This metabolite increases the turnover of epithelial cells and, in doing so, increases the probability of genetic damage occurring in the presence of oncogenic agents, because actively dividing cells are more susceptible to genetic damage.[107] The genotoxic effects of abnormally high ammonia concentrations (>100 mmol/l) are frequently seen in uterosigmoidostomy patients, since these individuals have a high risk of developing tumors distal to the site of ureteric implantation.[108,109]

Normally, ammonia is absorbed from the colon and detoxified in the liver, where it is converted to urea. However, in patients with liver disease, ammonia formation in the large intestine is responsible for the condition known as portal-systemic encephalopathy, or hepatic coma, in which ammonia accumulates in body fluids. Hepatic coma patients are treated with antibiotics and by administration of the synthetic disaccharide, lactulose. This carbohydrate is not absorbed in the small gut, but is fermented by bacteria in the colon. Its metabolism leads to an increased requirement for ammonia for use in biosynthetic reactions in the bacteria, thereby reducing ammonia concentrations in the gut lumin.[110,111] Ammonia is an important nitrogen source for a number of anaerobic bacteria and in the presence of fermentable carbohydrate, it is preferred to amino acids or peptides by many species.[112,113] Lactulose therapy serves as a particularly good example of the way in which fermentable carbohydrate is able to modify the metabolism of N-containing compounds in the large gut.

## D. AMINES

A wide range of amines are found in the colon, such as dimethylamine, putrescine, cadaverine, tyramine, agmatine, pyrrolidine, piperidine, and histamine.[13,114] The ability to form amines is widespread in human intestinal bacteria, including the clostridia,[79] bacteroides,[115] lactobacilli,[116] streptococci,[117] and enterobacteria.[118] These basic metabolites are produced in a variety of ways (see Figure 8). In the large bowel, however, decarboxylation of amino acids is probably the most important route. Tertiary and quaternary amines are also converted to simple amines in N-dealkylation reactions[119] and simple amines are also formed by the breakdown of polyamines.[120] It has been suggested that some bacteria decarboxylate amino acids to maintain intracellular $pCO_2$,[121] however, the early studies of Gale,[122] who mainly worked with enterobacteria, indicated that the production of basic metabolites served to regulate an organism's external environment during growth under acid conditions. This concept has gained widespread acceptance, and is supported to some extent by studies on amino acid decarboxylation in dental plaque bacteria,[123] studies with *B. fragilis*,[115] and in measurements of amines in the chicken digestive tract.[124] However, it is by no means universally applicable, since some amines are optimally produced at neutral pH by some intestinal bacteria. This was found to occur with putrescine, butylamine, and propylamine formation in batch cultures of *C. perfringens*.[115] Many other workers have also found that putrescine formation occurs maximally at neutral pH.[120,125,126] Decarboxylation of amino acids by both *C. perfringens* and *B. fragilis* occurs mainly during exponential growth, and is growth-rate associated in C-limited chemostats. Amine production by both species of bacteria in N-limited (carbon-excess) chemostats is, however, considerably diminished, again showing that carbohydrate strongly influences nitrogen metabolism in human gut bacteria.

Under normal circumstances, amines are almost certainly rapidly absorbed from the colon and detoxified by monoamine and diamine oxidases in the liver or gut mucosa. Some amines are excreted in urine, however, including the heterocyclic products of putrescine and cadaverine oxidative deamination, pyrrolidine and piperidine. Dimethylamine, which is produced by N-dealkylation of choline, is the predominant secondary amine in urine and about 20 mg/d are lost by this route, of which one-half is bacterial in origin.[127] Amine excretion in urine has been directly related to protein intake,[128,129] migraine,[130] and the onset of hypertensive symptoms.[131] Under certain circumstances, amines produced in the colon enter the general circulation because, like ammonia, they are implicated as causative agents in portal-systemic encephalopathy.[132]

More recent work has shown that N-acetyl and acetoxy derivatives of putrescine and cadaverine are produced in the infant gut, and by intestinal bacteria cultured *in vitro*. These substances (e.g., N-acetyl-1,5-diaminopentane, N-propionyl-1,5-diaminopentane) may be detrimental to the host since their concentrations are reportedly higher in the blood of patients with schizophrenia, whereas cancer sufferers excrete higher levels of these metabolites than healthy individuals.[133]

Many amines are pharmacologically active. As well as the polyamines spermine and spermidine, which are not excreted in significant amounts by colonic bacteria, putrescine is involved in regulation of cell growth and differentiation in the gastrointestinal epithelium.[134] A number of amines including putrescine, cadaverine, tyramine, and histamine function as pressor or depressor substances, variously acting as stimulators of gastric secretion or as vasodilators.[13] In some cases, amines have a traumatic effect on human metabolism as evidenced in cases where ingestion of high levels of tyramine in foodstuffs has been reported to cause heart failure and brain hemorrhage.[135]

Increased amine production by intestinal bacteria is seen in diarrhea in pigs, where levels of putrescine and cadaverine are particularly high,[136] and in humans, where children with gastroenteritis had significantly higher concentrations of tyramine and phenylethylamine in their feces.[137] This latter study also showed that in healthy infants, fecal tyramine levels were low in children fed on mother's milk, compared to those given cow's milk.

Amines also serve as reactants in N-nitrosation reactions, in which secondary amines such as dimethylamine, piperidine, or pyrrolidine undergo condensation with nitrite at acid pH in a chemical reaction, or at neutral pH in bacterial enzyme catalyzed reactions. Nitrosamines are mutagenic (see Chapter 7) and have been found in feces.[138,139] Although there are some studies which suggest that nitrosamine production may not be significant in the large gut,[140] several species of intestinal bacteria are known to possess the ability to carry out N-nitrosation of several secondary amines.[141,142]

## E. PHENOLS AND INDOLES

Human intestinal bacteria form a wide range of phenolic and indolic compounds from tyrosine, phenylalanine, and tryptophan. Complete breakdown of aromatic amino acids by microorganisms only occurs to a significant degree under aerobic conditions, when mono- and dioxygenases can incorporate molecular oxygen into the reactants.[143] This process is thermodynamically unfavorable in the absence of an inorganic electron acceptor. Consequently, despite the occurrence of extensive side-chain modifications, ring fission and complete breakdown of phenols and indoles cannot occur under normal circumstances in the human colon. A generalized scheme of aromatic amino acid breakdown, showing the major products and intermediates of metabolism is given in Figure 9. Phenol, p-cresol, 4-ethylphenol and hydroxylated phenol-substituted fatty acids are the main products of tyrosine fermentation. Phenylacetate and phenylpropionate are produced from phenylalanine, whereas indole (indican) and 3-methylindole (skatole) are the principal end products of tryptophan metabolism. Aromatic amino acids are metabolized by a wide range of intestinal anaerobes, including members of the genera *Bacteroides*, *Lactobacillus*, *Clostridium*, and *Bifidobacterium*. Studies by Elsden et al.[144] suggest that amino acid fermenting clostridia

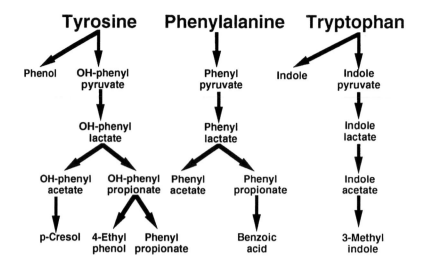

**Figure 9** Metabolic pathways involved in the formation of phenolic and indolic compounds by human colonic bacteria.

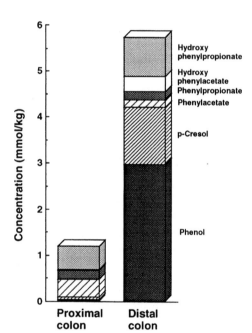

**Figure 10** Concentrations of phenol, p-cresol and phenol-substituted fatty acids in different areas of the large intestine. From *Holdeman, L. V., Cato, E. P., and Moore, W. E. C., Eds., Anaerobe Laboratory Manual, 4th Ed., Virginia Polytechnic Institute and State University, Blacksburg.*

are particularly active in metabolizing aromatic amino acids, especially species which produce phenylpropionate, such as *C. bifermentans*, *C. sporogenes*, and *C. sordelli*.

Phenolic compounds formed in the colon are usually absorbed and detoxified by glucuronide and sulfate conjugation in the mucosa of the large bowel, and in the liver, where they are excreted in urine.[145,146] Measurements of urinary phenols show that more than 90% consists of p-cresol, with most of the remainder being made up by phenol, and to a lesser extent, 4-ethylphenol.[147] Somewhere in the region of 50 to 160 mg of volatile phenols are excreted in urine per day,[148] but concentrations of individual phenolic compounds in gut contents seldom exceed more than a few millimolar, indicating that they are rapidly taken up by the gut mucosa. As with BCFA, there is a distinct regional trend with respect to the distribution of these metabolites in the gut, where overall concentrations increase by about fivefold in the distal colon (Figure 10), reinforcing the hypothesis that protein breakdown and amino acid fermentation become quantitatively more significant in the distal bowel. These data show qualitative as well as quantitative differences in that

**Table 8**  Effect of Diet on Excretion of Urinary Volatile Phenols and Fecal Ammonia Levels[149]

| Dietary regime | Urinary volatile phenol excretion (mg/d) | Fecal ammonia levels (mmol/kg) |
|---|---|---|
| Low protein (62.7 grams per day) | 74 ± 14.5 | 14.8 ± 1.3 |
| High protein (136 grams per day) | 108 ± 14.6 | 30.4 ± 1.1 |
| High protein plus 29.6 grams dietary fiber | 81 ± 4.8 | 28.4 ± 1.1 |

From Cummings, J. H., Hill, M. J., Bone, E. S., Branch, W. J., and Jenkins, D. J., *Am. J. Clin. Nutr.*, 32, 2094, 1979.

**Table 9**  Anaerobic Bacteria that Produce Indole in the Human Colon

| Indole producing species[a] | | Bacterial genera that do not contain indole producing species |
|---|---|---|
| *Peptostreptococcus asaccharolyticus* (9.7) | *Bacteroides thetaiotaomicron* (10.7) | *Lactobacillus* |
| *Fusobacterium naviforme* (-) | *Bacteroides uniformis* (-) | *Lachnospira* |
| *Fusobacterium nucleatum* (-) | *Porphyromonas asaccharolytica* (-) | *Butyrivibrio* |
| *Fusobacterium varium* (-) | *Eubacterium tenue* (8.8) | *Ruminococcus* |
| *Bacteroides ovatus* (10.0) | *Propionibacterium acnes* (9.4) | *Sarcina* |
| *Bacteroides putredinis* (10.7) | *Clostridium bifermentans* (8.7) | *Veillonella* |
| *Bacteroides eggerthii* (10.5) | *Clostridium cadaveris* (7.1) | *Megasphaera* |
| *Bacteroides coagulans* (10.5) | *Clostridium malenominatum* (6.4) | *Acidaminococcus* |
| *Bacteroides splanchnicus* (10.8) | | *Bifidobacterium* |

*Note:*  Data on indole producing species were obtained from Holdeman et al.[74] Where available, cell population densities of indole forming species (log bacterial numbers per gram dry weight of feces) are given in parentheses.[55]

whereas phenol-substituted fatty acids predominate in the proximal bowel, phenol and p-cresol are the major end products of aromatic amino acid metabolism that accumulate in the distal colon.

Although little is known of the factors that control the metabolism of aromatic amino acids in the large intestine, feeding studies with human volunteers have shown that increasing dietary protein intake results in higher levels of amino fermentation in the colon, as indicated by urinary phenol excretion and fecal ammonia concentrations (see Table 8). However, this effect was largely negated by increasing the amount of dietary fiber (fermentable carbohydrate) in the diet. Cummings et al.[149] related the reduction in formation of putrefactive products to greater demand for tyrosine and other amino acids for biosynthetic reactions by intestinal bacteria, due to the increase in carbohydrate availability. This study also showed that total N-excretion was unaffected by adding extra carbohydrate to the diet, the explanation for which was provided in subsequent work, where it was demonstrated that the majority of N in feces was associated with bacterial mass.[150]

More recent studies dealing with the effect of diet on amino acid fermentation in the colon indicated that urinary phenol and p-cresol excretion was markedly reduced when 18 volunteers were shifted from a conventional western diet to an uncooked vegetarian diet. When the subjects reverted to their original diet, the output of volatile phenols returned to the original level after about one month.[151]

The tryptophan metabolite indole is highly lipid soluble and as a consequence is rapidly absorbed from the colon.[152] A wide range of taxonomically diverse bacteria produce this metabolite (see Table 9), and its excretion in urine (indican) has often been used to provide an index of bacterial metabolism in the gut, especially in bacterial overgrowth syndromes.[153,154]

The formation of both phenols and indoles by intestinal bacteria has been associated with a variety of disease states in humans. Phenolic compounds have been implicated in schizophrenia,[155] and in animals p-cresol has a growth depressing effect.[156] Although phenols do not appear to be directly carcinogenic, there is evidence that they act as co-carcinogens.[157] In both active and quiescent ulcerative colitis, the ability of the colonic mucosa to detoxify these substances is impaired, which has been speculated to increase their toxic effects on the tissues.[146] Although the role of phenols in cancer is uncertain, nitrosation of dimethylamine by nitrite is stimulated by phenol, and the interaction of phenol with nitrite forms diazoquinone, which is mutagenic in the Ames test.[158] Like phenolic compounds, indole and other tryptophan metabolites have been linked to cancer,[159] and indole and skatole concentrations are significantly higher in the feces of cancer patients compared to healthy controls.[160] Nevertheless, these metabolites are believed to act as promoters of the disease rather than being directly carcinogenic by themselves.[161]

## F. GAS METABOLISM

In parallel with carbohydrate fermentation reactions, the metabolism of amino acids by human colonic bacteria frequently generates large quantities of $H_2$ gas, as observed in alanine, valine, leucine, and glutamate fermentations.[162,163] Hydrogen is also formed in Stickland reactions, as well as during the metabolism of individual amino acids, but in some bacteria, the fermentation of some amino acids depends on maintenance of low $pH_2$.[164] This situation would suggest that in the large intestine the activities of $H_2$ utilizing syntrophs may indirectly influence the way in which amino acids are metabolized.

## REFERENCES

1. **Metchnikoff, E.,** *The New Hygiene,* William Heinemann, London, 1906.
2. **Metchnikoff, E.,** *The Prolongation of Life,* William Heinemann, London, 1907.
3. **Arbuthnot Lane, W.,** *The Operative Treatment of Chronic Intestinal Stasis,* 4th ed., Hodder & Stoughton, London, 1918.
4. **Macfarlane, G. T. and Cummings, J. H.,** The colonic flora, fermentation, and large bowel function, in *The Large Intestine: Physiology, Pathophysiology and Disease,* Phillips, S. F., Pemberton, J. H., and Shorter, R. G., Eds., Raven Press, New York, 1991, 51.
5. **Macfarlane, G. T., Cummings, J. H., and Allison, C.,** Protein degradation by human intestinal bacteria, *J. Gen. Microbiol.,* 132, 1647, 1986.
6. **Chacko, A. and Cummings, J. H.,** Nitrogen losses from the human small bowel: obligatory losses and the effect of physical form of food, *Gut,* 29, 608, 1988.
7. **Gibson, J. A., Sladen, G. E., and Dawson, A. M.,** Protein absorption and ammonia production: the effects of dietary protein and removal of the colon, *Br. J. Nutr.,* 35, 61, 1976.
8. **Wrong, O. M.,** Bacterial metabolism of protein and endogenous nitrogen compounds, in *Role of the Gut Flora in Toxicity and Cancer,* Rowland, I. R., Ed., Academic Press, New York, 1988, 227.
9. **Christensen, J.,** Gross and microscopic anatomy of the large intestine, in *The Large Intestine: Physiology, Pathophysiology and Disease,* Phillips, S. F., Pemberton, J. H., and Shorter, R. G., Eds., Raven Press, New York, 1991, 13.
10. **Hutton, D. A., Pearson, J. P., Allen, A., and Foster, S. N. E.,** Mucolysis of the colonic mucus barrier by faecal proteinases: inhibition by interacting polyacrylate, *Clin. Sci.,* 78, 265, 1990.
11. **Slomiany, B. L., Piotrowski, J., and Slomiany, A.,** Effect of sucralfate on the degradation of human gastric mucus by *Helicobacter* protease and lipases, *Am. J. Gastroenterol.,* 87, 595, 1992.
12. **Bohe, M., Borgstrom, A., Genell, S., and Ohlson, K.,** Determination of immunoreactive trypsin, pancreatic elastase and chymotrypsin in extracts of human feces and ileostomy drainage, *Digestion,* 27, 8, 1983.
13. **Drasar, B. S. and Hill, M. J.,** *Human Intestinal Flora,* Academic Press, London, 1974.
14. **Goldberg, D. M., Campbell, R., and Roy, A. D.,** Binding of trypsin and chymotrypsin by human intestinal mucosa, *Biochim. Biophys. Acta,* 167, 613, 1968.

15. **Durr, H. K., Schneider, R., Bode, C., and Bode, J. C.,** Fecal chymotrypsin: study on some characteristics of the enzyme, *Digestion*, 17, 396, 1978.
16. **Catassi, C., Cardinalli, E., D'Angelo, G., Coppa, G. V., and Giorgi, P. L.,** Reliability of random fecal $a_1$-antitrypsin determination on non dried stools, *J. Pediatr.* (Sept.), 500, 1986.
17. **Bohe, M., Genell, S., and Ohlsson, K.,** Protease inhibitors in plasma and faecal extracts from patients with active inflammatory bowel disease, *Scand. J. Gastroenterol.*, 21, 598, 1986.
18. **Genell, S., Gustafsson, B. E., and Ohlsson, K.,** Immunochemical quantitation of pancreatic endopeptidases in the intestinal contents of germfree and conventional rats, *Scand. J. Gastroenterol.*, 12, 811, 1978.
19. **Macfarlane, G. T. and Macfarlane, S.,** Utilization of pancreatic trypsin and chymotrypsin by proteolytic and non-proteolytic *Bacteroides fragilis*-type bacteria, *Curr. Microbiol.*, 23, 143, 1991.
20. **Macfarlane, G. T., Cummings, J. H., Macfarlane, S., and Gibson, G. R.,** Influence of retention time on degradation of pancreatic enzymes by human colonic bacteria grown in a 3-stage continuous culture system, *J. Appl. Bacteriol.*, 67, 521, 1989.
21. **Macfarlane, G. T., Allison, C., Gibson, S. A. W., and Cummings, J. H.,** Contribution of the microflora to proteolysis in the human large intestine, *J. Appl. Bacteriol.*, 64, 37, 1988.
22. **Gibson, S. A. W., McFarlan, C., Hay, S., and Macfarlane, G. T.,** Significance of microflora in proteolysis in the colon, *Appl. Environ. Microbiol.*, 55, 679, 1989.
23. **Mangan, J. L.,** Quantitative studies on nitrogen metabolism in the bovine rumen, *Br. J. Nutr.*, 27, 261, 1972.
24. **Mahadevan, S., Erfle, D., and Sauer, F. D.,** Degradation of soluble and insoluble proteins by *Bacteroides amylophilus* protease and by rumen microorganisms, *J. Anim. Sci.*, 60, 723, 1980.
25. **Wallace, R. J.,** Adsorption of soluble proteins to rumen bacteria and the role of adsorption in proteolysis, *Br. J. Nutr.*, 53, 399, 1985.
26. **Morihara, K.,** Comparative specificity of microbial proteinases, in *Advances in Enzymology*, Vol. 41, Meister, A., Ed., John Wiley & Sons, New York, 1974, 179.
27. **Brock, F. M., Forsberg, C. W., and Buchanan-Smith, J. G.,** Proteolytic activity of rumen microorganisms and effects of proteinase inhibitors, *Appl. Environ. Microbiol.*, 44, 561, 1982.
28. **Lockwood, B. C., Coombs, G. H., and Williams, A. G.,** Proteinase activity in rumen ciliate protozoa, *J. Gen. Microbiol.*, 134, 2605, 1988.
29. **Englyst, H. N., Hay, S., and Macfarlane, G. T.,** Polysaccharide breakdown by mixed populations of human faecal bacteria, *FEMS Microbiol. Ecol.*, 95, 163, 1987.
30. **Macfarlane, G. T., Allison, C., and Gibson, G. R.,** Effect of pH on protease activities in the large intestine, *Lett. Appl. Microbiol.*, 7, 161, 1988.
31. **Allison, C. and Macfarlane, G. T.,** Protease production by *Clostridium perfringens* in batch and continuous culture, *Lett. Appl. Microbiol.*, 9, 45, 1989.
32. **Macfarlane, G. T. and Macfarlane, S.,** Physiological and nutritional factors affecting the synthesis and secretion of extracellular metalloproteases by *Clostridium bifermentans* NCTC 2914, *Appl. Environ. Microbiol.*, 58, 1195, 1992.
33. **Allison, C. and Macfarlane, G. T.,** Physiological and nutritional determinants of protease secretion by *Clostridium sporogenes*: characterization of six extracellular proteases, *Appl. Microbiol. Biotechnol.*, 37, 152, 1992.
34. **Cummings, J. H., Pomare, E. W., Branch, W. J., Naylor, C. P. E., and Macfarlane, G. T.,** Short chain fatty acids in human large intestine, portal, hepatic and venous blood, *Gut*, 28, 1221, 1987.
35. **Macfarlane, G. T., Gibson, G. R., and Cummings, J. H.,** Comparison of fermentation reactions in different regions of the human colon, *J. Appl. Bacteriol.*, 72, 57, 1992.
36. **Macfarlane, G. T. and Gibson, G. R.,** Metabolic activities of the normal colonic flora, in *Human Health: The Contribution of Microorganisms*, Gibson, S. A. W., Ed., Springer-Verlag, London, 1994, 17.
37. **Cummings, J. H., Macfarlane, G. T., and Drasar, B. S.,** The gut microflora and its significance, in *Gastrointestinal and Oesophageal Pathology*, Whitehead, R., Ed., Churchill Livingstone, Edinburgh, 1989, 201.
38. **Macy, J. M. and Probst, I.,** The biology of gastrointestinal *Bacteroides*, *Ann. Rev. Microbiol.*, 33, 561, 1979.
39. **Salyers, A. A.,** *Bacteroides* of the human lower intestinal tract, *Ann. Rev. Microbiol.*, 38, 293, 1984.
40. **Macfarlane, G. T., Gibson, S. A. W., and Gibson, G. R.,** Proteolytic activities of the fragilis group of *Bacteroides* spp., in *Medical and Environmental Aspects of Anaerobes*, Duerden, B. I., Brazier, J. S., Seddon, S. V., and Wade, W. G., Eds., Wrightson Biomedical Publishing Ltd., Petersfield,Hampshire, England, 1992, 159.
41. **Allison, C.,** *Nitrogen Metabolism of Human Large Intestinal Bacteria*, Ph.D. thesis, University of Cambridge, 1989.
42. **Ruseler-van Embden, J. G. H., van der Helm, R., and van Lieshout, L. M. C.,** Degradation of intestinal glycoproteins by *Bacteroides vulgatus*, *FEMS Microbiol. Lett.*, 58, 37, 1989.
43. **Macfarlane, G. T. and Gibson, G. R.,** Formation of glycoprotein degrading enzymes by *Bacteroides fragilis*, *FEMS Microbiol. Lett.*, 77, 289, 1991.
44. **Riepe, S. P., Goldstein, J., and Alpers, D. H.,** Effect of secreted *Bacteroides* proteases on human intestinal brush border hydrolases, *J. Clin. Invest.*, 66, 314, 1980.
45. **Gibson, S. A. W. and Macfarlane, G. T.,** Studies on the proteolytic activity of *Bacteroides fragilis*, *J. Gen. Microbiol.*, 134, 19, 1988.

46. **Macfarlane, G. T., Macfarlane, S., and Gibson, G. R.,** Synthesis and release of proteases by *Bacteroides fragilis*, *Curr. Microbiol.*, 24, 55, 1992.

47. **Gibson, S. A. W. and Macfarlane, G. T.,** Characterisation of proteases formed by *Bacteroides fragilis*, *J. Gen. Microbiol.*, 134, 2231, 1988.

48. **Smalley, J. W. and Birss, A. J.,** Trypsin-like enzyme activity of the extracellular membrane vesicles of *Bacteroides gingivalis* W50, *J. Gen. Microbiol.*, 133, 2883, 1987.

49. **Ando, Y.,** Collagenase, dipeptidyl peptidase IV and cathepsin D activities in gingival fluid and whole saliva from patients with periodontal disease, *J. Jpn. Assoc. Periodont.*, 22, 387, 1980.

50. **Abiko, Y., Hayakawa, M., Murai, S., and Takiguchi, H.,** Glycylprolyl dipeptidylaminopeptidase from *Bacteroides gingivalis*, *J. Dent. Res.*, 64, 106, 1985.

51. **Jousimies-Somer, H. R. and Finegold, S. M.,** Anaerobic gram negative bacilli and cocci, in *Manual of Clinical Microbiology*, 5th ed., Balows, A., Hausler, W. J., Herrmann, K. L., Isenberg, H. D., and Shadomy, H. J., Eds., American Society for Microbiology, Washington, D.C., 1991, 538.

52. **Macfarlane, G. T. and Gibson, G. R.,** Characteristics of protease synthesis in *Bacteroides splanchnicus*, *Appl. Microbiol. Biotechnol.*, 39, 506, 1993.

53. **Macfarlane, G. T., Macfarlane, S., and Gibson, G. R.,** Physiology and biochemistry of glycyl-prolyl aminopeptidase formation in *Bacteroides splanchnicus*, *Clin. Infect. Dis.*, 16, S408, 1993.

54. **Shone, C. C. and Hambleton, P.,** Toxigenic clostridia, in *Clostridia*, Minton, N. P. and Clarke, D. J., Eds., Plenum Publishing Corp., New York, 1989, 265.

55. **Finegold, S. M., Sutter, V. L., and Mathisen, G. E.,** Normal indigenous intestinal microflora, in *Human Intestinal Microflora in Health and Disease*, Hentges, D. J., Ed., Academic Press, London, 1983, 3.

56. **Hentges, D. J. and Smith, L. D. S.,** Hydrolytic enzymes as virulence factors of anaerobic bacteria, in *Bacterial Enzymes and Virulence*, Holder, I. A., Ed., CRC Press, Boca Raton, FL, 1985, 105.

57. **Sato, H., Yamakawa, Y., Ito, A., and Murata, R.,** Effect of zinc and calcium ions on the production of alpha-toxin and proteases by *Clostridium perfringens*, *Infect. Immun.*, 20, 325, 1978.

58. **Blaschek, H. P. and Solberg, M.,** Effect of nitrogen source on growth and exoprotease production by *Clostridium perfringens*, *J. Food Sci.*, 43, 1253, 1978.

59. **Curran, J. M., Solberg, M., Blaschek, H. P., and Rosen, D.,** Casein inhibition of *Clostridium perfringens* and exoprotease production, *J. Food Sci.*, 46, 169, 1981.

60. **Hapchuk, L. T., Pearson, A. M., and Price, J. F.,** Effect of a proteolytic enzyme produced by *Clostridium perfringens* upon porcine muscle, *Food Chem.*, 4, 213, 1979.

61. **MacLennan, J. D.,** The histotoxic clostridial infections of man, *Bact. Rev.*, 26, 177, 1962.

62. **Allison, C. and Macfarlane, G. T.,** Regulation of protease production in *Clostridium sporogenes*, *Appl. Environ. Microbiol.*, 56, 3485, 1990.

63. **Patterson-Curtis, S. I. and Johnson, E. A.,** Regulation of neurotoxin and protease formation in *Clostridium botulinum* Okra A and Hall A by arginine, *Appl. Environ. Microbiol.*, 55, 1544, 1989.

64. **Long, S., Mothibelli, M. A., Robb, F. T., and Woods, D. R.,** Regulation of extracellular alkaline protease activity by histidine in a collagenolytic *Vibrio alginolyticus* strain, *J. Gen. Microbiol.*, 127, 193, 1981.

65. **Pansare, A. C., Venugopal, C. V., and Lewis, N. F.,** A note on nutritional influence on extracellular protease synthesis in *Aeromonas hydrophila*, *J. Appl. Bacteriol.*, 58, 101, 1985.

66. **Whooly, M. A., O'Callaghan, J. A., and McLoughlin, A. J.,** Effect of substrate on the regulation of exoprotease production by *Pseudomonas aeruginosa* ATCC 10145, *J. Gen. Microbiol.*, 129, 981, 1983.

67. **Lwebuga-Mukasa, J. S., Harper, E., and Taylor, P.,** Collagenase enzymes from *Clostridium*: Characterization of individual enzymes, *Biochemistry*, 15, 4736, 1976.

68. **Priest, F. G.,** Extracellular enzyme synthesis in the genus *Bacillus*, *Bact. Rev.*, 41, 711, 1977.

69. **Strydom, E., MacKie, R. I., and Woods, D. R.,** Detection and characterization of extracellular proteases in *Butyrivibrio fibrisolvens*, *Appl. Microbiol. Biotechnol.*, 24, 214, 1986.

70. **Russell, J. B.,** Fermentation of peptides by *Bacteroides ruminicola* B14, *Appl. Environ. Microbiol.*, 45, 1566, 1983.

71. **Loesche, W. J. and Gibbons, R. J.,** Amino acid fermentation by *Fusobacterium nucleatum*, *Arch. Oral Biol.*, 13, 191, 1968.

72. **Robrish, S. A., Oliver, C., and Thompson, J.,** Amino acid-dependent transport of sugars by *Fusobacterium nucleatum* ATCC 10953, *J. Bacteriol.*, 169, 3891, 1987.

73. **Chalupa, W.,** Degradation of amino acids by the mixed rumen microbial population, *J. Anim. Sci.*, 43, 828, 1976.

74. **Holdeman, L. V., Cato, E. P., and Moore, W. E. C.,** Eds., *Anaerobe Laboratory Manual*, 4th ed., Virginia Polytechnic Institute and State University, Blacksburg, 1977.

75. **Pittman, K. A., Sitarama, S., and Bryant, M. P.,** Oligopeptide uptake by *Bacteroides ruminicola*, *J. Bacteriol.*, 93, 1499, 1967.

76. **Wahren, A. and Holme, T.,** Amino acid and peptide requirement of *Fusiformis necrophorus*, *J. Bacteriol.*, 116, 279, 1973.

77. **Harschild, A. H. W.,** Incorporation of $C^{14}$ from amino acids and peptides into protein by *Clostridium perfringens* type D, *J. Bacteriol.*, 90, 1569, 1965.

78. **Leach, F. R. and Snell, E. E.,** The absorption of glycine and alanine and their peptides by *Lactobacillus casei, J. Biol. Chem.*, 235, 3523, 1980.

79. **Payne, J. W. and Bell, G.,** Direct determination of the properties of peptide transport systems in *Escherichia coli* using a fluorescent labeling procedure, *J. Bacteriol.*, 137, 447, 1979.

80. **Matteuzzi, D., Crociani, F., and Emaldi, O.,** Amino acids produced by bifidobacteria and some clostridia, *Ann. Microbiol. Inst. Pasteur*, 129B, 175, 1978.

81. **Anraku, Y.,** Transport and utilization of amino acids by bacteria, in *Microorganism and Nitrogen Sources. Transport and Utilization of Amino Acids, Peptides, Proteins and Related Substrates*, Payne, J. W., Ed., John Wiley, Chichester, 1980, 9.

82. **Macfarlane, G. T. and Allison, C.,** Utilization of protein by human gut bacteria, *FEMS Microbiol. Letts.*, 38, 19, 1986.

83. **Chen, G., Strobel, H. J., Russell, J. B., and Sniffen, C. J.,** Effect of hydrophobicity on utilization of peptides by ruminal bacteria in vitro, *Appl. Environ. Microbiol.*, 53, 2021, 1987.

84. **Yang, C.-M. J. and Russell, J. B.,** Resistance of proline-containing peptides to ruminal degradation in vitro, *Appl. Environ. Microbiol.*, 58, 3954, 1992.

85. **Barker, H. A.,** Amino acid degradation by anaerobic bacteria, *Ann. Rev. Biochem.*, 50, 23, 1981.

86. **Mead, G. C.,** The amino acid fermenting clostridia, *J. Gen. Microbiol.*, 67, 47, 1971.

87. **Andreesen, J. R., Bahl, H., and Gottschalk, G.,** Introduction to the physiology and biochemistry of the genus *Clostridium*, in *Clostridia*, Minton, N. P. and Clarke, D. J., Eds., Plenum Press, London, 1989, 27.

88. **Allison, M. J.,** Production of branched-chain volatile fatty acids by certain anaerobic bacteria, *Appl. Environ. Microbiol.*, 35, 872, 1978.

89. **Saissac, R., Raynard, M., and Cohem, G.-N.,** Variation du type fermentaire des bacteries anaerobies du groupe de *Cl. sporogenes*, sous l' influence du glucose, *Ann. Inst. Pasteur*, 75, 305, 1948.

90. **Turton, L. J., Drucker, D. B., and Ganguli, L. A.,** Effect of glucose concentration in the growth medium upon neutral and acidic fermentation end-products of *Clostridium bifermentans*, *Clostridium sporogenes* and *Peptostreptococcus anaerobius*, *J. Med. Micro.*, 16, 61, 1983.

91. **Macfarlane, G. T., Gibson, G. R., Beatty, E., and Cummings, J. H.,** Estimation of short chain fatty acid production from protein by human intestinal bacteria, based on branched chain fatty acid measurements, *FEMS Microbiol. Ecol.*, 101, 81, 1992.

92. **Gottschalk, G.,** *Bacterial Metabolism*, 2nd ed., Springer-Verlag, Berlin, 1986.

93. **Elsden, S. R. and Hilton, M. G.,** Volatile acid production from threonine, valine, leucine and isoleucine by clostridia, *Arch. Microbiol.*, 117, 165, 1978.

94. **Wahren, A. and Gibbons, R. J.,** Amino acid fermentation by *Bacteroides melaninogenicus*, *Antonie van Leeuwenhoek; J. Microbiol. Serol.*, 36, 149, 1970.

95. **Miles, D. O., Dyer, J. K., and Wong, J. C.,** Influence of amino acids in the growth of *Bacteroides melaninogenicus*, *J. Bacteriol.*, 127, 899, 1976.

96. **Wong, J. C., Dyer, J. K., and Tribble, J. L.,** Fermentation of L-aspartate by a saccharolytic strain of *Bacteroides melaninogenicus*, *Appl. Environ. Microbiol.*, 33, 69, 1977.

97. **Sabbaj, J., Sutter, V. L., and Finegold, S. M.,** Urease and deaminase activities in fecal bacteria in hepatic coma, in *Antimicrobial Agents and Chemotherapy*, Hobby, G. I., Ed., American Society for Microbiology, Washington, D.C., 1970, 181.

98. **Summerskill, W. H. J. and Wolpert, E.,** Ammonia metabolism in the gut, *Am. J. Clin. Nutr.*, 22, 633, 1970.

99. **Varel, V. H., Bryant, M. P., Holdeman, L. V., and Moore, W. E. C.,** Isolation of ureolytoic *Peptostreptococcus productus* from feces using defined medium: failure of common urease tests, *Appl. Microbiol.*, 28, 394, 1974.

100. **Wolpert, E., Phillips, S. F., and Summerskill, W. H. J.,** Transport of urea and ammonia production in the human colon, *Lancet*, 2, 1387, 1971.

101. **Brown, R. L., Gibson, J. A., Fenton, J. C., Sneddon, W., Clark, M. L., and Sladen, G. E.,** Ammonia and urea transport by the excluded human colon, *Clin. Sci. Mol. Med.*, 48, 279, 1975.

102. **Wrong, O. M., Vince, A. J., and Waterlow, J. C.,** The contribution of endogenous urea to faecal ammonia in man, determined by $^{15}$N-labelling of plasma urea, *Clin. Sci.*, 68, 193, 1985.

103. **Wrong, O. M.,** Bacterial metabolism of protein and endogenous nitrogen compounds, in *Role of the Gut Flora in Toxicity and Cancer*, Rowland, I. R., Ed., Academic Press, London, 1988, 227.

104. **Wilson, D. R., Ing, T. S., Metcalfe-Gibson, A., and Wrong, O. M.,** In vivo dialysis of faeces as a method of stool analysis. III. The effect of intestinal antibiotics, *Clin. Sci.*, 34, 211, 1968.

105. **Visek, W. J.,** Effects of urea hydrolysis on cell life-span and metabolism, *Fed. Proc.*, 31, 1178, 1972.

106. **Visek, W. J.,** Diet and cell growth modulation by ammonia, *Am. J. Clin. Nutr.*, 31, S216, 1978.

107. **Warwick, G. P.,** Effect of the cell cycle on carcinogenesis, *Fed. Proc.*, 30, 1760, 1971.

108. **McConnel, J. B., Morison, J., and Steward, W. K.,** The role of the colon in the pathogenesis of hyperchloraemic acidosis in uterosigmoid anastomosis, *Clin. Sci.*, 57, 305, 1979.

109. **Tank, E. S., Krausch, D. N., and Lapides, J.,** Adenocarcinoma of the colon associated with uterosigmoidostomy, *Dis. Colon Rectum*, 16, 300, 1973.

110. **Weber, F. L.,** The effect of lactulose on urea metabolism and nitrogen excretion in cirrhotic patients, *Gastroenterology,* 77, 518, 1979.

111. **Weber, F. L., Banwell, J. G., Fresard, K. M., and Cummings, J. H.,** Nitrogen in fecal bacteria, fiber and soluble fractions of patients with cirrhosis: effects of lactulose and lactulose plus neomycin, *J. Lab. Clin. Med.,* 110, 259, 1987.

112. **Bryant, M. P. and Robinson, I. M.,** Some nutritional characteristics of predominant culturable ruminal bacteria, *J. Bacteriol.,* 84, 605, 1962.

113. **Pilgram, A. F., Gray, F. V., Weller, R. A., and Belling, C. B.,** Synthesis of microbial protein from ammonia in the sheep's rumen and the proportion of dietary nitrogen converted into microbial nitrogen, *Br. J. Nutr.,* 24, 589, 1970.

114. **Asatoor, A. M. and Simenhof, M. L.,** The origin of urinary dimethylamine, *Biochim. Biophys. Acta,* 3, 384, 1965.

115. **Allison, C. and Macfarlane, G. T.,** Influence of pH, nutrient availability and growth rate on amine production by *Bacteroides fragilis* and *Clostridium perfringens* in batch and continuous culture, *Appl. Environ. Microbiol.,* 55, 2894, 1989.

116. **Sumner, S., Speckland, M., Somers, E., and Taylor, S.,** Isolations of histamine-producing *Lactobacillus buchneri* from Swiss cheese implicated in food poisoning outbreak, *Appl. Environ. Microbiol.,* 50, 1094, 1985.

117. **Babu, S., Chandler, H., Batish, U. K., and Bhatia, K. L.,** Factors affecting amine production in *Streptococcus cremoris, Food Microbiol.,* 3, 359, 1986.

118. **Morris, D. R. and Boeker, E. A.,** Biosynthetic and biodegradative ornithine and arginine decarboxylase from *Escherichia coli, Meth. Enzymol.,* 94, 125, 1983.

119. **Johnson, K. A.,** The production of secondary amines by human gut bacteria and its possible relevance to carcinogenesis, *Med. Lab. Sci.,* 34, 131, 1977.

120. **White Tabor, C. and Tabor, H.,** Polyamines in microorganisms, *Microbiol. Rev.,* 49, 81, 1985.

121. **Morris, D. R. and Fillingame, R. H.,** Regulation of amino acid decarboxylation, *Ann. Rev. Biochem.,* 43, 303, 1974.

122. **Gale, E.,** The bacterial amino acid decarboxylases, *Adv. Enzymol.,* 6, 1, 1946.

123. **Hayes, M. L. and Hyatt, A. T.,** The decarboxylation of amino acids by bacteria derived from human dental plaque, *Arch. Oral. Biol.,* 19, 361, 1973.

124. **Nugon-Baudon, L., Szylit, O., Chaigneau, M., Dierick, N., and Raibaud, P.,** "In vitro" and "in vivo" production of amines by a *Lactobacillus* strain, *Ann. Inst. Pasteur Microbiol.,* 136, 63, 1985.

125. **Dierick, N. A., Vervaeke, I. J., Decuypere, J. A., and Hendrick, H. K.,** Influence of the gut flora and some growth-promoting food additives on nitrogen metabolism in pigs. I. Studies *in vitro, Livest. Prod. Sci.,* 14, 161, 1986.

126. **Goldschmidt, M. C., Lockhart, B. M., and Perry, K.,** Rapid methods for determining decarboxylase activity: ornithine and lysine decarboxylases, *Appl. Microbiol.,* 22, 344, 1971.

127. **Asatoor, A. M.,** The origin of urinary tyramine. Formation in tissues and by intestinal microorganisms, *Clin. Chem. Acta,* 22, 223, 1968.

128. **Irvine, W. T., Buthie, H. L., and Watson, W. G.,** Urinary output of free histamine after a meal, *Lancet,* 1, 1061, 1959.

129. **DeQuattro, V. L. and Sjoerdsma, A.,** Origin of urinary tyramine and tryptamine, *Clin. Chem. Acta,* 16, 227, 1967.

130. **Anon.,** Headache, tyramine, serotonin and migraine, *Nutr. Rev.,* 26, 40, 1968.

131. **Boulton, A. A., Cokkson, B., and Paulton, R.,** Hypertensive crisis in a patient on MAOI antidepressants following a meal of beef liver, *Can. Med. Assoc. J.,* 102, 1394, 1970.

132. **Phear, E. A. and Ruebner, B.,** The in vitro production of ammonium and amines by intestinal bacteria in relation to nitrogen toxicity as a factor in hepatic coma, *Br. J. Exp. Pathol.,* 37, 253, 1956.

133. **Murray, K. E., Shaw, K. J., Adams, R. F., and Conway, P. L.,** Presence of N-acyl and acetoxy derivatives of putrescine and cadaverine in the human gut, *Gut,* 34, 489, 1993.

134. **Seidel, E. R., Haddox, M. K., and Johnson, L. R.,** Polyamines in the response to intestinal obstruction, *Am. J. Physiol.,* 246, G649, 1984.

135. **Smith, T. A.,** Amines in food, *Food Chem.,* 6, 169, 1980.

136. **Porter, P. and Kenworthy, R.,** A study of intestinal and urinary amines in pigs in relation to weaning, *Res. Vet. Sci.,* 10, 440, 1969.

137. **Murray, K. E., Adams, R. F., Earl, J. W., and Shaw, K. J.,** Studies of the free faecal amines of infants with gastroenteritis and of healthy infants, *Gut,* 27, 1173, 1986.

138. **Fine, D. H., Ross, R., Roonbehler, D. P., Silvergleid, A., and Song, L.,** Formation *in vivo* of volatile N-nitrosmines in man after ingestion of cooked bacon and spinach, *Nature (London),* 265, 753, 1977.

139. **Leach, S. A., Cook, A. R., Challis, B. C., Hill, M. J., and Thompson, M. H.,** Reactions and endogenous formation of N-nitrosocompounds, in *Occurrence, Biological Effects and Relevance to Human Cancer,* Bartels, H., O'Neill, I. K., and Herman, R. S., Eds., International Agency for Research on Cancer Scientific Publications, Lyons, France, 1987, 396.

140. **Eisenbrand, G., Spiegelhalder, B., and Preussman, R.,** Analysis of human biological specimens for nitrosamine contents, in *Banbury Report No. 7,* Bruce, W. R., Correa, P., Lipkin, M., and Tannenbaum, S. R., Eds., Cold Spring Harbor Laboratory, New York, 1981, 275.

141. **Calmels, S. H., Ohshima, H., Vincent, P., Gounot, A. M., and Bartsch, H.,** Screening for nitrosation catalysis at pH 7 and kinetic studies on nitrosamine formation from secondary amines by *E. coli* strains, *Carcinogenesis,* 6, 911, 1985.

142. **Suzuki, K. and Mitsuoka, T.,** N-Nitrosamine formation by intestinal bacteria, in *IARC Publication No. 57*, O'Neill, I. K., Ed., International Agency for Research on Cancer Publications, Lyons, France, 1984, 275.

143. **Young, I. Y. and Rivera, M.,** Methanogenic degradation of four phenolic compounds, *Water Res.*, 19, 1325, 1985.

144. **Elsden, S. R., Hilton, M. G., and Waller, J. M.,** The end products of the metabolism of aromatic amino acids by clostridia, *Arch. Microbiol.*, 107, 283, 1976.

145. **Ramakrishna, B. S., Gee, D., Pannall, P., Roberts-Thomson, I. C., and Roediger, W. E. W.,** Estimation of phenolic conjugation by colonic mucosa, *J. Clin. Pathol.*, 42, 620, 1989.

146. **Ramakrishna, B. S., Roberts-Thomson, I. C., Pannal, P. R., and Roediger, W. E. W.,** Impaired sulphation of phenol by the colonic mucosa in quiescent and active ulcerative colitis, *Gut*, 32, 46, 1991.

147. **Tamm, A. and Villako, K.,** Urinary volatile phenols in patients with urinary obstruction, *Scand. J. Gastroenterol.*, 6, 5, 1971.

148. **Schmidt, E. G.,** Urinary phenols. Simultaneous determination of phenol and p-cresol in urine, *J. Biol. Chem.*, 179, 211, 1949.

149. **Cummings, J. H., Hill, M. J., Bone, E. S., Branch, W. J., and Jenkins, D. J. A.,** The effect of meat protein and dietary fiber on colonic function and metabolism. II. Bacterial metabolites in feces and urine, *Am. J. Clin. Nutr.*, 32, 2094, 1979.

150. **Stephen, A. M. and Cummings, J. H.,** The influence of dietary fibre on faecal nitrogen excretion in man, *Proc. Nutr. Soc.*, 38, 141A, 1979.

151. **Ling, W. H. and Hanninen, O.,** Shifting from a conventional diet to an uncooked vegan diet reversibly alters fecal hydrolytic activities in humans, *J. Nutr.*, 122, 924, 1992.

152. **Fordtran, J. S., Scroggie, J. B., and Polter, D. E.,** Colonic absorption of tryptophan metabolites in man, *J. Lab. Clin. Med.*, 64, 125, 1964.

153. **Greenberger, N. J., Saegh, S., and Ruppert, R. D.,** Urine excretion in malabsorptive disorders, *Gastroenterology*, 55, 204, 1968.

154. **Tomkin, G. H. and Weir, D. G.,** Indicanuria after gastric surgery, *Q. J. Med.*, 41, 191, 1972.

155. **Dalgliesh, C. E., Kelley, W., and Horning, E. C.,** Excretion of a sulphatoxyl derivative of skatole in pathological states in man, *Biochem. J.*, 70, 13P, 1958.

156. **Yokohama, M. T., Tabori, C., Miller, E. R., and Hogberg, M. G.,** The effects of antibiotics in the weanling pig diet on growth and excretions of volatile phenolic and aromatic bacterial metabolites, *Am. J. Clin. Nutr.*, 35, 1417, 1982.

157. **Bone, E., Tamm, A., and Hill, M. J.,** The production of urinary phenols by gut bacteria and their possible role in the causation of large bowel cancer, *Am. J. Clin. Nutr.*, 29, 1448, 1976.

158. **Kikugawa, K. and Kato, T.,** Formation of a mutagenic diazoquinone by interaction of phenol with nitrite, *Food Chem. Toxicol.*, 26, 209, 1986.

159. **Bryan, G. T.,** The role of bacterial tryptophan metabolites in the etiology of bladder cancer, *Am. J. Clin. Nutr.*, 24, 841, 1971.

160. **Zuccato, E., Venturi, M., Di, L., Colombo, L., Bertolo, C., Bressani, D. S., and Mussini, E.,** Role of bile acids and metabolic activity of colonic bacteria in increased risk of colon cancer after cholecystectomy, *Dig. Dis. Sci.*, 38, 514, 1993.

161. **Dunning, W. T., Curtis, M. R., and Mann, M. E.,** The effect of added dietary tryptophane on the occurrence of 2-acetylaminofluorene-induced liver and bladder cancer in rats, *Cancer Res.*, 10, 454, 1950.

162. **Nisman, B.,** The Stickland reaction, *Bacteriol. Rev.*, 1, 16, 1953.

163. **Stams, A. J. M. and Hansen, T. A.,** Fermentation of glutamate and other compounds by *Acidaminobacter hydrogenoformans* gen. nov. sp. nov., an obligate anaerobe isolated from black mud: studies with pure cultures and mixed cultures with sulfate-reducing and methanogenic bacteria, *Arch. Microbiol.*, 137, 329, 1984.

164. **Nagase, M. and Matsuo, T.,** Interactions between amino acid degrading bacteria and methanogenic bacteria in anaerobic digestion, *Biotech. Bioeng.*, 24, 2227, 1982.

165. **Matsubara, H. and Feder, J.,** Other bacterial, mold, and yeast proteases, in *The Enzymes*, Boyer, P. D., Ed., Academic Press, New York, 1971, 721.

166. **Siefter, S. and Harper, E.,** The collagenases, in *The Enzymes*, Boyer, P. D., Ed., Academic Press, New York, 1971, 649.

167. **Barrett, A. J.,** Introduction to the history and classification of tissue proteinases, in *Proteinases in Mammalian Cells and Tissues*, Barrett, A. J., Ed., North-Holland Publishing Co., New York, 1977, 1.

168. **Webb, J. L.,** *Enzyme and Metabolic Inhibitors*, Vol. 2, Academic Press, New York, 1966.

169. **Umezawa, H. and Aoyagi, T.,** Activities of proteinase inhibitors of microbial origin, in *Proteinases in Mammalian Cells and Tissues*, Barrett, A. J., Ed., Academic Press, New York, 1971, 637.

170. **Hazlewood, G. P. and Edwards, R.,** Proteolytic activities of a rumen bacterium, *Bacteroides ruminicola* R814, *J. Gen. Microbiol.*, 125, 11, 1981.

171. **Mihalyi, E.,** *Application of Proteolytic Enzymes to Protein Structure Studies*, Vol. 1, CRC Press, West Palm Beach, FL, 1978.

172. **Wahren, A., Bernholm, K., and Holme, T.,** Formation of proteolytic activity in continuous culture of *Sphaerophorus necrophorus*, *Acta. Path. Microbiol. Scand.*, 79, 391, 1971.

173. **Hazlewood, G. P., Jones, G. A., and Mangan, J. L.,** Hydrolysis of leaf fraction 1 protein by the proteolytic rumen bacterium *Bacteroides ruminicola*, R8/4, *J. Gen. Microbiol.*, 123, 223, 1981.

174. **Hausmann, E. and Kaufman, E.,** Collagenase activity in a particulate fraction of *Bacteroides melaninogenicus*, *Biochim. Biophys. Acta*, 194, 612, 1969.

175. **Killian, M., Thomsen, B., Petersen, T. E., and Bleeg, H. S.,** Occurrence and nature of IgA proteases, *Ann. N. Y. Acad. Sci.*, 409, 612, 1983.

176. **van Steenbergen, T. S. M. and de Graff, J.,** Proteolytic activity of black-pigmented *Bacteroides* strains, *FEMS Microbiol. Lett.*, 33, 219, 1986.

177. **El-Soda, M., Bergere, J. L., and Desmazeard, M. J.,** Detection and localization of peptide hydrolases in *Lactobacillus casei*, *J. Dairy Res.*, 45, 519, 1978.

178. **Kok, J. and de Vos, W. M.,** The proteolytic system of lactic acid bacteria, in *Genetics and Biotechnology of Lactic Acid Bacteria*, Gasson, M. J. and de Vos, W. M., Eds., Blackie Academic and Professional, London, 1994, 169.

179. **Senda, S., Fujiyama, Y., Ushijima, T., Hodohara, K., Bamba, T., Hosoda, S., and Kobayashi, K.,** *Clostridium ramosum*, an IgA protease-producing species and its ecology in the human intestinal tract, *Microbiol. Immunol.*, 29, 1019, 1985.

180. **Ingram, E., Holland, K. T., Gowland, G., and Cunliffe, W. J.,** Studies of the extracellular proteolytic activity produced by *Propionibacterium acnes*, *J. Appl. Bacteriol.*, 54, 263, 1983.

181. **Allison, C. and Macfarlane, G. T.,** Regulation of protease production in *Clostridium sporogenes*, *Appl. Environ. Microbiol.*, 56, 3485, 1990.

182. **Allison, C. and Macfarlane, G. T.,** Physiological and nutritional determinants of protease secretion by *Clostridium sporogenes*: characterization of six extracellular proteases, *Appl. Microbiol. Biotechnol.*, 37, 152, 1992.

# Chapter 5

# Short Chain Fatty Acids

*John H. Cummings*

## CONTENTS

I. Introduction ........................................................................................................................ 101
II. Occurrence in Hindgut ..................................................................................................... 102
    A. Substrates ................................................................................................................... 102
    B. Antibiotics ................................................................................................................. 105
    C. Stool Output and Diarrhea ........................................................................................ 105
    D. Diet ............................................................................................................................ 106
    E. Microflora .................................................................................................................. 106
    F. Transit Time ............................................................................................................... 107
III. Production Rates ............................................................................................................... 108
    A. Studies of Portal Blood ............................................................................................ 108
    B. Stoichiometry ............................................................................................................ 109
    C. Amino Acids as Substrates ....................................................................................... 110
    D. Stable Isotope Studies .............................................................................................. 110
IV. Absorption ........................................................................................................................ 111
    A. General Mechanisms ................................................................................................. 111
    B. Bicarbonate ............................................................................................................... 112
    C. Sodium ....................................................................................................................... 112
V. Cell Metabolism and Growth ........................................................................................... 112
    A. Energy Metabolism ................................................................................................... 113
    B. Mucosal Growth ....................................................................................................... 113
    C. Differentiation, Gene Expression, and Large Bowel Cancer ................................... 114
        1. Cellular Mechanisms ........................................................................................... 115
VI. Metabolism of SCFA ....................................................................................................... 116
VII. Clinical Conditions .......................................................................................................... 116
    A. Ulcerative Colitis ...................................................................................................... 116
        1. Sulfur Metabolism ............................................................................................... 117
        2. Treatment ............................................................................................................. 118
    B. Diversion Colitis ....................................................................................................... 119
    C. Pouchitis .................................................................................................................... 119
References ................................................................................................................................ 119

## I. INTRODUCTION

Short chain fatty acids (SCFA) are the major products of fermentation in the human colon. They arise from the breakdown of dietary and endogenous carbohydrates and proteins which reach the large bowel. As a result, they are a means whereby humans, and many other species, are able to obtain energy through symbiosis with anaerobic bacteria. The predominant short chain, and related fatty acid anions, found in the human colon are listed in Table 1.

The occurrence of SCFA in the human gut has been known for over 100 years,[1] ever since German and French groups began reporting amounts in feces in the early part of this century (summarized in Reference 2). The first English language papers on the subject were probably in 1929 from Olmsted and his colleagues in St. Louis, who were interested in the breakdown of dietary fiber in humans.[2-5] Their studies pointed to the fermentation of fiber as a major source of SCFA, and, although they ascribed laxative properties to SCFA, an idea subsequently thought to be incorrect,[6] they were pioneers in the development of concepts about SCFA. Interest in SCFA subsequently languished until the 1960s when a renal physician in London, Oliver Wrong, investigating the ionic composition of body fluids, developed

**Table 1**  Organic Anions of the Human Colon

| Common name | Chemical name | Formula | Mol wt | pKa | Range of concentrations found[a] |
|---|---|---|---|---|---|
| **Major fatty acids** | | | | | |
| Acetic acid | Ethanoic acid | $CH_3COOH$ | 60 | 4.75 | 40–100 |
| Propionic acid | Propanoic acid | $CH_3CH_2COOH$ | 74 | 4.87 | 15–40 |
| Butyric acid | Butanoic acid | $CH_3(CH_2)_2COOH$ | 88 | 4.81 | 10–30 |
| **Minor fatty acids** | | | | | |
| Formic acid | Methanoic acid | $HCOOH$ | 46 | 3.75 | 0–1 |
| Lactic acid | 2-Hydroxy-propanoic acid | $CH_3CH(OH)COOH$ | 90 | 3.08 | 0–5 |
| Isobutyric acid | 2-Methyl-propanoic acid | $(CH_3)_2CHCOOH$ | 88 | 4.84 | 0–3 |
| Valeric acid | Pentanoic acid | $CH_3(CH_2)_3COOH$ | 102 | 4.82 | 0–5 |
| Isovaleric acid | 3-Methyl-butanoic acid | $(CH_3)_2CHCH_2COOH$ | 102 | 4.77 | 0–4 |
| Caproic acid | Hexanoic acid | $CH_3(CH_2)_4COOH$ | 116 | 4.83 | 0–2 |

[a]  mmol/kg Contents: average values from various data sources and those in Table 2.

a novel technique for extracting SCFA from intestinal contents.[7] Using his dialysis bag method to study SCFA, Wrong produced a series of papers detailing the SCFA composition of fecal water and the influence of diet, antibiotics, and various oral electrolytes.[8-11]

The past 20 years have seen increasing interest in SCFA in human biology resulting in a number of reviews and books on the subject.[6,12-16] Today, there is considerable knowledge of their production by bacteria in the colon, epithelial transport, cellular metabolism, effects on cell growth and differentiation, and subsequent uptake by liver and muscle. Already there are clinical uses suggested for SCFA in the management of ulcerative colitis and diversion colitis, and as an energy source in enteral feeding. By producing SCFA, human colonic bacteria exert an indirect influence on many organs of the body, and on various aspects of metabolism.

## II. OCCURRENCE IN HINDGUT

Although there have been many reports of SCFA concentrations in human feces, few measurements have been made from material within different regions of the large bowel. Table 2 shows data from luminal contents obtained at autopsy from sudden death victims in the U.K., and also from right and left-sided colostomies of patients at Baragwanath Hospital in Soweto who were awaiting colostomy closure some months after surgery for abdominal trauma. In all cases acetate is the dominant SCFA, as it is in all other animal species studied. Propionate is equal to or greater than butyrate in humans and in most species[18] with the exception of the Dugong and West Indian manatee, where butyrate exceeds propionate substantially. In both the autopsy and colostomy cases shown in Table 2, SCFA concentrations are much higher in the right side of the colon, including the cecal area. It is here that colonic bacteria first encounter carbohydrate substrates leaving the small intestine and is thus the area of highest fermentative activity. Surprisingly, however, the molar ratios of acetate:propionate:butyrate are similar in both the right and left colon. Generally, concentrations in colostomy contents are higher but this may be because these subjects were eating a maize-based diet in which starchy carbohydrates accounted for 50 to 60% of their total energy intake, unlike in western countries such as the U.K., where carbohydrate amounts to around 45% of energy, of which only one-half comes from starchy sources (mainly bread and potatoes).

Weaver et al.[19] obtained colonic contents by enema from 35 healthy subjects and found molar ratios of 60.7:15.8:17.3 (acetate:propionate:butyrate) which is broadly similar to the values in Table 2. Less butyrate, however (11.2%), was found in patients with large bowel cancer. In a study of 11 transverse colostomy and 19 sigmoid colostomy patients, Mitchell et al.[20] reported higher total SCFA concentrations in the transverse region than in the sigmoid or feces, confirming maximal fermentative activity in the right colon.

### A. SUBSTRATES

There are many factors that determine the amount and proportion of end products produced during fermentation. These factors are summarized in Table 3. Substrate availability and type play a key role.

**Table 2** Short Chain Fatty Acids in Human Colonic Contents (mmol/kg) (molar ratios)

| | Right | | Left | |
|---|---|---|---|---|
| | Autopsy[a] | Colostomy[b] | Autopsy[a] | Colostomy[b] |
| Acetate | 66 (52) | 109 (62) | 47 (52) | 62 (56) |
| Propionate | 26 (20) | 36 (21) | 17 (19) | 30 (27) |
| Isobutyrate | 1.9 (1) | 0.8 (-) | 2.1 (2) | 1.4 (1) |
| Butyrate | 25 (20) | 25 (14) | 16 (18) | 11 (10) |
| Isovalerate | 2.7 (2) | 0.6 (-) | 3.6 (4) | 1.2 (1) |
| Valerate | 4.0 (3) | 2.1 (1) | 3.5 (4) | 2.2 (2) |
| Caproate | 1.6 (1) | 1.4 (1) | 0.9 (1) | 1.8 (2) |
| Lactate | 3.8 | — | 2.3 | — |
| Succinate | 2.0 | — | 2.0 | — |
| N | 6 | 8 | 6 | 8 |

[a] Data from Cummings et al.[17]
[b] Data from Cummings et al., unpublished.

**Table 3** Factors Affecting SCFA Yields

| Substrate | Amount |
|---|---|
| | Type (solubility, chemistry) |
| | Rate of breakdown |
| | Extent of breakdown |
| Microflora | Mix of species |
| | Pathogens |
| Host | Transit time (dilution rate) |
| | Mucus secretion |
| | Drugs (antibiotics, etc.) |
| | Immunoglobulin secretion |

The principal substrates in man are carbohydrates, i.e., resistant starch, nonstarch polysaccharides (NSP), nondigestible oligosaccharides such as inulin, fructans, and some nonabsorbed mono- and disaccharides.[21-23] Although it is not easy to compare experiments done in different laboratories, Table 4 summarizes 11 *in vitro* studies[24-34] which have reported SCFA yields using human fecal inocula. Total yields vary considerably, from as low as 10% g/g from some fiber preparations such as corn bran, pea and oat fiber[33] and from cellulose, to over 60% for starch and other polysaccharides. Low yields may equate with an incomplete fermentation, as is the case with bran NSP, or may mean other intermediates are being formed. In all cases, acetate is the major anion comprising 67% overall of the total SCFA. Pectin is a particularly good source of acetate (81% in 7 studies) while arabinogalactan (54%, n = 3) and guar (59%, n = 3) are the poorest sources. By contrast, guar and arabinogalactan are good sources of propionate (27 and 34%, respectively). Butyrate production varies over a wide range. Starch almost always gives high amounts of butyrate (63:14:22 acetate:propionate:butyrate, n = 6) followed closely by oat and wheat bran (60:16:20, n = 5). Pectin is the poorest source (81:11:8, n = 7) along with some of the very nondigestible corn, soya, sugar beet, pea, and oat sources used in Reference 33. With regard to overall production of butyrate, starch yields both relatively high molar proportions (22%) and has a high yield per g substrate fermented (about 60%).

Other products of fermentation include lactate, which is an intermediate in starch breakdown.[29,35] Although found in the human cecum,[17] lactate seldom accumulates in the colon of adults. It does, however, reach high concentrations in infants with diarrhea. Torres-Pinedo et al.[36] and Weijers et al.[37,38] noted lactate concentrations of around 20 mmol/kg in the stools of children with carbohydrate malabsorption. In both studies, there was also substantial acetate present, which accounted for about one-half of the total organic anion. In children who have recovered from diarrhea, lactate levels fall to 1 to 2 mmol/kg. The reason for lactate accumulation is partly its poor absorption and partly impaired further metabolism.

**Table 4**  Fermentation Products from Varying Carbohydrate Substrates

| Substrate | Total yield mg/g substrate[a] | Molar Ratios | | | Reference |
| --- | --- | --- | --- | --- | --- |
| | | Acetate | Propionate | Butyrate | |
| ***Starch*** | 590 | 50 | 22 | 29 | 24 |
| | 517 | 61 | 17 | 22 | 25 |
| | 356 | 80 | 10 | 10 | 27 |
| | — | 50 | 22 | 28 | 29 |
| | — | 69 | 9 | 20 | 28 |
| | — | 67 | 6 | 25 | 28 |
| | — | 58 | 17 | 25 | 30 |
| Mean | | 62 | 15 | 23 | |
| | | | | | |
| ***Nonstarch polysaccharides*** | | | | | |
| Arabinogalactan | 430 | 50 | 42 | 8 | 24 |
| | — | 55 | 32 | 11 | 28 |
| | — | 56 | 27 | 17 | 34 |
| | — | 68 | 24 | 8 | 30 |
| Psyllium | 266 | 77 | 10 | 12 | 27 |
| | — | 68 | 20 | 11 | 31 |
| Sterculia | 293 | 74 | 13 | 12 | 27 |
| Xylan | 540 | 82 | 15 | 3 | 24 |
| | — | 64 | 11 | 24 | 28 |
| Fibercon | 36 | 73[b] | 15 | 12 | 27 |
| Guar | 619 | 58 | 29 | 13 | 26 |
| | — | 60 | 27 | 13 | 31 |
| | — | 58 | 27 | 8 | 32 |
| Pectin | 350 | 84 | 14 | 2 | 24 |
| | — | 82 | 7 | 11 | 28 |
| | — | 78 | 12 | 10 | 31 |
| | — | 78 | 15 | 8 | 32 |
| (citrus) | 393 | 90 | 7 | 3 | 33 |
| (apple) | 434 | 82 | 11 | 7 | 33 |
| | — | 72 | 9 | 18 | 34 |
| | — | 75 | 16 | 10 | 30 |
| Tragacanth | — | 68 | 22 | 10 | 32 |
| | — | 67 | 19 | 8 | 32 |
| Cellulose | — | 62[b] | 23 | 15 | 31 |
| | — | 34[b] | 36 | 28 | 34 |
| Carboxy methylcellulose | — | 56 | 17 | 10 | 32 |
| Xanthan | — | 71 | 19 | 3 | 32 |
| Gellan | — | 62 | 20 | 7 | 32 |
| Karaya | — | 62 | 10 | 9 | 32 |
| Gum arabic | 451 | 66 | 25 | 8 | 33 |
| | — | 68 | 20 | 8 | 32 |
| Inulin | — | 72 | 19 | 8 | 30 |
| Mean (pectin) | | 80 | 12 | 8 | |
| Mean (others) | | 63 | 22 | 8 | |
| | | | | | |
| ***Brans*** | | | | | |
| Oat bran | 434 | 57 | 21 | 22 | 26 |
| | — | 64 | 12 | 19 | 28 |
| Wheat bran | 366 | 61 | 19 | 20 | 26 |
| | — | 64 | 16 | 19 | 28 |
| | — | 52 | 11 | 19 | 32 |
| | | | | | |
| ***Mixed & other sources*** | | | | | |
| Fiberall | 220 | 74 | 14 | 12 | 27 |
| Metamucil | 399 | 79 | 15 | 5 | 27 |
| Glucose | 491 | 62 | 16 | 22 | 25 |

**Table 4** Continued.

| Substrate | Total yield mg/g substrate[a] | Molar Ratios | | | Reference |
| | | Acetate | Propionate | Butyrate | |
|---|---|---|---|---|---|
| Kidney beans | 447 | 63 | 17 | 20 | 26 |
| Cabbage | — | 73 | 15 | 9 | 28 |
| Soya | — | 68 | 20 | 11 | 31 |
| | 125 | 71 | 21 | 8 | 33 |
| Corn bran | 42 | 63[a] | 30 | 7 | 33 |
| Sugar beet | 71 | 93[a] | 7 | 1 | 33 |
| Pea fiber | 49 | 74[a] | 17 | 9 | 33 |
| Oat fiber 1 | 21 | 88[a] | 12 | 1 | 33 |
| Oat fiber 2 | 9 | 51[a] | 41 | 8 | 33 |
| Polydextrose | — | 61 | 25 | 14 | 30 |
| Overall Mean | | 67 | 18 | 11 | |

*Note:* Data are values at 24 h (or 48 h in References 27 and 34). Molar percents include values for isobutyrate, isovalerate, and valerate in References 28 and 32.

[a]  Except Reference 33 where yield mg/g "Total Dietary Fiber" fermented.
[b]  Very low yields from substrate may not be representative.

Its major fate is oxidation by other fermenting bacteria. Lactate is produced mainly by bifidobacteria and lactobacilli. Lactate utilizing species include propionibacteria, some coliforms, and sulfate-reducing bacteria. Propionibacteria and sulfate reducers grow best at near neutral pH, and then only very slowly, so they may not be able to survive in the sudden rapid transit and low pH of an infant's gut in a state of diarrhea.

## B. ANTIBIOTICS
Some antibiotics impair fermentative activities in the colon and reduce fecal SCFA levels.[39-41] In the studies of Hoverstad et al.[40,41] in which groups of six healthy volunteers were given antibiotics by mouth for six d, the effects on fecal SCFA were closely related to the amounts of antibiotic detectable in feces. Reductions in total SCFA (mmol/kg feces) were seen with clindamycin (62.9 to 7.3), bacitracin (105.4 to 21.8), and vancomycin (69.3 to 19.4). Less striking changes were seen with ampicillin (62.4 to 47.8) and erythromycin (116.9 to 76.0), while metronidazole, co-trimoxazole, doxycycline, nalidixic acid, and ofloxacin had no effect. High concentrations of clindamycin, bacitracin, erythromycin, vancomycin, and nalidixic acid were found in feces. Molar ratios of butyrate decreased with clindamycin, ampicillin, bacitracin, and vancomycin.

Studies *in vitro* show that clindamycin,[42] erythromycin, and dicloxacillin reduce fermentative activity whereas penicillin and pivampicillin have no effect.[43] Lactate does not accumulate *in vitro* indicating a general suppression of fermentation rather than specific bacteria being affected. However, the studies by Wilson et al.[10] showed a pronounced increase in succinate concentrations in feces (4.2 to 40.0) in healthy subjects taking a combination of neomycin, bacitracin, colistin and nystatin,[11] together with reduction in SCFA (77.8 to 19.0), which suggests inhibition of succinate utilizing species.

## C. STOOL OUTPUT AND DIARRHEA
Diarrhea, whether associated with antibiotic administration[43] or not,[44-46] is also related to reduced SCFA concentrations in feces, although total daily excretion usually increases in proportion to rises in stool output. Studies which show increased SCFA outputs in feces in relation to normal stool weight include the effects of bran,[47] bran pentosan,[4] a variety of vegetables and cereal foods,[5,48] cellulose and pectin[49] and cellulose, xylan, corn bran, or pectin.[50] Similarly, in diarrhea due to mannitol, lactulose or raffinose,[44] magnesium sulfate,[46] and intestinal resection,[45] increased excretion with increased stool output is seen, but generally with lower concentrations.

In diarrhea, the underlying mechanism is failure of the colon to absorb water, whether because of increased volume and solute load from the small bowel, failure of solute absorption by a diseased mucosa, or the presence of bacterial toxins or nonabsorbable ions. In any event, fermentation continues until total gut transit time falls to about 18 h, below which SCFA levels decline to low levels.[45] Provided SCFA are

**Table 5**  Effect of Nonstarch Polysaccharides on Fecal SCFA (mmol/l and molar ratios)[a]

| Diet | | Acetate | Propionate | Butyrate | Total | Molar Ratios | | |
|---|---|---|---|---|---|---|---|---|
| | | | | | | Acetate | Propionate | Butyrate |
| Control | N = 18 | 50.2 | 16.3 | 20.2 | 87.0 | 58 | 19 | 23 |
| All subjects | | | | | | | | |
| Control | N = 6 | 47.6 | 14.8 | 17.7 | 80.8 | 59 | 18 | 22 |
| + Bran | | 45.5 | 10.8[b] | 20.6 | 77.0 | 59 | 14 | 27 |
| Control | N = 5 | 54.1 | 19.6 | 18.5 | 92.6 | 58 | 21 | 20 |
| + Cabbage | | 66.8 | 18.6 | 21.2 | 106.7 | 63 | 17 | 20 |
| Control | N = 5 | 57.3 | 16.1 | 24.3 | 97.7 | 59 | 16 | 25 |
| + Carrot | | 57.8 | 12.1[b] | 21.5 | 91.1 | 63 | 13 | 24 |
| Control | N = 6 | 44.2 | 13.2 | 18.1 | 75.8 | 58 | 17 | 24 |
| + Apple | | 44.6 | 19.2[b] | 24.4 | 88.1 | 51 | 22 | 28 |

[a]  See Reference 52 for details of study. Short chain fatty acids were measured by the dialysis bag technique.[8]
[b]  Significantly different from control diet.

being produced, and the mucosal surface is normal, the amount excreted each day will depend directly on stool weight, because SCFA are the principal anions in colonic contents. Thus, measurement of their output or concentration in feces will reflect stool weight, but their molar ratios can sometimes give useful information.

## D.  DIET

At least 95% of SCFA produced in the colon are absorbed, so it is not surprising if fecal SCFA measurements prove an insensitive guide to events going on more proximally in the large bowel. This situation has proved to be the case particularly when studying dietary change. In Fleming's review of 42 published studies of the effect of dietary fiber on fecal SCFA, the majority of human studies show no effect, although in the rat, pig, and monkey, some differences are seen.[51] Table 5 summarizes previously unpublished data on fecal SCFA concentrations in groups of healthy subjects fed a metabolically controlled diet, typical for the U.K. containing about 12 g/d nonstarch polysaccharides (NSP), and then the same diet with 12–20 g NSP added from carrot, cabbage, apple, or bran.[52] SCFA were measured using the dialysis bag technique of Wrong et al.[7] Diet periods were all three weeks. Total concentrations were around 80–100 mmol/kg with acetate always predominating. Neither the absolute concentration nor molar proportions of acetate changed with any of the diets. Propionate concentration went down significantly with bran and carrot, and increased with apple, but molar ratios were not different. Butyrate, which is perhaps the most important SCFA in terms of colonic function, went up with bran, cabbage, and apple and down with carrot, although the changes were relatively small. SCFA molar ratios changed with bran and apple but again were not statistically significant. Overall, there is no consistency in these results in that NSP added to the diet did not produce a characteristic change in SCFA pattern or excretion. Perhaps the changes seen with individual polysaccharides are important and reflect *in vitro* studies (Table 4). In a recent study of starch malabsorption induced by the α-glucosidase inhibitor acarbose, an increase in the relative and absolute amounts of butyrate in feces was seen.[53] Similarly, fecal butyrate concentrations increased on feeding resistant starch in the form of Hylon VII to healthy volunteers.[54] These latter findings are in keeping with the *in vitro* studies summarized in Table 4 where starch seems to be a good source of butyrate. Moreover, starch, not NSP, is probably the principal substrate for fermentation in the hindgut of many human populations, especially those with starchy staples as the main component of their diet[21,22] and may be important in the prevention of large bowel cancer[55] (see discussion later).

## E.  MICROFLORA

The type of bacteria present in the colon could theoretically make a contribution to the amount and type of SCFA present, because each genus has characteristically different fermentation end products. This may readily be seen when bacteria are grown in pure culture.[56] For example, *Clostridium perfringens* grown in carbon-limited cultures produces mainly acetate, butyrate, succinate, and lactate, while *Bacteroides ovatus* produces much less acetate, a lot of propionate, and some succinate. In the conditions of the colon, however, where many species co-exist, these individual differences do not show up and a constancy of SCFA production is much more evident.

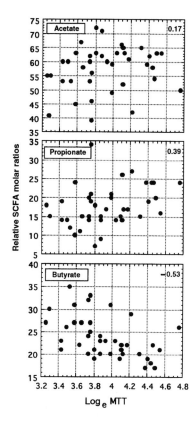

**Figure 1** Molar ratios of fecal short chain fatty acids vs. mean transit time in healthy subjects on diets with varying sources of non-starch polysaccharides (see Table 5 and Reference 52). Acetate, r = 0.17; propionate, ℓ = 0.39; butyrate, r = − 0.53.

Macfarlane and Gibson's studies using pure cultures have, however, shown clearly that substrate availability makes a significant difference to fermentation end product generation.[56] When *Bifidobacterium breve* was grown in carbohydrate-limited conditions, formate and acetate were the principal products, whereas with excess carbohydrate, lactate and acetate predominated. Lactate frequently appears where there is rapid fermentation of carbohydrate, where it acts as an electron sink to dispose of excess reducing power when substrate is plentiful. Similarly, with *Bacteroides ovatus*, acetate, succinate and propionate are the main products under carbohydrate-limited conditions, whereas acetate and succinate predominate with excess carbohydrate. These reactions are influenced by intracellular $pCO_2$. When sufficient $CO_2$ is available, as with carbohydrate excess, there is a reduced requirement to decarboxylate succinate and therefore this acid is produced in preference to propionate.[56]

## F. TRANSIT TIME

Time is another factor that affects microbial activity in the colon. In both ruminant and human studies,[57-59] bacterial growth and metabolism is much more efficient at fast turnover times, giving increased biomass yields from given amounts of substrate. This situation occurs because the maintenance energy requirements of bacteria (used for motility, active transport, and control of intracellular solute concentrations), is less at high dilution rates. *In vivo* studies in humans[59] show that speeding up transit through the gut from 64 to 35 h increases biomass excretion and reduces cellulose digestibility.

Time is also reflected in patterns of SCFA production. In studies of pure cultures of *C. perfringens* grown at different dilution rates SCFA production varies. At low dilution rates butyrate was between 12 and 17% of total organic anion, depending on carbon availability, whereas at faster rates it fell to 3 to 8%, with lactate production increasing significantly.[56] In an *in vitro* multichamber fermentation system, inoculated with mixed fecal bacteria, long retention times reduced bacterial viability, lactate and succinate production, and increased acetate and propionate, especially in the first vessel of the system, which was equivalent to the right colon.[60] Long retention times also increased protein degradation when studied in a multichamber system[61] giving rise to increased concentrations of phenols and ammonia.

There are no published studies in humans relating transit time to fecal SCFAs. Figure 1 shows the results of individual SCFA molar ratios as measured by the dialysis bag technique of Wrong et al.[8] from a study of 18 healthy individuals eating a variety of controlled diets.[52] Mean transit time (MTT) was measured in all subjects on each diet by the continuous marker method and, as can be seen in Figure 1, was significantly related to the proportion of butyrate in feces (r −0.53, p <0.002). The absolute concentration of butyrate was also significantly associated with MTT (r −0.356, p <0.03). No association was found between MTT and either concentration or molar ratios of acetate and propionate in this study. Thus, rapid transit, in the order of 30 to 40 h, is associated with a significantly higher fecal butyrate concentration and molar ratio than slow transit of 60 to 80 h.

The association of rapid transit with increased fecal butyrate might provide part of the explanation for Burkitt's proposed protective effect of dietary fiber against large bowel cancer.[62,63] Moreover, the greater efficiency of biomass production leads to increased stool output[59] again giving transit time a central role in determining colonic events.

Other changes in bacterial metabolism in the colon have been associated with transit time. Davignon et al.[64] showed that there was an inverse relationship between turnover time (equivalent to MTT) and formation of the neutral-steroid conversion products coprostanol and coprostanone in subjects on controlled diets. An inverse relationship was also seen between turnover and degradative losses of β-sitosterol.

In support of the *in vitro* findings of Macfarlane et al.[61] longer transit times are also associated with increasing fecal ammonia concentrations and urinary phenol outputs.[65,66] Both dietary protein intake and transit time affect urinary phenol excretion and together account for 64% of the variation in excretion.[66] Many years ago, Macy[67] reported a similar relationship between transit and urinary sulfate excretion in children. Both the ethereal sulfate found by Macy and urinary phenols are bacterial metabolites of dietary protein breakdown (see Chapter 4).

## III. PRODUCTION RATES

The importance of SCFA to humans lies not only in the relative amounts of each fatty acid produced, but more importantly, in the overall amount. Determining this amount is, however, much more difficult than simply measuring SCFA levels in gut contents. In ruminant animals, simultaneous catheterization of the portal vein and an artery, together with isotope infusions, allows estimates of production to be made.[68,69] Such studies are impossible in humans because of difficulty accessing the portal vein, and they pose difficulties in interpretation for the ruminant physiologist because each fatty acid is metabolized by different tissues in the body, e.g., acetate in muscle, propionate in the liver, and butyrate by the rumen epithelium (in general). The situation is further complicated by endogenous acetate production by the liver. In humans, a number of approaches are possible including direct measurement of arterio-venous (A/V) differences in SCFA concentrations across the gut at surgery or autopsy, *in vitro* fermentation models, estimation of substrate availability and breakdown, biomass production in the colon, and stable isotope infusions.

### A. STUDIES OF PORTAL BLOOD

Table 6 lists the few measurements of SCFA in human portal blood that have been made. These data have been acquired either at autopsy in sudden death victims (e.g., road accidents, other violent deaths, coronary heart disease) where the subject can be assumed to have been eating more or less normally until the time of death, during fasting in subjects undergoing surgery, or in surgical cases after "feeding" — either following installation of lactulose into the colon or during emergency surgery.[70-72] When taken together, these data give an average total SCFA concentration in portal blood of 288 mol/l, A/V differences of 192 mol/l and a daily production rate of about 277 mmol/d. If the data are separated into fasting (n = 2) and post meal (n = 3), then total concentrations are 167 mmol/l and 368 mmol/l and production rates are about 163 and 353 mmol/d, respectively.

Are these SCFA arising from the gut? The presence of acetate in arterial blood allows for the possibility of production at many sites. Some endogenous (noncolonic) production does occur because subjects who have had their large intestine removed (ileostomists) have 21 mol/l[73] acetate in peripheral blood. Moreover, during prolonged fasting (108 h), venous blood SCFA rise from 43.9 ± 4.4 (SEM) after 12 h fast, to 104.6 ± 13.2 mol/l after 60 h fast, and 114 ± 15.6 after 108 h fast in healthy subjects.[73] Skutches et al.[72] have also observed net synthesis in the liver and muscle of humans, when arterial blood levels fall below 80 mol/l. Nevertheless, in the fed state the gut is likely to be the major source of SCFA.

A number of assumptions are made when calculating production rates from A/V differences. First, portal blood flow over 24 h is estimated as 1 l/min. In practice, portal blood flow varies from less than

**Table 6** Short Chain Fatty Acid Production in the Human Gut

| Source | Portal blood (μmol/l) | | | Venous/Arterial (μmol/l) | | | Production[a] (mmol/d) | | | Reference |
|---|---|---|---|---|---|---|---|---|---|---|
| | A | P | B | A | P | B | A | P | B | |
| U.K. autopsy, nonfasting | 258 | 88 | 29 | 70 | 5 | 4 (V) | 270 | 119 | 36 | 17 |
| South Africa surgical, nonfasting | 271 | 96 | 56 | 134 | 16 | 11 (A) | 197 | 115 | 65 | 22 |
| Netherlands surgical, fasting | 114 | 32 | 9 | 35 | 2 | 1 (V) | 114 | 43 | 11 | 70 |
| New Zealand surgical, fasting | 128 | 34 | 18 | 67 | 4 | 0 (V) | 88 | 43 | 26 | 71 |
| post meal | 241 | 39 | 27 | 130 | 3 | 0 (V) | 160 | 52 | 39 | |
| USA[b] surgical, fasting | 133 | — | — | 60 | — | — (A) | 73 | — | — | 72 |

*Note:* A = acetate; P = propionate; B = butyrate.

[a] Production: A/V difference × 1440 (min/d) ÷ 1000 (mol to mmol).

[b] Only acetate measured.

1 l/min during fasting to about 1.5 l/min postprandially, but 1 l/min is a useful rule of thumb on average.[74-76] Blood SCFA levels also change between the fasting and fed state, but this variation is likely to be small since fermentation in the large intestine is a slow process and broad peak levels of SCFA are reached 8 to 12 h after a meal in peripheral blood.[77,78] Another factor to be taken into account is the partition of SCFA between red blood cells and plasma. The concentration of acetate in red cells is only 40 to 80% that in plasma,[79,80] and in whole blood is 80 to 85% of that in plasma. Of the values given in Table 6, all are measured in plasma except for the U.S. reference so a correction to production rates of 15 to 20% may be necessary. However, portal blood SCFA measurements do not account for fatty acid metabolism in the colonic epithelium. Almost certainly, 70 to 80% of butyrate is metabolized in the colonic epithelium, with some propionate and even a little acetate. Thus, A/V differences may underestimate total production by more than enough to compensate for partitioning between red cells and plasma.

## B. STOICHIOMETRY

Calculations from A/V differences give production rates of 300 to 400 mmol/d. How much substrate would be required to produce this amount of SCFA?

A number of attempts to draw up equations for SCFA production have been made[81-86] based both on rumen and human studies. Given that carbohydrate is the major substrate and SCFA, $CO_2$ and $H_2$ the principal products, the most valuable information needed to write an equation is the overall ratio in which SCFAs are produced. Table 7 shows molar ratios of SCFA from *in vitro* studies, portal blood data, and autopsy measurements.[17] Taking all these into account, and the probable metabolism of SCFA by the epithelium, a molar ratio of around 60:20:18 acetate:propionate:butyrate can be justifiably used which gives the following stoichiometry:

$$59\ C_6H_{12}O_6 + 38\ H_2O \rightarrow 60\ CH_3COOH + 22\ CH_3CH_2COOH +$$

$$18\ CH_3(CH_2)_2COOH + 96\ CO_2 + 256\ [H^+] \tag{1}$$

This avoids making assumptions about routes of $H^+$ disposal[87] but takes note of the fact that different substrates may produce very different relative amounts of the three SCFA when incubated alone (Table 4). The above stoichiometry gives a yield (weight/weight) of SCFA from carbohydrate of 63%. This is close to the figure of 61% which Livesey and Elia[81] suggest fits best with experimental data used to calculate the energy value of fermentable carbohydrate in mixed diets fed to humans and also with ruminant studies.[57,58,83] However, if the data in Table 4 are examined it will be apparent that yields of around 60% are seen only with starch fermentation. Cell wall polysaccharides such as arabinogalactans and pectins give yields of 35 to 54%. Therefore, a theoretical yield of 63% from the equation is either an upper limit, or starch has to be the major substrate for fermentation.

**Table 7**  Molar Ratios of SCFA
From *In Vitro* and *In Vivo* Studies[a]

|                          | Acetate | Propionate | Butyrate |
|--------------------------|---------|------------|----------|
| Autopsy – R. Colon       | 56      | 22         | 22       |
| L. Colon                 | 59      | 21         | 20       |
| Colostomy – R. Colon     | 64      | 21         | 15       |
| L. Colon                 | 60      | 29         | 11       |
| *In vitro* – Starch      | 63      | 14         | 22       |
| Other substrates         | 68      | 18         | 14       |
| Feces – Various diets    | 69      | 17         | 24       |
| Portal Blood – Fasting   | 72      | 20         | 8        |
| Fed                      | 70      | 20         | 10       |
| Overall average[b]       | 60      | 22         | 18       |

[a]  Data from Tables 2, 4, 5, and 6.
[b]  This average takes into account some butyrate uptake by epithelium, and is also weighted in favor of starch as major substrate worldwide and *in vivo* data.

Equation 1 therefore requires 32 to 42 g of carbohydrate to be fermented in the human colon each day to produce 300 to 400 mmol of SCFA. Estimates of substrate availability[21,22] show that at least this amount is available for fermentation.

## C.  AMINO ACIDS AS SUBSTRATES

Amino acid fermentation also gives rise to SCFA and may therefore be an additional source in humans.[88] Amino acid fermentation yields not only SCFA, but the branched chain fatty acids isobutyrate, isovalerate, and 2-methylbutyrate, which arise, respectively, from valine, leucine and isoleucine. Table 2 shows the concentration of these branched chain fatty acids found in the human cecum. (Isovalerate and 2-methylbutyrate are combined in Table 2 because they run together in the chromatography system used.) Although the amounts are much lower than the three major SCFA, the presence of branched chain fatty acids indicate that amino acid fermentation is taking place. A fraction of total SCFA must therefore come from protein breakdown. Using batch culture studies of human fecal inocula, Macfarlane et al.[88] have shown that SCFA are the principal end products formed during the degradation of protein by human colonic bacteria. Approximately 30% of protein is converted to SCFA. Branched chain fatty acids constitute 16% of SCFA produced from bovine serum albumin and 21% of SCFA generated when casein is the substrate. Branched chain fatty acid concentrations in gut contents taken from the human proximal and distal colons were on average, 4.6 and 6.3 mmol/kg respectively, corresponding to 3.4 and 7.5% of total SCFA. These results suggest that protein fermentation could potentially account for about 17% of the SCFA found in the cecum, and 38% of the SCFA produced in the sigmoid/rectum. Measurements of branched chain fatty acids in portal and arterial blood taken from individuals undergoing emergency surgery indicated that net production of branched chain fatty acids by the gut was in the region of 11 mmol/d, which would require the fermentation of about 12 g of protein. These data show that protein is a significant source of SCFA, and, with the known amounts of carbohydrate entering the colon, we have enough substrate available to account for 300 to 400 mmol/d of SCFA.

## D.  STABLE ISOTOPE STUDIES (Table 8)

Undoubtedly one of the best ways of measuring SCFA production rates in humans is to use stable isotope technology. The first published attempts to do this were by Walter et al.[89] who were investigating the sources of propionate in children with methylmalonicaciduria. Propionate production was measured by a continuous intravenous infusion of $^{13}C$ propionate, with breath and blood sampling for $^{13}C$ measurement. Total propionate production was 55 to 186 mmol/kg/h in the children and 17 mmol/kg/h in 5 healthy adults. Only 5 to 40% of the propionate produced in the children could be ascribed to protein catabolism leaving a large role for colonic fermentation. Using the data of Walter et al. for the children and the stoichiometry of Equation 1, total SCFA production in a 70 kg person would be approximately 380–760 mmol/d. These calculations are not possible for the healthy controls because of propionate uptake by the healthy liver.

**Table 8**  Acetate Production or
Turnover as Measured in
Isotope Infusion Studies (mol/kg/min)

| Subjects | Rate | Reference |
|---|---|---|
| Premature infants | 64 | 92 |
| Starving obese adults | 1 | 93 |
| Fasting adults | 5–8 | 72 |
| One adult fasting | 3 | 94 |

Breves and colleagues[90] and Freeman et al.[91] applied stable isotope technology to studies of pigs and showed that 7 to 40% of maintenance energy requirements come from hindgut fermentation and that the gut itself, including the liver, may utilize a substantial portion of acetate production. In premature infants, Kien et al.[92] used $^{13}C$ acetate intragastric infusions together with blood and breath sampling to show that 24 to 74% of ingested lactose was fermented to acetic acid. This process allows the conclusion that premature babies, who are known to digest lactose poorly, may salvage substantial energy from unabsorbed carbohydrate through fermentation.

Overall therefore, SCFA production in humans has been estimated only approximately at present from static A/V difference studies across the gut or indirectly using knowledge of substrate availability, and stoichiometry derived from *in vitro* animal and human colon fermentation studies. Net yields vary greatly depending on the type of substrate, its rate of fermentation, and transit time through the colon. Stable isotope technology offers a more promising approach to this problem.[93,94] On the whole, these lines of evidence all point to a daily production in adults on western diets of 300 to 400 mmol/d derived from fermentation of 30 to 40 g carbohydrate and 12 g protein. These production rates are for the western colon and are low, contributing much less to energy needs than those observed in the pig and particularly in ruminant species. The amounts and types of SCFA produced in the human colon can be readily increased, however, by changing the pattern of the diet, with the introduction of more cereal-based starchy foods, fruits, vegetables, or other sources of resistant starch, NSP and nondigestible oligosaccharides such as inulin and other fructans. This type of diet is found in many Third World populations where SCFA production may be much greater.

## IV. ABSORPTION

SCFA are rapidly taken up from the human colon[95,96] and absorption rates in the rectum, descending and transverse colon are comparable to those observed in other mammals. The first demonstration of SCFA absorption in humans was made in 1964 by Dawson et al.[97] using an isolated human large bowel. Early mechanistic studies[95,96,98] showed absorption of all three SCFA which was largely unaffected by pH, with associated bicarbonate appearance in the lumin, consistent with part nonionic diffusion and part transport of the anion.

### A. GENERAL MECHANISMS

SCFA absorption has been extensively studied in the rumen. They are rapidly absorbed, stimulate sodium absorption, and are associated with bicarbonate exchange. pH has little effect but absorption rates are concentration dependent and show little variation due to chain length.[99] Magnesium absorption may also be stimulated.[100]

In humans, as in ruminants, a large concentration gradient for SCFA exists between the colonic lumin and blood, at around 100 mmol/l, thereby favoring its movement into blood. At the pH of colonic contents (around 5.5),[17] SCFA are largely present as anions. A transepithelial potential difference exists, with the lumin positive, also favoring absorption of the anion.

Interpretation of experimental studies is complicated by the variety of methods used to investigate absorption. Techniques include short circuit current experiments with isolated epithelia, apical and basolateral membrane vesicles, and whole organ perfusion. This situation coupled with the clear regional, and possible interspecies differences which occur as well as the differential metabolism of SCFA by colonic epithelial cells, makes it difficult to find a consistent pattern.

Some generalizations may, however, be made. SCFA are absorbed from the human gut and concentration dependence can be shown;[96] pH has little effect.[95] One major route for absorption is likely to be by passive diffusion of the protonated acid.[101] von Engelhardt and colleagues[102-104] suggested that, in the guinea pig, SCFA anions may permeate by a paracellular pathway. However, subsequent studies in both horse and guinea pig using voltage clamping have shown that this makes no difference to unidirectional fluxes, and largely rules out paracellular transport and any associated bulk water flow.

For nonionic diffusion to occur, SCFA anions have to be protonated. $H^+$ is probably made available via a $Na^+-H^+$ exchange in the proximal colon and $K^+-H^+$ ATPase in the distal colon.[105,106] In his studies on the guinea pig colon, von Engelhardt has suggested that about 35% of SCFA are absorbed in the undissociated form in the cecum, 30 to 50% in the proximal colon, and 60 to 80% in the distal colon.[107]

## B. BICARBONATE

A relationship between SCFA and bicarbonate secretion has often been observed.[95,96,108-113] The consistent appearance in the lumin of bicarbonate during SCFA absorption is independent of chloride-bicarbonate exchange because it occurs in the absence of luminal chloride and is independent of chloride absorption. The amount of bicarbonate that accumulates is equivalent to about one-half of the acetate that is absorbed. Bicarbonate appearance in the lumin is probably the result of a SCFA-bicarbonate exchange at the cell surface.

Since SCFA transport is nonelectrogenic, they must be absorbed either by anion exchange or co-transport with a cation. However, the associated changes in luminal $pCO_2$ and pH do not accord with a simple anion exchange. Carbon dioxide and bicarbonate in body fluids are related through the equation:

$$H_2O + CO_2 \xleftrightarrow[\text{anhydrase}]{\text{carbonic}} H_2CO_3 \xleftrightarrow{\substack{pKa \\ 6.4}} H^+ + HCO_3^- \qquad (2)$$

Any increase in luminal bicarbonate due to secretion from the mucosa will push the reaction to the left resulting in a rise in pH and $pCO_2$. In experimental studies in the pig,[109] rumen,[114] and humans,[96] pH rises, but $pCO_2$ falls during acetate absorption. The explanation that has been given for this is that luminal or juxtamucosal hydration of $CO_2$ occurs, and that hydrogen ion is used to protonate SCFA anion prior to crossing the mucosa as undissociated acid. Thus, absorption of the acid leads to bicarbonate accumulation, a rise in pH, and a fall in $pCO_2$. The colonic epithelium, in contrast to the small bowel, is rich in carbonic anhydrase,[115,116] and its inhibition reduces SCFA transport.[117] Consequently, bicarbonate secretion by the colonic mucosa, either from the cytosol or a juxta-mucosal carbonic anhydrase, is important in SCFA absorption and provides the principal intraluminal buffer for fermentation.

## C. SODIUM

Stimulation of sodium absorption is seen in many studies of SCFA transport.[96,110,118-121] The stimulatory effect of fatty acids on sodium absorption is considerable. In a study by Roediger and Moore[121] on the isolated perfused human colon, net sodium absorption (nmol/min/cm$^2$ ± SEM) increased from 320 ± 10 in the control perfusion to 1960 ± 480, with the addition of 20 mmol/l butyrate to the perfusate. In the rat colon, neither succinate nor lactate, which are poorly absorbed anions, stimulate sodium absorption, whereas acetate[111,122] and nitrate[123] do.

von Engelhardt[107] believes that interaction of SCFA with $Na^+$ absorption indicates that electroneutral transport of $Na^+$ stimulates SCFA absorption but not the reverse. Therefore, SCFA absorption itself affects $Na^+$ transport very little. SCFA appear to be coupled in some way and thus provide a powerful mechanism for the movement of sodium and water out of the colonic lumin. In this context, SCFA must be seen as antidiarrheal and failure in their production leads to disturbances of salt and water absorption which are most notable in the germ-free animal.

Thus, SCFA absorption probably involves a number of mechanisms including both nonionic, ionic, and ion exchange processes. There are secondary effects on electrolyte transport and, perhaps most importantly for fermentation, on water movement out of the colon, and on pH through $HCO_3^-$ and $H^+$ transport at the epithelial surface.

## V. CELL METABOLISM AND GROWTH

Probably the single most important interaction between bacteria in the large intestine and their host lies in the metabolic effects of SCFA in the colonic epithelial cell. All three major SCFA are metabolized to

**Table 9**  Butyrate Clearance by Colonic Mucosa

| | Mol/L | | |
| --- | --- | --- | --- |
| | **Acetate** | **Propionate** | **Butyrate** |
| Portal blood (autopsy)    a | 258 | 88 | 29 |
| Arterial blood (autopsy)  b | 70 | 5 | 5 |
| SCFA added   (a-b) | 188 | 83 | 25 |
| Molar ratio | 63 | 28 | 8 |
| Molar ratio colonic contents | 57 | 22 | 21 |
| Portal blood (surgery)     c | 271 | 96 | 56 |
| Arterial blood (surgery)  d | 134 | 16 | 11 |
| SCFA added   (c-d) | 137 | 80 | 45 |
| Molar ratio | 52 | 30 | 17 |
| Molar ratio colostomy contents (from Table 2) | 62 | 25 | 13 |

some extent by the epithelium to provide energy, but butyrate is especially important as a fuel for these cells and may also play a critical role in moderating cell growth and differentiation.

## A. ENERGY METABOLISM

The colonic epithelium derives 60 to 70% of its energy from bacterial fermentation products.[124,125] Using the isolated rat colonocyte, Roediger,[126] and subsequently confirmed by others,[127-130] showed that SCFA, particularly butyrate, are metabolized and suppress glucose oxidation. Studies of $CO_2$ production using mixtures of SCFA indicate that cellular activation is in the order butyrate>propionate>acetate. SCFA are metabolized to $CO_2$ and ketone bodies and are precursors for mucosal lipid synthesis.[131] In the intact rat, about 12% of butyrate is transformed to ketone bodies[132] in the wall of the cecum. No ketogenesis was seen in the colon, a finding similar to that observed in the rabbit and guinea pig.[128,133,134] In the rabbit, studies of [14]C-SCFA metabolism have shown label being transferred to amino acids, carboxylic acids, and sugars in the cecal wall.[127,128] The human colonocyte also metabolizes glucose and glutamine.[135] However, more than 70% of oxygen consumption in isolated colonocytes is due to butyrate oxidation, although there are regional differences. Carbon dioxide production from butyrate is similar in both proximal and distal colon in humans, but ketone body appearance is less in the distal gut. This situation implies that more butyrate enters the TCA cycle in the distal colon and is more important in this region. Conversely, glutamine metabolism is greater proximally, as is glucose oxidation. Thus, the proximal colon more resembles the small bowel, while the distal colon relies on butyrate as a primary energy source. Roediger[136] has suggested that the health of colonic epithelial cells is largely dependent on the availability and efficient metabolism of butyrate (see discussion later).

Uptake and utilization of butyrate by the colonic epithelium *in vivo* can be demonstrated from study of levels of SCFA in portal and arterial blood and in colonic contents. Table 9 shows the amounts and molar ratios of SCFA in material taken from sudden death victims,[17] and at emergency surgery in trauma patients.[22] In the autopsy study, the molar proportion of butyrate falls from 21% within the colonic lumin to 8% in portal blood, indicating substantial butyrate clearance by the mucosa (assuming no mucosal production of acetate and propionate). Table 7 shows that *in vitro*, about 18% of total SCFA produced is butyrate, and *in vivo*, 21% of SCFA found in the colon is butyrate. The fall to 8% in portal blood indicates a clearance of 65% of the butyrate by the mucosa. If either acetate or propionate are also metabolized by the mucosa, then 65% is a minimum estimate of clearance. However, the surgical data in Table 9 show that portal blood butyrate molar ratio is 17%, whereas in colostomy contents of patients from an identical population in the same hospital, butyrate is only 13% of total SCFA. This difference means that either the mucosa of this population is selectively metabolizing acetate and propionate, an unlikely event, producing butyrate, which is equally unlikely, or that all three acids are being taken up. These data all point to substantial butyrate, and probably some propionate and acetate utilization by the mucosa in humans.

## B. MUCOSAL GROWTH

A number of studies have shown that the presence in the small bowel of dietary fiber stimulates mucosal growth and function[137-141] and likewise in the large bowel.[142-146] This trophic effect on the epithelium is

**Table 10** Functions of Butyrate in the Colonic Epithelial Cell

1. Energy source
2. Substrate for membrane lipid synthesis
3. Arrests cell growth early in $G_1$
4. Induces differentiation
5. Affects gene expression
6. Associated changes in cytoskeletal architecture
7. Increases histone acetylation
8. Induces apoptosis

almost certainly mediated by SCFA.[147-151] Sakata has shown, by instilling a mixture of SCFA into the colon daily in rats with ileal fistulae, an increase in crypt cell production rates within 2 d, which was independent of luminal pH.[150,151] Where SCFA production is reduced in either large or small bowel, epithelial cell proliferation is depressed. This situation is seen with bypass surgery,[152] low fiber diets,[153,154] and in germ-free animal studies.[148] The mechanism of this effect is intriguing, particularly the trophism, which is also seen in the jejunum during SCFA instillation into the colon.[151] Inert bulk is not sufficient to mediate the effect.[147,150] Rombeau and colleagues have shown that the jejunal response to SCFA occurs after either intravenous or colonic infusions[155] and that butyrate is more related to trophic effects within the colon.[156] The effect is independent of pH.[155] The effects of SCFA at a site distant from their production suggests a systemic mediator of the trophic effect. Recent studies by Rombeau and Sakata[157] have shown that the mechanism may require the autonomic nervous system. Using rats with microsurgical denervation of the cecum, these authors showed that an intact nervous system was needed for the jejunal trophic effect. Local effects are also possible involving increases in blood flow, SCFA as an energy source, or locally released growth factors.

The trophic effect of SCFA on the large bowel mucosa has led to the suggestion that it may be a factor favoring tumor development.[158-160] However, Scheppach and colleagues[161,162] have shown, using biopsy material from the human colon, that this is unlikely to be true. Using [$^3$H]-thymidine and bromodeoxyuridine to label incubated crypts, they have calculated the labeling index (a measure of crypt cell growth rate) in whole crypts and five equal compartments of the crypt. Butyrate and propionate both increased proliferation rates, whereas acetate did not. However, cell growth was stimulated only in the basal three compartments, not those near the surface as is characteristic of preneoplastic conditions.[163] Moreover, butyrate is well established as a growth inhibitor and inducer of differentiation in many cell lines (see discussion later).

## C. DIFFERENTIATION, GENE EXPRESSION, AND LARGE BOWEL CANCER

Apart from being an important respiratory fuel for the colonocyte, butyrate has remarkably diverse properties in a wide range of cells. These properties include arrest of cell growth early in G1, induction of differentiation, stimulation of cytoskeletal organization, and alteration in gene expression (Table 10). The effects of butyrate on the cell were first highlighted in 1976 in a review by Prasad and Sinha.[164] A large number of papers have been written since then and various mechanisms of action have been proposed.[165-171]

The slowing or arrest of cell growth is seen in many cell lines including chick fibroblasts and HeLa cells,[172] ovarian cells,[173] mouse fibroblasts,[174,175] hepatoma,[176] bladder,[177] colon,[178-184] pancreas,[185] cervix,[186] melanoma,[187,188] neuroblastoma,[189] prostate,[190,191] and breast.[192] Changes in cell growth are associated with differentiation, as indicated for example by the expression of alkaline phosphatase activity. Kim and co-workers[193-195] have shown that in normal colonic mucosa, intestinal alkaline phosphatase activity is increased by butyrate, while treatment of tumor tissue produces placental-like alkaline phosphatase (PLAP). They have also shown that butyrate induction of PLAP in colon cancer cells is regulated by increased message levels which occur with differentiation. In LS174T cells, the 5′ flanking regions of the PLAP gene were found to contain *cis*-acting elements which regulate PLAP expression. The suppressor effect for gene expression of the PLAP promoter was removed by butyrate, which suggests that butyrate-induced factors lead to the release of a repressor band in this region before induction.[195] The importance of specific DNA sequences in butyrate sensitivity of particular genes has also been shown for two cytotoxic cell protease genes from T-lymphocytes[168] and the embryonic globulin gene.[170]

The effect of butyrate on differentiation is therefore related to the control of gene expression. Toscani et al.[175,196] have shown with Swiss 3T3 cells that the arrest of cell growth and differentiation is not a generalized shutting off of the expression of growth-associated genes, but rather a specific reduction of c-*myc*, p53, thymidine kinase, and induction of c-*fos* and aP2. These events ultimately lead to adipocyte differentiation in Swiss 3T3 cells, when combined with either insulin or dexamethasone.

Butyrate alters the expression of many genes, including the induction of hemoglobin synthesis in murine erythroleukemia cells,[197] EGF receptors[198-200] in hepatocytes, plasminogen activator synthesis in endothelial cells,[201] thyroid hormone receptors in the pituitary,[202] metallothionein in hepatoma cells,[203] estrogen, prolactin, and EGF receptors in breast tissue cells[204,205] and many others. In colo-rectal cancer cells a number of changes in gene expression have been observed including induction of c-*fos*,[206] PLAP[207] and carcinoembryonic antigen,[208] inhibition of urokinase and release of plasminogen activator inhibitor,[209,210] expression of brush border glycoprotein,[184] and P-glycoprotein phosphorylation.[211] The induction of differentiation in tumor cell lines is associated with changes in cytoskeletal architecture and adhesion properties of cells.[212-214]

## 1. Cellular Mechanisms

Kruh and colleagues have suggested a number of cellular mechanisms for the action of butyrate.[167,215] The best known mechanism is the effect on histone acetylation which has been shown in many cell types. Smith[169] has demonstrated that by inhibiting histone deacetylase, butyrate allows hyperacetylation of histones to occur. In turn, this "opens up" DNA structure facilitating access of DNA repair enzymes. Smith has grown human adenocarcinoma cells (HT29) at a range of butyrate concentrations up to 10 mmol/l and shown inhibition of cellular proliferation, which was approximately concentration dependent. Cells were blocked in GI phase. Cells grown in 5 mmol/l butyrate and exposed to various DNA damaging agents showed increased resistance to ultraviolet radiation damage and to adriamycin, but surprisingly, increased sensitivity to X-irradiation. Accessibility of DNA to endonucleases was increased and a compensatory increase in ultraviolet repair incision rates seen. In an animal study in which wheat bran was fed, Boffa et al.[216] demonstrated that butyrate levels in the colonic lumin are positively related to colonic epithelial cell histone acetylation and inversely related to cell proliferation. Butyrate thus appears to be able to modulate DNA synthesis *in vivo*.

However, increase in histone acetylation may not be the entire explanation for the specificity of butyrate in modulating gene expression. Evidence is now accumulating that butyrate acts by a mechanism that involves specific regulatory DNA sequences. Glauber and colleagues[170] have shown, with adult erythroid cells, that 5′ flanking sequences upstream of the embryonic p-gene are a major determinant of p-gene expression and mediate the stimulatory action of butyrate on p-gene transcription. Similarly, in T-lymphocytes, Fregeau et al.[168] have shown a butyrate sensitive region in the 5′ flanking sequence of a cytotoxic cell protease gene. A specific effect of butyrate on transcriptional regulatory proteins and promoter regions seems likely.[195,206,217] Furthermore, Klehr et al.[218] have shown that the stimulatory effect of butyrate on gene expression is greatest if one, or especially two, scaffold/matrix-attached regions are present adjacent to the gene. They suggest that butyrate induced inhibition of histone deacetylase may have consequences for chromatin structure which would, in turn, promote transcription.

Other possible mechanisms of action suggested by Kruh[166,167] include inhibition of chromatin protein phosphorylation, hypermethylation of DNA, and chromatin structure. Recently, Paraskeva and colleagues[171] have suggested that butyrate causes apoptosis. In cell lines from colorectal cancers, sodium butyrate in 1 to 4 mmol/l concentrations induced apoptosis, while TGF- did not. Apoptosis in the colon may therefore be triggered as cells migrate up the crypt and are exposed to luminal growth factors.

Does all this convert into a mechanism whereby anaerobic bacteria in the gut produce, from dietary carbohydrate, factors which lead to protection from large bowel cancer? Butyrate can induce transformed cells to acquire the phenotype of more differentiated cells, but as Young and Gibson[184,219] point out butyrate has paradoxical effects on normal and transformed colon epithelial cells. Some of the conflicting results in the literature may be related to the concentration of butyrate used in the *in vitro* studies. The most effective concentration for inducing differentiation is usually no more than 5 mmol/l while high concentrations lead to cell death. However, luminal concentrations of butyrate in the human colon can exceed 20 mmol/l (Table 2) and are usually in the range 10 to 30 mmol/l. It is difficult to believe that a naturally occurring fatty acid, ubiquitous in the mammalian hind gut, could promote or select for tumor growth. Its origin from fermentation of dietary NSP and RS does however provide a link between epidemiological studies which show these carbohydrates to be protective against large bowel cancer and the cellular mechanisms that occur. In the 14 years that have elapsed since it was first proposed that butyrate was the link between fiber and protection against large bowel cancer,[220] it has remained the most likely candidate to fulfill this role. In addition, it is an important energy source for the colonic epithelium, substrate for lipid synthesis and its production by bacteria is therefore of major importance to health.

## VI. METABOLISM OF SCFA

Acetate is an essential metabolic fuel in ruminant animals because all glucose reaching the rumen is fermented by the resident bacteria. In humans, its importance is less certain because it derives from the hind gut and as such must represent a secondary fuel for the tissues, after carbohydrate absorbed from the small bowel. Acetate is always found in human venous blood[77,221,222] with fasting levels around 50 mol/l rising to 100 to 300 mol/l after meals containing fermentable carbohydrate.[77,221-224] Oral metronidazole has little effect, despite suppressing $H_2$ and $CH_4$ production.[222] Acetate is rapidly cleared from the blood with a half-life of only a few minutes,[221,225] and is metabolized by skeletal[222,226] and cardiac muscle[227,228] and brain.[229] Blood acetate is derived primarily from the gut.[222,230] Patients who have no large intestine have very low blood acetate levels. However, liver synthesis probably occurs when blood levels fall below a certain critical level[221,230,231] as seen in starvation for example.[222] Acetate spares fatty acid oxidation in humans. Free fatty acid levels fall when oral acetate, or alcohol which is an immediate precursor of acetate, is given.[226,232,233] Acetate either orally or intravenously has little effect on glucose metabolism and does not stimulate insulin release in humans.[234,235]

Propionate metabolism has been extensively studied in ruminants where it is a major glucose precursor.[236,237] Much less is known about its role in humans. Propionate can be found in portal blood, although some may be metabolized in the colonic epithelium and may be a differentiating factor, but with less power than butyrate.[178] Propionate supplemented diets have been shown to lower blood cholesterol in rats[238,239] and pigs,[240-243] but in humans the effects are less clear. In a double-blind placebo controlled study of 10 female volunteers fed 7.5 g sodium propionate daily for 7 weeks, there was no change in serum cholesterol but HDL cholesterol increased as did triglycerides. There were decreases in fasting blood glucose levels and maximum insulin responses to standard glucose loads.[244] In a similar study, lasting one week, again no effect on cholesterol was seen but glucose levels were lowered.[245] In another study nine healthy males, whose initial cholesterols were all >5.5 mmol/l, were given 5.4 g propionate daily for 15 d and showed a fall in total cholesterol (to $5.28 \pm 0.23$ mmol/l) and in LDL cholesterol, while acetate was without effect.[246] To overcome the objections that oral propionate does not reflect the true physiology of propionate produced by fermentation in the hind gut, two studies using rectal or colonic infusions of propionate have been undertaken.[247,248] Propionate had no effect, or slightly raised serum cholesterol in short-term studies. Other experiments have shown inhibition of hepatic cholesterol synthesis by propionate and redistribution of cholesterol from plasma to liver.[238,239,249]

There is also extensive hepatic metabolism of SCFA.[250] The metabolic role of SCFA in humans is summarized in Figure 2.

## VII. CLINICAL CONDITIONS

### A. ULCERATIVE COLITIS

A role for intestinal bacteria in the etiology of ulcerative colitis (UC) has been suggested on many occasions, but few studies have been able to identify the specific species involved.[251] No mechanism has yet been identified but it will be very surprising if bacteria are not involved in the initiation and/or maintenance of inflammation in this condition.

Roediger's work in establishing that SCFA, particularly butyrate, were important fuels for the human colonic epithelium led him to examine their metabolism in disease, and in ulcerative colitis in particular. He was the first to show that butyrate oxidation was significantly impaired in UC, both during the active phase of the disease, and in remission.[252] This observation has been confirmed by a number of other groups.[253-255] It has now been shown that butyrate oxidation is substantially reduced even in uninvolved segments of the large bowel in patients with active colitis.[253] One study, in UC patients undergoing routine colonoscopy, did not show any difference in butyrate oxidation either regionally in the colon or between

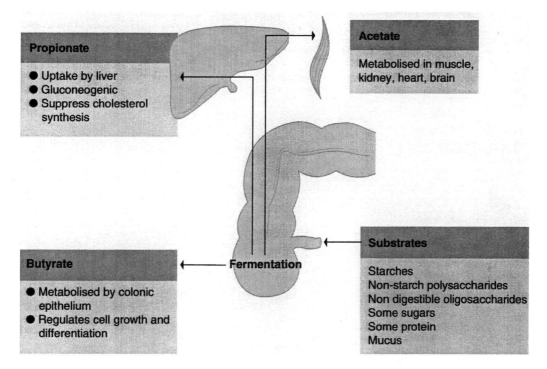

**Figure 2**   Metabolic fate of short chain fatty acids in humans.

UC and controls.[256] In this investigation ileal butyrate metabolism was greater than colonic. A number of methodological differences exist between this latter and the earlier work. Roediger and his colleagues[257] have also shown reduced butyrate oxidation *in vivo*, using a dialysis bag technique and bicarbonate production as a marker, and have gone on to show that in rats, inhibiting β-oxidation of fatty acids in colonocytes with 2-bromooctanoate produces lesions resembling those of UC.[258]

## 1.  Sulfur Metabolism

If butyrate oxidation is diminished in UC is this secondary to other cellular events? What is the intracellular mechanism of cell damage? A number of lines of evidence point to bacterial sulfur metabolism being involved.

In experimental animals, UC can be induced by feeding polysaccharides such as carageenan, sulfated dextran, and amylopectin. All these materials contain sulfate[259-262] and will be fermented in the colon. Intestinal bacteria are necessary for this model to succeed since it does not occur in germ-free animals,[263] nor those treated with antibiotics.[264]

Which particular bacteria may be involved? Gibson and colleagues[265-267] have shown that sulfate-reducing bacteria (SRB) occur in the human gut and are carried principally by people who do not excrete $CH_4$ in breath.[268,269] In *in vitro* studies, SRB outcompete methanogens for $H_2$.[266] SRB require sulfate as an electron acceptor and produce $SH^-$. Feeding sulfate to healthy methanogenic individuals causes inhibition of methane excretion and the growth of SRB in those subjects carrying low numbers of SRB initially.[270] *In vitro*, SRB growth is modulated by sulfate availability[269,271] even in the form of mucus.[271,272] In UC, nearly all subjects (about 95%) carry SRB[272,273] the majority of which (92%) belong to the genus *Desulfovibrio*.[272] Figure 3 shows that these SRB are adapted to live at high dilution rates such as might be found during acute episodes of diarrhea in UC. Furthermore, rates of $H_2S$ production are higher from UC patients than controls [controls $30 \pm 12$ (SD) mmol/g feces/h vs. UC $72 \pm 12$].[272]

So does $SH^-$, or do other reduced sulfur compounds produce the observed metabolic defect in UC cells? Roediger has shown that fatty acid oxidation can be inhibited by a range of S-containing compounds, including mercaptoacetate and sulfite, which specifically inhibit butyrate oxidation.[274] Mercaptoacetate is produced during fermentation in the colon.[275] In studies using rat colonocytes, sulfide

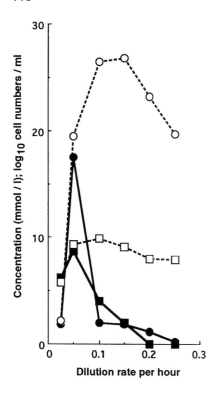

**Figure 3** Influence of dilution rate on viable counts and $H_2S$ production by *Desulfovibrio desulfuricans* grown in continuous culture. Dsv 18 isolated from colitic feces. Cell counts (□), $H_2S$ production (○). Dsv 25 isolated from healthy feces. Cell counts (■), $H_2S$ production (●). Data from *Gibson, G. R., Cummings, J. H., and Macfarlane, G. T., FEMS Microbiol. Ecol., 86, 103, 1991.*

impaired β-oxidation but this could be reversed by methionine. Methionine increases S-adenosyl-L-methionine levels in enterocytes and its beneficial effect, Roediger suggests, could be due to methylation of sulfide to methanethiol which is much less toxic. He postulates that sulfides form persulfides with butyryl-CoA which in turn inhibit short chain acyl dehydrogenase and β-oxidation. In human colonocytes, sulfide inhibits β-oxidation more significantly in the recto-sigmoid area than the ascending colon, which closely mirrors the metabolic and pathological abnormalities seen in UC.[276,277] Other disorders of the colonocyte involving S-metabolism have been shown in UC, including impaired sulfation of phenols[278] and reduced sulfation of mucus.[279]

## 2. Treatment

The suggestion that diminished butyrate oxidation in UC might be important in the inflammatory process, coupled to observations of low SCFA in the stool of UC patients with severe disease,[280-282] has prompted a number of groups to try using SCFA to treat UC. The first such report[283] was of an open trial using a solution of 80 mmol/l acetate, 30 mmol/l propionate, and 40 mmol/l butyrate in twice daily enemas in patients with distal disease. Of the 10 patients who completed the 6-week study, 9 were improved. Breuer et al. are now conducting a multicenter randomized placebo controlled study in the U.S. Midwest using the same SCFA mixture vs. saline. Preliminary results show 53% of UC patients improved compared to 26% of those on placebo.[284]

The effect of enemas containing butyrate alone (100 mmol/l) has been reported by Scheppach et al.[285] Ten patients with left-sided UC who were unresponsive to, or intolerant of standard treatment, were given either butyrate or sodium chloride enemas twice daily for two weeks, each in a randomized single blind cross-over study. After butyrate administration, stool frequency, rectal inflammation, and upper crypt labeling index all fell significantly. Other reports of similar studies in progress also look equally promising.[286,287]

Thus, it appears that SCFA will prove to be an effective treatment for many UC patients. Where does this leave the sulfur story? If S-compounds inhibit β-oxidation in the way that Roediger suggests, then high concentrations of butyrate might be able to overcome the block, simply by mass action. A direct attack on SRB and attempts to reduce $SH^-$ production in the colon might also be effective unless there is an inherited disorder of S-metabolism in the colonocyte of UC patients. This might explain why not everyone who carries SRB gets UC. Nevertheless, reduction of $SH^-$ in the colonic lumin should offer an effective therapeutic approach.

## B. DIVERSION COLITIS

In 1981, Glotzer et al.[288] described a condition in which mucosal inflammation occurred in segments of the large intestine from which the fecal stream had been diverted by proximal colostomy. This "diversion colitis" resembled UC in many ways but the inflammation is cured by reanastomosing the bowel and thus allowing the fecal stream back into the defunctioned colon.[288,289] Various causes have been postulated, including changes in the bacterial flora,[290,291] but the most likely cause is loss of an essential trophic or nutritional factor from the fecal stream, which is probably SCFA.[291] Diversion colitis parallels a number of other colitides in which SCFA production is reduced, such as occurs in starvation and famine, or utilization impairment as in UC.[292]

Harig et al.[293] tested these theories in four patients with diversion colitis. They showed by breath $H_2$ testing that no fermentative flora were present in the excluded segment, and neither were there any recognized pathogenic bacteria. Total SCFA concentration in material aspirated from the excluded segments was 0, 4.0, and 0.8 mmol/l for acetic, propionic, and butyric acids, respectively. They then treated the patients with SCFA enemas comprising 60 mmol/l acetate, 30 mmol/l propionate, and 40 mmol/l butyrate, twice daily for 2 to 6 weeks. All subjects showed clinical improvement and disappearance of histological changes of inflammation. Control periods using either no enemas or saline did not lead to any improvement.

A double-blind study was then undertaken by Guillemot et al.[294] in 13 patients who were given the same enema mixture as in Harig's study, but for two weeks only. No improvement in either endoscopic appearances or histology of the mucosa was observed. It is therefore probable that treatment for longer than 2 weeks is needed. Further trials are awaited, but at this stage the evidence suggests that in these patients absence of SCFA from the bowel lumin is damaging.

## C. POUCHITIS

Pouchitis is another inflammatory condition of the intestine which occurs in the distal ileal loops that are used to fashion an artificial rectal reservoir in patients who have had their large bowel and rectum removed for UC. It is very common after colectomy and is associated with reduced SCFA levels in luminal contents[295,296] [56 ± 13 (SEM) mmol/l pouchitis versus 139 ± 8 mmol/l healthy pouch in 6 patients of Clausen et al].[296,298] SCFA production from fecal homogenates incubated *in vitro* is also reduced, but L-lactate levels are increased threefold. No bacteriological differences have been observed, and the condition resolves rapidly with metronidazole treatment.

There are sporadic reports of the use of SCFA to treat pouchitis. One group used SCFA enemas in 2 patients for 4 weeks without benefit,[297] whereas another group used butyrate suppositories[298] also without great benefit. In these same patients, however, glutamine suppositories led to improvement in 60% of patients. Glutamine is the primary fuel for the ileal mucosa, rather than butyrate or other SCFA.

## REFERENCES

1. **Brieger, L.,** Ueber die fluchtiger Bestandthiele der menschlichen excremente, *Journal fur Praktische Chemie*, 17, 124, 1878.
2. **Grove, E. W., Olmsted, W. H., and Koenig, K.,** The effect of diet and catharsis on the lower volatile fatty acids in the stools of normal men, *J. Biol. Chem.*, 85, 127, 1929–30.
3. **Olmsted, W. H., Duden, C. W., Whitaker, W. M., and Parker, F. F.,** A method for the rapid distillation of the lower fatty acids from the stools, *J. Biol. Chem.*, 85, 115, 1929–30.
4. **Olmsted, W. H., Curtis, G., and Timm, O. K.,** Stool volatile fatty acids. IV. The influence of feeding bran pentosan and fiber to man, *J. Biol. Chem.*, 108, 645, 1935.
5. **Williams, R. D. and Olmsted, W. H.,** The effect of cellulose, hemicellulose and lignin on the weight of the stool: a contribution to the study of laxation in man, *J. Nutr.*, 11, 433, 1936.
6. **Cummings, J. H.,** Short chain fatty acids in the human colon, *Gut*, 22, 763, 1981.
7. **Wrong, O., Morrison, R. B. I., and Hurst, P. E.,** A method of obtaining faecal fluid by in vivo dialysis, *Lancet*, 1, 1208, 1961.
8. **Wrong, O., Metcalfe-Gibson, A., Morrison, R. B. I., Ng, S. T., and Howard, A. V.,** In vivo dialysis of faeces as a method of stool analysis. I. Technique and results in normal subjects, *Clin. Sci.*, 28, 357, 1965.
9. **Wilson, D. R., Ing, T. S., Metcalfe-Wilson, A., and Wrong, O. M.,** The chemical composition of faeces in uraemia, as revealed by in-vivo dialysis, *Clin. Sci.*, 35, 197, 1968.
10. **Wilson, D. R., Ing, T. S., Metcalfe-Gibson, A., and Wrong, O. M.,** In vivo dialysis of faeces as a method of stool analysis. III. The effect of intestinal antibiotics, *Clin. Sci.*, 34, 211, 1968.

11. **Rubinstein, R., Howard, A. V., and Wrong, O. M.,** In vivo dialysis of faeces as a method of stool analysis. IV. The organic anion component, *Clin. Sci.*, 37, 549, 1969.

12. **Roche, A. F.,** Ed., *Short Chain Fatty Acids: Metabolism and Clinical Importance*, Report of the Tenth Ross Conference on Medical Research, Ross Laboratories, Columbus, OH, 1991.

13. **Binder, H. J., Cummings, J. H., and Soergel, K. H.,** Eds., *Short Chain Fatty Acids*, Kluwer Academic Publishers, Lancaster, U.K., 1994.

14. **Cummings, J. H., Rombeau, J. L., and Sakata, T.,** Eds., *Physiological and Clinical Aspects of Short Chain Fatty Acids*, Cambridge University Press, Cambridge, U.K., 1995.

15. **Bugaut, M.,** Occurrence, absorption and metabolism of short-chain fatty acids in the digestive tract of mammals, *Comp. Biochem. Physiol.*, 86, 439, 1987.

16. **Bugaut, M. and Bentejac, M.,** Biological effects of short-chain fatty acids in nonruminant mammals, *Ann. Rev. Nutr.*, 13, 217, 1993.

17. **Cummings, J. H., Pomare, E. W., Branch, W. J., Naylor, C. P. E., and Macfarlane, G. T.,** Short chain fatty acids in human large intestine, portal, hepatic and venous blood, *Gut*, 28, 1221, 1987.

18. **Breves, G. and Stuck, K.,** Short chain fatty acids (SCFA) in the hindgut, in *Physiological and Clinical Aspects of Short Chain Fatty Acids*, Cummings, J. H., Rombeau, J. L., and Sakata, T., Eds., Cambridge University Press, Cambridge, U.K., 1995, chap. 5.

19. **Weaver, G. A., Krause, J. A., Miller, T. L., and Wolin, M. J.,** Short chain fatty acid distributions of enema samples from a sigmoidoscopy population: an association of high acetate and low butyrate ratios with adenomatous polyps and colon cancer, *Gut*, 29, 1539, 1988.

20. **Mitchell, B. L., Lawson, M. J., Davies, M., Kerr Grant, A., Roediger, W. E. W., Illman, R. J., and Topping, D. L.,** Volatile fatty acids in the human intestine: studies in surgical patients, *Nutr. Res.*, 5, 1089, 1985.

21. **Cummings, J. H. and Englyst, H. N.,** Fermentation in the human large intestine and the available substrates, *Am. J. Clin. Nutr.*, 45, 1243, 1987.

22. **Cummings, J. H., Gibson, G. R., and Macfarlane, G. T.,** Quantitative estimates of fermentation in the hind gut of man, in *Proc. Int. Symp. Comp. Aspects Physiol. Dig. Rumin. Hindgut Ferm.*, Skadhauge, E. and Norgaard, P., Eds., Acta Veterinaria Scandinavica, Copenhagen, Denmark, Suppl. 86, 76, 1989.

23. **McBurney, M. I.,** Passage of starch into the colon of humans: quantitation and implications, *Can. J. Physiol. Pharmacol.*, 69, 130, 1991.

24. **Englyst, H. N., Hay, S., and Macfarlane, G. T.,** Polysaccharide breakdown by mixed populations of human faecal bacteria, *FEMS Microbiol. Ecol.*, 95, 163, 1987.

25. **Weaver, G. A., Krause, J. A., Miller, T. L., and Wolin, M. J.,** Constancy of glucose and starch fermentations by two different human faecal microbial communities, *Gut*, 30, 19, 1989.

26. **McBurney, M. I. and Thompson, L. U.,** Effect of human faecal inoculum on in vitro fermentation variables, *Br. J. Nutr.*, 58, 233, 1987.

27. **Gibson, G. R., Macfarlane, S., and Cummings, J. H.,** The fermentability of polysaccharides by mixed human faecal bacteria in relation to their suitability as bulk-forming laxatives, *Lett. Appl. Microbiol.*, 11, 251, 1990.

28. **Weaver, G. A., Krause, J. A., Miller, T. L., and Wolin, M. J.,** Cornstarch fermentation by the colonic microbial community yields more butyrate than does cabbage fiber fermentation: cornstarch fermentation rates correlate negatively with methanogenesis, *Am. J. Clin. Nutr.*, 55, 70, 1992.

29. **Macfarlane, G. T. and Englyst, H. N.,** Starch utilization by the human large intestinal microflora, *J. Appl. Bacteriol.*, 60, 195, 1986.

30. **Wang, X. and Gibson, G. R.,** Effects of the *in vitro* fermentation of oligofructose and inulin by bacteria growing in the human large intestine, *J. Appl. Bacteriol.*, 70, 373, 1993.

31. **McBurney, M. I. and Thompson, L. U.,** In vitro fermentabilities of purified fiber supplements, *J. Food Sci.*, 54, 347, 1989.

32. **Adiotomre, J., Eastwood, M. A., Edwards, C. A., and Brydon, G. W.,** Dietary fiber in vitro methods that anticipate nutrition and metabolic activity in humans, *Am. J. Clin. Nutr.*, 52, 128, 1990.

33. **Titgemeyer, E. C., Bourquin, L. D., Fahey, G. C., Jr., and Garleb, K. A.,** Fermentability of various fiber sources by human fecal bacteria in vitro, *Am. J. Clin. Nutr.*, 53, 1418, 1991.

34. **Vince, A. J., McNeil, N. I., Wager, J. D., and Wrong, O. M.,** The effect of lactulose, pectin, arabinogalactan and cellulose on the production of organic acids and metabolism of ammonia by intestinal bacteria in a faecal incubation system, *Br. J. Nutr.*, 63, 17, 1990.

35. **Etterlin, C., McKeown, A., Bingham, S. A., Elia, M., Macfarlane, G. T., and Cummings, J. H.,** D-Lactate and acetate as markers of fermentation in man, *Gastroenterology*, 102, A551, 1992.

36. **Torres-Pinedo, R., Lavastida, M., Rivera, C. L., Rodriguez, H., and Ortiz, A.,** Studies on infant diarrhea. I. A comparison of the effects of milk feeding and intravenous therapy upon the composition and volume of the stool and urine, *J. Clin. Invest.*, 45, 469, 1966.

37. **Weijers, H. A. and Van de Kamer, J. H.,** Aetiology and diagnosis of fermentative diarrhoeas, *Acta Paediatr. Scand.*, 52, 329, 1963.

38. **Weijers, H. A., Van de Kamer, J. H., Dicke, W. K., and Ijsseling, J.,** Diarrhoea caused by deficiency of sugar splitting enzymes, *Acta Paediatr. Scand.*, 50, 55, 1961.

39. **Heimdahl, A. and Nord, C. E.,** Effect of erythromycin and clindamycin on the indigenous human anaerobic flora and new colonization of the gastrointestinal tract, *Eur. J. Clin. Microbiol.*, 1, 38, 1982.

40. **Hoverstad, T., Carlstedt-Duke, B., Lingaas, E., Midtvedt, T., Norin, K. E., Saxerholt, H., and Steinbakk, M.,** Influence of ampicillin, clindamycin, and metronidazole on faecal excretion of short-chain fatty acids in healthy subjects, *Scand. J. Gastroenterol.*, 21, 621, 1986.

41. **Hoverstad, T., Carlstedt-Duke, B., Lingaas, E., Norin, E., Saxerholt, H., Steinbakk, M., and Midtvedt, T.,** Influence of oral intake of seven different antibiotics on faecal short-chain fatty acid excretion in healthy subjects, *Scand. J. Gastroenterol.*, 21, 997, 1986.

42. **Edwards, C. A., Duerden, B. I., and Read, N. W.,** Effect of clindamycin on the ability of a continuous culture of colonic bacteria to ferment carbohydrate, *Gut*, 27, 411, 1986.

43. **Clausen, M. R., Bonnen, H., Tvede, M., and Mortensen, P. B.,** Colonic fermentation to short-chain fatty acids is decreased in antibiotic-associated diarrhea, *Gastroenterology*, 101, 1497, 1991.

44. **Saunders, D. R. and Wiggins, H. S.,** Conservation of mannitol, lactulose, and raffinose by the human colon, *Am. J. Physiol.*, 241, G397, 1981.

45. **Cummings, J. H., James, W. P. T., and Wiggins, H. S.,** Role of the colon in ileal-resection diarrhoea, *Lancet*, 1, 344, 1973.

46. **Grove, E. W., Olmsted, W. H., and Koenig, K.,** The effect of diet and catharsis on the lower volatile fatty acids in the stools of normal men, *J. Biol. Chem.*, 85, 127, 1929.

47. **Cummings, J. H., Hill, M. J., Jenkins, D. J. A., Pearson, J. R., and Wiggins, H. S.,** Changes in fecal composition and colonic function due to cereal fibre, *Am. J. Clin. Nutr.*, 29, 1468, 1976.

48. **Williams, R. D. and Olmsted, W. H.,** The manner in which food controls the bulk of faeces, *Ann. Intern. Med.*, 10, 717, 1936.

49. **Spiller, G. A., Chernoff, M. C., Hill, R. A., Gates, J. E., Nassar, J. J., and Shipley, E. A.,** Effect of purified cellulose, pectin, and a low-residue diet on faecal volatile fatty acids, transit time and fecal weight in humans, *Am. J. Clin. Nutr.*, 33, 754, 1980.

50. **Fleming, S. E. and Rodriguez, M. A.,** Influence of dietary fibre in faecal excretion of volatile fatty acids by human adults, *J. Nutr.*, 113, 1613, 1983.

51. **Fleming, S. E.,** Influence of dietary fiber on the production, absorption, or excretion of short chain fatty acids in humans, in *CRC Handbook of Dietary Fiber in Human Nutrition*, 2nd ed., Spiller, G. A., Ed., CRC Press, Boca Raton, FL, 1992, chap. 6.

52. **Cummings, J. H., Southgate, D. A. T., Branch, W., Houston, H., Jenkins, D. J. A., and James, W. P. T.,** The colonic response to dietary fibre from carrot, cabbage, apple, bran and guar gum, *Lancet*, 1, 5, 1978.

53. **Scheppach, W., Fabian, C., Sachs, M., and Kasper, H.,** The effect of starch malabsorption on fecal short chain fatty acid excretion in man, *Scand. J. Gastroenterol.*, 23, 755, 1988.

54. **Van Munster, I. P., Nagengast, F. M., and Tangerman, A.,** The effect of resistant starch on fecal bile acids, cytotoxicity and colonic mucosal proliferation, *Gastroenterology*, 104, A460, 1993.

55. **Cassidy, A., Bingham, S. A., and Cummings, J. H.,** Starch intake and colorectal cancer risk: an international comparison, *Br. J. Cancer*, 69, 937, 1994.

56. **Macfarlane, G. T. and Gibson, G. R.,** Microbiological aspects of short chain fatty acid production in the large bowel, in *Physiological and Clinical Aspects of Short Chain Fatty Acids*, Cummings, J. H., Rombeau, J. L., and Sakata, T., Eds., Cambridge, University Press, Cambridge, U.K., 1995, chap. 6.

57. **Isaacson, H. R., Hinds, F. C., Bryant, M. P., and Owens, F. N.,** Efficiency of energy utilisation by mixed rumen bacteria in continuous culture, *J. Dairy Sci.*, 58, 1645, 1975.

58. **Owens, F. N. and Isaacson, H. R.,** Ruminal microbial yields: factors influencing synthesis and bypass, *Fed. Proc.*, 36, 198, 1977.

59. **Stephen, A. M., Wiggins, H. S., and Cummings, J. H.,** Effect of changing transit time on colonic microbial metabolism in man, *Gut*, 28, 601, 1987.

60. **Allison, C., McFarlan, C., and Macfarlane, G. T.,** Studies on mixed populations of human intestinal bacteria grown in single-stage and multistage continuous culture systems, *Appl. Environ. Microbiol.*, 55, 672, 1989.

61. **Macfarlane, G. T., Cummings, J. H., Macfarlane, S., and Gibson, G. R.,** Influence of retention time on degradation of pancreatic enzymes by human colonic bacteria grown in a 3-stage continuous culture system, *J. Appl. Bacteriol.*, 67, 521, 1989.

62. **Burkitt, D. P.,** Epidemiology of cancer of the colon and rectum, *Cancer*, 28, 3, 1971.

63. **Burkitt, D. P., Walker, A. R. P., and Painter, N. S.,** Effect of dietary fibre on stools and transit times, and its role in the causation of disease, *Lancet*, 2, 1408, 1972.

64. **Davignon, J., Simmonds, W. J., and Ahrens, E. H.,** Usefulness of chromic oxide as an internal standard for balance studies in formula-fed patients and for assessment of colonic function, *J. Clin. Invest.*, 47, 127, 1968.

65. **Cummings, J. H.,** Diet and transit through the gut, *J. Plant Foods*, 3, 83, 1978.

66. **Cummings, J. H., Hill, M. J., Bone, E. S., Branch, W. J., and Jenkins, D. J. A.,** The effect of meat protein and dietary fiber on colonic function and metabolism. Part II. Bacterial metabolites in feces and urine, *Am. J. Clin. Nutr.*, 32, 2094, 1979.

67. **Macy, I. G.,** *Nutrition and Chemical Growth in Childhood*, Vol. 1, *Evaluation*, C. C. Thomas, Baltimore, 122, 1942.

68. **Bergman, E. N.,** Production and utilization of metabolites by the alimentary tract as measured in portal and hepatic blood, in *Digestion and Metabolism in the Ruminant*, McDonald, I. M. and Warner, A. C. I., Eds., University of New England, Sydney, Australia, 1975, 292, 305.

69. **Bergman, E. N.,** Energy contributions of volatile fatty acids from the gastrointestinal tract in various species, *Physiol. Rev.*, 70, 567, 1990.

70. **Dankert, J., Zijlstra, J. B., and Wolthers, B. G.,** Volatile fatty acids in human peripheral and portal blood: quantitative determination by vacuum distillation and gas chromatography, *Clin. Chim. Acta*, 110, 301, 1981.

71. **Peters, S. G., Pomare, E. W., and Fisher, C. A.,** Portal and peripheral blood short chain fatty acid concentrations after caecal lactulose instillation at surgery, *Gut*, 33, 1249, 1992.

72. **Skutches, C. L., Holroyde, C. P., Myers, R. N., Paul, P., and Reichard, G. A.,** Plasma acetate turnover and oxidation, *J. Clin. Invest.*, 64, 708, 1979.

73. **Scheppach, W., Pomare, E. W., Elia, M., and Cummings, J. H.,** The contribution of the large intestine to blood acetate in man, *Clin. Sci.*, 80, 177, 1991.

74. **Schenk, W. G., McDonald, J. C., McDonald, K., and Draponas, T.,** Direct measurement of hepatic blood flow in surgical patients, *Ann. Surg.*, 156, 463, 1962.

75. **Norryd, C., Dencker, H., Lunderquist, A., Olin, T., and Tylen, U.,** Superior mesenteric blood flow during digestion in man, *Acta Chir. Scand.*, 141, 197, 1975.

76. **Qamar, M. I. and Read, A. E.,** Effects of ingestion of carbohydrate, fat, protein and water on the mesenteric blood flow in man, *Scand. J. Gastroenterol.*, 23, 26, 1988.

77. **Pomare, E. W., Branch, W. J., and Cummings, J. H.,** Carbohydrate fermentation in the human colon and its relation to acetate concentration in venous blood, *J. Clin. Invest.*, 75, 1448, 1985.

78. **Bridges, S. R., Anderson, J. W., Deakins, D. A., Dillon, D. W., and Wood, C. L.,** Oat bran increases serum acetate of hypercholesterolemic men, *Am. J. Clin. Nutr.*, 56, 4555, 1992.

79. **Neilsen, L. G., Owen-Ash, K., and Thor, E.,** Gas chromatographic method for plasma acetate analysis in acetate intolerance studies, *Clin. Chem.*, 24, 348, 1978.

80. **Akanji, O. A.,** *Measurement of Plasma Acetate Concentrations in Humans with Reference to Diabetes, Dietary Composition and Bowel Function*, D. Phil. thesis, University of Oxford, 1988.

81. **Livesey, G. and Elia, M.,** Short chain fatty acids as an energy source in the colon: metabolism and clinical implications, in *Physiological and Clinical Aspects of Short Chain Fatty Acids*, Cummings, J. H., Rombeau, J. L., and Sakata, T., Eds., Cambridge University Press, Cambridge, U.K., 1995.

82. **Wolin, M. J. and Miller, T. L.,** Carbohydrate fermentation, in *Human Intestinal Microflora in Health and Disease*, Hentges, D. J., Ed., Academic Press, New York, 1983, 147.

83. **Wolin, M. J.,** A theoretical rumen fermentation balance, *J. Dairy Sci.*, 43, 1452, 1960.

84. **Wolin, M. J.,** Fermentation in the rumen and human large intestine, *Science*, 213, 1463, 1981.

85. **Miller, T. L. and Wolin, M. J.,** Fermentations by saccharolytic intestinal bacteria, *Am. J. Clin. Nutr.*, 32, 164, 1979.

86. **Mathers, J. C. and Annison, E. F.,** Stoichiometry of polysaccharide fermentation in the large intestine, in *Dietary Fibre and Beyond — Australian Perspectives*, Samman, S. and Annison, G., Eds., Nutrition Society of Australia, Occasional Publications, Adelaide, 1993, 123.

87. **Gibson, G. R., Macfarlane, G. T., and Cummings, J. H.,** Leading article: sulphate-reducing bacteria and hydrogen metabolism in the human large intestine, *Gut*, 34, 437, 1993.

88. **Macfarlane, G. T., Gibson, G. R., Beatty, E. R., and Cummings, J. H.,** Estimation of short-chain fatty acid production from protein by human intestinal bacteria based on branched chain fatty acid measurements, *FEMS Microbiol. Ecol.*, 101, 81, 1992.

89. **Walter, J. H., Leonard, J. V., Thompson, G. N., Bartlett, K., and Halliday, D.,** Contribution of amino acid catabolism to propionate production in methylmalonic acidaemia, *Lancet*, 1, 1298, 1989.

90. **Breves, G., Schulze, E., Sallmann, H. P., and Gadeken, D.,** The application of $^{13}$C labelled short chain fatty acids to measure acetate and propionate production rates in the large intestines. Studies in a pig model, *Zeitschrift fur Gastroenterologie*, 31, 179, 1993.

91. **Freeman, K., Foy, T., Feste, A. S., Reeds, P. J., and Lifschitz, C. L.,** Contribution of colonic acetate to the circulating acetate pool in the infant pig, *Paediatr. Res.*, (in press).

92. **Kien, C. L., Kepner, J., Grotjohn, K., Ault, K., and McLead, R. E.,** Stable isotope model for estimating colonic acetate production in premature infants, *Gastroenterology*, 102, 1458, 1992.

93. **Seufert, C. D., Mewes, W., and Soeling, H. D.,** Effect of long-term starvation on acetate and ketone body metabolism in obese patients, *Eur. J. Clin. Invest.*, 14, 163, 1984.

94. **Rocchiccioli, F., Lepetit, N., and Bougneres, P. F.,** Capillary gas-liquid chromatographic/mass spectrometric measurement of plasma acetate content and (2-13C) acetate enrichment, *Biomed. Environ. Mass Spectrom.*, 18, 816, 1989.

95. **McNeil, N. I., Cummings, J. H., and James, W. P. T.,** Short chain fatty acid absorption by the human large intestine, *Gut,* 19, 819, 1978.

96. **Ruppin, H., Bar-Meir, S., Soergel, K. H., Wood, C. M., and Schmitt, M. G.,** Absorption of short chain fatty acids by the colon, *Gastroenterology,* 78, 1500, 1980.

97. **Dawson, A. M., Holdsworth, C. D., and Webb, J.,** Absorption of short chain fatty acids in man, *Proc. Soc. Exp. Biol. Med.,* 117, 97, 1964.

98. **McNeil, N. I., Cummings, J. H., and James, W. P. T.,** Rectal absorption of short chain fatty acids in the absence of chloride, *Gut,* 20, 400, 1979.

99. **Gabel, G. and Martens, H.,** Transport of Na$^+$ and Cl$^-$ across the forestomach epithelium: mechanisms and interactions with short chain fatty acids, in *Physiological Aspects of Digestion and Metabolism in Ruminants: Proceedings of the Seventh International Symposium on Ruminant Physiology,* Tsuda, T., Sasaki, Y., and Kawashima, R., Eds., Academic Press, New York, 1991, 129.

100. **Martens, H., Leonhard, S., and Gabel, G.,** Minerals and digestion: exchanges in the digestive tract, in *Rumen Microbial Metabolism and Ruminant Digestion,* Jouany, J. P., Ed., INRA Editions, Paris, 1991, 199.

101. **Soergel, K. H., Harig, J. M., Loo, F. D., Ramaswamy, K., and Wood, C. M.,** Colonic fermentation and absorption of SCFA in man, in *Proc. Int. Symp. Comp. Aspects Physiol. Dig. Rumin. Hindgut Ferm.,* Skadhauge, E. and Norgaard, P., Eds., Acta Veterinaria Scandinavica, Copenhagen, Denmark, 107, 1989.

102. **Ronnau, K., Guth, D., and Engelhardt, W. v.,** Absorption of dissociated and undissociated short-chain fatty acids across the colonic epithelium of guinea pig, *Q. J. Exp. Physiol.,* 74, 511, 1989.

103. **Luciano, L., Reale, E., Rechkemmer, G., and Engelhardt, W. v.,** Structure of zonulae occludentes and the permeability of the epithelium to short chain fatty acids in the proximal and distal colon of guinea pig, *J. Membrane Biol.,* 82, 145, 1984.

104. **von Engelhardt, W. and Rechkemmer, G.,** Segmental differences of short-chain fatty acid transport across guinea pig large intestine, *Exp. Physiol.,* 77, 491, 1992.

105. **von Engelhardt, W., Burmester, M., Hansen, K., Becker, G., and Rechkemmer, G.,** Effects of amiloride and ouabain on short-chain fatty acid transport in guinea pig large intestine, *J. Physiol.,* 460, 455, 1993.

106. **Rechkemmer, G. and von Engelhardt, W.,** Absorption and secretion of electrolytes and short-chain fatty acids in the guinea pig large intestine, in *Ion Transport in Vertebrate Colon,* Clauss, W., Ed., Adv. Comp. Environ. Physiol., 16, Springer Verlag, New York, 1993, 139.

107. **von Engelhardt, W.,** Absorption of short-chain fatty acids from the large intestine, in *Physiological and Clinical Aspects of Short Chain Fatty Acids,* Cummings, J. H., Rombeau, J. L., and Sakata, T., Eds., Cambridge University Press, Cambridge, U.K., 1995.

108. **Argenzio, R. A., Southworth, M., Lowe, J. E., and Stevens, C. E.,** Interrelationship of Na, HCO$_3$ and volatile fatty acid transport by equine large intestine, *Am. J. Physiol.,* 233, E469, 1977.

109. **Argenzio, R. A. and Whipp, S. C.,** Inter-relationship of sodium, chloride, bicarbonate and acetate transport by the colon of the pig, *J. Physiol.,* 295, 365, 1979.

110. **Rubsamen, K. and Engelhardt, W. V.,** Absorption of Na$^+$, H$^+$ ions and short chain fatty acids from the sheep colon, *Pflüegers Arch.,* 391, 141, 1981.

111. **Umesaki, Y., Yajima, T., Yokokura, T., and Mutai, M.,** Effect of organic acid absorption on bicarbonate transport in rat colon, *Pflüegers Arch.,* 379, 43, 1979.

112. **Mascolo, N., Rajendran, V. M., and Binder, H. J.,** Mechanism of short-chain fatty acid uptake by apical membrane vesicles of rat distal colon, *Gastroenterology,* 101, 331, 1991.

113. **Harig, J. M., Knaup, S. M., Shoshara, J., Dudeja, P. K., Ramaswamy, K., and Brasitus, T. A.,** Transport of N-butyrate into human colonic luminal membrane vesicles, *Gastroenterology,* 98, A543, 1990.

114. **Stevens, C. E. and Stettler, B. K.,** Factors affecting the transport of volatile fatty acids across rumen epithelium, *Am. J. Physiol.,* 210, 365, 1966.

115. **Lonnerholm, G., Selking, O., and Wistrand, P. J.,** Amount and distribution of carbonic anhydrases CA I and CA II in the gastrointestinal tract, *Gastroenterology,* 88, 1151, 1988.

116. **Charney, A. N., Wagner, J. D., Birnbaum, G. J., and Johnstone, J. N.,** Functional role of carbonic anhydrase in intestinal electrolyte transport, *Am. J. Physiol.,* 251, G682, 1986.

117. **Engelhardt, W. v., Gros, G., Burmester, M., Hansen, K., Becker, G., and Rechkemmer, G.,** Functional role of bicarbonate in propionate transport across guinea pig large intestine, *J. Physiol.,* 1994 (in press).

118. **Parsons, D. S. and Paterson, C. R.,** Fluid and solute transport across rat colonic mucosa, *Q. J. Exp. Physiol.,* 50, 220, 1965.

119. **Argenzio, R. A., Miller, N., and Engelhardt, W. v.,** Effect of volatile fatty acids on water and ion absorption from the goat colon, *Am. J. Physiol.,* 229, 997, 1975.

120. **Crump, M. H., Argenzio, R. A., and Whipp, S. C.,** Effects of acetate on absorption of solute and water from the pig colon, *Am. J. Vet. Res.,* 41, 1565, 1980.

121. **Roediger, W. E. W. and Moore, A.,** Effect of short chain fatty acid on sodium absorption in isolated human colon perfused through the vascular bed, *Dig. Dis. Sci.,* 26, 100, 1981.

122. **Umesaki, Y., Yajima, T., Tohyama, K., and Mutai, M.,** Characterization of acetate uptake by the colonic epithelial cells of the rat, *Pflüegers Arch.*, 388, 205, 1980.

123. **Roediger, W. E. W., Deakin, E. J., Radcliffe, B. C., and Nance, S.,** Anion control of sodium absorption in the colon, *J. Exp. Physiol.*, 71, 195, 1986.

124. **Roediger, W. E. W.,** Short chain fatty acids as metabolic regulators of ion absorption in the colon, *Acta Vet. Scand.*, 86, 116, 1989.

125. **Roediger, W. E. W.,** Role of anaerobic bacteria in the metabolic welfare of the colonic mucosa in man, *Gut*, 21, 793, 1980.

126. **Roediger, W. E. W.,** Utilization of nutrients by isolated epithelial cells of the rat colon, *Gastroenterology*, 83, 424, 1982.

127. **Marty, J. and Vernay, M.,** Absorption and metabolism of the volatile fatty acids in the hind-gut of the rabbit, *Br. J. Nutr.*, 51, 265, 1984.

128. **Vernay, M.,** Effects of plasma aldosterone on butyrate absorption and metabolism in the rabbit proximal colon, *Comp. Biochem. Physiol.*, 86A, 657, 1987.

129. **Ardawi, M. S. M. and Newsholme, E. A.,** Fuel utilization in colonocytes of the rat, *Biochem. J.*, 231, 713, 1985.

130. **Fleming, S. E., Fitch, M. D., DeVries, S., Liu, M. L., and Kight, C.,** Nutrient utilization by cells isolated from rat jejunum, cecum and colon, *J. Nutr.*, 121, 869, 1991.

131. **Roediger, W. E. W., Kapaniris, O., and Millard, S.,** Lipogenesis from n-butyrate in colonocytes. Action of reducing agent and 5-aminosalicylic acid with relevance to ulcerative colitis, *Mol. Cell. Biochem.*, 118, 113, 1992.

132. **Remesy, C. and Demigne, C.,** Partition and absorption of volatile fatty acids in the alimentary canal of the rat, *Ann. Rech. Vet.*, 7, 39, 1976.

133. **Henning, S. J. and Hird, F. J. R.,** Concentration and metabolism of volatile fatty acids in the fermentative organs of 2 species of kangaroo and the guinea-pig, *Br. J. Nutr.*, 24, 146, 1970.

134. **Henning, S. J. and Hird, F. J. R.,** Ketogenesis from acetate and butyrate by the caecum and the colon of rabbits, *Biochem. J.*, 130, 785, 1972.

135. **Roediger, W. E. W.,** Role of anaerobic bacteria in the metabolic welfare of the colonic mucosa in man, *Gut*, 21, 793, 1980.

136. **Roediger, W. E. W.,** Cellular metabolism of short-chain fatty acids in colonic epithelial cells, in *Short Chain Fatty Acids: Metabolism and Clinical Importance*, Report of the Tenth Ross Conference on Medical Research, Roche, A. F., Ed., Ross Laboratories, Columbus, OH, 1991, 67.

137. **Brown, R. C., Kelleher, J., and Losowsky, M. S.,** The effect of pectin on the structure and function of the rat small intestine, *Br. J. Nutr.*, 42, 357, 1979.

138. **Jacobs, L. R.,** Effects of dietary fiber on mucosal growth and cell proliferation in the small intestine of the rat: a comparison of oat bran, pectin, and guar with total fiber deprivation, *Am. J. Clin. Nutr.*, 37, 954, 1983.

139. **Sigleo, S., Jackson, M. J., and Vahouny, G. V.,** Effects of dietary fiber constituents on intestinal morphology and nutrient transport, *Am. J. Physiol.*, 246, G34, 1984.

140. **Johnson, I. T., Gee, J. M., and Mahoney, R. R.,** Effect of dietary supplements of guar gum and cellulose on intestinal cell proliferation, enzyme levels and sugar transport in the rat, *Br. J. Nutr.*, 52, 477, 1984.

141. **Sakata, T. and Yajima, T.,** Influence of short chain fatty acids on the epithelial cell division of digestive tract, *Q. J. Exp. Physiol.*, 69, 639, 1984.

142. **Cassidy, M. M., Lightfoot, F. G., Grau, L. E., Story, J. A., Kritchevsky, D., and Vahouny, G. V.,** Effect of chronic intake of dietary fibers on the ultrastructural topography of rat jejunum and colon: a scanning electron microscopy study, *Am. J. Clin. Nutr.*, 34, 218, 1981.

143. **Jacobs, L. R. and White, F. A.,** Modulation of mucosal cell proliferation in the intestine of rats fed a wheat bran diet, *Am. J. Clin. Nutr.*, 37, 945, 1983.

144. **Jacobs, L. R. and Lupton, J. R.,** Effect of dietary fibers on rat large bowel mucosal growth and cell proliferation, *Am. J. Physiol.*, 246, G378, 1984.

145. **Jacobs, L. R. and Schneeman, B. O.,** Effects of dietary wheat bran on rat colonic structure and mucosal cell growth, *J. Nutr.*, 111, 798, 1981.

146. **Sakata, T. and von Engelhardt, W.,** Stimulatory effect of short chain fatty acids on the epithelial acell proliferation in rat large intestine, *Comp. Biochem. Physiol.*, 74A, 459, 1983.

147. **Goodlad, R. A., Lenton, W., Ghatei, M. A., Adrian, T. E., Bloom, S. R., and Wright, N. A.,** Effects of an elemental diet, inert bulk and different types of dietary fibre on the response of the intestinal epithelium to refeeding in the rat and relationship to plasma gastrin, enteroglucagon, and PYY concentrations, *Gut*, 28, 171, 1987.

148. **Goodlad, R. A., Ratcliffe, B., Fordham, J. P., and Wright, N. A.,** Does dietary fibre stimulate intestinal epithelial cell proliferation in germ free rats?, *Gut*, 30, 820, 1989.

149. **Sakata, T.,** Short chain fatty acids as the luminal trophic factor, *Can. J. Anim. Sci.*, 64, 189, 1984.

150. **Sakata, T.,** Effects of indigestible dietary bulk and short chain fatty acids on the tissue weight and epithelial cell proliferation rate of the digestive tract in rats, *J. Nutr. Sci. Vitaminol.*, 32, 355, 1986.

151. **Sakata, T.,** Stimulatory effect of short chain fatty acids on epithelial cell proliferation in the rat intestine: a possible explanation for trophic effects of fermentable fibre, gut microbes and luminal trophic factors, *Br. J. Nutr.*, 58, 95, 1987.

152. **Sakata, T.,** Depression of intestinal epithelial cell production rate by hindgut bypass in rats, *Scand. J. Gastroenterol.*, 23, 1200, 1988.

153. **Janne, P., Carpenter, Y., and Willems, G.,** Colonic mucosal atrophy induced by a liquid elemental diet in rats, *Am. J. Dig. Dis.*, 22, 808, 1977.

154. **Goodlad, R. A. and Wright, N. A.,** The effects of addition of cellulose or kaolin to an elemental diet on intestinal cell proliferation in the mouse, *Br. J. Nutr.*, 50, 91, 1983.

155. **Kripke, S. A., Fox, A. D., Berman, J. M., Settle, R. G., and Rombeau, J. L.,** Stimulation of intestinal mucosal growth with intracolonic infusion of short-chain fatty acids, *J. Parent. Ent. Nutr.*, 13, 109, 1989.

156. **Koruda, M. J., Rolandelli, R. H., Bliss, D. Z., Hastings, J., Rombeau, J. L., and Settle, R. G.,** Parenteral nutrition supplemented with short-chain fatty acids: effect on the small-bowel mucosa in normal rats, *Am. J. Clin. Nutr.*, 51, 685, 1990.

157. **Frankel, W. L., Zhang, W., Singh, A., Klurfeld, D., Don, T., Sakata, T., Modlin, I., and Rombeau, J. L.,** Mediation of the trophic effects of short-chain fatty acids on the rat jejunum and colon, *Gastroenterology*, 106, 375,1994.

158. **Jacobs, L. R. and Lupton, J. R.,** Relationship between colonic luminal pH, cell proliferation, and colon carcinogenesis in 1,2-dimethylhydrazine-treated rats fed high fiber diets, *Cancer Res.*, 46, 1727, 1986.

159. **Freeman, H. J.,** Effects of differing concentrations of sodium butyrate on 1,2-dimethylhydrazine-induced rat intestinal neoplasia, *Gastroenterology*, 91, 596, 1986.

160. **Watanabe, K., Reddy, B. S., Weisburger, J. H., and Kritchevsky, D.,** Effect of dietary alfalfa, pectin and wheat bran on azoxymethane-or methylnitrosourea-induced colon carcinogenesis in F344 rats, *J. Natl. Cancer Inst.*, 63, 141, 1979.

161. **Scheppach, W., Bartram, P., Richter, A., Richter, F., Liepold, H., Dusel, G., Hofstetter, G., Ruthlein, J., and Kasper, H.,** Effect of short-chain fatty acids on the human colonic mucosa in vitro, *J. Parent. Ent. Nutr.*, 16, 43, 1992.

162. **Scheppach, W. M.,** Short-chain fatty acids are a trophic factor for the human colonic mucosa in vitro, in *Short-Chain Fatty Acids: Metabolism and Clinical Importance*, Report of the Tenth Ross Conference on Medical Research, Roche, A. F., Ed., Ross Laboratories, Columbus, OH, 1991, 90.

163. **Lipkin, M., Blattner, W. A., Fraumeni, J. F., Jr., Lynch, H. T., Deschner, E. E., and Winawar, S.,** Tritiated thymidine labeling distribution in the identification of hereditary predisposition to colon cancer, *Cancer Res.*, 43, 1899, 1983.

164. **Prasad, K. N. and Sinha, P. K.,** Effect of sodium butyrate on mammalian cells in culture: a review, *In Vitro*, 12, 125, 1976.

165. **Prasad, K. N.,** Butyric acid: a small fatty acid with diverse biological functions, *Life Sci.*, 27, 1351, 1980.

166. **Kruh, J.,** Effects of sodium butyrate, a new pharmacological agent, on cells in culture, *Mol. Cell. Biochem.*, 42, 65, 1982.

167. **Kruh, J., Defer, N., and Tichonicky, L.,** Action moleculaire et cellulaire du butyrate, *C. R. Soc. Biol.*, 186, 12, 1992.

168. **Fregeau, C. J., Helgason, C. D., and Bleackley, R. C.,** Two cytotoxic cell proteinase genes are differentially sensitive to sodium butyrate, *Nucleic Acids Res.*, 20, 3113, 1992.

169. **Smith, P. J.,** n-Butyrate alters chromatin accessibility to DNA repair enzymes, *Carcinogenesis*, 7, 423, 1986.

170. **Glauber, J. G., Wandersee, N. J., Little, J. A., and Ginder, G. D.,** 5′ Flanking sequences mediate butyrate stimulation of embryonic globin gene expression in adult erythroid cells, *Mol. Cell. Biol.*, 11, 4690, 1991.

171. **Hague, A., Manning, A. M., Hanlon, K. A., Huschischa, L. I., Hart, D., and Paraskeva, C.,** Sodium butyrate induces apoptosis in human colonic tumour cell lines in a p53-independent pathway: implications for the possible role of dietary fibre in the prevention of large bowel cancer, *Int. J. Cancer*, 55, 498, 1993.

172. **Hagopian, H. K., Riggs, M. G., Swartz, L. A., and Ingram, V. M.,** Effect of n-butyrate on DNA synthesis in chick fibroblasts and HeLa cells, *Cell*, 12, 855, 1977.

173. **D'Anna, J. A., Tobey, R. A., and Gurley, L. R.,** Concentration-dependent effects of sodium butyrate in Chinese hamster cells: cell-cycle progression, inner histone acetylation, histone H1 dephosphorylation and induction of an H1-like protein, *Biochemistry*, 19, 2656, 1980.

174. **Wintersberger, E., Mudrak, I., and Wintersberger, U.,** Butyrate inhibits mouse fibroblasts at a control point in the G1 phase, *J. Cell. Biochem.*, 21, 239, 1983.

175. **Toscani, A., Soprano, D. R., and Soprano, K. J.,** Molecular analysis of sodium butyrate-induced growth arrest, *Oncogene Res.*, 3, 223, 1988.

176. **Van Wijk, R., Tichonicky, L., and Kruh, J.,** Effect of sodium butyrate on the hepatoma cell cycle: possible use for cell synchronization, *In Vitro*, 17, 859, 1981.

177. **Flatow, U., Rabson, A. B., Hand, P. H., Willingham, M. C., and Rabson, A. S.,** Characterization and tumorigenicity of a butyrate-adapted T24 bladder cancer cell line, *Cancer Invest.*, 7, 423, 1989.

178. **Gamet, L., Daviaud, D., Denis-Pouxviel, C., Remesy, C., and Murat, J.-C.,** Effects of short-chain fatty acids on growth and differentiation of the human colon-cancer cell line HT29, *Int. J. Cancer*, 52, 286, 1992.

179. **Dexter, D. L., Lev, R., McKendall, G. R., Mitchell, P., and Calabres, P.,** Sodium butyrate-induced alteration of growth properties and glycogen levels in cultured human colon carcinoma cells, *Histochem. J.*, 16, 137, 1984.

180. **Kim, Y. S., Tsao, D., Siddiqui, B., Whitehead, J. S., Arnstein, P., Bennett, J., and Hicks, J.,** Effects of sodium butyrate and dimethylsulfoxide on biochemical properties of human colon cancer cells, *Cancer*, 45, 1185, 1980.

181. **Whitehead, R. H., Young, G. P., and Bhathal, P. S.,** Effects of short chain fatty acids on a new human colon carcinoma cell line (LIM1215), *Gut*, 27, 1457, 1986.

182. **Tsao, D., Morita, A., Bella, A., Luu, P., and Kim, Y. S.,** Differential effects of sodium butyrate, dimethyl sulfoxide, and retinoic acid on membrane-associated antigen, enzymes and glycoproteins of human rectal adenocarcinoma cells, *Cancer Res.*, 42, 1052, 1982.

183. **Chung, Y. S., Song, I. S., Erickson, R. H., Sleisinger, M. H., and Kim, Y. S.,** Effect of growth and sodium butyrate on brush border membrane associated hydrolases in human colorectal cancer cell lines, *Cancer Res.*, 45, 2976, 1985.

184. **Gibson, P. R., Moeller, I., Kagelari, O., Folino, M., and Young, G. P.,** Contrasting effects of butyrate on the expression of phenotypic markers of differentiation in neoplastic and non-neoplastic colonic epithelial cells in vitro, *J. Gastroenterol. Hepatol.*, 7, 165, 1992.

185. **Bloom, E. J., Siddiqui, B., Hicks, J. W., and Kim, Y. S.,** Effect of sodium butyrate, a differentiating agent, on cell surface glycoconjugates of a human pancreatic cell line, *Pancreas*, 4, 59, 1989.

186. **Dyson, J. E. D., Daniel, J., and Surrey, C. R.,** The effect of sodium butyrate on the growth characteristics of human cervix tumour cells, *Br. J. Cancer*, 65, 803, 1992.

187. **Nordenberg, J., Wasserman, L., Beery, E., Aloni, D., Malik, H., Stenzel, K. H., and Novogrodsky, A.,** Growth inhibition of murine melanoma by butyric acid and dimethylsulfoxide, *Exp. Cell Res.*, 162, 77, 1986.

188. **Nordenberg, J., Wasserman, L., Peled, A., Malik, Z., Stenzel, K. H., and Novogrodsky, A.,** Biochemical and ultrastructural alterations accompany the anti-proliferative effect of butyrate on melanoma cells, *Br. J. Cancer*, 55, 493, 1987.

189. **Rama, B. N. and Prasad, K. N.,** Modification of the hyperthermic response on neuroblastoma cells by cAMP and sodium butyrate, *Cancer*, 58, 1448, 1986.

190. **Reese, D. H., Gratzner, H. G., Block, N. L., and Politano, V. A.,** Control of growth, morphology and alkaline phosphatase activity by butyrate and related short-chain fatty acids in the retinoid-responsive 9-1C rat prostatic adenocarcinoma cell, *Cancer Res.*, 45, 2308, 1985.

191. **Halgunset, J., Lamvik, T., and Esspevik, T.,** Butyrate effects on growth, morphology, and fibronectin production in PC-3 prostatic carcinoma cells, *The Prostate*, 12, 65, 1988.

192. **Guilbaud, N. F., Gas, N., Dupont, M. A., and Valette, A.,** Effects of differentiation-inducing agents on maturation of human MCF-7 breast cancer cells, *J. Cell. Physiol.*, 145, 162, 1990.

193. **Morita, A., Tsao, D., and Kim, Y. S.,** Effect of sodium butyrate on alkaline phosphatase in HRT-18, a human rectal cancer cell line, *Cancer Res.*, 42, 4540, 1982.

194. **Gum, J. R., Kam, W. K., Byrd, J. C., Hicks, J. W., Sleisenger, M. H., and Kim, Y. S.,** Effects of sodium butyrate on human colonic adenocarcinoma cells: induction of placental-like alkaline phosphatase, *J. Biol. Chem.*, 262, 1092, 1987.

195. **Kim, Y. S.,** Colonic cell differentiation and proliferation, in *Short Chain Fatty Acids*, Binder, H. J., Cummings, J. H., and Soergel, K. H., Eds., Proc. of 73rd Falk Symposium, Strasbourg, France, Kluwer Academic Publishers, Norwell, MA, 1994, 19.

196. **Toscani, A., Soprano, D. R., and Soprano, K. J.,** Sodium butyrate in combination with insulin or dexamethasone can terminally differentiate actively proliferating Swiss 3T3 cells into adipocytes, *J. Biol. Chem.*, 265, 5722, 1990.

197. **Leder, A. and Leder, P.,** Butyric acid, a potent inducer of erythroid differentiation in cultured erythroleukemia cells, *Cell*, 5, 319, 1975.

198. **Gladhaug, I. P., Refsnes, M., and Christoffersen, T.,** Regulation of hepatocyte epidermal growth factor receptors by n-butyrate and dimethyl sulfoxide: sensitivity to modulation by the tumor promoter TPA, *Anticancer Res.*, 9, 1587, 1989.

199. **Gladhaug, I. P. and Christoffersen, T.,** n-Butyrate and dexamethasone synergistically modulate the surface expression of epidermal growth factor receptors in cultured rat hepatocytes, *FEBS Lett.*, 243, 21, 1989.

200. **Gladhaug, I. P., Refsnes, M., Sand, T.-E., and Christoffersen, T.,** Effects of butyrate on epidermal growth factor receptor binding, morphology, and DNA synthesis in cultured rat hepatocytes, *Cancer Res.*, 48, 6560, 1988.

201. **Kooistra, T., Van Den Berg, J., Tons, A., Platenburg, G., Rijken, D. C., and Van den Berg, E.,** Butyrate stimulates tissue-type plasminogen-activator synthesis in cultured human endothelial cells, *Biochem. J.*, 247, 605, 1987.

202. **Cattini, P. A., Kardami, E., and Eberhardt, N. L.,** Effect of butyrate on thyroid hormone-mediated gene expression in rat pituitary tumour cells, *Mol. Cell. Endocrinol.*, 56, 263, 1988.

203. **Birren, B. W. and Herschman, H. R.,** Regulation of the rat metallothionein-I gene by sodium butyrate, *Nucleic Acids Res.*, 14, 853, 1986.

204. **Ormandy, C. J., de Fazio, A., Kelly, P. A., and Sutherland, R. L.,** Coordinate regulation of oestrogen and prolactin receptor expression by sodium butyrate in human breast cancer cells, *Biochem. Biophys. Res. Comm.*, 182, 740, 1992.

205. **de Fazio, A., Chiew, Y.-E., Donoghue, C., Lee, C. S. L., and Sutherland, R. L.,** Effect of sodium butyrate on estrogen receptor and epidermal growth factor receptor gene expression in human breast cancer cell lines, *J. Biol. Chem.*, 267, 18008, 1992.

206. **Souleimani, A. and Asselin, C.,** Regulation of c-fos expression by sodium butyrate in the human colon carcinoma cell line Caco-2, *Biochem. Biophys. Res. Comm.*, 193, 330, 1993.

207. **Wice, B. M., Trugnan, G., Pinto, M., Rousset, M., Chevalier, G., Dussaulx, E., Lacroix, B., and Zweibaum, A.,** The intracellular accumulation of UDP-N-acetyl hexosamines is concomitant with the inability of human colon cancer cells to differentiate, *J. Biol. Chem.*, 260, 139, 1985.

208. **Saini, K., Steele, G., and Thomas, P.,** Induction of carcinoembryonic-antigen-gene expression in human colorectal carcinoma by sodium butyrate, *Biochem. J.*, 272, 541, 1990.

209. **Gibson, P. R., Rosella, O., Rosella, G., and Young, G. P.,** Butyrate is a potent inhibitor of urokinase secretion by colonic epithelium, *In Vitro*, submitted.

210. **Gibson, P. R., Folino, M., McIntyre, A., Rosella, O., Finch, C., and Young, G. P.,** Dietary modulation of colonic mucosal urokinase activity in rats, *Gut*, submitted.

211. **Bates, S. E., Currier, S. J., Alvarez, M., and Fojo, A. T.,** Modulation of p-glycoprotein phosphorylation and drug transport by sodium butyrate, *Biochemistry*, 31, 6366, 1992.

212. **Higgins, P. J. and Ryan, M. P.,** Cytoarchitecture of ras oncogene-expressing tumor cells: butyrate modulation of substrate adhesion, cytoskeletal actin content and subcellular microfilament distribution, *Int. J. Biochem.*, 21, 1143, 1989.

213. **Malik, H., Nordenberg, J., Novogrodsky, A., Fuchs, A., and Malik, Z.,** Chemical inducers of differentiation, dimethylsulfoxide, butyric acid, and dimethylthiourea, induce selective ultrastructural patterns in B16 melanoma cells, *Biol. Cell*, 60, 33, 1987.

214. **Wilson, J. R. and Weiser, M. M.,** Colonic cancer cell (HT29) adhesion to laminin is altered by differentiation: adhesion may involve galactosyltransferase, *Exp. Cell Res.*, 201, 330, 1992.

215. **Kruh, J., Defer, N., and Tichonicky, L.,** Effects of butyrate on cell proliferation and gene expression, in *Physiological and Clinical Aspects of Short Chain Fatty Acids*, Cummings, J. H., Rombeau, J. L., and Sakata, T., Eds., Cambridge University Press, Cambridge, U.K., 1995, chap. 18.

216. **Boffa, L. C., Lupton, J. R., Mariani, M. R., Ceppi, M., Newmark, H. L., Scalmati, A., and Lipkin, M.,** Modulation of colonic epithelial cell proliferation, histone acetylation, and luminal short chain fatty acids by variation of dietary fiber (wheat bran) in rats, *Cancer Res.*, 52, 5906, 1992.

217. **Johnston, L. A., Tapscott, S. J., and Eisen, H.,** Sodium butyrate inhibits myogenesis by interfering with the transcriptional activation function of MyoD and Myogenin, *Mol. Cell. Biol.*, 12, 5123, 1992.

218. **Klehr, D., Schlake, T., Maass, K., and Bode, J.,** Scaffold-attached regions (SAR elements) mediate transcriptional effects due to butyrate, *Biochemistry*, 31, 3222, 1992.

219. **Young, G. P. and Gibson, P. R.,** Butyrate and the human cancer cell, in *Physiological and Clinical Aspects of Short Chain Fatty Acids*, Cummings, J. H., Rombeau, J. L., and Sakata, T., Eds., Cambridge University Press, Cambridge, U.K., 1995, chap. 21.

220. **Cummings, J. H., Stephen, A. M., and Branch, W. J.,** Implications of dietary fiber breakdown in the human colon, in *Banbury Report 7: Gastrointestinal Cancer: Endogenous Factors*, Bruce, R., Correa, P., Lipkin, M., Tannenbaum, S. R., and Wilkins, T. D., Eds., Cold Spring Harbor Laboratory, New York, 1981, 71.

221. **Skutches, C. L., Holroyde, C. P., Myers, R. N., Paul, P., and Reichard, G. A.,** Plasma acetate turnover and oxidation, *J. Clin. Invest.*, 64, 708, 1979.

222. **Scheppach, W., Pomare, E. W., Elia, M., and Cummings, J. H.,** The contribution of the large intestine to blood acetate in man, *Clin. Sci.*, 80, 177, 1991.

223. **Bridges, S. R., Anderson, J. W., Deakins, D. D., Dillon, D. W., and Wood, C. L.,** Oat bran increases serum acetate of hypercholesterolemic men, *Am. J. Clin. Nutr.*, 56, 455, 1992.

224. **Riddell-Mason, S., Geil, P. B., and Anderson, J. W.,** Dry bean consumption increases serum acetate in hypercholesterolemic men, *FASEB J.*, 7, A740, 1993.

225. **Akanji, A. O. and Hockaday, T. D. R.,** Acetate tolerance and the kinetics of acetate utilization in diabetic and nondiabetic subjects, *Am. J. Clin. Nutr.*, 51, 112, 1992.

226. **Lundquist, F., Sestoft, L., Damgaard, S. E., Clausen, J. P., and Trap-Jensen, J.,** Utilization of acetate in the human forearm during exercise after ethanol ingestion, *J. Clin. Invest.*, 52, 3231, 1973.

227. **Lindeneg, O., Mellemgaard, K., Fabricius, J., and Lundquist, F.,** Myocardial utilization of acetate, lactate and free fatty acids after ingestion of ethanol, *Clin. Sci.*, 27, 427, 1964.

228. **Williamson, J. R.,** Effects of insulin and starvation on the metabolism of acetate and pyruvate by the perfused rat heart, *Biochem. J.*, 93, 97, 1964.

229. **Juhlen-Dannfelt, A.,** Ethanol effects of substrate utilization by the human brain, *Scand. J. Clin. Lab. Invest.*, 37, 443, 1977.

230. **Buckley, B. M. and Williamson, D. H.,** Origins of blood acetate in the rat, *Biochem. J.*, 166, 539, 1977.

231. **Seufert, C. D., Graf, M., Janson, G., Kuhn, A., and Soling, H. D.,** Formation of free acetate by isolated perfused livers from normal, starved and diabetic rats, *Biophys. Res. Comm.*, 57, 901, 1974.

232. **Crouse, J. R., Gerson, C. D., DeCarli, L., and Lieber, C. S.,** Role of acetate in the reduction of plasma free fatty acids produced by ethanol in man, *J. Lipid Res.*, 9, 509, 1968.

233. **Bouchier, I. A. D. and Dawson, A. M.,** The effect of infusions of ethanol on the plasma free fatty acids in man, *Clin. Sci.*, 26, 47, 1964.

234. **Scheppach, W., Cummings, J. H., Branch, W. J., and Schrezenmeir, J.,** Effect of gut-derived acetate on oral glucose tolerance in man, *Clin. Sci.*, 75, 355, 1988.

235. **Scheppach, W., Wiggins, H. S., Halliday, D., Self, R., Howard, J., Branch, W. J., Schrezenmeir, J., and Cummings, J. H.,** Effect of gut-derived acetate on glucose turnover in man, *Clin. Sci.,* 75, 363, 1988.

236. **Bergman, E. N.,** Production and utilization of metabolites by the alimentary tract as measured in portal and hepatic blood, in *Digestion and Metabolism in the Ruminant,* McDonald, I. W. and Warner, A. C. I., Eds., University of New England, Sydney, Australia, 1975, 295.

237. Propionate and cholesterol homeostasis in animals, *Nutr. Rev.,* 45, 188, 1987.

238. **Chen, W.-J. L., Anderson, J. W., and Jennings, D.,** Propionate may mediate the hypocholesterolemic effects of certain soluble plant fibres in cholesterol-fed rats, *Proc. Soc. Exp. Biol. Med.,* 175, 215, 1984.

239. **Illman, R. J., Topping, D. L., McIntosh, G. H., Trimble, R. P., Storer, G. B., Taylor, M. N., and Cheng, B.-Q.,** Hypocholesterolaemic effects of dietary propionate studies in whole animals and perfused rat liver, *Ann. Nutr. Metab.,* 32, 97, 1988.

240. **Boila, R. J., Salomons, M. D., Milligan, L. P., and Aherne, F. X.,** The effect of dietary propionic acid on cholesterol synthesis in swine, *Nutr. Rep. Int.,* 23, 1113, 1981.

241. **Thacker, P. A. and Bowland, J. P.,** Effects of dietary propionic acid on serum lipids and lipoproteins of pigs fed diets supplemented with soybean meal or canola meal, *Can. J. Anim. Sci.,* 61, 439, 1981.

242. **Thacker, P. A., Bowland, J. P., and Fenton, M.,** Effects of vitamin B12 on serum lipids and lipoproteins of pigs fed diets supplemented with propionic acid or calcium propionate, *Can. J. Anim. Sci.,* 62, 527, 1982.

243. **Thacker, P. A., Salomons, M. O., Aherne, F. X., Milligan, L. P., and Bowland, J. P.,** Influence of propionic acid on the cholesterol metabolism of pigs fed hypercholesterolemic diets, *Can. J. Anim. Sci.,* 61, 969, 1981.

244. **Venter, C. S., Vorster, H. H., and Cummings, J. H.,** Effects of dietary propionate on carbohydrate and lipid metabolism in healthy volunteers, *Am. J. Gastroenterol.,* 85, 549, 1990.

245. **Todesco, T., Rao, V. A., Bosello, O., and Jenkins, D. J. A.,** Propionate lowers blood glucose and alters lipid metabolism in healthy subjects, *Am. J. Clin. Nutr.,* 54, 860, 1991.

246. **Amaral, L., Hoppel, C., and Stephen, A. M.,** Propionate — sources and effect on lipid metabolism, in *Short Chain Fatty Acids,* Binder, H. J., Cummings, J. H., and Soergel, K. H., Eds., Proc. of 73rd Falk Symposium, Strasbourg, France, Kluwer Academic Publishers, Norwell, MA, 1994.

247. **Wolever, T. M. S., Brighenti, F., Royall, D., Jenkins, A. L., and Jenkins, D. J. A.,** Effect of rectal infusion of short chain fatty acids in human subjects, *Am. J. Gastroenterol.,* 84, 1027, 1989.

248. **Wolever, T. M. S., Spadafora, P., and Eshuis, H.,** Interaction between colonic acetate and propionate in humans, *Am. J. Clin. Nutr.,* 53, 681, 1991.

249. **Anderson, J. W. and Bridges, S. R.,** Plant fiber metabolites alter hepatic glucose and lipid metabolism, *Diabetes,* 30 (Suppl. 1), 133A, 1981.

250. **Remesy, C., Demigne, C., and Morand, C.,** Metabolism of short chain fatty acids in the liver, in *Physiological and Clinical Aspects of Short Chain Fatty Acids,* Cummings, J. H., Rombeau, J. L., and Sakata, T., Eds., Cambridge University Press, Cambridge, U.K., 1995.

251. **Burke, D. A. and Axon, A. T. R.,** Ulcerative colitis and *Escherichia coli* with adhesive properties, *J. Clin. Pathol.,* 40, 782, 1987.

252. **Roediger, W. E. W.,** The colonic epithelium in ulcerative colitis: an energy-deficiency disease?, *Lancet,* 2, 712, 1980.

253. **Ireland, A. and Jewell, D. P.,** 5-Aminosalicylic acid (5-ASA) has no effect on butyrate metabolism in human colonic epithelial cells, *Gastroenterology,* 98, A176, 1989.

254. **Chapman, M. A. S., Grahn, M. F., Boyle, M. A., Hutton, M., Rogers, J., and Williams, N. S.,** Butyrate oxidation is impaired in the colonic mucosa of sufferers of quiescent ulcerative colitis, *Gut,* 35, 73, 1994.

255. **Williams, N. N., Branigan, A., Fitzpatrick, J. M., and O'Connell, P. R.,** Glutamine and butyric acid metabolism measurement in biopsy specimens (in vivo): a method of assessing treatment on inflammatory bowel conditions, *Gastroenterology,* 102, A713, 1992.

256. **Finnie, I. A., Taylor, B. A., and Rhodes, J. M.,** Ileal and colonic epithelial metabolism in ulcerative colitis: increased glutamine metabolism in distal colon but no defect in butyrate metabolism, *Gut,* 1994 (in press).

257. **Roediger, W. E. W., Lawson, M. J., Kwok, V., Kerr Grant, A., and Pannall, P. R.,** Colonic bicarbonate output as a test of disease activity in ulcerative colitis, *J. Clin. Pathol.,* 37, 704, 1984.

258. **Roediger, W. E. W. and Nance, S.,** Metabolic induction of experimental ulcerative colitis by inhibition of fatty acid oxidation, *Br. J. Exp. Pathol.,* 67, 773, 1986.

259. **Watt, J. and Marcus, R.,** Ulcerative colitis in the guinea pig caused by seaweed extract, *J. Pharm. Pharmac.,* 21, 1875, 1969.

260. **Watt, J. and Marcus, R.,** Carageenan-induced ulceration of the large intestine in the guinea pig, *Gut,* 12, 164, 1971.

261. **Marcus, R. and Watt, J.,** Ulcerative disease of the colon in laboratory animals induced by pepsin inhibitors, *Gastroenterology,* 67, 473, 1974.

262. **Okayasu, I., Hatakeyama, S., Yamada, M., Onkusa, T., Inagaki, Y., and Nakaya, R.,** A novel method in the induction of reliable experimental acute and chronic ulcerative colitis in mice, *Gastroenterology,* 98, 694, 1990.

263. **Onderdonk, A. B., Hermos, J. A., and Bartlett, J. G.,** The role of the intestinal microflora in experimental colitis, *Am. J. Clin. Nutr.,* 30, 1819, 1977.

264. **Onderdonk, A. B., Hermos, J. A., Dzink, J. L., and Bartlett, J. G.,** Protective effect of metronidazole in experimental ulcerative colitis, *Gastroenterology*, 74, 521, 1978.

265. **Gibson, G. R., Macfarlane, G. T., and Cummings, J. H.,** Occurrence of sulphate-reducing bacteria in human faeces and the relationship of dissimilatory sulphate reduction to methanogenesis in the large gut, *J. Appl. Bacteriol.*, 65, 103, 1988.

266. **Gibson, G. R., Cummings, J. H., and Macfarlane, G. T.,** Competition for hydrogen between sulphate-reducing bacteria and methanogenic bacteria from the human large intestine, *J. Appl. Bacteriol.*, 65, 241, 1988.

267. **Gibson, G. R.,** A review: physiology and ecology of the sulphate-reducing bacteria, *J. Appl. Bacteriol.*, 69, 769, 1990.

268. **Christl, S. U., Gibson, G. R., and Cummings, J. H.,** Methanogenesis in the human large intestine, *Gut*, 34, 573, 1993.

269. **Gibson, G. R., Macfarlane, S., and Macfarlane, G. T.,** Metabolic interactions involving sulphate-reducing and methanogenic bacteria in the human large intestine, *FEMS Microbiol. Ecol.*, 12, 117, 1993.

270. **Christl, S. U., Gibson, G. R., and Cummings, J. H.,** Role of dietary sulphate in the regulation of methanogenesis in the human large intestine, *Gut*, 33, 1234, 1992.

271. **Gibson, G. R., Cummings, J. H., and Macfarlane, G. T.,** Use of a three stage continuous culture system to study the effect of mucin on dissimilatory sulfate reduction and methanogenesis by mixed populations of human gut bacteria, *Appl. Environ. Microbiol.*, 54, 2750, 1988.

272. **Gibson, G. R., Cummings, J. H., and Macfarlane, G. T.,** Growth and activities of sulphate-reducing bacteria in gut contents of healthy subjects and patients with ulcerative colitis, *FEMS Microbiol. Ecol.*, 86, 103, 1991.

273. **Florin, T. H. J., Gibson, G. R., Neale, G., and Cummings, J. H.,** A role for sulfate-reducing bacteria in ulcerative colitis?, *Gastroenterology*, 98, A170, 1990.

274. **Roediger, W. E. W. and Nance, S.,** Selective reduction of fatty acid oxidation in colonocytes: correlation with ulcerative colitis, *Lipids*, 25, 646, 1990.

275. **Duncan, A., Kapaniris, O., and Roediger, W. E. W.,** Measurement of mercaptoacetate levels in anaerobic batch culture of colonic bacteria, *FEMS Microbiol. Ecol.*, 74, 303, 1990.

276. **Roediger, W. E. W., Duncan, A., Kapaniris, O., and Millard, S.,** Sulphide impairment of substrate oxidation in rat colonocytes: a biochemical basis for ulcerative colitis?, *Clin. Sci.*, 85, 623, 1993.

277. **Roediger, W. E. W., Duncan, A., Kapaniris, O., and Millard, S.,** Reducing sulfur compounds of the colon impair colonocyte nutrition: implications for ulcerative colitis, *Gastroenterology*, 104, 802, 1993.

278. **Ramakrishna, B. S., Roberts-Thomson, K., Pannall, P. R., and Roediger, W. E. W.,** Impaired sulphation of phenol by the colonic mucosa in quiescent and active ulcerative colitis, *Gut*, 32, 46, 1991.

279. **Raouf, A. H., Tsai, H. H., Parker, N., Hoffman, J., Walker, R. J., and Rhodes, J. M.,** Sulphation of colonic and rectal mucin in inflammatory bowel disease: reduced sulphation of rectal mucus in ulcerative colitis, *Clin. Sci.*, 83, 623, 1992.

280. **Fukushima, T., Kawamoto, M., Kubo, A., Takemura, H., and Tsuchiya, S.,** Fecal bacteria and short-chain fatty acid in patients with ulcerative colitis, in *Inflammatory Bowel Disease*, Shiratori, T. and Nakano, H., Eds., University of Tokyo Press, Tokyo, 1982, 59.

281. **Vernia, P., Caprilli, R., Latella, G., Barbetti, F., Magliocca, F. M., and Cittadini, M.,** Fecal lactate and ulcerative colitis, *Gastroenterology*, 95, 1564, 1988.

282. **Roediger, W. E. W., Heyworth, M., Willoughby, P., Piris, J., Moore, A., and Truelove, S. C.,** Luminal ions and short chain fatty acids as markers of functional activity of the mucosa in ulcerative colitis, *J. Clin. Pathol.*, 35, 323, 1982.

283. **Breuer, R. I., Buto, S. K., Christ, M. L., Bean, J., Vernia, P., Paoluzi, P., Di Paolo, M. C., and Caprilli, R.,** Rectal irrigation with short-chain fatty acids for distal ulcerative colitis, *Dig. Dis. Sci.*, 36, 185, 1991.

284. **Breuer, R., Lashner, B., Soergel, K. H., Hanauer, S. B., Harig, J., Vanagunas, A., Robinson, M., and Keshvarzian, A.,** Short chain fatty acid rectal irrigation therapy of left sided ulcerative colitis: a randomized placebo controlled clinical trial, in *Short Chain Fatty Acids*, Falk Symposium, Strasbourg, France, Kluwer Academic Publishers, Norwell, MA, 1994, 214.

285. **Scheppach, W., Sommer, H., Kirchner, T., Paganelli, G. -M., Bartram, P., Christl, S., Richter, F., Dusel, G., and Kasper, H.,** Effect of butyrate enemas on the colonic mucosa in distal ulcerative colitis, *Gastroenterology*, 103, 51, 1992.

286. **Steinhart, A. H., Baker, J. P., and Brzezinski, A.,** Butyrate enemas in the treatment of refractory ulcerative proctosigmoiditis, in *Short Chain Fatty Acids*, Falk Symposium, Strasbourg, France, Kluwer Academic Publishers, Norwell, MA, 73, D8, 1993.

287. **Vernia, P., Cittadini, M., Marcheggiano, A., Frieri, G., Caprilli, R., Valpiani, D, Miglo, F., and Torsoli, A.,** Short chain fatty acid (SCFA) enemas in distal ulcerative colitis: a double blind placebo controlled study, in *Short Chain Fatty Acids*, Falk Symposium, Strasbourg, France, Kluwer Academic Publishers, Norwell, MA, 73, D8, 1993.

288. **Glotzer, D. J., Glick, M. E., and Goldman, H.,** Proctitis and colitis following diversion of the fecal stream, *Gastroenterology*, 80, 438, 1981.

289. **Korelitz, B. I., Cheskin, L. J., Sohn, N., and Sommers, S. C.,** Proctitis after fecal diversion in Crohn's disease and its elimination with reanastomosis: implications for surgical management, *Gastroenterology*, 87, 710, 1984.

290. **Neut, C., Colombel, J. F., Guillemot, F., Cortot, A., Gower, P., Quandalle, P., Ribet, M., Romond, C., and Paris, J. C.,** Impaired bacterial flora in human excluded colon, *Gut*, 30, 1094, 1989.

291. **Lusk, L. B., Reichen, J., and Levine, J. S.,** Aphthous ulceration in diversion colitis, *Gastroenterology*, 87, 1171, 1984.

292. **Roediger, W. E. W.,** The starved colon - diminished mucosal nutrition, diminished absorption, and colitis, *Dis. Colon Rectum*, 33, 858, 1990.

293. **Harig, J. M., Soergel, K. H., Komorowski, R. A., and Wood, C. M.,** Treatment of diversion colitis with short chain fatty acid irrigation, *N. Engl. J. Med.*, 320, 23, 1989.

294. **Guillemot, F., Colombel, J. F., Neut, C., Verplanck, N., Lecomte, M., Romond, C., Paris, J. C., and Cortot, A.,** Treatment of diversion colitis by short chain fatty acids, *Dis. Colon Rectum*, 34, 861, 1991.

295. **Wischmeyer, P. E., Tremaine, W. J., Haddad, A. C., Ambroze, W. L., Pemberton, J. H., and Phillips, S. F.,** Fecal short chain fatty acids in patients with pouchitis after ileal pouch anal anastomosis, *Gastroenterology*, 100, A848, 1991.

296. **Clausen, M. R., Tvede, M., and Mortensen, P. B.,** Short chain fatty acids in pouch contents from patients with and without pouchitis after ileal pouch - anal anastomosis, *Gastroenterology*, 103, 1144, 1992.

297. **De Silva, H. J., Ireland, A., Kettlewell, M., Mortensen, N., and Jewell, D. P.,** Short chain fatty acid irrigation in severe pouchitis (Letter), *N. Engl. J. Med.*, 321, 1416, 1989.

298. **Wischmeyer, P., Grotz, R. L., Pemberton, J. H., and Phillips, S. F.,** Treatment of pouchitis after ileo-anal anastomosis with glutamine and butyric acid, *Gastroenterology*, 102, A617, 1992.

## Chapter 6

# Gas Metabolism in the Large Intestine

*Michael D. Levitt, Glenn R. Gibson, and Stefan U. Christl*

## CONTENTS

I. Introduction and Methods of Study ........................................................................................ 131
   A. *In Vivo* Study of Bacterial Gas Production ................................................................... 132
      1. Flatus Measurements ................................................................................................ 132
      2. Breath Measurements ............................................................................................... 133
   B. *In Vitro* Study of Bacterial Gas Production .................................................................. 135
II. Gas Production in the Human Large Intestine ..................................................................... 136
   A. Hydrogen .......................................................................................................................... 136
      1. Hydrogen Production ................................................................................................ 136
      2. Hydrogen Consumption ........................................................................................... 136
   B. Carbon Dioxide ................................................................................................................ 139
   C. Trace Gases ...................................................................................................................... 140
III. Methanogenesis ..................................................................................................................... 140
IV. Dissimilatory Sulfate Reduction ........................................................................................... 142
V. Other Bacterial Mechanisms of Hydrogen Disposal ........................................................... 145
   A. Dissimilatory Nitrate Reduction .................................................................................... 145
   B. Acetogenesis .................................................................................................................... 145
VI. Clinical Applications of Gas Metabolism ............................................................................ 146
   A. Breath Hydrogen ............................................................................................................. 146
      1. Carbohydrate Malabsorption ................................................................................... 146
      2. Transit Time Measurements ..................................................................................... 147
      3. Bacterial Overgrowth of the Small Intestine .......................................................... 147
   B. Flatulence and Meteorism ............................................................................................... 147
   C. Pneumatosis Cystoides Intestinalis ................................................................................ 147
   D. Toxicity of Sulfide and Ulcerative Colitis ..................................................................... 148
   E. Methane and Colorectal Cancer ..................................................................................... 148
VII. Conclusions ........................................................................................................................... 148
References .................................................................................................................................... 149

## I. INTRODUCTION AND METHODS OF STUDY

Although intestinal bacteria produce a variety of volatile metabolites in trace concentrations, most scientific interest has been directed towards the three gases that are produced in quantitatively important volumes (i.e., >1% of flatus): $H_2$, methane ($CH_4$), and $CO_2$. While $H_2$ and $CH_4$ are solely of bacterial origin,[1] $CO_2$ may originate directly from bacterial metabolism, or from the interaction of bicarbonate and acids (which may be of bacterial origin), or via diffusion from the blood to the intestinal lumin.[2] Since high flatus $CO_2$ concentrations are almost invariably associated with high $H_2$ concentrations,[3] it seems likely that most flatus $CO_2$, like $H_2$, originates from bacterial metabolism.

Gases accumulating in the intestinal lumin are handled differently from the nonvolatile bacterial metabolites. First, gas is propelled through the bowel much more rapidly than liquid; consequently the stomach to anus transit time of gas may be as short as 20 min, compared to several days for compounds that remain in the fecal stream.[4] As a result, analysis of the anal excretion of gas provides more of a moment-to-moment scenario of colonic bacterial metabolism than does analysis of other metabolites excreted in feces. Second, bacterial gases produced in the intestine are absorbed into blood perfusing the mucosa, carried to the lungs, and then excreted in expired air.[5] Analysis of expired air provides a simple, noninvasive, "clean" means of assessing the ongoing metabolism of gas-producing bacteria. Thus, gases represent a unique type of intestinal bacterial metabolite whose production can be studied *in situ* in the

0-8493-4524-3/95/$0.00+$.50

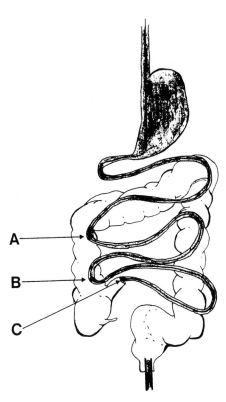

**Figure 1** Schematic representation of the position of the triple-lumin tube during infusion study. The distal opening (C) is the terminal ileum. The central opening (B) is 60 cm upstream in the proximal ileum, while the proximal opening (A) is another 60 cm upstream in the midjejunum. Gas is infused through orifice A and samples collected at B, C, and also from a rectal tube.

gut, via breath or flatus measurements, as well as in conventional culture techniques that are used to assess the bacterial production of nonvolatile metabolites. Because some bacterial gases, such as $H_2$, are catabolized by other bacteria,[6] what appear to be "production" measurements actually represent the net of absolute production minus consumption.

## A. *IN VIVO* STUDY OF BACTERIAL GAS PRODUCTION
### 1. Flatus Measurements

Study of gas excretion in flatus is more difficult than one might suppose, due to problems with the long-term, quantitative collection of rectal gas. The first analyses of flatus composition were published more than 100 years ago by Ruge,[7] a Viennese physician, who used a flatus collection system consisting of a glass tube placed in the anus that led through a hole in the bottom of a chair to a water displacement system. Over the years, variations of this technique have been employed, including flexible rectal tubes leading to gas impermeable bags or syringes.[3,8,9] However, the tubes are uncomfortable and tend to plug with mucus and fecal material. Ileostomy patches over the anus have been used, but the high $O_2$ levels observed with this technique suggest that there is appreciable leakage, since flatus usually contains negligible amounts of this gas.[3,10] Thus, there have been no reliable long-term (>24 h) studies of flatus composition utilizing direct collection of gas passed.

Short-term flatus collections (via rectal tubes) suggest that healthy subjects excrete a total of 400 to 2000 ml of gas per day.[9] The percentage composition of bacterially produced gases in flatus varies widely, with $H_2$ concentrations ranging from 1 to 40%, $CH_4$ from <1 to 20%, and $CO_2$ from 5 to 50%.[10]

Intubation of the intestine makes it possible to quantitate the rate of gas production at specific sites in the bowel. For example, in an early study of $H_2$ and $CH_4$ production in the human intestine, a mercury-weighted, triple lumin tube was swallowed and allowed to pass until the distal orifice was in the terminal ileum (Figure 1).[11] The tube was constructed so that the middle orifice was then in the proximal ileum and the most proximal orifice in the jejunum. The gut was then constantly perfused, at a rate of 30 ml/min via the proximal orifice, with nitrogen containing $SF_6$, a virtually nonabsorbable dilutional marker gas. The infused gas was rapidly propelled through the gut, and when a steady state became established, gas samples were collected from the proximal and distal ileum via the intestinal tube and the rectum via

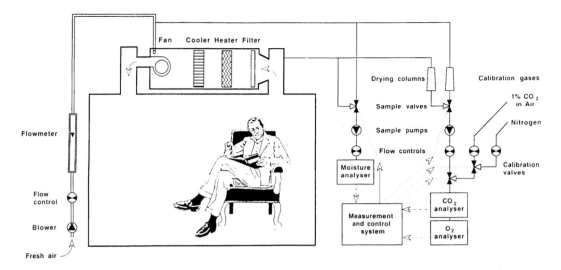

**Figure 2**  A whole body calorimeter as used to measure total excretion of $H_2$ and $CH_4$ in human subjects for up to 36 h. Figure courtesy of P. Murgatroyd.

a rectal tube. The total rate that gas passed a collection site was determined from the change in concentration of $SF_6$ and infusion rate:

$$\text{Rate}_{total} = (\text{Infusion rate}) \cdot \left( [SF_6]_{infusate} / [SF_6]_{aspirate} \right) \qquad (1)$$

The rate of flow of $H_2$, $CH_4$, or $CO_2$ past a collection site was then calculated from the concentration of the specific gas and the total flow rate of gas:

$$\text{Rate } H_2 \text{ or } CH_4 \text{ or } CO_2 = \text{Rate}_{total} \cdot [\text{Specific gas}] \qquad (2)$$

These experiments demonstrated that virtually all $H_2$ and $CH_4$ production in the normal human gut occurred in the large intestine. Studies carried out after the tube had passed into the proximal colon of a few subjects suggested that $CH_4$ was produced primarily in the distal colon while $H_2$ was liberated throughout the colon.[11]

An important advance in the study of intestinal gas production was provided by Christl et al.[12] who housed subjects in a whole body calorimeter. This system is shown in Figure 2. The total rates of excretion of $H_2$ and $CH_4$ were determined from the difference between the concentration of these gases entering and leaving the calorimeter and the flow rate throughout (100 L/min). Pulmonary excretion of $H_2$ and $CH_4$ (determined by measuring concentrations in alveolar air), and alveolal ventilation rates calculated from individual $CO_2$ production measured in the calorimeter, was subtracted from total excretion to determine the anal excretion rate. This technique permits long-term, quantitative measurements of total intestinal gas excretion under physiological and comfortable conditions, and allows measurements to be made that are impossible with any other technique. However, a drawback to this methodology is the need for very sensitive and accurate assay techniques, since the excreted gases are diluted in a large volume of air. In addition, assessment of $CO_2$ produced in the intestine is not possible because production of this gas in the gut cannot be differentiated from that of host cellular metabolism.

## 2.  Breath Measurements

Because of the wide use of breath measurements to assess intestinal gas production, it is necessary to understand the physiology underlying the pulmonary excretion of bowel gases, and why breath measurements should be considered to be only a semi-quantitative indicator of intestinal production.

All gases have appreciable aqueous and lipid solubility and thus are able to readily diffuse across the epithelial cell that separates the gut lumin from mucosal blood flow. The direction of the net movement of a gas is determined by the partial pressure difference between the lumin and blood. Because of differences in the partial pressure, the net movement of $H_2$, $CH_4$, and trace gases produced by bacteria is from the lumin into the bloodstream. When $CO_2$ is being rapidly produced in the lumin, it behaves similarly. However, in some situations where gut $CO_2$ production is low, it is possible that the net direction of movement is the reverse.

Since the diffusion process is entirely passive, there is no saturation of this transport mechanism and a constant fraction of gas should be transported into the blood per unit time. The rate of diffusion (D) of a gas between lumin and blood is determined by a series of factors including the partial pressure difference between lumin ($P_L$) and blood ($P_B$), the aqueous diffusion coefficient (D) of the gas, the aqueous solubility coefficient ($\alpha$) of the gas, and the surface area (A) and thickness (T) of the gut epithelium:

$$D = \left(P_L - P_B\right) \cdot \left(\alpha\right) \cdot \left(D\right) \cdot \left(A/T\right) \tag{3}$$

All gases have sufficient lipid solubility to render the resistance of the very thin lipid membrane negligible, relative to the resistance of the aqueous body of the epithelial cell or an unstirred water layer over the mucosa. Thus, the rate limiting step in diffusion is movement through a water phase of the cell, which is determined by the aqueous solubility and diffusion coefficient of the gas. Lipid solubility appears to be irrelevant. The ratios of aqueous diffusion coefficients of $H_2$, $CH_4$, and $CO_2$ are about 1:0.7:0.3, respectively, while the ratio of aqueous solubilities is about 1:2:36.[13] Since diffusion rate is proportional to [(D)·($\alpha$)], $CO_2$ diffuses much more rapidly than do the other major intestinal gases.

Equation 3 does not mention the rate of intestinal mucosal blood flow or the solubility of the gas in blood. These factors indirectly influence the uptake of gases from the blood via their effect on $P_B$, the partial pressure of gas in the blood. When blood flow is sufficiently slow, $P_B$ rises and begins to approach $P_L$, and uptake becomes blood flow limited. In contrast, absorption is diffusion limited when the gas diffuses slowly, relative to the capacity of the blood to carry away the gas, i.e., $P_B$ remains negligible relative to $P_L$. Carbon monoxide is a gas whose uptake is totally diffusion limited, since the binding of CO to hemoglobin maintains a negligible CO tension in blood.[14] For this reason, CO has been used as a probe to measure diffusion barriers in the gut and lung. The gas with the most rapid diffusion relative to its blood solubility is $H_2$; hence, this gas is most likely to manifest blood flow limitation to uptake. Although data are unavailable for the human intestine, studies in rats suggest that uptake of the major bacterial gases from the colon is largely diffusion limited, i.e., the rate of blood flow plays a minor role.[15]

The rate that intestinal gases are cleared from the blood in expired air is a function of the solubility of the gas in air relative to blood, the higher the air:blood solubility ratio, the more efficient the clearance. Gases such as $H_2$ and $CH_4$ have a very high ratio of gas:blood solubility (approximately 50:1), and therefore these gases are almost 100% cleared from the blood during a single passage through the lungs. Carbon dioxide has a much lower solubility ratio and the $pCO_2$ of blood only falls from about 46 to 40 torr during passage through the lungs, i.e., only about 13% is removed in a single passage.

No clearance mechanism exists to remove $H_2$ and $CH_4$ from blood except pulmonary excretion. Thus, the rate of pulmonary excretion of these gases is a quantitative indicator of the rate that they are absorbed from the gut. With lung disease, a somewhat higher blood concentration of $H_2$ or $CH_4$ may be required to yield a given rate of pulmonary excretion. In the steady state, however, breath excretion equals absorption rate from the gut. As discussed earlier, a number of factors influence the absorption rate of these gases from the intestine, but only luminal partial pressure is related to the rate of production of the gas. As a result, breath excretion of $H_2$ and $CH_4$ should be considered to be only a semi-quantitative indicator of intestinal production of these gases.

Recent studies by Christl et al.[12] using a whole body calorimeter have helped to clarify the relationship between breath and flatus excretion of $H_2$ and $CH_4$. When the total excretion rate of these gases (flatus plus breath) was less than 200 ml/24 h, about 65% of the $H_2$ was excreted on the breath, while at higher production rates, the percentage excreted via the lungs decreased to about 25%. Apparently, rapid production of bowel gas results in increased propulsion of gas through the gut, and hence a greater fractional excretion via the anus.

Initial studies of the breath excretion of $H_2$ and $CH_4$ involved analyses of quantitative collections of expired air.[16] However, since sedentary subjects have a relatively constant alveolar ventilation (roughly

**Figure 3** Growth of SRB in a single-stage chemostat inoculated with mixed fecal bacteria. Note the black appearance of the fermentor, due to a production of sulfide and formation of FeS in the culture medium.

5000 ml/70 kg), the simple measurement of the concentration of $H_2$ or $CH_4$ in alveolar air can be substituted for the far more complicated measurement of excretion rate.[17]

A number of techniques have been used to collect alveolar air specimens. Many investigators now use a commercial device that automatically discards the first 500 ml of expired air (dead space) and then collects the subsequent exhalation in a foil bag. However, it is seldom recognized that this technique is far from perfect, since minor degrees of hyperventilation or breath holding can appreciably influence the concentration of $H_2$ or $CH_4$ in alveolar air.[18] Thus, reproducible measurements require attention to the details of alveolar air collection.

Storage of gas samples without loss or contamination is far more difficult than is storage of other bacterial products. Syringes represent the easiest means of handling gas; however, plastic syringes are permeable to most gases. The most satisfactory system for storing gas samples for up to several weeks appears to be glass syringes lubricated with mineral oil (unlubricated glass syringes leak at a prodigious rate).[19]

Concentrations of the major bacterial gases in flatus are sufficiently high that a variety of chemical as well as more sensitive analytical techniques can be employed. In contrast, the relatively low concentrations of $H_2$ and $CH_4$ in breath (usually between 1 and 200 ppm) require much more sensitive technology. Most studies have used relatively simple gas chromatographic methods for this purpose.

## B. *IN VITRO* STUDY OF BACTERIAL GAS PRODUCTION

A number of fermentation systems exist which are used to investigate microbiological aspects of colonic function, including gas metabolism (for a review see Rumney and Rowland[20]). A widely used form of batch culture involves the incubation of fecal slurries in air tight bottles, with a headspace of anaerobic gas and fermentable substrate(s) added. Removal of gas samples periodically, either through butyl rubber seals or suba seals, gives some idea of net production. In these simple systems, it is often difficult to control medium pH, which is likely to decrease as fermentation proceeds. This situation may adversely affect the activities of some gas metabolizing species such as methanogens or sulfate reducers (see discussion later). A more sophisticated approach involves the use of semi-continuous[21-23] or single-stage continuous culture sytems.[24-26] Here, bacterial growth can be more adequately controlled in either pure culture, defined mixed populations, or fecal slurries. Figure 3 shows the growth of a mixed inoculum of fecal bacteria, containing active sulfate-reducing bacteria (SRB), in a single-stage chemostat.

The chemostat is particularly useful for investigating bacterial activities under steady-state conditions. These have been further developed into multiple-stage continuous culture systems, which attempt to simulate the varying physicochemical environments of the proximal and distal colons.[27-29] Inocula for these systems is usually feces and the bacteria are left to establish under varying conditions of pH, growth rate, and substrate availability which are imposed. Gas sampling can be achieved by input from a vessel port in the fermentors into vacuolated bottles, football bladders, or other appropriate collection systems. Subsequent analyses are usually performed using gas chromatography. However, a preferable alternative is continuous gas monitoring by, for example, mass spectroscopy.

## II. GAS PRODUCTION IN THE HUMAN LARGE INTESTINE

### A. HYDROGEN
### 1. Hydrogen Production

Studies carried out immediately after birth in the bacteria-free human newborn showed no detectable $H_2$ production, while after 24 h of life (when the infant had begun to develop an intestinal flora) copious excretion of this gas was observed.[1] Similarly, germ-free rats did not excrete $H_2$, but within 24 h of contamination with feces from conventional rats, production was readily detectable.[30] Thus, all $H_2$ production in humans (and rats) appears to be of bacterial origin. $H_2$ is an important fermentation intermediate which serves as an efficient mechanism for disposing of reducing power (electrons) generated during the bacterial metabolism of carbohydrate and protein.

The biochemistry of $H_2$ production by isolates of the human intestinal flora has received extensive study. During the fermentation of sugars and amino acids, bacteria use $H_2$ as an electron sink rather than reducing $O_2$ to $O^{-2}$, the electron disposal mechanism in aerobic metabolism. Many types of enterobacteria produce $H_2$ from pyruvate.[31] The initial step in this reaction is the cleavage of pyruvate to formate. Formate is then metabolized to $H_2$ and $CO_2$ via the action of formate hydrogen lyase, a reaction that yields equivalent amounts of these gases. Clostridia generate $H_2$ from pyruvate via ferredoxin. In addition, many bacteria form $H_2$ in the process of oxidizing the pyridine nucleotides $NADH^+$ and $FADH^+$ to $NAD^+$ and $FAD^+$. Since high $PH_2$ values inhibit this regeneration of oxidized nucleotides, the removal of $H_2$ from the system is necessary for maximum efficiency of the entire fermentation reaction.[32] In contrast, the liberation of $H_2$ from pyruvate is not influenced by $H_2$ tension. Wolin[33] has proposed that in the absence of $H_2$ oxidizing bacteria, the overall stoichiometry of carbohydrate fermentation in the human colon takes the following form:

$$34.5 \ C_6H_{12}O_6 \rightarrow 48 \ \text{acetate} + 58 \ CO_2 + 95 \ H_2 + 11 \ \text{propionate} + 5 \ \text{butyrate} + 10.5 \ H_2O \qquad (4)$$

A variety of short chain organic acids are produced in the above reaction. Since the optimal pH of the enzymes for $H_2$ production is approximately 7.0,[34] the reaction is self-inhibitory unless acid is removed, or base is added, to the reaction mixture.

The potential for gas production from the above reaction is considerable. In terms of carbohydrate load likely to reach the colon, malabsorption of the 12.5 g of lactose in a cup of milk would be expected to yield 4300 ml of $H_2$ if fermented as described above. Such enormous quantities of $H_2$ not only would create major problems with distention and flatulence, but would place the individual in imminent danger of a Hindenburg-like explosion, if a flame were anywhere in the vicinity when gas was passed. Fortunately, studies of human subjects enclosed in the whole body calorimeter[12] have demonstrated that malabsorption of carbohydrate yields a total excretion (breath plus flatus) that was only a small fraction of that predicted by the equation of Wolin.[33] For example, ingestion of 15 g of the totally nonabsorbable disaccharide, lactulose, resulted in the excretion of only about 227 ml of $H_2$ (average of 10 subjects) as opposed to the theoretical value of 5100 ml, a 20-fold discrepancy between observed and predicted values.

### 2. Hydrogen Consumption

Theoretically, the approximately 20-fold discrepancy between the expected and observed $H_2$ excretion could reflect a failure to produce this gas in the quantities predicted by Equation 4 and/or its consumption (oxidation). Studies in fecal homogenates in which $H_2$ consumption was thought to be totally inhibited[35] showed that liberation during the fermentation of carbohydrate was about one-fourth that predicted by Equation 4. The bulk of the 20-fold discrepancy between observed and predicted $H_2$ excretion, therefore, presumably represents its utilization by fecal bacteria.

The existence of $H_2$ consumption in the colon was first clearly demonstrated in rat studies in which $H_2$ and helium were co-instilled into segments of intestine isolated between ligatures.[6] While the bulk of the helium instilled into the cecum of conventional rats was absorbed and recovered in expired air, only about 10% of the $H_2$ was recovered. This disappearance was not observed when $H_2$ was instilled into the small intestine or into the cecum of germ-free rats. Thus, cecal bacteria appeared to be responsible for the consumption of $H_2$. Subsequent incubation studies demonstrated that human fecal homogenates consumed $H_2$ at a very rapid rate. Therefore, previous measurements of $H_2$ "production" *in vivo*, or by fecal homogenates, *in vitro*, actually measured the net of absolute production minus consumption, and not the true (absolute) production rate.

The human colon has been shown to contain at least three types of $H_2$ consuming (oxidizing) microorganisms: methanogenic, sulfate-reducing, and acetogenic bacteria.[36-38] The major type of methanogen in human feces is *Methanobrevibacter smithii* which utilizes $H_2$ to reduce $CO_2$ to $CH_4$:

$$4H_2 + CO_2 \rightarrow CH_4 + 2H_2O \tag{5}$$

While the methanogen manages to extract a small amount of energy out of this reaction, the host benefits markedly via a dramatic reduction in colonic gas. With efficient methanogenesis, therefore, Equation 4 becomes:

$$34.5\ C_6H_{12}O_6 \rightarrow 48\ \text{acetate} + 34\ CO_2 + 24\ CH_4 + 11\ H_2O + 11\ \text{propionate} + 5\ \text{butyrate} \tag{6}$$

There are a variety of SRB in the human colon, some of which are capable of utilizing $H_2$ to reduce sulfate to sulfide:[39]

$$4H_2 + SO_4^{2-} + H^+ \rightarrow HS^- + 4H_2O \tag{7}$$

Acetogenic bacteria utilize $H_2$ to reduce $CO_2$ to acetate:[28]

$$4H_2 + 2CO_2 \rightarrow CH_3COOH + 2H_2O \tag{8}$$

In order to determine the relative roles of absolute $H_2$ production vs. consumption in the determination of an individual's net excretion, it is necessary to be able to measure both parameters independently. Several techniques have been used to inhibit consumption, thereby permitting absolute measurements of production. One method is based on the observation that utilization rate is directly proportional to $pH_2$.[35] If $H_2$ tension was maintained at a very low level, consumption became negligible relative to the absolute production rate. To maintain such a low $pH_2$ in the homogenate, it was necessary to incubate a very thin layer (0.1 ml) of well-stirred homogenate in the bottom of a 1 L flask. If inhibitors of the two major $H_2$ consuming mechanisms, methanogenesis[40] and sulfate reduction,[41] were added to the homogenate, a somewhat thicker layer of homogenate could be used, since acetogenesis and possibly other $H_2$ consuming reactions, seemingly require a relatively high $H_2$ tension for appreciable activity.

Studies of fecal homogenates in which $H_2$ consumption had been inhibited, suggested that fecal samples from different individuals all had similar absolute production per gram of carbohydrate fermented.[35] Therefore, differences in $H_2$ consumption presumably explain the large individual differences observed in excretion following malabsorption of similar carbohydrate loads. It is apparent from the foregoing discussion that an understanding of $H_2$ consumption is critical for interpreting excretion measurements.

Since $CH_4$ is primarily produced from the reduction of $CO_2$ by $H_2$, and there appears to be no catabolism of $CH_4$ in the colon, the importance of methanogenesis as a $H_2$ consuming mechanism is readily quantified in the intact individual, i.e., four times the total $CH_4$ excretion (see Equation 5) equals the $H_2$ that is being metabolized via this route. Of interest is the common observation that fasting subjects may excrete large quantities of $CH_4$ on the breath while $H_2$ concentrations are only a few ppm. Thus, the vast majority of $H_2$ produced in this situation appears to be consumed via methanogenesis and only a small fraction is excreted intact.

Unfortunately, the *in situ* consumption of $H_2$ via sulfate reduction and acetogenesis cannot be measured. The product of sulfate reduction (sulfide) appears to be rapidly absorbed and/or metabolized, since very little is present in feces.[36] Measurement of acetogenesis in the colon would require administration of $^{14}CO_2$ and the seemingly impossible knowledge of its specific activity in the fecal mass. Thus, the importance of these two $H_2$ consuming reactions in the colon is limited to extrapolation from incubation studies carried out with fecal homogenates. In such studies, sulfate reduction rate is determined using measurements of the production of sulfide or the conversion of $[^{35}S]$ sulfate to $[^{35}S]$ sulfide.[36] Acetogenesis can be determined from the rate of incorporation of $^{13}CO_2$ or $^{14}CO_2$ into acetate.[38] Methanogenesis is readily measured by production rate of the gas. The role of these reactions has also been indirectly assessed in studies in which $H_2$ disappearance is measured in the presence and absence of specific metabolic inhibitors of the $H_2$ consuming reactions. Molybdate,[41] chloroform,[42] and 2-bromoethanesulfonic acid[40] are relatively specific inhibitors of sulfate reduction, acetogenesis, and methanogenesis, respectively.

The extent to which the results of $H_2$ consumption measurements with fecal homogenates truly represent what is occurring in the complex ecosystem of the colon is problematic, since a variety of intracolonic factors may not be simulated in the homogenate. For example, while methanogenic and acetogenic bacteria reduce a ubiquitous substrate ($CO_2$), the activity of the SRB is limited by sulfate availability. Appreciable sulfate clearly is delivered to the colon, as judged from studies with ileostomy effluent;[43] additional sulfate may be provided from the fermentation of host glycoproteins (see Chapter 8). However, sulfate is apparently used by a series of bacterial reactions, and by the time the fecal stream reaches the rectum, there may be insufficient quantities of sulfate to support $H_2$ consumption.[44] Thus, because we lack knowledge of sulfate availability throughout the colon, it is not possible at present to predict the importance of sulfate reduction as a $H_2$ consuming mechanism.

The pH of the colonic environment is another factor that is often not appropriately simulated in homogenates. Methanogenesis and sulfate reduction occur optimally at neutral to slightly alkaline pH values, whereas acetogenesis appears to be maximal at an acidic pH.[36] Of particular importance are recent observations that cell population densities of the $H_2$ consuming bacteria are not constant throughout the colon, but vary longitudinally along its length.[45,46] Thus, the common use of fecal homogenates (right colonic contents are obtained with extreme difficulty) to simulate the colonic bacterial environment may markedly overestimate the role played by methanogens in $H_2$ consumption. Here, gut modeling systems may have an increasingly important role.

Intestinal intubation studies of healthy subjects showed that virtually all $H_2$ was liberated in the large bowel.[11] Colonic $H_2$ production was low in subjects who had fasted for 12 h, but increased markedly when a small amount of lactose was infused into the small intestine and colon. From these observations, it was concluded that bacteria which produce $H_2$ were limited to the colon of healthy subjects and that they produced very little $H_2$ from substrates endogenous to the intestine. Copious $H_2$ production was dependent on the delivery of ingested fermentable substrate (carbohydrate and protein) that was malabsorbed in the small gut, and thus became available for $H_2$-releasing fermentation reactions in the colon. Despite much discussion, it is still unclear to what extent bacterial overgrowth of the small bowel may result in appreciable $H_2$ production from substrates that are totally absorbed in the small intestine.[11,47]

Colonic bacteria are capable of producing $H_2$ during the fermentation of a wide variety of dietary components, including sugars, sugar-alcohols (mannitol or sorbitol), complex carbohydrates, and some forms of fiber. In addition, glycoproteins have been shown to support $H_2$ production, generally in proportion to their carbohydrate content.[48] Mucin, a particularly carbohydrate rich protein, yielded the greatest $H_2$ production per gram of substrate fermented.

Since $H_2$ producing bacteria are primarily limited to the colon, only carbohydrates that are incompletely absorbed by the small bowel can serve as substrates for production of this gas. In healthy subjects, such malabsorption may occur for a variety of reasons. Many foods contain carbohydrates that are inherently indigestible and hence, nonabsorbable. Examples include oligosaccharides such as raffinose and stachyose, which are present in high concentrations in legumes and beans.[3] The simple sugars of these oligosaccharides are linked by $\alpha$-galactosidic bonds that cannot be split by the enzymes of the small gut, but which are readily fermented by colonic bacteria. Lactulose, an indigestible disaccharide, is widely used to treat constipation and hepatic encephalopathy. Some fruits such as unripe bananas contain starch that is resistant to amylase digestion and hence is malabsorbed.[12] The refrigeration of products made from wheat flour or potato, for example, causes some of the starch to crystallize into what is known as retrograded starch, which is a material that is extremely resistant to pancreatic amylase digestion, but which can be slowly broken down by bacteria in the colon.[49] In addition, sugar-alcohols such as sorbitol or mannitol require no digestion but are so slowly transported by the intestinal mucosa that a large amount escapes absorption in the normal small intestine.[50] Finally, most forms of fiber are more or less slowly fermented, with the production of $H_2$.[51]

Patients with various digestive-absorptive abnormalities may malabsorb carbohydrates that are completely absorbed by normal subjects. The most common example is lactase deficiency, a condition in which there is insufficient small bowel β-galactosidase to digest even relatively small quantities of lactose. Other conditions in which carbohydrate may be malabsorbed include intestinal disease states with diffuse mucosal injury, such as celiac sprue, giardiasis, gastroenteritis, and conditions with inadequate digestion, such as pancreatic insufficiency.

It has recently been demonstrated that, although completely fermented, malabsorbed carbohydrate that is slowly broken down (such as resistant starch) results in much less $H_2$ excretion than does soluble, rapidly fermented carbohydrate such as lactose.[12] There are several likely explanations for this diminished $H_2$ excretion with slowly fermented carbohydrates. First, there is an increasing concentration of some $H_2$-consuming bacteria, such as the methanogens, as fecal material moves distally in the colon.[45,46] Second, gut contents become more solidified in the distal colon and, as a result, stirring becomes less efficient. Consequently, $H_2$ produced by the bacteria tends to be trapped, maintaining a high gas tension in the fecal material. Since the ability of bacteria to utilize this gas is directly proportional to $H_2$ tension,[35] consumption is more efficient, and only limited quantities of gas escapes to be excreted by the anus or breath.

Multiple factors can influence an individual's $H_2$ excretion status:

1. The quantity and type (rapidly vs. slowly fermentable) of carbohydrate and protein malabsorbed.
2. The ability of the colonic microflora to ferment the malabsorbed carbohydrate.
3. The numbers, activity, type, and location of the $H_2$ producing and consuming microorganisms.
4. The efficiency of stirring of gut contents.
5. Environmental factors such as pH and sulfate availability.

## B. CARBON DIOXIDE

Large quantities of $CO_2$ are formed in the large intestine, and this gas is often the predominant component of flatus.[3] However, production of $CO_2$ in the gut has received relatively little scientific study. As discussed earlier, breath measurements cannot distinguish between the minor fraction of this gas that originates in the intestine versus the amount resulting from cellular metabolism of the host. As a result, *in vivo* studies have been limited to reports of flatus composition, and one study in which $CO_2$ tension was measured in the duodenum.[52]

There are three potential sources of $CO_2$ in the gut: diffusion from the blood, acidification of bicarbonate, and bacterial metabolism. The maximal $CO_2$ concentration that can be achieved by diffusion is about 5% (45 torr), the $CO_2$ tension in the blood. Since $CO_2$ concentrations much greater than 5% have frequently been observed in flatus[3] and upper gastrointestinal tract contents,[10] it is apparent that diffusion accounts for only a minor proportion of gut $CO_2$.

The interaction of acid from a gastric or meal origin with duodenal $HCO_3^-$ releases large quantities of $CO_2$. The $pCO_2$ of duodenal contents has been found to average 300 and 500 torr in controls and subjects with duodenal ulcer, respectively.[52] Thus, the bulk of upper gastrointestinal gas is frequently $CO_2$. However, this gas is very rapidly absorbed across the gut mucosa and little, if any, reaches the rectum. Studies of the $CO_2$ concentration of flatus demonstrate values ranging from 3 to greater than 50%.[3,10] In general, high flatus $CO_2$ concentrations occur in association with high concentrations of $H_2$. Since the sole source of $H_2$ in flatus is from bacterial metabolism, the great bulk of rectal $CO_2$ similarly results from fermentation rather than an interaction of acid and bicarbonate in the upper gastrointestinal tract.

Carbon dioxide may thus be a direct metabolic product of bacterial reactions, or an indirect result of the interaction of organic acids produced during fermentation, and bicarbonate secreted by the gut. It has been proposed[33] that the major overall fermentation reaction in the human colon yields about 124 ml and 212 ml of $CO_2$ per g of carbohydrate fermented in $CH_4$ producers and nonproducers, respectively (see Equations 4 and 6). About 64 mol of these acids are produced during the fermentation of 6.2 kg of carbohydrate (see Equation 4) and the interaction with bicarbonate yields 22,400 ml of $CO_2$ per mol. Thus, the maximal $CO_2$ that might be released via this mechanism is about 231 ml/g of carbohydrate fermented. To the extent that some of these organic acids are excreted or absorbed in their protonated form, $CO_2$ release will be less than the maximal value cited above. Given the extremely rapid absorption of $CO_2$ from the gut lumin, high concentrations of this gas in flatus indicate very rapid bacterial production in the colon.

## C. TRACE GASES

Sensitive gas chromatographic techniques have demonstrated that human feces generate a vast array of gases in trace quantities. Examples of these gases include volatile amines, $NH_3$, and mercaptans, while many remain to be identified.[53]

The five major gut gases (i.e., $H_2$, $CH_4$, $CO_2$, $O_2$, and $N_2$) do not have an odor. Thus, the unpleasant smell of feces and flatus is attributable to gases present in trace concentrations. While older studies suggested that indole and skatole were responsible for this odor,[54] recent work has indicated that the offending agents are methyl sulfides of bacterial origin such as methanethiol and dimethyl sulfide.[53]

## III. METHANOGENESIS

As discussed earlier, a major means of disposing of $H_2$ in the colon is methanogenesis. Methane is produced by several species of archaebacteria in various anaerobic ecosystems such as sludge, sediments, the rumen, and the large intestine of animals.[55,56] Some important observations regarding gut methanogenesis are given in Table 1. In humans, only two different methanogenic bacteria have been described: *Methanosphaera stadtmaniae* reduces methanol with $H_2$ to form $CH_4$, but is found only in relatively low numbers;[57,58] the numerically predominant species is *Methanobrevibacter smithii*,[59,60] a slow growing strictly anaerobic microorganism capable of reducing $CO_2$ with $H_2$ (Equation 5).[61,62] Thus, both species have an obligate growth requirement for $H_2$. A distinctive feature of methanogens, including gut species, is their fluorescence at 420 nm under ultraviolet light. This feature makes a rapid, qualitative detection in feces relatively straightforward (Plate 3). In the rumen, methanogenesis always occurs, but this is not the case in humans. For example, only between 30 and 60% of persons in western populations are methanogenic[63-66] compared to about 90% in some rural African populations.[67,68] The reasons for these differences in carriage rates of methanogens are not yet fully understood, although there is an increasing awareness of competitive bacterial processes that may influence methanogenesis in humans (see discussion later). While young children are usually $CH_4$-negative, excretor status is gradually acquired with age.[64,69] This fact has led to the suggestion that environmental contamination might be responsible for colonization with methanogens. Further evidence for this hypothesis is provided by the high coincidence of $CH_4$ excretion among family members (70% in twins),[70] and in a group of institutionalized, mentally retarded patients (over 90%).[64]

Similar to $H_2$, $CH_4$ is expelled with flatus. It also partly diffuses from the bowel lumin to the bloodstream and is eventually excreted in breath, where it can be detected using gas-chromatographic methods.[64,65,71] In most studies, a cutoff point of 1 or 2 ppm excretion above room air in end-expiratory breath samples is used to separate $CH_4$-producers and nonproducers. Excretion kinetics correspond to those for $H_2$, with most of the gas (60 to 90%) being exhaled in breath at low excretion rates, while this proportion decreases to 20% when large total volumes are excreted.[12] Unlike $H_2$, which is not used by mammalian cells but may be further metabolized by certain bacteria, $CH_4$ is thought to be inaccessible to both the host and the gut microflora and is therefore regarded as physiologically inert.

In the rumen, virtually all available $H_2$ is consumed in methanogenic reactions.[72] In the human large intestine, however, $H_2$ usually accumulates, even in presence of methanogenic bacteria. This situation may be explained by the varying anatomy and physiology of these organs, which can result in different retention times and stirring mechanisms. Thus, although present, $H_2$ may not be completely available to human colonic methanogens. Quantitative gas measurements have shown that about 75% of the $H_2$ excreted in methanogenic volunteers was as $CH_4$.[12] Most studies show that in methanogenic subjects $H_2$ and $CH_4$ excretions are inversely related. Thus, subjects with high rates of methanogenesis usually excrete little, or even no, $H_2$.[63,65] Using intubation techniques, it has been shown that methanogenesis is much more active in the left colon as compared to the right side where most of the saccharolytic activity occurs.[45,73] This has been confirmed by *in vitro* studies using a multichamber fermentation system,[37] and incubation of colonic contents from sudden death victims.[46] A possible explanation is provided by the pH profile in the large intestine, with the more acidic environment of the right colon being less favorable to

**Table 1** Historical Perspective of Methanogenesis in the Human Large Intestine

| Author(s) | Date | Observation(s) | Reference |
|---|---|---|---|
| Kirk | 1949 | Early report of methanogenesis in up to 30% of persons tested. | 162 |
| Calloway | 1966 | $CH_4$ excreted in breath. Concentrations show a daily cyclic pattern. | 5 |
| Levitt and Ingelfinger | 1968 | $CH_4$ in the gut measured by a constant infusion technique. Methanogenesis confined to the large bowel. | 163 |
| Bond et al. | 1971 | Populations separated into $CH_4$-producers and nonproducers. Infants generally negative, while methanogenesis is acquired with age. High prevalence of $CH_4$ excretors in institutionalized mentally ill persons (93%). High concordance in twins (96%) and within families. | 64 |
| Haines et al. | 1977 | 80% $CH_4$ excretors in patients with colorectal cancer, against 40% in healthy persons. | 158 |
| Pitt et al. | 1980 | Prevalence of methanogenesis in females (49%) compared to males (33%) and in Caucasians (48%) vs. Indians (32%) and Orientals (24%). Stimulation of methanogenesis occurs by a single dose of lactulose. | 66 |
| Perman and Modler | 1982 | $CH_4$ production shown to be stimulated by endogenous glycoproteins. | 48 |
| Miller et al. | 1982 | Isolation of *Methanobrevibacter smithii* from human feces. | 60 |
| Miller and Wolin | 1982 | Enumeration of *M. smithii* as the predominant methanogen in feces. | 59 |
| Bjornkelett and Jenssen | 1982 | Negative correlation between breath $H_2$ and $CH_4$ production. | 63 |
| Pique et al. | 1984 | High (91%) proportion of $CH_4$ excretors in colorectal cancer patients compared to controls (43%). After tumor resection, a reduction in methanogenesis was demonstrated. | 159 |
| Peled et al. | 1985 | No methanogenesis occurs before the age of 3 years. | 69 |
| Miller and Wolin | 1985 | *Methanosphaera stadtmaniae* isolated from feces. A methanogen that reduces methanol with $H_2$. | 58 |
| McKay et al. | 1985 | Incidence of $CH_4$ production not different between males and females. Low prevalence of methanogenesis in patients with Crohn's disease (13%), ulcerative colitis (15%), and pneumatosis cystoides intestinalis (11%). | 65 |
| Peled et al. | 1987 | Methanogenesis in Crohn's disease (6.1%) and ulcerative colitis (34%) patients confirmed as low compared to healthy persons (50%). Little effect of antimicrobial treatment on $CH_4$ excretion. | 164 |
| Allison and Macfarlane | 1988 | Inhibition of $CH_4$ production in fecal slurries by the addition of $NO_3^-$. | 110 |
| Segal et al. | 1988 | High prevalence of $CH_4$ production in rural black Africans (84%) compared to white persons (52%). Colon cancer risk lower in the black Africans, however. | 68 |
| Kashtan et al. | 1989 | No significant difference in $CH_4$ excretion between colon cancer patients and healthy persons. | 160 |
| Flourie et al. | 1991 | Methanogenesis predominantly occurs in the distal colon. Stimulation of methanogenesis by exogenous and endogenous substrates demonstrated. | 73 |
| Sivertsen et al. | 1992 | No association between colon cancer and breath $CH_4$ excretion. | 161 |
| Christl et al. | 1992 | Total $CH_4$ excretion measured in a whole body calorimeter. 70% of the total $H_2$ excreted in methanogenic subjects is as $CH_4$. 60% of $CH_4$ produced is exhaled in breath. | 12 |
| Pochart et al. | 1993 | Pyxigraphic technique developed for the purpose of sampling methanogens in the proximal colon. | 165 |

methanogenic bacteria, which grow best at approximately neutral pH values.[36] Stephen et al.[74] have shown a positive correlation between intestinal transit time and methanogenesis, which seems logical because low turnover rates should favor slow growing microorganisms. Nevertheless, rural Africans have short transit times, but up to 90% may be $CH_4$-producers, with excretion rates usually much higher than those observed in Caucasian subjects.[68]

It is generally believed that $CH_4$ excretor status is relatively constant.[64] Under certain circumstances, however, $CH_4$ may be absent in otherwise methanogenic subjects. This phenomenon is still poorly understood. In the past, it has been more or less accepted that rates of methanogenesis are independent of dietary influence, but recent studies, including quantitative measurements of total excretion, have shown that $CH_4$ excretion may substantially increase after administrating fermentable substrates such as lactulose, with excretion profiles being similar to those of $H_2$.[12,70,75] As will be discussed later, the dietary influence of sulfate may also be important.

Significant $CH_4$ excretion may persist,[64] unlike breath $H_2$ where excretion may be almost completely depressed in the fasting state. When total gas excretion was studied in subjects given a diet free of nonabsorbable and fermentable carbohydrate, an average rate of $H_2$ excretion of 35 ml/d was measured, with a $CH_4$ production of 149 ml/d.[12] Results from *in vitro* studies suggest that $CH_4$ may then originate from fermentation of endogenous substrates in the colon.[48]

Methane production is certainly an efficient means of disposing of $H_2$ from the large intestine thereby effectively reducing the volume of free gas. Methanogenesis does not seem to be physiologically as important, however, for the colonic fermentation as it is for the rumen. A high percentage of humans live comfortably without excreting detectable $CH_4$ in breath. While in the rumen, the composition of fermentation products is strongly dependent on methanogenesis, no difference in short chain fatty acid production pattern has been found between methanogenic and nonmethanogenic persons.[33,61] Thus, the ecological and physiological importance of $CH_4$ production in the human colonic ecosystem remains unclear.

## IV. DISSIMILATORY SULFATE REDUCTION

In 1895, Beijerinck[76] first described bacteria that were able to utilize the sulfate anion as an electron acceptor for the dissimilation of organic and inorganic compounds. In subsequent years, a number of significant findings have ensured that SRB are often in the forefront of microbiological research. In particular, since the 1970s, these bacteria have been the subject of a number of taxonomic, biochemical and physiological advances. Although sulfate reducers are often considered to be ubiquitous microorganisms that are able to grow well in anaerobic ecosystems where sulfate is readily available, the majority of studies have used inocula from aquatic sediments and oil reservoirs.[77,78] For those bacteriologists familiar with their activities in sewage-related areas, the presence of SRB in human feces will probably come as no surprise. The fact that human gut contents may contain large populations of SRB is by no means a new concept. Studies by Moore et al.[79] and Beerens and Romond[80] first demonstrated their existence in stool samples. The ecological significance of dissimilatory sulfate reduction in the large bowel has only recently been recognized however and, as is usual with these bacteria, is currently an area of some debate. Table 2 summarizes some of the recent findings concerning colonic sulfate reduction.

In fecal samples from persons living in the U.K., high numbers of SRB are common and these bacteria occur in approximately 65% of the healthy population.[81] In Americans, however, the percentage is somewhat lower.[35,82] From British studies, it seems clear that these bacteria play an important role in $H_2$ metabolism, thereby contributing towards a reduction in gas distension problems (see Table 2 and references therein). The consequence of this process, however, is production of potentially toxic $H_2S$ which is the normal end product of SRB metabolism. This may have relevance towards host health (see discussion later). The ability of gut SRB to outcompete methanogens for the mutual growth substrate $H_2$ has been demonstrated.[83] This factor possibly explains why breath $CH_4$ production only occurs in a certain percentage of human populations.[64,68] Variations in the flux level of sulfate available in the colon may select for either sulfate reducers (when present in high concentrations), or methanogens (when available in limited amounts only). American and French studies have indicated that dissimilatory sulfate reduction may not be quantitatively or (from the viewpoint of gas metabolism) physiologically as important, however.[82,84] For example, in Minneapolis, the production of sulfide and methane was assessed in the presence and absence of specific inhibitors, and indicated that about one-third of the population consumed $H_2$ via methanogenesis, one-third via sulfate reduction, and one-third via some unidentified reaction.[44] Although there is no definitive explanation for these discrepancies, a possibility is variations in the supply of electron acceptor (sulfate) to the indigenous microbiota or variant methodologies in different laboratories.

The failure to find high levels (activities) of methanogens and SRB in the same fecal sample suggests that the two types of bacteria are in competition, presumably for $H_2$. Once again there is a minor controversy between Cambridge and Minneapolis investigators over which bacterial type predominates, the former arguing that the sulfate reducers win the competition for $H_2$[83] while the latter group claiming that methanogens predominate.[82]

**Table 2**  Historical Perspective of Dissimilatory Sulfate Reduction in the Human Large Intestine

| Author(s) | Date | Observation(s) | Reference |
|---|---|---|---|
| Moore et al. | 1976 | New genus of SRB isolated using human feces as the inoculum (*Desulfomonas* gen. nov.). | 79 |
| Beerens and Romond | 1977 | Approximately $10^7$/g feces SRB detected in stool samples. | 80 |
| Gibson et al. | 1988 | Inverse relationship between carriage of SRB and MB suggested. Five different genera isolated in 14 of 20 British subjects, but only 3 of 20 South Africans. | 83 |
| Gibson et al. | 1988 | The ability of colonic SRB to outcompete MB for the mutual growth substrate $H_2$ demonstrated. | 166 |
| Gibson et al. | 1988 | Utilization of mucin by cultures of fecal bacteria able to stimulate sulfate reduction at the expense of methanogenesis in mixed culture. | 37 |
| Gibson et al. | 1990 | Sulfate reduction and methanogenesis more or less mutually exclusive in South African volunteers. Availability of $SO_4^{2-}$ postulated as a regulatory mechanism involved in $H_2$ disposal. | 36 |
| Gibson et al. | 1991 | Microbiological features of SRB from the healthy and colitis gut shown to be different. | 154 |
| Christl et al. | 1992 | Feeding of 15 mmol/d $SO_4^{2-}$ to $CH_4$ producing volunteers may regulate methanogenesis. | 109 |
| Macfarlane et al. | 1992 | Studies with gut contents from sudden death victims show that sulfate reduction is favored by physicochemical conditions found in the distal colon. | 46 |
| Strocchi and Levitt | 1992 | Critical role of $H_2$ tension and fecal stirring on gas consumption shown. | 35 |
| Strocchi et al. | 1992 | Reliable method developed for the quantification of $H_2S$ in feces. | 167 |
| Pochart et al. | 1992 | SRB and MB shown to exert a competitive relationship but without being mutually exclusive. Hypothesis that $H_2$ metabolism is regulated by $SO_4^{2-}$ availability contradicted. | 84 |
| Tsai et al. | 1992 | Bacterially produced; novel mucin sulfatase isolated from feces. | 168 |
| Strocchi et al. | 1992 | Methanogens in feces of Minnesotans able to outcompete other $H_2$ utilizing bacteria, even when supplemented with $SO_4^{2-}$. | 82 |
| Strocchi et al. | 1993 | Low carriage of SRB in fecal samples from American persons demonstrated (4 of 14). | 44 |
| Gibson et al. | 1993 | Metabolic diversity of gut SRB demonstrated. | 81 |
| Roediger et al. | 1993 | Reduced sulfur compounds shown to impair colonocyte function and potentially be important in the etiology of ulcerative colitits. | 155 |
| Strocchi et al. | 1995 | Shifts in the balance of the $H_2$ uptake microflora in respone to time demonstrated. | 85 |

Sulfate can be supplied by dietary means or by the depolymerization and desulfation of endogenously produced sulfated glycoproteins such as mucins, which may release the anion in high concentrations. Both processes are likely to be highly variable. It is difficult to estimate the dietary supply of sulfate to the colon, but studies with ileostomists have indicated that values range from 2 to 9 mmol/d.[43] Alternatively, $H_2$ availability may be of some significance. If this substrate is in excess, both hydrogenotrophic methanogenesis and sulfate reduction could occur concomantly. Recent findings by Strocchi et al.[44,85] have indicated that this may indeed be the case in some people. Other factors such as gut pH, or transit time of contents may also be of some relevance.

In a competitive environment where the substrate, in this case $H_2$, is limited, the scenario might be quite different. If sulfate reducers, methanogens, and other $H_2$ utilizers such as acetogenic bacteria are all present, then given adequate sulfate the SRB should theoretically win the competition for this mutual growth substrate. The explanation for this lies in the relative affinities of SRB and methanogens for $H_2$ and, to a lesser extent, energetic considerations of the bacteria.[86] This simple ecological theory should also be applicable to the human large intestine. Studies from the U.K. indicate that this may indeed be the case. In sediments, sulfate concentrations above 30 μmol/l allow the majority of $H_2$ uptake to occur through sulfate reduction. Methanogens become much more important at lower concentrations.[87-94] However, care should be made when extrapolating these findings to the gut ecosystem, because a great deal of variation in $H_2$ availability, both inter- and intraindividually, is likely to occur. However, should gut methanogens

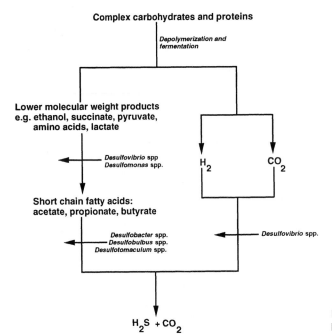

**Complex carbohydrates and proteins**

*Depolymerization and fermentation*

**Lower molecular weight products
e.g. ethanol, succinate, pyruvate,
amino acids, lactate**

*Desulfovibrio* spp
*Desulfomonas* spp.

$H_2$      $CO_2$

**Short chain fatty acids:
acetate, propionate, butyrate**

*Desulfobacter* spp.
*Desulfobulbus* spp.
*Desulfotomaculum* spp.

*Desulfovibrio* spp.

$H_2S + CO_2$

**Figure 4**  Postulated mechanisms of colonic fermentation which may involve SRB.

ever be shown to outcompete SRB for mutual growth substrates without sub-optimal growth conditions for each, then the enigmatic nature of colon microbiology will be much enhanced.

Until now, only a restricted number of colonic SRB have been described.[83] They are species belonging to the genera *Desulfovibrio*,[95] *Desulfobacter*,[96] *Desulfobulbus*,[97] *Desulfotomaculum*,[98] and *Desulfomonas*.[99] Not all of these bacteria are able to utilize $H_2$ as an electron acceptor, however. In contrast to colonic methanogens, SRB show a great deal of metabolic diversity. Figure 4 shows the potential electron donors available for sulfate reduction in the gut. The limiting factor concerning which substrate predominates will be the amount of sulfate available for biosynthesis. Stoichiometrically, however, $H_2$ is likely to be particularly important because 4 mol of this gas can be converted to $H_2S$ by only 1 mol of sulfate (Equation 7). In contrast, the oxidation of other electron donors such as acetate, lactate, propionate, and butyrate require higher concentrations of sulfate.[32,100-102] In the gut, the major hydrogenotrophic SRB belong to the *Desulfovibrio* genus, which always seems to be the numerically predominant group in feces.[39] Some species belonging to this genus are able to grow fermentatively in the absence of sulfate in many anaerobic ecosystems.[77,78,103] Bearing in mind, however, the high availability of organic material in the gut, it is unlikely that these SRB will be of any quantitative significance in this respect.

The microbiological factors that influence the growth of SRB in the human colon are far from understood. Apart from their involvement in $H_2$ disposal, there is very little information on how these bacteria interact with other members of the large intestinal microbiota. It is possible that further studies will increase the taxonomic range of gut isolates so that new genera may be isolated, some of which may perhaps be involved in gas metabolism. Many species may remain in a nonculturable form in the gut. For example, the colonic mucosal layer supplies a rich source of bound sulfate. Considering the preference of some SRB for surface growth (e.g., desulfonemas[104,105]), it is possible that some bacteria may reside within the semi-solid mucin layer that lines the mucosa. These microorganisms are virtually impossible to isolate using conventional microbiological approaches. In this context, recent advances in genotypic methods for studying sulfate reducer growth[106-108] may eventually have some applicability to the large intestine.

The composition of the gut ecosystem is thought to be relatively stable on a long-term basis but notoriously difficult to manipulate externally, i.e., through dietary addition. However, this may not entirely be true for methanogens and SRB. A recent study by Christl et al.[109] showed that addition of 15 mmol/d sodium sulfate to the diet of six $CH_4$-producing volunteers, rapidly inhibited methanogenesis in three of these people. A stimulation of sulfate reducing activities in gut contents from these individuals was observed concomitantly. These results indicate that in some people, SRB are apparently inactive, in

**Table 3**  Historical Perspective of Acetogenesis in the Human Large Intestine

| Author(s) | Date | Observation(s) | Reference |
|---|---|---|---|
| Wolin and Miller | 1983 | $H_2$ dependent acetogenesis suggested as an important source of acetate in the colon. | 62 |
| Lajoie et al. | 1988 | Homoacetogenic reactions confirmed in human fecal slurries. Highest activities recorded where methanogenesis was low. | 38 |
| DeGraeve and Demeyer | 1989 | Addition of $H_2$ gas to fecal incubations increases acetogenesis. | 169 |
| Henderson and Demeyer | 1989 | Potential for acetate production from a combination of $H_2$ and $CO_2$ confirmed. | 116 |
| Gibson et al. | 1990 | In U.K. and South African samples, negligible acetogenesis was detected whereas methanogenesis and sulfate reduction were active. | 36 |
| Duncan and Henderson | 1990 | Approximately 20% of the acetate formed *in vitro* during fermentation studies was provided by homoacetogenesis. | 170 |
| Doré et al. | 1993 | Acetogenic bacteria enumerated in feces using an MPN method (up to 8.5 log N/g) recorded. | 121 |
| Ducros et al. | 1993 | A low pH in conjunction with a lactic acid fermentation lowers acetogenesis. | 171 |
| Bernalier et al. | 1993 | $^{13}$C NMR studies used to demonstrate a role for acetogenesis by the human colonic microflora in a competitive association with methanogenesis for $H_2$. | 172 |

terms of $H_2$ metabolism, and are probably limited by a reduced supply of electron acceptor. However, these bacteria are able to outcompete methanogens for substrate when sulfate becomes more available. Further studies by Strocchi et al.[85] have shown that fluctuations in the methanogenic or sulfate reducing status of particular individuals occur over time. Obviously, explanations of the full microbiological impact of dissimilatory sulfate reduction in the gut are far from complete.

## V. OTHER BACTERIAL MECHANISMS OF HYDROGEN DISPOSAL

British studies of fecal samples from healthy persons living in either the U.K. (n = 81)[81] or South Africa (n = 30)[36] indicate that methanogens or dissimilatory SRB are predominant in terms of colonic $H_2$ disposal. Using metabolic inhibitors of these processes, other studies have shown that alternative pathways of $H_2$ consumption may be operative.[44] Possibilities include dissimilatory nitrate reduction, or the most frequently touted mechanism, acetogenesis.

### A. DISSIMILATORY NITRATE REDUCTION

In fecal slurries, $H_2$ is the preferred electron donor for dissimilatory nitrate reduction.[110] Energetically, and from the viewpoint of substrate affinity, this mechanism of gas disposal would be a favorable method of operation.[86] Therefore a question remains: are sufficient amounts of nitrate available in the colon for this process to be of any quantitative relevance? The standard western diet may contain in excess of 100 mg of $NO_3^-$ per day,[111,112] from a variety of sources including fresh vegetables and drinking water. Studies with ileostomists indicate that an insignificant amount reaches the colon from dietary origin, however.[113] An alternative source is the secretion of $NO_3^-$ from some cells involved in immunological reactions,[114] although this is almost impossible to quantify.

### B. ACETOGENESIS

Acetate production from a combination of $H_2$ and $CO_2$ has been suggested to be an important $H_2$ sink in the colon (Table 3).[38,115,116] This reaction is energetically less favorable than either methanogenesis or dissimilatory sulfate reduction.[86] Acetogenesis has been demonstrated in the rat cecum,[42] termite gut,[117] and pig bowel.[118] It is not surprising that efforts have been made to demonstrate similar activities in the human large intestine. A study by Gibson et al.[36] concluded that acetogenesis was insignificant when either methanogenesis or sulfate reduction were involved in $H_2$ disposal. These results are hardly novel findings since it is recognized that both processes usually dominate $H_2$ dependent acetate production in many anaerobic ecosystems. It probably should not be discounted, however, that environmental microniches in the colon may harbor populations of acetogenic bacteria. For example, sulfate reducers and methanogens are both pH sensitive, possibly allowing for increased acetogenesis under acidic conditions, e.g., in the cecum. In this context, Macfarlane et al.[46] showed, using gut contents taken at autopsy from sudden death victims, that both methanogenesis and sulfate reduction were quantitatively more important distally rather than in proximal regions.

A further problem exists, however. A number of acetogenic bacteria which are likely to grow in the colon are also saccharolytic, e.g., *Eubacterium limosum*[119] and *Peptostreptococcus productus*.[120] *In vitro*, these species may reduce $CO_2$ with $H_2$, however in the large gut, it is probable that they prefer to utilize carbohydrates or amino acids that provide increased energy yields. Moreover, the fact that fairly high numbers of acetogens may be detected in gut contents is unsurprising after it is considered that the enumeration methods used are forms of enrichment culture.[121]

Studies by Lajoie et al.[38] have indicated that in some people, acetogenic bacteria may in fact be of some importance for $H_2$ disposal. Significant rates occurred *in vitro* when fermenters contained high levels of $H_2$. A recent paper by Pochart et al.[84] concluded that reliable enumeration methods need to be developed for a better understanding of colonic acetogenesis. This is probably equally as important as optimization of rate measurements. The study by Gibson et al.[36] involved the use of $^{14}$C-labeled $HCO_3^-$ into acetate, whereas Lajoie et al.[38] used a (probably more accurate) $^{13}$C-labeled approach with nuclear magnetic resonance detection.

## VI. CLINICAL APPLICATIONS OF GAS METABOLISM

### A. BREATH HYDROGEN

Breath $H_2$ measurements have been used for a variety of diagnostic purposes, including carbohydrate malabsorption, bowel transit time, and bacterial overgrowth of the small intestine.

### 1. Carbohydrate Malabsorption

Breath excretion has been used to detect various malabsorption syndromes, assuming that certain carbohydrates not absorbed in the small intestine are fermented by colonic bacteria, and that they therefore cause a rise in $H_2$ concentrations. To test for lactose intolerance (i.e., deficiency of lactase, small intestinal disaccharidase), patients are fed a test dose of lactose, and breath $H_2$ concentrations monitored. This procedure is a simple, safe, and noninvasive test which is particularly valuable in pediatric medicine.[17,122] Similarly, fructose and sucrose intolerance can be tested for.[75,123] The method relies on a relatively constant proportion of $H_2$ being absorbed and excreted by the lungs and assumes that net production resulting from fermentation of malabsorbed carbohydrate is sufficient to increase breath levels of sufficient magnitude to be differentiated from background variability. Despite the widespread use of breath $H_2$ testing, limited attention has been directed toward the appropriate criterion for a positive test. The accuracy of breath $H_2$ for detecting malabsorption of dietary carbohydrate cannot be assessed by comparison with some "gold standard," because none exists. Independent of this, the accuracy of the test is maximized when the fasting breath $H_2$ concentration is very low. To this end, the final meal on the day before testing should contain no poorly absorbed, slowly fermented carbohydrates. This factor is often overlooked.

Other than breath $H_2$ testing, the only clinically useful technique for assessing carbohydrate absorption is the tolerance test. Here, a large dose of carbohydrate is ingested and the normality of absorption determined by the finding of an appreciable increase in blood glucose over the next 1 to 2 h. Despite the various pitfalls of the $H_2$ breath test, it does offer a number of advantages over the absorption approach:

1. It is less invasive in that expired air rather than blood samples are used.
2. It is not dependent on normal intermediary glucose metabolism for accuracy.
3. It is not influenced by variable rates of gastric emptying.
4. It employs lower, more physiological doses of carbohydrate.
5. It detects malabsorption of a small fraction of the test sugar, whereas about 50% of the dose must be malabsorbed to produce a positive tolerance test.

Quantification of carbohydrate malabsorption using breath $H_2$ determinations is difficult, if not impossible. Some of its problems include variability in the proportion of $H_2$ excreted, which is dependent on whether the test carbohydrate is slowly or rapidly fermented. Moreover, as outlined earlier, a number of bacteria in the gut are able to metabolize $H_2$, which results in a nonlinear correlation between its production and carbohydrate fermentation. A further problem exists in the handling of highly variable fasting breath $H_2$ meaurements. The conventional approach is to subtract the fasting value from all subsequent measurements obtained after the test meal. This procedure would be correct if the basal level remained constant throughout the breath testing period. However, high basal levels fall rapidly over several hours. Thus, the conventional maneuver of subtracting the basal $H_2$ from subsequent measurements markedly underestimates excretion from the meal in subjects with high fasting levels.

## 2. Transit Time Measurements

The time elapsing between the ingestion of a nonabsorbed carbohydrate (such as lactulose), and an increase in breath $H_2$ excretion should theoretically be a measure of the time required for the "head" of the carbohydrate load to move from the stomach to the cecum.[124] This technique has been used for many years in the absence of good experimental evidence of its accuracy. The recent development of scintillation methods to assess transit time has made it possible to simultaneously measure transit time with lactulose and a radioactive probe.[125] Such studies have shown that comparable transit times can be recorded with the two techniques provided that early unsustained and, usually small, increases in breath $H_2$ are neglected. These early $H_2$ peaks are frequently observed in breath studies and can be attributed to bacterial overgrowth of the small bowel, bacteria in the mouth, movement of ileal carbohydrate from a previous meal into the cecum, and increased colonic motility with release of $H_2$ trapped in colonic contents.

## 3. Bacterial Overgrowth of the Small Intestine

A rise in breath $H_2$ excretion shortly after ingesting a carbohydrate (before the carbohydrate could reach the cecum) might be expected to be a good indicator of bacterial overgrowth in the small bowel.[126,127] Although the test is used clinically, there is still controversy about its limited correlation with other methods to detect small gut overgrowth.[128,129] Problems with the technique include the common tendency to find small early breath $H_2$ peaks in healthy people. In addition, in conditions such as gastrectomy states, small bowel transit time of lactulose may be extremely rapid, e.g., 5 to 10 min.[130] Thus, utilization of the test sugar in the cecum cannot be differentiated from the small bowel fermentation. Although the simple, noninvasive, nonradioactive nature of breath $H_2$ measurements makes it an appealing technique for some diagnostic purposes, it appears to be of limited value in the diagnosing bacterial overgrowth syndromes.

## B. FLATULENCE AND METEORISM

Although gas is generally produced in the human large intestine it does not usually cause symptoms. First, this may depend on the volume of free gas present in the colon. A proportion of this gas is nitrogen (20 to 90%),[131] the origin of which is mostly swallowed air. As mentioned earlier, the remainder consists of the fermentation products $H_2$, $CO_2$, and $CH_4$. Production of these gases depends on the amount and nature of fermentable substrate delivered to the colon.[5] This is particularly obvious when rapidly fermentable substrates such as oligosaccharides from legumes are ingested, since they are notoriously flatogenic with flatus excretions as high as 176 ml/h being reported.[3] Most nonabsorbable carbohydrate in the average western diet consists of nonstarch polysaccharides (dietary fiber) or starch, which yield substantially lower gas volumes, as shown *in vivo* and *in vitro*.[12,132,133] Thus, the ability of the bacterial population to reduce the volume of gas produced is potentially as important as substrate supply to the colonic flora.

Studies using colonic washout techniques[131] to assess intestinal gas volumes have shown that the large bowel usually contains less than 200 ml of gas. Thus, individual tolerance to increased intraluminal gas volumes may also be a factor which determines meteorism and flatulence. Reliable quantitative studies of total gas excretion in flatulent people have not been done except in calorimetric studies of patients with pneumatosis cystoides intestinalis where $H_2$ excretion can be as high as 1700 ml/d.[134] In clinical medicine, meteorism and flatulence may be a problem in malabsorption syndromes such as lactase deficiency, but also when nonabsorbable sugars or sugar-alcohols (such as lactulose or lactitol) are given to patients with hepatic encephalopathy.[135] Disaccharidase inhibitors, (e.g., acarbose) which are used to delay carbohydrate assimilation in diabetic patients may also cause significant malabsorption and fermentation making flatulence a major complaint during such treatment.[136]

## C. PNEUMATOSIS CYSTOIDES INTESTINALIS

Pneumatosis cystoides intestinalis is a disease whose clinical features and symptoms are dominated by intestinal gas disorder. It is a rare condition characterized by the presence of multiple gas filled cysts in the bowel wall.[137,138] These cysts are shown in Plate 3 and contain mostly $N_2$ and $H_2$. Excessive flatulence and very high fasting breath $H_2$ levels are symptomatic.[139,140] The etiology is uncertain and the prognosis benign. While previously the cysts were presumed to originate from ruptured emphysematous cysts[141] or from gas-forming bacteria within the gut wall,[142] it has been suggested recently that an unusually high $H_2$ tension in the bowel lumin might be responsible.[140-144] This hypothesis has been strengthened by a study showing that patients with this disease experience excessive $H_2$ excretion in breath and flatus.[134] Although only three patients were studied, they had neither significant numbers of sulfate-reducing nor methanogenic bacteria in fecal samples. The absence of important $H_2$ metabolizing bacteria in the colon may help explain the accumulation of gas with flatulence and possibly the formation of cysts. The disease may prove to be a useful model for studying gas excretion *in vivo* in the absence of significant $H_2$ consumption. On a polysaccharide-free diet, pneumatosis patients excreted about 400 ml/d of $H_2$ compared to only 35 ml/d in control subjects. $H_2$ from lactulose was 100 ml/g in the patients compared to 13 ml/g in controls.[134] Thus, the total volume of $H_2$ excreted by these patients was much closer to values predicted from stoichiometric calculations,[62,134] than values measured in normal subjects.

## D. TOXICITY OF SULFIDE AND ULCERATIVE COLITIS

Hydrogen sulfide and $CH_4$ are the major products of $H_2$ metabolism in humans. $CH_4$ is a fairly inert gas not further metabolized by mammalian cells and is thought to be free of any adverse effects on humans (apart from its combustibility). Hydrogen sulfide, however, is a highly toxic substance[145] which can cause cellular hypoxia, presumably by inhibiting mitochondrial cytochrome oxidases.[146] The amount of sulfide produced in the human colon, mainly through dissimilatory sulfate reduction, is unknown. With the assumption that most dietary sulfate not absorbed in the small intestine can be reduced to sulfide, as much as 10 mmol of this metabolite per day can be theoretically produced,[12,43,109] which is a potentially lethal dose. However, as sulfide is rapidly metabolized within the gut mucosa by oxidation and methylation[147,148] systemic effects do not seem to occur. More likely is an effect on the colonic mucosa itself which is indicated by several lines of evidence: When rats are fed highly sulfated polysaccharides, such as certain carrageenans, they develop a condition very similar to ulcerative colitis in humans.[149,150] Moreover, even dysplasia and adenocarcinoma (both observed in longstanding ulcerative colitis) can be induced in animals fed carrageenan or dextran sulfate.[151,152] It may be speculated that this is caused by elevated sulfate reduction and accumulation of toxic sulfide. However, no direct measurements are available to support this hypothesis. In humans, despite the fact that only about 50% of a healthy population harbor SRB, and display active sulfate reduction, this is the case in more than 90% of patients with ulcerative colitis.[153] Moreover, sulfate reducing activity is higher in these patients.[154] Further support for the hypothesis of sulfide somehow being involved in the pathogenesis of ulcerative colitis was provided in a study by Roediger and his co-workers[155] which showed that butyrate dependent energy metabolism of colonocytes in cell culture was severely depressed in the presence of sodium-hydrogen sulfide, methanethiol, or mercaptoacetate. This is particularly interesting as pathophysiology of ulcerative colitis seems to include a deficiency in the metabolism of butyrate, the preferred "fuel" of colonocytes.[156,157]

## E. METHANE AND COLORECTAL CANCER

Haines et al.[158] have reported a considerably increased incidence of methanogenesis in patients with colorectal cancer (80 vs. 40% in controls). These results have been confirmed by Pique et al.[159] who also showed that the proportion of $CH_4$ producers in the cancer group (92%) fell to that of a control population after resection. However, more recent studies were unable to reproduce these findings.[160,161] Moreover it is known that rural African populations are mostly methanogenic (over 90%) but have a very low incidence of colorectal cancer compared to western populations.[68] Thus, the link between methanogenesis and colorectal cancer is still open to discussion. It is more likely that any association between methanogenesis and bowel cancer can be attributed to slow growth of the bacteria and a decreased transit time of the patient. Physical obstructions such as tumors would contribute to a prolonged residence time of gut contents (including colonic bacteria).

## VII. CONCLUSIONS

This chapter has given an overview of bacterial gas metabolism in the human large intestine. The mechanisms involved are sometimes complex and often controversial. For example, the full influence of dissimilatory sulfate reduction both in colonic ecophysiology and, possibly, pathology has yet to be clarified. As research progresses, it becomes clear that some accepted mechanisms of study should be reconsidered. In particular, breath $H_2$ testing as a means of quantifying carbohydrate fermentation in the hindgut should now be regarded as reliable on a semi-quantifiable basis only. Although the full composition of flatus gas is still not wholly understood, it is thought that bacterial mechanisms of production seem to be central in this. Most importantly, gas metabolism of any nature has never been shown to be involved in the etiology of gut disease, although interesting studies with pneumatosis cystoides intestinalis, ulcerative colitis, and colon cancer have been carried out.

## REFERENCES

1. **Engel, R. R. and Levitt, M. D.,** Intestinal tract gas formation in newborns, in *Program for Meeting of American Pediatric Society and Society for Pediatric Research*, 1970, 255.
2. **Levitt, M. D. and Bond, J. H.,** Volume, composition and source of intestinal gas, *Gastroenterology*, 59, 921, 1970.
3. **Steggerda, F. R.,** Gastrointestinal gas following food consumption, *Ann. N.Y. Acad. Sci.*, 150, 57, 1968.
4. **Lasser, R. B., Bond, J. H., and Levitt, M. D.,** The role of intestinal gas in functional abdominal pain, *N. Engl. J. Med.*, 293, 524, 1975.
5. **Calloway, D. H.,** Respiratory hydrogen and methane as affected by consumption of gas-forming foods, *Gastroenterology*, 51, 383, 1966.
6. **Levitt, M. D., Berggren, T., Hastings, J., and Bond, J. H.,** Hydrogen ($H_2$) catabolism in the colon of the rat, *J. Lab. Clin. Med.*, 84, 163, 1974.
7. **Ruge, E.,** Beitrag zur kennuness der darmgase, *Sitsber. Kaiserlicken Akad.*, 44, 739, 1861.
8. **Tomlin, J., Lewis, C., and Read, N. W.,** Investigation of normal flatus production in healthy volunteers, *Gut*, 32, 665, 1991.
9. **Beazell, J. M. and Ivy, A. C.,** The quantity of colonic flatus excreted by the normal individual, *Am. J. Digest. Dis.*, 8, 128, 1941.
10. **Calloway, D. H.,** Gas in the alimentary canal, in *Handbook of Physiology*, Code, C. F., Ed., American Physiological Society, Washington, D.C., 2839, 1967.
11. **Levitt, M. D.,** Production and excretion of hydrogen gas in man, *N. Engl. J. Med.*, 281, 122, 1969.
12. **Christl, S. U., Murgatroyd, P. R., Gibson, G. R., and Cummings, J. H.,** Quantitative measurement of hydrogen and methane from fermentation using a whole body calorimeter, *Gastroenterology*, 102, 1269, 1992.
13. **Altman, P. L. and Dittmer, D. S.,** Respiration and circulation, *Fed. Am. Societ. Exp. Biol.*, 3, 1571, 1971.
14. **Levitt, M. D., Aufderheide, T., Fetzer, C. A., Bond, H. H., and Levitt, D. G.,** Use of carbon monoxide to measure luminal stirring in the rat gut, *J. Clin. Invest.*, 74, 2056, 1984.
15. **Levitt, D. G., Bond, J. H., and Levitt, M. D.,** Use of model of small bowel mucosa to predict passive absorption, *Am. J. Physiol.*, 239, G23, 1980.
16. **Levitt, M. D. and Donaldson, R. M.,** Use of respiratory hydrogen excretion to detect carbohydrate malabsorption, *J. Lab. Clin. Med.*, 75, 937, 1970.
17. **Metz, G., Jenkins, D. J. A., Peters, T. J., Newman, A., and Blendis, L. M.,** Breath hydrogen as a diagnostic method for hypolactasia, *Lancet*, 1, 1155, 1975.
18. **Strocchi, A., Ellis, C. J., and Levitt, M. D.,** Reproducibility of measurements of trace gas concentrations in expired air, *Gastroenterology*, 101, 175, 1991.
19. **Ellis, C. J., Kneip, J. M., and Levitt, M. D.,** Storage of breath samples for hydrogen analysis, *Gastroenterology*, 94, 822, 1988.
20. **Rumney, C. J. and Rowland, I. R.,** *In vivo* and *in vitro* models of the human colonic microflora, *Crit. Rev. Food Sci. Technol.*, 31, 299, 1992.
21. **Edwards, C. A., Duerden, B. L., and Read, N. W.,** Metabolism of mixed human colonic bacteria in a continuous culture mimicking the human cecal contents, *Gastroenterology*, 88, 1903, 1985.
22. **Miller, T. L. and Wolin, M. J.,** Fermentation by the human large intestinal community in an *in vitro* semicontinuous culture system, *Appl. Environ. Microbiol.*, 42, 400, 1981.
23. **Manning, B. W., Federle, T. W., and Cerniglia, C. E.,** Use of a semicontinuous culture system as a model for determining the role of human intestinal microflora in the metabolism of xenobiotics, *J. Microbiol. Meth.*, 6, 81, 1987.
24. **Mallet, A. K., Bearne, C. A., and Rowland, I. R.,** The use of continuous flow systems for studying the metabolic activity of the hindgut microflora *in vitro*, *Food Chem. Toxicol.*, 24, 743, 1986.
25. **Freter, R., Stauffer, E., Cleve, D., Holdeman, L. V., and Moore, W. E. C.,** Continuous flow cultures as *in vitro* models of the ecology of large intestinal flora, *Infect. Immun.*, 39, 666, 1983.
26. **Bearne, C. A., Mallett, A. K., Rowland, I. R., and Brennan-Craddock, W. E.,** Continuous culture of human faecal bacteria as an *in vitro* model for the colonic microflora, *Toxicol. In Vitro*, 4, 522, 1990.
27. **Bearne, C. A., Mallett, A. K., and Rowland, I. R.,** The use of continuous flow methodology to study human gut flora metabolism, *Proc. IX Int. Cong. Gnotobiology*, Versailles, 1987.

28. **Macfarlane, G. T., Hay, S., and Gibson, G. R.,** Influence of mucin on glycosidase, protease and arylamidase activities of human gut bacteria grown in a 3-stage continuous culture system, *J. Appl. Bacteriol.,* 66, 407, 1989.

29. **Macfarlane, G. T., Cummings, J. H., Macfarlane, S., and Gibson, G. R.,** Influence of retention time on degradation of pancreatic enzymes by human colonic bacteria grown in a 3-stage continuous culture system, *J. Appl. Bacteriol.,* 67, 521, 1989.

30. **Levitt, M. D., French, P., and Donaldson, R. M.,** Use of hydrogen and methane excretion in the study of the intestinal flora, *J. Lab. Clin. Med.,* 72, 988, 1968.

31. **Macfarlane, G. T.,** Fermentation reactions in the large intestine, in *Short Chain Fatty Acids: Metabolism and Clinical Importance,* Cummings, J. H., Rombeau, J. I., and Sakata, T., Eds., Ross Laboratories Press, Columbus, OH, 1991, 5.

32. **Gibson, G. R., Macfarlane, G. T., and Cummings, J. H.,** Sulphate-reducing bacteria and hydrogen metabolism in the human large intestine, *Gut,* 34, 437, 1993.

33. **Wolin, M. J.,** Interactions between $H_2$-producing species, in *Microbial Formation and Utilization of Gases,* Schlegel, H. G. and Pfennig, N., Eds., Goltze Press, Göttingen, Germany, 1976, 141.

34. **Perman, J. A., Modler, S., and Olson, A. C.,** Role of pH in production of hydrogen from carbohydrates by colonic bacterial flora, *J. Clin. Invest.,* 67, 643, 1981.

35. **Strocchi, A. and Levitt, M. D.,** Factors affecting hydrogen production and consumption by human fecal flora: the critical role of hydrogen tension and methanogenesis, *J. Clin. Invest.,* 89, 1304, 1992.

36. **Gibson, G. R., Cummings, J. H., Macfarlane, G. T., Allison, C., Segal, I., Vorster, H. H., and Walker, A. R. P.,** Alternative pathways for hydrogen disposal during fermentation in the human colon, *Gut,* 31, 679, 1990.

37. **Gibson, G. R., Cummings, J. H., and Macfarlane, G. T.,** Use of a three-stage continuous culture system to study the effect of mucin on dissimilatory sulfate reduction and methanogenesis by mixed populations of human gut bacteria, *Appl. Environ. Microbiol.,* 54, 2750, 1988.

38. **Lajoie, S. F., Bank, S., Miller, T. L., and Wolin, M. J.,** Acetate production from hydrogen and [$^{13}$C] carbon dioxide by the microflora of human feces, *Appl. Environ. Microbiol.,* 54, 2723, 1988.

39. **Gibson, G. R.,** A review. Physiology and ecology of the sulphate-reducing bacteria, *J. Appl. Bacteriol.,* 69, 769, 1990.

40. **Smith, M. R.,** Reversal of 2-bromoethanesulfonate inhibition of methanogenesis in *Methanosarcina* sp., *J. Bacteriol.,* 156, 516, 1983.

41. **Taylor, B. F. and Oremland, R. S.,** Depletion of adenosine triphosphate in desulfovibrio by oxyanions of group IV elements, *Curr. Microbiol.,* 3, 101, 1979.

42. **Prins, R. A. and Lankhorst, A.,** Synthesis of acetate from $CO_2$ in the cecum of some rodents, *FEMS Microbiol. Lett.,* 1, 255, 1977.

43. **Florin, T. H. J., Neale, G., Gibson, G. R., Christl, S. U., and Cummings, J. H.,** Metabolism of dietary sulphate: Absorption and excretion in humans, *Gut,* 32, 766, 1991.

44. **Strocchi, A., Ellis, C. J., and Levitt, M. D.,** Use of metabolic inhibitors to study $H_2$ consumption by human feces: Evidence for a pathway other than methanogenesis and sulfate reduction, *J. Lab. Clin. Med.,* 121, 320, 1993.

45. **Flourie, B. F., Etanchaud, F., Florent, C., Pellier, P., Bouhnik, Y., and Rambaud, J. C.,** Comparative study of hydrogen and methane production in the human colon using caecal and faecal homogenates, *Gut,* 31, 684, 1990.

46. **Macfarlane, G. T., Gibson, G. R., and Cummings, J. H.,** Comparison of fermentation reactions in different regions of the human colon, *J. Appl. Bacteriol.,* 72, 57, 1992.

47. **Kerlin, P. and Wong, L.,** Breath hydrogen testing in bacterial overgrowth of the small intestine, *Gastroenterology,* 95, 982, 1988.

48. **Perman, J. A. and Modler, S.,** Glycoproteins as substrates for production of hydrogen and methane by colonic bacterial flora, *Gastroenterology,* 82, 911, 1982.

49. **Tomlin, J. and Read, N. W.,** The effect of resistant starch on colon function in humans, *Br. J. Nutr.,* 64, 589, 1990.

50. **Jain, N. K., Rosenberg, D. B., Ulahannan, M. J., Glasser, M. J., and Pitchumoni, C. S.,** Sorbitol intolerance in adults, *Am. J. Gastroenterol.,* 80, 678, 1985.

51. **Tadesse, K. and Eastwood, M. A.,** Metabolism of dietary fiber components in man assessed by breath hydrogen and methane, *Br. J. Nutr.,* 40, 393, 1978.

52. **Rune, S. J.,** Acid-base parameters of duodenal contents in man, *Gastroenterology,* 62, 533, 1972.

53. **Moore, J. G., Jessop, L. D., and Osborne, D. N.,** A gas chromatographic and mass spectrometric analysis of the odor of human feces, *Gastroenterology,* 93, 1321, 1987.

54. **Orton, J. M. and Neuhaus, O. W.,** Nutrition: Digestion, absorption and energy metabolism, in *Human Biochemistry,* 9th ed., Orton, J. M. and Neuhaus, O. W., Eds., C. V. Mosby, St. Louis, 1975, 471.

55. **Balch, W. E., Fox, G. E., Magrum, L. J., Woese, C. L., and Wolfe, R. S.,** Methanogens: reevaluation of a unique biological group, *Microbiol. Rev.,* 43, 260, 1979.

56. **Bryant, M. P.,** The microbiology of anaerobic degradation and methanogenesis with special reference to sewage, in *Microbial Energy Conservation,* Schlegel, H. G. and Barnea, J., Eds., Goltze Press, Göttingen, Germany, 1976, 107.

57. **Miller, T. L. and Wolin, M. J.,** Oxidation of hydrogen and reduction of methanol to methane is the sole energy source for a methanogen isolated from human feces, *J. Bacteriol.,* 153, 1051, 1983.

58. **Miller, T. L. and Wolin, M. J.,** *Methanosphaera stadtmaniae* gen. nov., sp. nov. a species that forms methane by reducing methanol with hydrogen, *Arch. Microbiol.,* 141, 116, 1985.

59. **Miller, T. L. and Wolin, M. J.,** Enumeration of *Methanobrevibacter smithii* in human feces, *Arch. Microbiol.,* 45, 317, 1982.

60. **Miller, T. L., Wolin, M. J., de Macario, E. C., and Macario, A. J. L.,** Isolation of *Methanobrevibacter smithii* from human feces, *Appl. Environ. Microbiol.,* 43, 227, 1982.

61. **Wolin, M. J.,** Metabolic interactions among intestinal microorganisms, *Am. J. Clin. Nutr.,* 27, 1320, 1974.

62. **Wolin, M. J. and Miller, T. L.,** Carbohydrate fermentation, in *Human Intestinal Microflora in Health and Disease,* Hentges, D. J., Ed., Academic Press, London, 1983, 147.

63. **Bjornkelett, A. and Jenssen, E.,** Relationship between hydrogen ($H_2$) and methane ($CH_4$) production in man, *Scand. J. Gastroenterol.,* 17, 985, 1982.

64. **Bond, J. H., Engel, R. R., and Levitt, M. D.,** Factors influencing pulmonary methane excretion in man. An indirect method of studying the *in situ* metabolism of the methane-producing colonic bacteria, *J. Exp. Med.,* 133, 572, 1971.

65. **McKay, F. L., Eastwood, M. A., and Brydon, W. G.,** Methane excretion in man: a study of breath, flatus and feces, *Gut,* 26, 69, 1985.

66. **Pitt, P., de Bruijn, K. M., Beeching, M. F., Goldberg, E., and Blendis, L. M.,** Studies on breath methane: the effect of ethnic origin and lactulose, *Gut,* 21 951, 1980.

67. **Drasar, B. S., Tomkins, A. M., Wiggins, H., and Hudson, M.,** Breath methane levels and intestinal methanogenesis among rural Nigerians on a local diet, *Proc. Nutr. Soc.,* 43, 86A, 1984.

68. **Segal, I., Walker, A. R. P., Lord, S., and Cummings, J. H.,** Breath methane and large bowel cancer risk in contrasting African populations, *Gut,* 29, 608, 1988.

69. **Peled, Y., Gilat, T., Libermann, T., and Bujanover Y.,** The development of methane production in childhood and adolescence, *J. Pediatr. Gastroenterol. Nutr.,* 4, 575, 1985.

70. **Flatz, G., Czeizel, A., Metneki, J., Flatz, S. D., Kühnau, W., and Jahn, D.,** Pulmonary hydrogen and methane excretion following ingestion of an unabsorbable carbohydrate: a study of twins, *J. Pediatric. Gastroenterol. Nutr.,* 4, 936, 1985.

71. **Calloway, D. H. and Murphy, E. L.,** The use of expired air to measure intestinal gas formation, *Ann. N. Y. Acad. Sci.,* 159, 82, 1968.

72. **Hungate, R. E.,** *The Rumen and its Microbes,* Academic Press, New York, 1966.

73. **Flourie, B., Pellier, P., Florent, C., Marteau, P., Pochart, P., and Rambaud, J. C.,** Site and substrates for methane production in human colon, *Am. J. Physiol.,* 260, G752, 1991.

74. **Stephen, A. M., Haddad, A. C., and Phillips, S. F.,** Passage of carbohydrate in the colon. Direct measurements in humans, *Gastroenterology,* 85, 589, 1983.

75. **Perman, J. R., Barr, R. G., and Watkins, J. B.,** Sucrose malabsorption in children: non-invasive diagnosis by interval breath hydrogen determination, *J. Pediatr.,* 93, 17, 1978.

76. **Beijerinck, M. W.,** Uber *Spirillum desulfuricans* als ursache von sulfatreduction, *Zentralblatt Bakteriol.,* 1, 1, 1895.

77. **Postgate, J. R.,** *The Sulphate-Reducing Bacteria,* 2nd ed., Cambridge University Press, Cambridge, 1984.

78. **Odom, J. M. and Singleton, R.,** *The Sulfate-Reducing Bacteria: Contemporary Perspectives,* Springer-Verlag, New York, 1993.

79. **Moore, W. E. C., Johnson, J. L., and Holdeman, L. V.,** Emendation of *Bacteroidaceae* and *Butyrivibrio* and descriptions of *Desulfomonas* gen. nov. and ten new species in the genera *Desulfomonas, Butyrivibrio, Eubacterium, Clostridium* and *Ruminococcus, Int. J. Sys. Microbiol.,* 26, 238, 1976.

80. **Beerens, H. and Romond, C.,** Sulfate-reducing anaerobic bacteria in human feces, *Am. J. Clin. Nutr.,* 30, 1770, 1977.

81. **Gibson, G. R., Macfarlane, S., and Macfarlane, G. T.,** Metabolic interactions involving sulphate-reducing and methanogenic bacteria in the human large intestine, *FEMS Microbiol. Ecol.,* 12, 117, 1993.

82. **Strocchi, A., Furne, J. K., Ellis, C. J., and Levitt, M. D.,** Competition for hydrogen by human fecal bacteria: evidence for the predominance of methanobacteria, *Gut,* 32, 1498, 1992.

83. **Gibson, G. R., Macfarlane, G. T., and Cummings, J. H.,** Occurrence of sulphate-reducing bacteria in human faeces and the relationship of dissimilatory sulphate reduction to methanogenesis in the large gut, *J. Appl. Bacteriol.,* 65, 103, 1988.

84. **Pochart, P., Dore, J., Lemann, F., Goderel, I., and Rambaud, J. C.,** Interrelations between populations of methanogenic archaea and sulfate-reducing bacteria in the human colon, *FEMS Microbiol. Lett.,* 98, 225, 1992.

85. **Strocchi, A., Ellis, C. J., Furne, J. K., and Levitt, M. D.,** Study of the constancy of the hydrogen consuming flora of the human colon, *Dig. Dis. Sci.,* in press.

86. **Thauer, R. K., Jungermann, K., and Decker, K.,** Energy conservation in chemotrophic anaerobic bacteria, *Bacteriol. Rev.,* 41, 100, 1977.

87. **Lovley, D. R., Dwyer, D. F., and Klug, M. J.,** Kinetic analysis of competition between sulfate reducers and methanogens for hydrogen in sediments, *Appl. Environ. Microbiol.,* 43, 1373, 1982.

88. **Lovley, D. R. and Klug, M. J.,** Model for the distribution of sulfate reduction and methanogenesis in freshwater sediments, *Geochim. et Cosmochim. Acta,* 50, 11, 1986.

89. **Kristjansson, J. K., Schonheit, P., and Thauer, R. K.,** Different Ks values for hydrogen of methanogenic bacteria and sulfate-reducing bacteria: an explanation for the apparent inhibition of methanogenesis by sulfate, *Arch. Microbiol.,* 131, 278, 1982.

90. **Oremland, R. S. and Polcin, S.,** Methanogenesis and sulfate reduction: competitive and non-competitive substrates in estuarine sediments, *Appl. Environ. Microbiol.,* 44, 1270, 1982.

91. **Capone, D. G. and Kiene, R. P.,** Comparison of microbial dynamics in marine and freshwater sediments: contrasts, in anaerobic carbon catabolism, *Limnol. Oceanogr.,* 33, 725, 1988.

92. **Winfrey, M. R. and Zeikus, J. G.,** Effect of sulfate on carbon and electron flow during microbial methanogenesis in freshwater sediments, *Appl. Environ. Microbiol.,* 37, 244, 1979.

93. **Abram, J. W. and Nedwell, D. B.,** Inhibition of methanogenesis by sulphate-reducing bacteria competing for transferred hydrogen, *Arch. Microbiol.,* 117, 89, 1978.

94. **Robinson, J. and Tiedje, J. M.,** Competition between sulfate-reducing and methanogenic bacteria for hydrogen under resting and growing conditions, *Arch. Microbiol.,* 137, 26, 1984.

95. **Postgate, J. R. and Campbell, L. L.,** Classification of *Desulfovibrio* species, the non-sporulating sulfate-reducing bacteria, *Bacteriol. Rev.,* 30, 732, 1966.

96. **Widdel, F. and Pfennig, N.,** Studies on dissimilatory sulfate-reducing bacteria that decompose fatty acids. I. Isolation of new sulfate-reducing bacteria enriched with acetate from saline environments. Description of *Desulfobacter postgatei* gen. nov., sp. nov., *Arch. Microbiol.,* 129, 395, 1981.

97. **Widdel, F. and Pfennig, N.,** Studies on dissimilatory sulfate-reducing bacteria that decompose fatty acids. II. Incomplete oxidation of propionate by *Desulfobulbus propionicus* gen. nov., sp. nov., *Arch. Microbiol.,* 131, 360, 1982.

98. **Widdel, F. and Pfennig, N.,** Sporulational and further nutritional characteristics of *Desulfotomaculum acetoxidans,* *Arch. Microbiol.,* 129, 401, 1981.

99. **Widdel, F.,** Microbiology and ecology of sulfate- and sulfur-reducing bacteria, in *Biology of Anaerobic Microorganisms,* Zehnder, A. J. B., Ed., John Wiley & Sons, New York, 1988, 469.

100. **Sørensen, J., Christensen, D., and Jørgensen, B. B.,** Volatile fatty acids and hydrogen as substrates for sulfate-reducing bacteria in anaerobic marine sediments, *Appl. Environ. Microbiol.,* 42, 5, 1981.

101. **Christensen, D.,** Determination of substrates oxidized by sulphate reduction in intact cores of marine sediments., *Limnol. Oceanogr.,* 29, 192, 1984.

102. **Parkes, R. J., Gibson, G. R., Mueller-Harvey, I., Buckingham, W. J., and Herbert, R. A.,** Determination of the substrates for sulphate-reducing bacteria within marine and estuarine sediments with different rates of sulphate reduction, *J. Gen. Microbiol.,* 135, 175, 1989.

103. **Pfennig, N., Widdel, F., and Truper, H. G.,** The dissimilatory sulfate-reducing bacteria, in *The Prokaryotes: A Handbook on Habitats, Isolation and Identification of Bacteria,* Starr, M. P., Stolp, H., Truper, H. G., Balows, A., and Schlegel, H. G., Eds., Springer-Verlag, Berlin, 1981, 926.

104. **Widdel, F., Kohring, G. W., and Mayer, F.,** Studies on dissimilatory sulfate-reducing bacteria that decompose fatty acids. III. Characterization of the filamentous gliding *Desulfomena limicola* gen. nov., sp. nov. and *Desulfonema magnum* sp. nov., *Arch. Microbiol.,* 134, 286, 1983.

105. **Widdel, F.,** *Anaeroberabbau von fettsäuren und benzoesäure durch neu iosolerte arten sulfat-reduzierender bakterien,* Ph.D. thesis, University of Göttingen, Göttingen, Germany, 1980.

106. **Wall, J. D.,** Genetics of the sulfate-reducing bacteria, in *The Sulfate-Reducing Bacteria: Contemporary Perspectives,* Odom, J. M. and Singleton, R., Eds., Springer-Verlag, New York, 1993, 77.

107. **Voordouw, G.,** Molecular biology of the sulfate-reducing bacteria, in *The Sulfate- Reducing Bacteria: Contemporary Perspectives,* Odom, J. M. and Singleton, R., Eds., Springer-Verlag, New York, 1993, 88.

108. **Voordouw, G. and Wall, J. D.,** Genetics and molecular biology of sulfate-reducing bacteria, in *Genetics and Molecular Biology of Anaerobic Bacteria,* Sebald, M., Ed., Springer-Verlag, New York, 1993, 456.

109. **Christl, S. U., Gibson, G. R., and Cummings, J. H.,** Role of dietary sulphate in the regulation of methanogenesis in the human large intestine, *Gut,* 33, 1234, 1992.

110. **Allison, C. and Macfarlane, G. T.,** Effect of nitrate on methane production by slurries of human faecal bacteria, *J. Gen. Microbiol.,* 134, 1397, 1988.

111. **Walker, R.,** Naturally occurring nitrate/nitrite in foods., *J. Sci. Food Agr.,* 26, 1735, 1975.

112. **Knight, T. M., Forman, D., Al-Dabbagh, S. A., and Doll, R.,** Estimation of dietary intake of nitrate and nitrite in Great Britain, *Food Chem. Toxicol.,* 25, 277, 1987.

113. **Florin, T. H. J., Neale, G., and Cummings, J. H.,** The effect of dietary nitrate on nitrate and nitrite excretion in man, *Br. J. Nutr.,* 64, 387, 1990.

114. **Beeken, W., Northwood, I., Beliveau, C., and Gump, D.,** Phagocytes in cell suspensions of human colon mucosa, *Gut,* 28, 976, 1987.

115. **Demeyer, D. I., DeGraeve, K., Durand, M., and Stevani, J.,** Acetate: a hydrogen sink in hindgut fermentation as opposed to rumen fermentation, *Acta Vet. Scand.,* 86, 88, 1989.

116. **Henderson, C. and Demeyer, D. I.,** The rumen as a model of the microbiology of fibre digestion in the human large intestine, *Anim. Feed Sci. Technol.,* 23, 227, 1989.

117. **Breznak, J. A. and Switzer, J. M.,** Acetate synthesis from $H_2$ plus $CO_2$ by termite gut microbes, *Appl. Environ. Microbiol.*, 52, 623, 1986.

118. **Greening, R. C. and Leedle, J. A. Z.,** Enrichment and isolation of *Acetitomaculum ruminis*, gen nov., sp. nov.: acetogenic bacteria from the bovine rumen, *Arch. Microbiol.*, 151, 399, 1989.

119. **Genthner, B. R. S. and Bryant, M. P.,** Growth of *Eubacterium limosum* with carbon monoxide as the energy source, *Appl. Environ. Microbiol.*, 43, 70, 1982.

120. **Lorowitz, W. H. and Bryant, M. P.,** *Peptostreptococcus productus* strain that grows rapidly with $CO_2$ as the energy source, *Appl. Environ. Microbiol.*, 47, 961, 1984.

121. **Doré, J., Morvan, B., Pochart, P., Goderel, I., Rieu-Lesme, F., Fonty, G., Rambaud, J. C., and Gouet, P.,** Enumeration of bacteria forming acetate from $H_2$ and $CO_2$ and other $H_2$-utilizing micro-organisms from the digestive tract of animals and man, *Proc. Nutr. Soc.*, 52, 117A, 1993.

122. **Newcome, A. D., McGill, B., Thomas, P. J., and Hofmann, A. F.,** Prospective comparison of indirect methods for lactase deficiency, *N. Engl. J. Med.*, 293, 232, 1975.

123. **Rumessen, J. J. and Gudmand-Hoyer, E.,** Absorption capacity of fructose in healthy adults. Comparison with sucrose and its constituent monosaccharides, *Gut*, 27, 1161, 1986.

124. **Read, N. W., Miles, C. A., and Fisher, D.,** Transit of a meal through the stomach, small intestine, and colon in normal subjects and its role in the pathogenesis of diarrhea, *Gastroenterology*, 79, 1276, 1980.

125. **Caride, V. J., Propkop, E. K., Troncale, F. J., Buddoura, W., Winchenbach, K., and McCallum, R. W.,** Scintigraphic determination of small intestinal transit time: comparison with the hydrogen breath technique, *Gastroenterology*, 86, 714, 1984.

126. **Metz, G., Gassull, M. A., Drasar, B. S., Jenkins, D. J. A., and Blendis, L. M.,** Breath hydrogen test for small-intestinal bacterial colonisation, *Lancet*, 1, 668, 1976.

127. **Rhodes, J. M., Middleton, P., and Jewell, D. P.,** The lactulose hydrogen breath test as a diagnostic test for small-bowel bacterial overgrowth, *Scand. J. Gastroenterol.*, 14, 333, 1979.

128. **Read, N. W.,** Small bowel transit time of food in man: measurement, regulation and possible importance, *Scand. J. Gastroenterol.*, 96, 77, 1984.

129. **King, C. E. and Toskes, P. P.,** Comparison of the 1-gram ($^{14}C$) xylose, 10-gram lactulose-$H_2$, and 80-gram glucose-$H_2$ breath tests in patients with small intestine bacterial overgrowth, *Gastroenterology*, 91, 1447, 1986.

130. **Ambrecht, U., Bosaeus, I., Gillberg, R., Seeberg, S., and Stockbrügger, R. W.,** Hydrogen ($H_2$) breath test and gastric bacteria in acid-secreting subjects and in achlorhydric and postgastrectomy patients before and after antimicrobial treatment, *Scand. J. Gastroenterol.*, 20, 805, 1985.

131. **Levitt, M. D.,** Volume and composition of human intestinal gas determined by means of an intestinal washout technique, *N. Engl. J. Med.*, 284, 1394, 1971.

132. **Gibson, G. R., Macfarlane., S., and Cummings, J. H.,** The fermentability of polysaccharides by mixed human faecal bacteria in relation to their suitability as bulk-forming laxatives, *Lett. Appl. Microbiol.*, 11, 251, 1990.

133. **Flourie, B., Pellier, P., Florent, C., Marteau, P., Pochart, P., and Rambaud, J. C.,** Site and substrates for methane production in human colon, *Am. J. Physiol.*, 260, G752, 1991.

134. **Christl, S. U., Gibson, G. R., Murgatroyd, P., Scheppach, W., and Cummings, J. H.,** Impaired hydrogen metabolism in pneumatosis cystoides intestinalis, *Gastroenterology*, 104, 392, 1993.

135. **Blanc, B., Daures, J. P., Rouillon, J. M., Peray, P., Pirrugues, R., Larrey, D., Gremy, F., and Michel, H.,** Lactitol or lactulose in the treatment of chronic hepatic encephalopathy: results of a meta-analysis, *Hepatology*, 15, 222, 1992.

136. **Hollander, P.,** Safety profile of acarbose, an alpha-glucosidase inhibitor, *Drugs*, 44, 47, 1992.

137. **Koss, L. G.,** Abdominal gas cysts (pneumatosis cystoides intestinalis hominis). An analysis with a report of a case and a critical review of the literature, *Arch. Pathol.*, 53, 523, 1952.

138. **Jamart, J.,** Pneumatosis cystoides intestinalis. A statistical analysis of 919 cases, *Acta Hepato-Gastroenterol.*, 26, 419, 1979.

139. **Ecker, J. A., Williams, R. G., and Clay, K. L.,** Pneumatosis cystoides intestinalis-bullous emphysema of the intestine, *Am. J. Gastroenterol.*, 56, 125, 1971.

140. **Forgacs, P., Wright, P. H., and Wyatt, A. P.,** Treatment of intestinal gas cysts by oxygen breathing, *Lancet*, 1, 579, 1973.

141. **Keyting, W. S., McCarver, R. R., Kovarik, J. L., and Daywitt, A. L.,** Pneumatosis intestinalis: a new concept, *Radiology*, 76, 733, 1961.

142. **Yale, C. E., Balish, E., and Wu, J. P.,** The bacterial etiology of pneumatosis cystoides intestinalis, *Arch. Surg.*, 109, 89, 1974.

143. **Read, N. W., Al-Janabi, M. N., and Cann, P. A.,** Is raised breath hydrogen related to the pathogenesis of pneumatosis coli?, *Gut*, 25, 839, 1984.

144. **Van der Linder, W. and Marsell, R.,** Pneumatosis cystoides coli associated with high $H_2$ excretion. Treatment with an elemental diet, *Scand. J. Gastroenterol.*, 14, 173, 1979.

145. **Karrer, P.,** *Organic Chemistry*, Elsevier Press, New York, 1960.

146. **Gossel, T. A. and Bricker, J. D.,** *Principles of Clinical Toxicology*, Raven Press, New York, 107, 1990.

147. **Bremer, J. and Greenberg, D. M.,** Enzymatic methylation of foreign sulfydryl compounds, *Biochem. Biophy. Acta*, 46, 217, 1960.
148. **Weiseger, R. A., Pinkus, L. M., and Jacoby, W. B.,** Thiol S-methyltransferase: suggested role in detoxication of intestinal hydrogen sulfide, *Biochem. Pharmacol.*, 29, 2885, 1980.
149. **Marcus, R. and Watt, J.,** Seaweeds and ulcerative colitis in laboratory animals, *Lancet*, 2, 489, 1969.
150. **Watt, J. and Marcus, R.,** Carrageenan induced ulceration of the large intestine in the guinea pig, *Gut*, 12, 164, 1971.
151. **Kitano, A., Matsumoto, T., Hiki, M., Hashimura, H., Yoshiyasu, K., Okawa, K., Kuwajiama, S., and Kobayashi, K.,** Epithelial dysplasia of the rabbit colon induced by degraded carrageenan, *Cancer Res.*, 46, 1374, 1986.
152. **Yamada, M., Ohkusa, T., and Okayasu, I.,** Occurrence of dysplasia and adenocarcinoma after experimental chronic ulcerative colitis in hamsters induced by dextran sodium sulphate, *Gut*, 33, 1521, 1992.
153. **Florin, T. H. J., Gibson, G. R., Neale, G., and Cummings, J. H.,** A role for sulfate-reducing bacteria in ulcerative colitis?, *Gastroenterology*, 98, A170, 1990.
154. **Gibson, G. R., Cummings, J. H., and Macfarlane, G. T.,** Growth and activities of sulphate-reducing bacteria in gut contents of healthy subjects and patients with ulcerative colitis, *FEMS Microbiol. Ecol.*, 86, 103, 1991.
155. **Roediger, W. E. W., Duncan, A., Kapaniris, O., and Millard, S.,** Reducing sulfur compounds of the colon impair colonocyte nutrition: implications for ulcerative colitis, *Gastroenterology*, 104, 802, 1993.
156. **Roediger, W. E. W.,** The colonic epithelium in ulcerative colitis: an energy-deficient disease?, *Lancet*, 2, 712, 1980.
157. **Scheppach, W., Sommer, H., Kirchner, T., Paganelli, G. M., Bartram, P., Christl, S. U., Richter, F., Dusel, G., and Kasper, H.,** Effect of butyrate enemas on the colonic mucosa in distal ulcerative colitis, *Gastroenterology*, 103, 51, 1992.
158. **Haines, A., Metz, G., Dilwari, J., Blendis, L., and Wiggins, H.,** Breath methane in patients with cancer of the large bowel, *Lancet*, 2, 481, 1977.
159. **Pique, J. M., Pallares, M., Cuso, E., Vilar-Bonet, J., and Gassull, M. A.,** Methane production and colon cancer, *Gastroenterology*, 87, 601, 1984.
160. **Kashtan, H., Rabau, M., Peled, Y., Milstein, A., and Witznitzer, T.,** Methane production in patients with colorectal cancer, *Isr. J. Med. Sci.*, 25, 614, 1989.
161. **Sivertsen, S. H., Bjornkelett, A., Gullestad, H. P., and Nygaard, K.,** Breath methane and colorectal cancer, *Scand. J. Gastroenterol.*, 27, 25, 1992.
162. **Kirk, E.,** The quantity and composition of of human colonic flatus, *Gastroenterology*, 12, 782, 1949.
163. **Levitt, M. D. and Ingelfinger, F. D.,** Hydrogen and methane production in man, *Ann. N. Y. Acad. Sci.*, 150, 68, 1968.
164. **Peled, Y., Weinberg, D., Hallak, A., and Gilat, T.,** Factors affecting methane production in humans, *Dig. Dis. Sci.*, 32, 267, 1987.
165. **Pochart, P., Lemann, F., Flourie, B., Pellier, P., Goderel, I., and Rambaud, J.-C.,** Pyxigraphic sampling to enumerate methanogens and anaerobes in the right colon of healthy humans, *Gastroenterology*, 105, 1281, 1993.
166. **Gibson, G. R., Cummings, J. H., and Macfarlane, G. T.,** Competition for hydrogen between sulphate-reducing bacteria and methanogenic bacteria from the human large intestine, *J. Appl. Bacteriol.*, 65, 241, 1988.
167. **Strocchi, A., Furne, J. K., and Levitt, M. D.,** A modification of the methylene blue method to measure bacterial sulfide production in feces, *J. Microbiol. Meth.*, 15, 75, 1992.
168. **Tsai, H. H., Sunderland, D., Gibson, G. R., Hart, C. A., and Rhodes, J. M.,** A novel mucin sulphatase from human faeces: its identification, purification and characterization, *Clin. Sci.*, 82, 447, 1992.
169. **DeGraeve, K. and Demeyer, D. I.,** Rumen and hindgut fermentation: differences for possible exploitation, *Forum Appl. Biotechnol. Ghent.*, 44, 1989.
170. **Duncan, A. J. and Henderson, C.,** A study of the fermentation of dietary fibre by human colonic bacteria grown in vitro in semi-continuous culture, *Microbiol Ecol. Health Dis.*, 3, 87, 1990.
171. **Ducros, V., Durand, M., Beaumatin, P., Hannequart, G., Cordelet, C., and Grivet, J. P.,** Adaptation to two doses of lactulose by human colonic flora in continuous culture, *Proc. Nutr. Soc.*, 52, 156A, 1993.
172. **Bernalier, A., Doisneau, E., Cordelet, C., Beaumatin, P., Durand, M., and Grivet, J. P.,** Competition for hydrogen between methanogenesis and hydrogenotrophic acetogenesis in human colonic flora studied by $^{13}$C NMR, *Proc. Nutr. Soc.*, 52, 118A, 1993.

# Chapter 7

# Toxicology of the Colon: Role of the Intestinal Microflora

*Ian R. Rowland*

## CONTENTS

I. Introduction ................................................................................................................... 155
    A. Interactions of the Microflora and Its Host ........................................................... 155
    B. Bacterial Metabolism — General Considerations ................................................. 156
II. Methods for Studying Metabolism and Toxicity ..................................................... 157
III. Toxicological Consequences of Gut Microflora Metabolism ................................... 157
    A. Activation of Chemicals to Toxic, Mutagenic, and Carcinogenic Derivatives ................. 158
        1. Plant Glycosides ........................................................................................... 158
        2. Azo Compounds ........................................................................................... 160
        3. Nitro Compounds ......................................................................................... 161
        4. IQ ................................................................................................................. 161
    B. Synthesis of Carcinogens ...................................................................................... 162
        1. Nitrate Reduction and *N*-Nitroso Compound Synthesis ............................. 162
        2. Fecapentaenes ............................................................................................... 163
    C. Synthesis of Tumor Promoters .............................................................................. 163
        1. Secondary Bile Acids ................................................................................... 163
        2. Protein and Amino Acid Metabolites ........................................................... 164
        3. Fecapentaenes ............................................................................................... 164
    D. Biliary Conjugates and Enterohepatic Circulation ............................................... 164
    E. Detoxication and Protective Effects of the Gut Microflora ................................... 165
        1. Methylmercury .............................................................................................. 165
        2. Lignans and Phyto-Estrogens ....................................................................... 166
        3. Flavonoids .................................................................................................... 167
        4. Short Chain Fatty Acids ................................................................................ 168
        5. Carcinogen Binding by Intestinal Bacteria .................................................. 168
    F. Modification of Mammalian Metabolism and Tumor Incidence by Gut Bacteria ............. 168
References ............................................................................................................................ 170

## I. INTRODUCTION

### A. INTERACTIONS OF THE MICROFLORA AND ITS HOST

The colonic microflora is a dynamic population which is influenced by its host, and in turn influences its host. Interactions between the microflora and human host have implications for nutrition, infection, metabolism, toxicity, and cancer. At the nutritional level, the bacterial population in the colon obtains all of its nutrients from the host through either undigested dietary residues or intestinal secretions. The host receives some nutrients in the form of certain vitamins[1] and short chain fatty acids (SCFA). Although the microflora can be a source of infection of wounds and the urogenital tract, and some component members can generate toxins, the flora plays a very important role in preventing many enteric pathogens from establishing in the gut and causing disease. This process is known as colonization resistance and is a major beneficial effect of the normal gut flora.[2]

From the standpoint of its role in toxicity and cancer, bacteria in the colon can have both beneficial and detrimental influences. The involvement of the microflora in toxic events is often mediated by metabolism, for example, the conversion of an ingested compound into a form which is more or less toxic than the parent compound, resulting in activation or detoxification, respectively. There are, however, examples of more subtle effects of the flora on the toxicity of chemicals, in particular, modification by gut bacteria of mammalian metabolic processes that are involved in the activation of toxic chemicals.

**Table 1**  Evidence Implicating Intestinal Bacteria in the Etiology of Cancer

1. In the human gut, the region with highest bacterial numbers (large intestine) is the most common site of tumors.
2. Bacterial colonization of the stomach (e.g., in pernicious anemia patients) increases cancer risk.[119]
3. Human feces contain genotoxic substances thought to be of bacterial origin.[120]
4. 1,2-Dimethylhydrazine induces fewer colon tumors in germ-free rats than in conventional flora rats.[80]
5. Intestinal bacteria can produce genotoxins and carcinogens from dietary components.
6. Diets identified as having a colon cancer risk can modify gut microflora metabolism, particularly enzymes and reactions that generate potential carcinogens.

**Table 2**  Bacterial Enzymes that Generate Toxic, Genotoxic, or Carcinogenic Products

| Enzyme | Substrates |
| --- | --- |
| β-Glycosidase | Plant glycosides |
| | Amygdalin |
| | Rutin |
| | Franguloside |
| Azoreductase | Azo compounds |
| | Benzidine-based dyes |
| Nitroreductase | Nitro compounds |
| | Dinitrotoluene |
| | Dinitrobenzene |
| | Nitrochrysene |
| β-Glucuronidase | Biliary glucuronides |
| | Benzo(a)pyrene |
| | IQ |
| | Benzidine |
| IQ "hydratase-dehydrogenase" | IQ, MeIQ |
| Nitrate/nitrite reductase | Nitrate, nitrite |

A number of factors can affect both metabolism of the gut microflora and its involvement in toxic events. These factors include age, drugs, diet, and species differences. They have been reviewed previously.[3]

A wide variety of sources have provided evidence, albeit circumstantial, that intestinal bacteria involved in the etiology of cancer, although somewhat circumstantial. The main pieces of supporting evidence are summarized in Table 1.

## B. BACTERIAL METABOLISM — GENERAL CONSIDERATIONS

The importance of gut microflora metabolism in toxicological events requires a general introduction before the more detailed discussions of specific reactions described below.

There is little information about the ability of the individual species that comprise the microflora to metabolize nutrients and foreign compounds. Even if more information were known, it would be difficult to predict whether a reaction that occurred *in vitro* with a pure culture of a gut organism would proceed when that organism was surrounded by other members of the ecosystem within the mammalian gut. Information on the reactions performed by individual gut species are therefore usually of little benefit. A more valid and valuable approach for understanding the role of colonic bacteria in nutrition and toxic events in humans is to consider the whole flora as an additional "organ" within the host, ignoring its multi-organism composition. This approach has been used with considerable success by a number of research groups.[4-6]

Table 2 lists the major classes of reactions catalyzed by intestinal bacteria. It is constructive to compare, in broad terms, the metabolic activities of the mammalian liver and the microflora (Table 3). Whereas hepatic reactions are largely oxidative, reactions of the flora are reductive. This clearly expands the range of reactions that are possible *in vivo*. The liver has an important function in conjugating many drugs and xenobiotics prior to their excretion. The polar nature of such conjugated materials should facilitate their elimination in feces after excretion into the gut via the bile. The hydrolytic activities of the

**Table 3**  Comparison of Mammalian and Gut Microflora Metabolism

| Mammalian | Gut microflora |
|---|---|
| Oxidation common (phase I) (cytochromes P450-dependent mixed-function oxidase, $O_2$-dependent) | Reduction common |
| Synthesis common (phase 2) | Degradation and hydrolysis common |
| Conjugation with glucuronic acid sulfate amino acids | Hydrolysis of glucuronides sulfates amino acid conjugates |
| Acetylation | N-acetyl and O-acetyl compounds |

flora, however, can cleave such conjugates, permitting reabsorption of the parent compound and resulting in an enterohepatic circulation (see discussion below).

An important factor to be taken into account when considering foreign compound metabolism by gut bacteria is the location of the main population in the alimentary tract. The colon is the region of the gut which harbors the greatest number of bacteria. Indeed other areas of the human gastrointestinal tract are very sparsely populated.[7] This does not mean, however, that only poorly absorbed chemicals encounter the colonic flora. Substances and their metabolites may enter the colon across the intestinal wall from the blood or may reach the colon after excretion in the bile. Thus there is ample opportunity for a wide variety of materials in diet to encounter and be metabolized by colonic bacteria.

## II. METHODS FOR STUDYING METABOLISM AND TOXICITY

The ideal method for studying the metabolic activities and the role in toxic events of the intestinal bacteria is to use human volunteers. Unfortunately, this method is not always practical or ethical, so the problem has to be addressed by other means, each of which has intrinsic advantages and disadvantages, but which together provide a reasonably accurate view of the ecosystem. A summary of the advantages and disadvantages of the available *in vitro* and *in vivo* methods for studying the intestinal microflora is given in Table 4. A detailed description and evaluation and examples for many of the uses of these models has been provided previously.[8-10]

In general, *in vitro* methods can demonstrate whether a particular reaction can be catalyzed by intestinal bacteria, and what are its likely products and kinetics. However, because such methods do not take into account absorption, excretion, and distribution of a chemical in the body, they do not provide information on whether the reaction actually occurs after ingestion of the substance by man. Consequently, a combination of *in vitro* and *in vivo* assays is used when assessing the involvement of intestinal bacteria in metabolism and toxicity. It should also be remembered that although static cultures of gut organisms (fecal slurries) are simple to use, they often do not simulate the reactions of the microflora *in vivo* as accurately as the more complex continuous or semi-continuous culture techniques (reviewed by Rumney and Rowland).[10] Similarly, although laboratory animals are convenient for studying metabolism and toxic events and also enable monitoring of the flora in different regions of the gut, as well as feces, their use as models for the human fecal microflora and its metabolism may be questioned. It is known that there are differences in the bacterial composition of the gut floras of animals and man,[7] and in gut microbial enzymic activities.[11] Consequently, even though they are more difficult and expensive to undertake, studies using human volunteers or germ-free animals associated with a complete human microflora provide more relevant data.

## III. TOXICOLOGICAL CONSEQUENCES OF GUT MICROFLORA METABOLISM

Metabolism of foreign or endogenously produced substances by colonic bacteria can have wide ranging implications for human health because both beneficial and detrimental effects can result. These effects are summarized in Table 5 and are discussed in the following sections.

**Table 4**    Methods for Investigating Involvement of Gut Microflora in Metabolism and Toxicity

| Method | Description | Advantages | Disadvantages |
|---|---|---|---|
| *In vitro* | | | |
| Static (batch) culture | Feces or colonic contents suspended in buffer are incubated for short periods under anaerobic conditions | Useful for short-term metabolism studies and studies of enzyme activities | Bacterial composition of suspension changes with time, so only useful for limited periods |
| Continuous culture | | | |
| Defined bacteria | Chemostat culture system is inoculated with one or more kinds of bacteria; medium addition and culture removal continuous | Useful for studies of bacterial interactions | Relevance of results to gut ecosystem questionable |
| Complex flora | Culture system is inoculated with a nondefined flora, usually feces | System of continuous flow has some analogies to *in vivo* gut; metabolism and ecological studies over long periods (days, weeks) facilitated | More complex and difficult to set up than static culture; ignores host inputs, e.g., gut secretions, immunology |
| *In vivo* | | | |
| Human volunteers | Fresh fecal samples taken and analyzed bacteriologically, chemically, or metabolically | Fresh samples easily obtained; dietary studies can be performed; useful for clinical studies | Toxicological studies not usually possible on ethical grounds; assumes feces are representative of colonic material |
| Laboratory animals | Fresh fecal samples or gut contents removed from conventional flora (CV) animals and analyzed | Wide range of studies (dietary, metabolic, toxicological, clinical) possible and relatively simple to perform; not restricted to fecal material | Major differences between animals and humans in microflora composition and metabolism |
| Germ-free (GF) and antibiotic-treated (AB) animals | Comparison of metabolism and toxicity in animals with and without a gut microflora | Provides direct evidence for involvement of microflora | Antibiotics may react with test substance. GF and CV rats have physiological and morphological differences which can affect metabolism. |
| Gnotobiotic animals Defined strains | Germ-free rats or mice inoculated with specific microorganisms | Specific host-bacterial, bacteria-bacteria interactions can be studied | Relevance to situation of normal conventional flora gut ecosystem is questionable |
| Human flora-associated animals | Germ-free rats or mice inoculated with human feces | Flora retains many characteristics of human gut microflora; facilitate studies of diet, toxic chemicals, carcinogens on human gut flora; circumvents ethical problems associated with such studies in humans | Expensive; gut physiology (secretions, pH, peristalsis) of host animal may not be same as that of humans |

## A. ACTIVATION OF CHEMICALS TO TOXIC, MUTAGENIC, AND CARCINOGENIC DERIVATIVES

### 1. Plant Glycosides

Plants produce a wide variety of substances which are linked to sugar moieties resulting in glycosides.[12] Their wide distribution, often in large quantities, in edible fruits and vegetables and in beverages such as tea and wine, derived from plants, results in human intake of around 1 g/d.[12] In their glycosidic form, most

**Table 5** Toxicological Consequences
of Gut Flora Activities

**A. Activation**
    to toxicants
    to mutagens
    to carcinogens

**B. Synthesis of carcinogens**
    $N$-nitroso compounds
    fecapentaenes

**C. Synthesis of promoters**
    bile acids
    ammonia
    phenols, cresols
    fecapentaenes (as promoters)

**D. Enterohepatic circulation**
    carcinogens
    steroid hormones and drugs

**E. Detoxification/protection**
    methyl mercury
    lignans and phytosterols
    mutagen binding
    SCFA

**F. Modification of mammalian metabolism/activation**
    hepatic activation of aflatoxin and cooked-food mutagens
    hepatic DNA adducts
    modification of hepatoma incidence

of these substances are harmless, and poorly absorbed, thus passing into the colon. Once in the colon, they are subject to the action of β-glycosidases associated with the resident microbial flora that cleave the sugar moieties releasing aglycones, which may be toxic, carcinogenic, or mutagenic. Amygdalin, a glycoside found in several drupes and pomes including apricot, apple, plum, and almond is hydrolyzed in the gut to mandelonitrile which subsequently decomposes to cyanide. After exposure to oral doses of amygdalin, germ-free rats have lower thiocyanate levels in the blood and do not exhibit cyanide toxicity, unlike their counterparts with a conventional microflora. This indicates the critical role played by colonic bacteria in the metabolism and toxicity of amygdalin. Cyanogenic glycosides are present in cassava, a staple crop in many countries in the tropics. In periods of food shortage, the plants are sometimes consumed without appropriate treatments for removing the glycosides, which leads to cyanide exposure and acute poisoning or chronic neurological disease.[12]

When fed to conventional flora rats, cycasin (a component of cycad nuts) is hydrolyzed by bacterial β-glucosidase in the colon releasing the aglycone methylazoxymethanol (MAM) which induces colon tumors. No such tumors are found in germ-free rats fed the glycoside.[13]

Many of the glycosides found in plants are mutagenic on hydrolysis by bacterial glycosidases (Table 6). Some of most common glycosides are quercetin, including rutin (quercetin-$O$-rutinoside) and quercetrin (quercetin-$O$-rhamnoside) which are found in lettuce and many other edible plants. When tested for mutagenicity using the *Salmonella*/microsome assay, mutagenic activity in these plant materials can only be detected when an extract of fecal bacteria is added to the incubation mixture (Table 7).[14] Similarly, potent mutagenic activity can be detected in red wine and tea when the beverages are incubated with an extract of human feces, a finding attributed to the hydrolysis of plant glycosides by bacterial β-glycosidases.[15] Apart from MAM, the carcinogenicity of aglycones derived from plant glycosides is debatable. Quercetin has been investigated for carcinogenicity in a number of animal models, and in general has not been found to exhibit tumorigenic properties (reviewed by Brown).[12] In two-stage carcinogenicity studies with quercetin, no evidence for initiating activity for rat liver tumors (using phenobarbital and partial hepatectomy as promoters), or for promoting activity (using MAM as an initiator) was found.[16,17] Similarly, a lack of initiating and promoting activity towards rat bladder carcinogenesis has been reported.[18]

Assessment of the toxicological significance of glycoside hydrolysis by the intestinal microflora is further complicated by reports of potential anticarcinogenic and antimutagenic effects of flavonoid aglycones.

**Table 6** Some Plant Glycosides that Become Mutagenic
After Transformation by Intestinal Bacteria

| Type | Glycoside | Aglycone | Source |
|---|---|---|---|
| Flavonol | Rutin | Quercetin | Citrus fruits |
| | Quercetrin | Quercetin | Berries |
| | Robinin | Kaempferol | Legumes |
| | Astragazin | Kaempferol | Leafy vegetables |
| | Tiliroside | Kaempferol | Herbs/spices |
| Diterpenoid | Stevioside | Steviol | *Stevia* |
| | Rebaudioside A | Steviol | |
| Anthraquinone | Chrysazin glucoside | Chrysazin | |
| | Franguloside | Emodin | Rhubarb |
| | Quinizarin glucoside | Quinizarin | |
| | Lucidin 3-*O*-primveroside | Lucidin | *Rubia tinctorum* |
| Azoxy | Cycasin | Methylazoxymethanol | Cycad plants |

**Table 7** Influence of Gut Microflora Metabolism
on Mutagenicity of Extracts of Vegetables

| Vegetable (cultivar) | His+ mutants/plate[a] | |
|---|---|---|
| | −GFE | +GFE |
| Lettuce (Renate)[b] | 10 | 346[f] |
| Lettuce (Dandy)[b] | 6 | 169[f] |
| Runner Bean (Romore)[c] | 24 | 262[f] |
| Paprika (Goldstar)[d] | 17 | 90[f] |
| Paprika (Bruinsma green)[d] | 8 | 198[f] |
| Rhubarb (Paragon)[e] | 14 | 126[f] |
| Rhubarb (Goliath)[e] | 13 | 13 |

[a] Methanol extracts of the freeze-dried vegetables were tested in *Salmonella typhimurium* TA98 in
the presence and absence of an extract of rat cecal contents (GFE); values for negative controls
are subtracted.
[b] Amount of methanol extract used per plate was equivalent to 150 mg of freeze-dried vegetable.
[c] Amount of methanol extract used per plate was equivalent to 400 mg of freeze-dried vegetable.
[d] Amount of methanol extract used per plate was equivalent to 200 mg of freeze-dried vegetable.
[e] Amount of methanol extract used per plate was equivalent to 132 mg of freeze-dried vegetable.
[f] Significant mutagenic effect.

From Van der Hoeven, J. C., Lagerweij, W. J., Bruggeman, I. M., Voragen, F. G., and Koeman, J. H.,
*J. Agric. Food Chem.*, 31, 1020, 1983.

It is clear therefore that hydrolysis of plant glycosides in the gut can lead, potentially, to both adverse
and beneficial consequences for humans. Further studies are needed to establish the repercussions for
colon cancer of flavonoid release from glycosides by β-glycosidases in the gut.

## 2. Azo Compounds

A number of dyes used in food, cosmetics, textiles, leather, and paper printing are based on azo
compounds. These substances are reduced by the intestinal flora in the gut to produce, ultimately, amines.
The reduction products are often toxic; for example, after bacterial reduction the food dye Brown FK
causes vacuolar myopathy in rats given high doses.[19]

Workers exposed to Direct Black 38, a dye used in the leather and textile industry, have an elevated
risk of bladder cancer which has been attributed to reduction of the dye by the gut microflora to benzidine,
a human bladder carcinogen.[20,21] Studies in man and a number of laboratory animal species demonstrate
that exposure to benzidine-based dyes is followed by urinary excretion of aromatic amines and their *N*-
acetylated derivatives.[22] When Direct Black 38 is incubated with cultures of intestinal bacteria from man,
rats, or monkeys, benzidine is rapidly formed, indicating that the microflora is the major site of

metabolism of the dye.[21] Studies with more sophisticated, continuous-culture systems have revealed that several other urinary metabolites of the dye, including the *N*-acetylated derivative can be attributed to bacterial metabolism in the colon.[23]

Acetylation is a crucial step in the activation of aromatic amines to their ultimate carcinogenic metabolite. Consequently, these studies implicate gut microflora not only in the overall metabolism in the body of benzidine-based azo dyes but also in the generation of reactive, carcinogenic species.

## 3. Nitro Compounds

Heterocyclic and aromatic nitro compounds are extensively used in industry, being important intermediates in the manufacture of thousands of consumer products[24] and also an important class of drugs with uses as antibiotic, antiparasitic, and radiosensitizing agents. Nitroaromatics are ubiquitous environmental contaminants which result from combustion of fossil fuels and are found in diesel exhaust, cigarette smoke, and airborne particulates. These compounds often possess toxic, mutagenic, and carcinogenic activity which may contribute to the environmental cancer risk in man.[25] Reduction of the nitro group occurs in a series of steps involving nitroso and hydroxylamino intermediates, leading ultimately to an amine. Reduction is usually required for the pharmaceutical and toxicological activity of these compounds to be expressed. Examples are for the antitrichomonad activity and mutagenicity of metronidazole[26] and the induction of methemoglobinemia by nitrobenzenes.[27] Although the nitro group on nitroaromatic compounds can be reduced by both hepatic and bacterial reductases, in most cases nitroreduction by the gut microflora appears to play a more important role than mammalian enzymes.

Evidence for the importance of bacterial reductases in the induction of methemoglobinemia by nitrobenzene has been obtained by exposing conventional flora rats, antibiotic-treated rats, and germ-free rats to the compound. Methemoglobin levels of 30 to 40% were induced in the conventional rats within 2 h of exposure to nitrobenzene, whereas no increase was detected in the animals without a gut flora.[27] Levin and Dent[28] further confirmed the role of intestinal microflora by comparing the *in vitro* metabolism of nitrobenzene by rat hepatic microsomes and cecal contents. Nitrobenzene was readily reduced via nitrosobenzene and phenylhydroxylamine (the presumed methemoglobinogenic metabolites) to aniline by the cecal contents. Aniline was also produced in the presence of microsomes under anaerobic conditions, but at less than 1% of the rate of bacterial reduction. Therefore, there is conclusive evidence that the intestinal microflora is a major determinant of both the metabolism and acute toxicity of nitrobenzene.

The toxicity of nitrotoluenes, which are important intermediates in the manufacture of plastics and dyes, is similarly dependent on the reductive activity of the intestinal microflora. In this case, genotoxicity and the ability to bind covalently to macromolecules (thought to be an early step in tumorigenesis) was lower in germ-free rats than in conventional rats.[29,30]

Gut bacterial nitroreduction is also believed to be crucial for the metabolism of nitrated polycyclic aromatic hydrocarbons, such as 1-nitropyrene and 6-nitrochrysene. El-Bayoumy et al.[31] detected 1-aminopyrene in the feces of conventional flora rats, but not germ-free rats given 1-nitropyrene. Studies in a semi-continuous flow culture of the human colonic microflora revealed that 1-nitropyrene was virtually completely reduced to 1-aminopyrene and *N*-formyl-1-aminopyrene over 24 h.[32] Using the same model system, rapid reduction of 6-nitrochrysene to 6-aminochrysene and some other minor metabolites, including 6-nitrosochrysene, was seen.[33] It is significant that 6-aminochrysene is known to induce liver and lung tumors in mice,[34] which suggests that nitroreduction by the gut microflora plays an important role in activation and tumorigenesis by nitrochrysene.

## 4. IQ

2-Amino-3-methyl-3*H*-imidazo[4,5-*f*]quinoline (IQ) is one of several heterocyclic amine compounds which are produced in small amounts when proteinaceous foods, particularly meat and fish, are grilled or fried.[35] These compounds are highly mutagenic in *Salmonella typhimurium*, but require activation by hepatic cytochrome P450-dependent mixed function oxidases for their activity.[36,37] IQ is also carcinogenic in rodent bioassays and induces tumors at various sites including the large intestine, which suggests that it may play a role in the etiology of colon cancer in man.[38]

Incubation of IQ with a suspension of human feces yields the 7-keto derivative, 2-amino-3,6-dihydro-3-methyl-7*H*-imidazo[4,5-*f*]quinoline-7-one (7-OHIQ).[39] A species of *Eubacterium* has been shown to be particularly active in catalyzing the hydration-dehydrogenation reaction involved.[40] This reaction has not been shown to occur in gut contents obtained from germ-free rats.[41] The 7-keto metabolite has been found in feces of individuals consuming a diet containing a high level of fried meat, indicating that the formation of 7-OHIQ can occur *in vivo* in humans.[42] Unlike IQ, the bacterial metabolite is a direct-acting mutagen in *Salmonella typhimurium*, inducing up to 10 revertants per ng in the absence of hepatic S-9.[40,41] Thus, there is strong evidence for the bacterial formation in the human gut of a directly genotoxic derivative of a dietary carcinogen.

Although there are many structurally related heterocyclic amine compounds in cooked foods,[35] apart from IQ, only 2-amino-3,4-dimethyl-3H-imidazo[4,5-*f*]quinoline (MeIQ) appears to undergo metabolism by fecal suspensions to a genotoxin (C. Rumney and I.R. Rowland, unpublished observations, 1993).

## B. SYNTHESIS OF CARCINOGENS
### 1. Nitrate Reduction and *N*-Nitroso Compound Synthesis

Nitrate is widely distributed in the environment and is present in both the diet and drinking water. Although of very low toxicity, it is readily converted by the nitrate reductase activity of the bacterial population of the mammalian gastrointestinal tract to its more reactive and toxic reduction product, nitrite.

One of the consequences of nitrate reduction is the reaction of nitrite with nitrogenous compounds such as amines, amides, and methylureas (nitrosation) in the body to produce *N*-nitroso compounds, many of which are highly carcinogenic.[43] Acidic conditions are required for the reaction to occur chemically, and so the main site of nitrosation in humans is considered to be the stomach which normally has a pH of 1 to 2. However, nitrosation can also be catalyzed at a neutral pH by bacteria.[44,45] Recently, we have explored the possibility that intestinal bacteria are involved in nitrosation of nitrogenous compounds *in vivo*.[46] A particular problem in assessing the overall extent of endogenous nitrosation is the large variety of different classes of N-nitroso compounds that may be formed. These compounds include derivatives of dialkylamines, guanidines, urethanes, amides, and amino acids. In order to obtain an overall view of N-nitroso compound synthesis in the body, we have used the group selective procedure devised by Walters et al.[47] which measures the concentration of all compounds containing the *N*-nitroso grouping, although it does not give information on the levels of individual compounds. This method is termed apparent total *N*-nitroso compound (ATNC) analysis.

To determine the role of the gut microflora in nitrosation reactions in the body, we investigated the concentration of *N*-nitroso compounds in tissues and gut contents of germ-free rats and conventional microflora rats.[46] To prevent the results being confounded by ingestion of preformed *N*-nitroso compounds, the animals were fed diets containing no detectable ATNC. They were given drinking water containing high amounts of nitrate. ATNC were detected in significant amounts only in gut contents of the conventional flora rats, with the largest quantities being found in colonic contents. There was a significant positive correlation between *N*-nitroso compound formation and nitrate concentration in the drinking water. In the germ-free rats, little or no ATNC were detected in gut contents. These results demonstrate that the compounds were being formed by microbial action, probably through bacterial reduction of the ingested nitrate, and subsequent bacterial catalysis of the nitrosation reaction. Although the enzyme involved in the nitrosation process has not been isolated, it has been suggested from studies in pure cultures of bacteria, that nitrosation may be associated with the activity of nitrate and nitrite reductases.[45,48]

Recently, we have extended our investigations of nitrosation to man. *N*-nitroso compounds were detectable at concentrations between 40 and 590 µg N-NO/kg of feces of male and female volunteers consuming their own free-choice diets.[49] Considerable variation in fecal ATNC concentration was seen between individuals and also between samples from the same individual sampled at weekly intervals for nine weeks. No attempt was made in these studies to control the diet, so the fecal nitroso compounds could have been derived from the diet or from endogenous synthesis. To resolve this uncertainty, we provided volunteers with drinking water and food with undetectable ATNC levels and very low nitrate content. Under these dietary conditions, fecal ATNC concentrations decreased to almost undetectable values over 5 d. When nitrate was added to the drinking water a rapid rise in fecal ATNC concentration was seen, reaching a level similar to that recorded for the volunteers on their own diets. The results of this study suggest that the ATNC in feces can be increased by nitrate consumption and that the majority of the ATNC in feces is a consequence of endogenous synthesis, rather than coming from residues of *N*-nitroso compounds ingested with the diet.

## 2. Fecapentaenes

Although technically difficult, it is possible to test human fecal samples for mutagenic activity. It is usually done with the Ames *Salmonella* assay. Using this technique, feces from individuals excreting large amounts of mutagenic activity were extracted and concentrated in an attempt to identify the agent or agents involved. Two groups[50,51] performed this work and structural analysis revealed a glyceryl ether compound containing a pentaene moiety with a chain length of 12 or 14 (fecapentaenes 12 or 14). Both types are found in feces, although the ratio varies considerably.[52] Although fecapentaenes occur in feces of the majority of western populations, more detailed epidemiological studies have revealed some anomalies. For example, lower fecapentaene levels were found in feces from colorectal cancer patients than in controls[53] and fecal excretion of fecapentaenes is higher in vegetarians, a population at low risk from colon cancer.[54] These and other studies suggest that the more fecapentaene excreted the lower the risk of colon cancer. This reasoning has a certain logic since fecapentaenes are very potent, direct-acting mutagens which would be expected to react extremely rapidly with DNA and other macromolecules in the colonic mucosa. Thus, increased fecal excretion may reflect lower endogenous exposure to the genotoxin. Recent studies demonstrating binding of fecapentaenes to dietary fiber support this hypothesis.[55] Future epidemiological studies on fecapentaenes may have to utilize more sophisticated measurements of exposure to the genotoxin such as DNA adduct determination.

The gut microflora has been implicated in fecapentaene synthesis by the demonstration of this process *in vitro* by fecal suspensions, under anaerobic conditions, and the inhibition of that synthesis by antibiotics and heat sterilization.[51,56] Further work has established that *Bacteroides* species are the organisms mainly responsible for fecapentaene production.[57] Production of mutagens *in vitro* is stimulated by addition of bile acids[58] although they are not thought to be precursors (these have not been identified).

The initial mutagenicity studies on fecapentaenes were performed in *Salmonella*, and showed that the compounds were potent, direct-acting mutagens (i.e., did not require metabolic activation for their activity), although they were present in only low concentrations (about 5 µg/kg) in feces. Subsequently, synthetic fecapentaene-12 has become available, enabling the mutagen to be tested in several toxicological assays. The compound has been shown to cause a range of genetic damage in human and rodent cells *in vitro*. At low concentrations (0.6 to 10 µg/ml), it induced single-strand DNA breaks, gene mutations, sister chromatid exchanges and unscheduled DNA synthesis (a response to DNA damage) in human fibroblasts.[59] Induction of chromosome aberrations in human lymphocytes and increases in neoplastic transformation in mouse cells have also been reported.[60,61] Results of *in vivo* studies have yielded more equivocal results. When fecapentaene-12 was injected into rat colon, increased proliferation of mucosal cells and DNA single strand breaks were seen.[62] Similarly, intrarectal administration to mice increased mitosis in colonic crypt cells, but did not induce nuclear aberrations in these cells.[63] Nuclear aberrations were detected after exposure to colon carcinogens such as dimethylhydrazine and have also been reported when fecal extracts were tested in the assay.[64] However, activity in the latter was not attributed to presence of fecapentaenes in the extracts since the UV spectrum characteristic of fecapentaenes was not seen. *In vivo* rodent bioassays have indicated that fecapentaene-12 does not have carcinogenic or tumor-initiating activity.[65-67] However, a recent study has provided evidence for tumor-promoting activity of fecapentaene.[68] This evidence is consistent with some of the data described above which indicates that the compound induces mucosal cell proliferation in the colon.

## C. SYNTHESIS OF TUMOR PROMOTERS

Investigations of possible tumor promoters generated by bacterial activity in the colon have focused on areas related to fat and animal protein in an attempt to provide mechanistic explanations for the epidemiological studies which indicate a link between these macronutrients and colon cancer. Since this is an area which has been the subject of numerous reviews,[69-73] only a brief summary will be provided here.

## 1. Secondary Bile Acids

Population studies indicate a strong positive correlation between the incidence of colon cancer and a high intake of fat. Such high risk populations are reported to excrete larger amounts of bile acids, particularly secondary bile acids, in feces.[69] Case control studies and experiments in human volunteers have not usually confirmed these observations[71] although *in vitro* and animal studies provide some evidence for tumor-promoting activity.

The liver secretes two major bile acids, cholic and chenodeoxycholic acids, which are subject to extensive metabolism by the intestinal microflora. These reactions have been reviewed by MacDonald et al.[74] and Hill.[69] The most important reaction *in vivo* is 7-α-dehydroxylation which converts cholic to deoxycholic acid and chenodeoxycholic to lithocholic acid. These two secondary bile acids, which comprise over 80% of total fecal bile acids, are postulated to play an important role in the etiology of colon cancer by acting as promoters of the tumorigenic process. There is considerable evidence to indicate that acid steroids, in particular secondary bile acids, can exert a range of biological effects. They induce cell necrosis, hyperplasia, metabolic alterations and DNA synthesis in intestinal mucosal cells,[75] enhance the genotoxicity of a number of mutagens in *vitro* assays,[76] and exhibit tumor-promoting activity in the colon (reviewed by Rowland et al).[72]

## 2. Protein and Amino Acid Metabolites

Epidemiological studies indicating a link between high intake of protein and incidence of colon cancer have led to investigations of the relationship, and of possible mechanisms in laboratory animals. Such studies suggest that the amount of protein, rather than its source, may be the most important factor[73] and have led to the hypothesis that colonic ammonia, produced by bacterial catabolism of amino acids and other nitrogenous compounds, may participate in tumor promotion. An increase in the protein content of the diet increases the colonic luminal ammonia concentration.[77] Ammonia exhibits a number of effects which suggest that it may be involved in promotion, including increasing mucosal cell turnover and altering DNA synthesis. It has also been shown to increase the incidence of colon carcinomas induced by N-methyl-N-nitro-N-nitrosoguanidine in rats.[73] Less well studied than ammonia has been the formation of phenols and cresols by bacterial deamination of certain aromatic amino acids in the colon. The generation of phenols is higher when high protein diets are consumed and there is some evidence for tumor-promoting activity in a mouse skin assay (reviewed by Hill[70]).

## 3. Fecapentaenes

Fecapentaenes have been shown to be highly genotoxic, but their carcinogenic activity has not been established (see discussion earlier). Recently, however, evidence has been obtained that fecapentaenes may possess tumor-promoting activity in a rat colon carcinogenesis model, using N-methyl-N-nitrosurea (MNU) as an initiating agent.[68] The number of carcinoma bearing rats and the number of carcinomas per rat were significantly higher in the animals given MNU and fecapentaenes, as compared with those given MNU alone.

## D. BILIARY CONJUGATES AND ENTEROHEPATIC CIRCULATION

Many potentially toxic substances, and also endogenously produced compounds such as steroids are metabolized in the liver and conjugated to glucuronic acid, sulfate, glutathione, or amino acids before being excreted via the bile into the small intestine.[78] In the colon, the bacterial enzymes β-glucuronidase (an enzyme associated with many common gut organisms), sulfatase, and chondroitin sulfate lyase can hydrolyze the conjugates releasing the parent compound, or its hepatic metabolite. This process may result in an enterohepatic circulation as the compound is reabsorbed, and returns to the liver where it can be subjected to further metabolism and conjugation. Enterohepatic circulation results in xenobiotics and steroids being retained in the body with a concomitant potentiation of their pharmacological and physiological effects.[79] This interplay between mammalian and bacterial metabolism of foreign compounds can have important consequences for humans, particularly when carcinogenic or mutagenic metabolites are produced. The role of β-glucuronidase in such reactions has been the most extensively studied. In the case of the colon carcinogen 1,2-dimethylhydrazine (DMH), small amounts of procarcinogenic glucuronide metabolites formed in the liver are excreted in the bile and hydrolyzed by bacteria, releasing the carcinogen methylazoxymethanol (MAM) in the colon. As might be expected, germ-free rats have fewer colon tumors after treatment with DMH or with MAM-glucuronic acid conjugate than do conventional flora rats.[80] The carcinogen benzo(a)pyrene (BaP), a common contaminant of the human diet, undergoes a similar sequence of reactions to that of DMH. Polar conjugates, including the glucuronide (about 35% of the dose), are excreted in bile following administration of BaP to rodents[81] and are hydrolyzed to BaP diols by gut bacteria. These products, unlike the conjugates from which they are derived, bind to DNA and are mutagenic in *in vitro* assays.[82,83] It is possible therefore that bacterial products of BaP formed in the colon may be initiators of carcinogenesis. Other carcinogens in which β-glucuronidase activity is important for their action include benzidine, nitropyrene, and dinitrotoluenes.

A number of drugs such as chloramphenicol, diflunisal, fenclofenac, phenacetin, and phenobarbital also undergo enterohepatic circulation which influences their pharmaceutical and toxicological effects.[84] For example, chloramphenicol is excreted as a glucuronide conjugate in the bile of rats, and is subsequently hydrolyzed in the gut and reduced to an arylamine. It is then reabsorbed and exerts a toxic effect on the thyroid.[85]

Enterohepatic circulation appears to be of particular importance in the absorption and metabolism of endogenously produced and synthetic steroids including estrone, estriol, pregnanolone, tetrahydrocortisone, and diethylstilbestrol. Bacterial β-glucuronidase and sulfatase acting on the conjugates of these steroids can thus affect hormonal activity and bioavailability of these compounds and their metabolites.[71]

## E. DETOXICATION AND PROTECTIVE EFFECTS OF THE GUT MICROFLORA
### 1. Methylmercury
The high toxicity of methylmercury (MeHg) is due to a combination of its efficient absorption from the gut, long retention time in the body, and its ability to penetrate the blood-brain barrier and accumulate in the lipid-rich neuropil of the central nervous system where it exerts its major toxic effects.[86] Consumption of fish contaminated with MeHg resulted in the disastrous outbreaks of mercury poisoning in the populations of Minimata and Niigata in Japan in the 1950s and 1960s.[87]

MeHg is rapidly, and virtually completely, absorbed from the gut.[88] As a result, the cumulative body-burden of mercury after methylmercury exposure is determined not only by the quantity taken in, but also, critically, by its rate of elimination. The main route of excretion occurs via feces in man and laboratory animals.[89] It is significant that in rats and mice given MeHg, the majority (50 to 90%) of the mercury in feces is in the mercuric form.[89-91] Therefore, demethylation of MeHg would appear to be an important step in the excretion of the organomercurial from the body.

MeHg gains access to the gut flora via its secretion as a MeHg-glutathione complex in bile.[92] The methylmercury-glutathione complex is largely reabsorbed leading to an enterohepatic circulation of mercury.[93] Conversion of MeHg in bile to the poorly absorbed, mercuric form would interrupt this enterohepatic circulation resulting in increased mercury excretion in feces. Evidence for this hypothesis is derived from *in vitro* and *in vivo* studies. Incubation of $^{203}$Hg-labeled methylmercuric chloride with gut contents from the rat or mouse, or with suspensions of human feces, results in extensive metabolism of the organomercurial[91,94] with a variety of metabolites being produced including elemental mercury and mercuric ion. Further evidence comes from studies in germ-free rats and animals treated with antibiotics to suppress their gut bacteria. Suppression or absence of the gut microflora was associated with decreased excretion of total mercury, and lower amounts of mercuric ion in feces as compared with conventional flora rats.[90,95-97] Rowland et al.[96] demonstrated that the decreased fecal excretion of Hg in mice treated with antibiotics was reflected in significantly higher body-burdens of mercury. The half-time of mercury elimination increased from 10 d in the conventional flora mice to more than 100 d in the antibiotic-treated animals. In general, suppression of the gut microflora in mice results in higher total mercury levels and higher proportions of MeHg in most tissues. The concentration of mercury in the brain (the target organ for MeHg) after MeHg exposure was found to be 25 to 45% greater in germ-free or antibiotic-treated animals than in controls. The consequences of this altered level of Hg in the central nervous system were elucidated by Rowland et al.[90] who investigated the effect of eliminating the microflora on the neurotoxicity of MeHg. MeHg-induced behavioral signs of neurotoxicity and the severity of the histopathological lesions in the cerebellum were much greater in antibiotic-treated rats than in conventional rats.

Changes in the composition of the gut microflora with age may influence susceptibility to MeHg toxicity. Neonatal and weaned mice exhibit dramatic differences in the rate of excretion of MeHg. Suckling mice absorb and retain the majority of an oral dose of MeHg (half-time of mercury elimination is greater than 100 d), whereas older mice excrete the mercurial much more rapidly (half time 6 to 10 d).[98] This developmental change in the rate of excretion coincides with the time of weaning to a pelleted rodent diet, and has been linked to a number of possible mechanisms (reviewed by Rowland et al.[91]). The most likely of these mechanisms is a change in the rate of biliary secretion of MeHg into the gut at weaning[99] and an increase in the demethylation activity of the gut microflora. The latter has been demonstrated *in*

**Table 8**  Effect of Age and Diet on MeHg
Demethylation by Mouse Cecal and
Human Fecal Suspensions

| Species | Age | Diet | % MeHg remaining |
|---------|-----|------|------------------|
| Mouse | 10 d | Milk | 94 |
| | 20 d | Stock | 50 |
| | 3 mos. | Stock | 46 |
| Human | 2 d | Milk | 97 |
| | 4.5 mos. | Milk | 90 |
| | 10 mos. | Milk | 88 |
| | 8 mos. | Solid mixed diet | 29 |
| | 4.5 y | Solid mixed diet | 18 |

*Note:*  The fecal suspensions were incubated with MeHg for 24 h at
37°C and % MeHg remaining determined by benzene
extraction.

From Rowland, I. R., Robinson, R. D., Doherty, R. A., and Landry,
T. D., *Reproductive and Developmental Toxicity of Metals*, Clarkson,
T. W., Nordberg, G. F., and Sager, P. R., Eds., Plenum Press, New
York, 1983, 745.

*vitro.* Cecal contents from adult (3-month-old) mice demethylated MeHg rapidly, but in suckling (10-day-old) mice the reaction was very slow. The rate of demethylation by the cecal microflora of 20-day-old mice was similar to that in adult mice (Table 8)[91] indicating that the underlying metabolic change occurred in mice during weaning (15 to 18 d), which is a time of major alteration in the bacterial composition of the gut microflora.[100]

These *in vitro* observations are supported by results from *in vivo* studies where it has been shown that the mercuric mercury excretion after MeHg exposure is much greater in 20-day-old mice than in 4- or 10-day-old mice.[91] Corresponding developmental changes in demethylating capacity occur in the human gut microflora since weaned and unweaned children of similar ages exhibit markedly different fecal demethylation capacities (Table 8). The virtual absence of demethylating capacity in fecal suspensions from unweaned human babies implies that they would absorb and retain more of an oral dose of MeHg than adults, making them more susceptible to MeHg neurotoxicity.

It is also tempting to speculate that the wide range of mercury elimination rates seen in man[101] may be related to the variations in the composition of the gut flora between individuals.

## 2.  Lignans and Phyto-Estrogens

The human diet, particularly grain and other fiber containing foods is a source of plant lignans such as secoisolariciresinol-diglycoside and matairesinol. When ingested, these substances are converted to enterolactone and enterodiol (the major lignans found in human urine) by hydrolysis, dehydroxylation, demethylation, and oxidation reactions catalyzed by the facultative anaerobes of the intestinal microflora.[102]

Similarly, phyto-estrogens such as glycosides of genistein, daidzein, biochanin A, and formonometin are modified by the intestinal microflora with the isoflavan equol being a major metabolite which is excreted in urine.[103] Soy-protein derived foods are particularly high in phyto-estrogens.

Numerous studies of the biological activity of lignans and isoflavones indicate that they possess hormonal activity mediated by the intestinal microflora. They are diphenolic compounds which are structurally similar to the potent synthetic estrogens diethylstilbestrol and hexestrol. In general, however, lignans and isoflavones are weakly estrogenic, and may also exhibit antiestrogenic properties in animals (reviewed by Setchell and Adlercreutz).[102] Over the last 10 years, many studies have focused on the potential beneficial effects of lignans and phyto-estrogens for colon and breast cancers. Epidemiological studies of populations with different risks of breast cancer have shown that the excretion of enterolactone is usually highest in the population with the greatest breast cancer risk.[104] A case control study of women with breast cancer in Singapore indicated that a decreased risk of the disease was associated with high intake of soya products which are known to be high in phyto-estrogens.[105] Studies in laboratory animals have yielded equivocal results. Supplementation of a

**Table 9**  Anti-Mutagenic and Anti-Carcinogenic Effects of Flavonoids

| Flavonoid | Effect |
|---|---|
| Quercetin | Inhibits tumor promotion by phorbol ester and teleocidin in mice.[121] |
|  | Inhibits DAB-induced hepatic tumors in rats.[111] |
|  | Inhibits B(a)P-induced lung tumors.[110] |
| Quercetin/Myricetin | Inhibit PAH-DNA adducts in skin and lungs of mice.[122] |
| Quercetin/Myricetin/Morin | Inhibit mutagenicity of cooked food mutagens *in vitro*.[112] |
| Myricetin | Inhibits PAH diol-epoxide mutagenicity.[123] |
| Quercetin/Rutin | Inhibit azoxymethane-induced colon tumors and mucosal cell proliferation.[113] |

high-fat diet with 5% flaxseed flour (a rich source of lignans), which resulted in a large increase in urinary enterolactone and enterodiol concentration, reduced cell proliferation and nuclear aberrations (considered to be early markers for carcinogenesis) in the rat mammary gland.[106] However, a subsequent study in which the effects of flaxseed flour on initiation and promotional stages of mammary tumorigenesis in rats showed no consistent influence on tumor incidence.[107] Furthermore, although the flaxseed flour fed during the promotion phase (i.e., after dosing with the initiating agent 7,12-dimethylbenzanthracene) reduced the size of tumors, the number of tumors per rat was increased. Evidence that lignans may be protective against colon cancer has been obtained from a study in which rats, treated with azoxymethane as an inducer of colon tumors, were fed a basal diet with or without supplementation with flaxseed oil.[108] The incidence of aberrant crypt foci (an early marker of neoplasia) in the colons of the rats after 4 weeks was reduced by about 50% in the rats fed the flaxseed oil diet.

A number of possible mechanisms may be proposed to account for the anticarcinogenic effects of lignans and phyto-estrogens. Clearly, in cancers which are hormonally dependent, such as breast cancer, it is possible that lignans may reduce the sensitivity of estrogen receptors to estradiol, or that the anti-estrogenic properties of enterodiol and enterolactone may inhibit the action of estrogen-stimulated tumor growth. Similar mechanisms could be operating in the inhibition of colon cancer since androgen and estrogen receptors have been identified in some primary colon tumors.[109] However, enterolactone exhibits antimitotic effects *in vitro* so this is an alternative mode of action.

## 3.  Flavonoids

In addition to the capacity of flavonoid glycosides to exert toxic and genotoxic effects after hydrolysis by gut microbial enzymes (see discussion earlier), there is evidence that they may also afford protection against genotoxic and carcinogenic activity of other chemicals (Table 9). Quercetin has been shown to antagonize tumor induction by benzo(a)pyrene and 4-dimethylaminoazobenzene[110,111] and in common with other flavonoids has been shown to inhibit the mutagenic activity of benzo(a)pyrene and its directly genotoxic diol-epoxide metabolite. The mode of action in the latter case appears to be the binding of the flavonoid to the active metabolite. Quercetin and two other flavonoids, myricetin and morin, exhibit potent antimutagenic effects *in vitro* against a range of different cooked food mutagens including IQ, MeIQ, and MeIQx.[112] These heterocyclic amine mutagens are produced when meat or fish is grilled or fried, and so are consumed by the majority of the population on a regular basis, albeit in small quantities. As well as being mutagenic, some heterocyclic amines induce tumors in the large intestine of laboratory animals, and so are potential inducers of colon cancer in man. In the case of the heterocyclic amines, the mechanism of action of the flavonoids appears to be inhibition of the activity of the hepatic enzymes responsible for activating the amines to their ultimate genotoxic species.[112] The enzymes involved (cytochrome P450 IA-dependent mixed function oxidases), participate in the bioactivation of a wide range of carcinogens. As well as being present in the liver, these enzymes are also found in the intestinal tract. It is possible therefore that flavonoids released from their glycosidic forms by bacterial β-glycosidases in the gut may inhibit the bioactivation by gut mucosal mixed function oxidases of dietary carcinogen precursors to their reactive species.

Quercetin and rutin have been shown to decrease the incidence of colon tumors induced by azoxymethane in mice.[113] The effect appeared to be on the promotional phase of the carcinogenic process. In particular, a decrease in epithelial cell proliferation was noted which suggests that the flavonoids were not acting on the generation of reactive carcinogenic metabolites involved in initiation.

## 4. Short Chain Fatty Acids

In recent years, a number of beneficial effects on the colonic mucosa have been ascribed to SCFA produced by fermentation of carbohydrates in the colon. These effects include amelioration of the symptoms of ulcerative colitis and anticancer effects which are discussed in more detail in Chapter 5.

## 5. Carcinogen Binding by Intestinal Bacteria

Morotomi and Mutai[114] investigated the *in vitro* binding of seven carcinogens produced during cooking of meat and fish to intestinal bacteria and freeze-dried feces. Two of the mutagens, 3-amino-1,4-dimethyl-5*H*-pyrido[4,3-*b*]indole (Trp-P-1) and 3-amino-1-methyl-5*H*-pyrido[4,3-*b*]indole (Trp-P-2), were effectively bound by most of the 22 bacterial strains tested and by freeze-dried feces. The binding resulted in a decrease in *vitro* mutagenic activity. Other carcinogens including IQ, MeIQ and MeIQx, which are present in much larger quantities than Trp-P-1 and Trp-P-2 in cooked foods, were in general much less effectively bound. It could be hypothesized that carcinogens bound to intestinal bacteria would have reduced ability to damage the colonic mucosa or to be absorbed which may represent a nonspecific protective effect of colonic bacteria against carcinogenesis. However, it remains to be established whether binding can occur *in vivo*, whether the physicochemical conditions are very different from those *in vitro*, and whether such binding has any toxicological consequences.

## F. MODIFICATION OF MAMMALIAN METABOLISM AND TUMOR INCIDENCE BY GUT BACTERIA

In addition to the involvement of the microflora in synthesis or activation of carcinogens in the gut, there is also evidence that intestinal bacteria can have indirect effects on carcinogen metabolism by affecting enzymic activity in other parts of the body, notably the liver. We have explored this hypothesis by comparing the ability of hepatic cell free extracts (postmitochondrial supernatants or S-9) from conventional flora and germ-free animals to activate genotoxins from several procarcinogens present in the human diet. The dietary carcinogens included 2-amino-3,4-dimethylimidazo[4,5-*f*]quinoline (MeIQ) and 3-amino-1-methyl-5H-pyrido[4,3-*b*]indole (Trp-P-2), examples of substances formed when meat or fish are cooked, and the fungal toxin aflatoxin $B_1$. All these carcinogens require metabolism by cytochrome P450-dependent mixed function oxidase enzymes to generate reactive species before they can damage DNA. The presence of a gut microflora significantly influenced the activation of the carcinogens to genotoxins detectable in the Ames *Salmonella* mutagenicity assay.[115] In the case of MeIQ and Trp-P-2, activation was somewhat greater with germ-free rat S-9, whereas aflatoxin $B_1$ mutagenicity was twofold higher in the presence of conventional flora liver fraction. To assess the significance of the latter *in vivo*, we performed a host-mediated assay in which germ-free and conventional flora mice were given *Salmonella* indicator bacteria intravenously and simultaneously, a *per os* dose of the carcinogen. After 1 h, the mice were killed, the Salmonella recovered from the liver by centrifugation, and the number of mutant bacteria counted. The number of mutants induced by aflatoxin in the conventional mice (141 per liver) was significantly greater ($p < 0.001$) than those in the germ-free mice (67 per liver). The mechanism by which the flora influences the hepatic activation of aflatoxin has not been elucidated. It may be envisaged, however, that the need to deal with microbial products absorbed from the gastrointestinal tract of the conventional flora rodent might result in an overall increase in metabolic capacity of the liver. It should be noted in this regard that the conventional rat liver is larger than the organ in the germ-free rat.

Clearly, long-term studies would have to be carried out using the two types of mice to confirm that tumorigenesis by aflatoxin can be affected by the presence of gut bacteria. There is, however, other evidence suggesting that the types of bacteria in the gut can influence liver tumor incidence. Roe and Grant[116] reported that the spontaneous incidence of hepatomas in germ-free C3H mice was lower than conventional microflora mice. Differences between germ-free and conventional flora mice were also observed in the incidence of liver and lung tumors induced by a single dose of the carcinogen dimethylbenzanthracene. The incidence of liver tumors being 81 and 13% in germ-free and conventional mice, respectively. The corresponding figures for lung tumors were 48 and 4%.

These results for spontaneous hepatoma incidence were confirmed in C3H/He mice by Mizutani and Mitsuoka,[117] who reported 30% incidence in germ-free mice and 75% incidence in conventional flora mice after 1 year. The investigation by these authors was carried further by associating the mice with one or more human gut bacteria. The hepatoma incidence could be manipulated between 46 and 95% by

**Table 10** Incidence of Hepatomas in Germ-free, Conventional Microflora, and Gnotobiotic C3H/He Mice

| Intestinal microflora | No. of mice | Mice with hepatomas after 1 year |
|---|---|---|
| Germ-free | 139 | 42 (30%) |
| Conventional flora | 56 | 42 (75%) |
| *Lactobacillus acidophilus* | 27 | 13 (48%) |
| *Bifidobacterium longum* | 17 | 8 (47%) |
| *Bacteroides multiacidus* | 28 | 21 (75%) |
| *Escherichia coli* + *Enterococcus faecalis* + *Clostridium paraputrificum* | 20 | 19 (95%) |
| *E. coli* + *Ent. faecalis* + *C. paraputrificum* + *Bif. longum* | 13 | 6 (46%) |

From Mizutani, T. and Mitsuoka, T., *Recent Advances in Germfree Research*, Sakai, S., Ed., Tokai University Press, Tokai, 1981, 639.

**Table 11** Diet-Microflora Interactions and Hepatic DNA Adducts

| Diet | Microbiological status | Adduct levels |
|---|---|---|
| Low fat (15% of total energy) | GF | 1.38 + 0.30 |
| Low fat | HFA | 3.50 + 0.31[a] |
| High fat (45% of total energy) | HFA | 5.44 + 1.15[a] |

*Note:* Covalently-bound DNA adducts in liver were quantitated by the $^{32}$P-postlabeling technique. Values shown are mean relative adduct levels per $10^9$ DNA bases ±SD for 4 rats. The rats were fed freeze-dried, isocaloric human diets.

[a] Value is significantly different from GF group fed the low fat diet.

From Rumney, C. J., Rowland, I. R., Coutts, C. M., Randerath, K., Reddy, R., Shah, A. B., Ellul, A., and O'Neill, I. K., *Carcinogenesis*, 14, 79, 1993.

altering the bacterial association (Table 10). The mechanism by which this occurs has not been elucidated. It should be noted that the C3H strain of mouse has a very high spontaneous incidence of hepatic neoplasia, and that the relevance of the hepatoma to the process of carcinogenesis in general, and liver cancer in particular, has been a subject of controversy. It is possible therefore that synthesis, or activation, of carcinogens by the bacteria may not be involved, but that production of promoters such as phenols and cresols, or changes in hormone metabolism may play a part. Whatever the mechanism, the flora must have a major effect on physiological and metabolic events in the liver.

Evidence that intestinal bacteria may be involved in induction of tumors in the liver has been obtained by comparing levels of DNA adducts in hepatic tissue from germ-free and HFA rats. The formation of DNA adducts is considered to be a crucial early stage in the process of carcinogenesis. The levels of covalently-bound DNA adducts in the livers of bacterially colonized animals was significantly higher than in the germ-free animals (Table 11).[118] In the same experiment, we investigated the role of diet by feeding the rats freeze-dried, human diets of varying composition. The feeding of a diet containing fat at a level (45% of calories) and of a type typical of that consumed by the UK population, was associated with increased levels of hepatic DNA adducts by comparison to rats fed a low-fat diet (Table 11). These results imply that the human intestinal microflora can produce, from naturally occurring substances in the human diet, chemical species capable of reacting with, and covalently binding to, hepatic DNA.

The increased adduct levels in rats fed the high-fat diet suggest that dietary fat may be acting as a precursor of these DNA-reactive species, or is increasing the rate of their production.

# REFERENCES

1. **Coates, M. E.,** Vitamins, in *The Germ-Free Animal in Biomedical Research*, Coates, M. E. and Gustafsson, B. E., Eds., Laboratory Animals Ltd, London, 1984, 269.

2. **Van der Waaij, D.,** Colonization resistance of the digestive tract, in *The Germ-Free Animal in Biomedical Research*, Coates, M. E. and Gustafsson, B. E., Eds., Laboratory Animals Ltd., London, 1984, 155.

3. **Mallett, A. K. and Rowland, I. R.,** Factors affecting the gut microflora, in *Role of the Gut Flora in Toxicity and Cancer*, Rowland, I. R., Ed., Academic Press, London, 1988, 348.

4. **Goldin, B. R. and Gorbach, S. L.,** The relationship between diet and rat fecal enzymes implicated in colon cancer, *J. Natl. Cancer Inst.*, 57, 371, 1976.

5. **Midtvedt, T.,** Monitoring the functional state of the microflora, in *Recent Advances in Microbial Ecology*, Hattori, T., Ishida, Y., Maruyama, Y., Morita, R. Y., and Uchida, A., Eds., Japan Scientific Societies Press, Tokyo, 1989, 515.

6. **Rowland, I. R.,** Metabolic profiles of intestinal floras, in *Recent Advances in Microbial Ecology*, Hattori, T., Ishida, Y., Maruyama, Y., Morita, R. Y., and Uchida, A., Eds., Japan Scientific Societies Press, Tokyo, 1989, 510.

7. **Drasar, B. S.,** The bacterial flora of the intestine, in *Role of the Gut Flora in Toxicity and Cancer*, Rowland, I. R., Ed., Academic Press, 1988, 23.

8. **Coates, M. E., Drasar, B. S., Mallett, A. K., and Rowland, I. R.,** Methodological considerations for the study of bacterial metabolism, in *Role of the Gut Flora in Toxicity and Cancer*, Rowland, I. R., Ed., Academic Press, London, 1988, 1.

9. **Corpet, D. E.,** Microbial ecology of the intestine. *In vitro, in vivo* and mathematical models, *Rev. Sci. Off. Int. Epiz.*, 8, 391, 1989.

10. **Rumney, C. J. and Rowland, I. R.,** *In vivo* and *in vitro* models of the human colonic flora, *Crit. Rev. Food Sci. Nutr.*, 31, 299, 1992.

11. **Rowland, I. R., Mallett, A. K., Bearne, C. A., and Farthing, M. J. G.,** Enzyme activities of the hindgut microflora of laboratory animals and man, *Xenobiotica*, 16, 519, 1983.

12. **Brown, J. P.,** Hydrolysis of glycosides and esters, in *Role of the Gut Flora in Toxicity and Cancer*, Rowland, I. R., Ed., Academic Press, London, 1988, 109.

13. **Laqueur, G. L. and Spatz, M.,** The toxicology of cycasin, *Cancer Res.*, 28, 2262, 1968.

14. **Van der Hoeven, J. C., Lagerweij, W. J., Bruggeman, I. M., Voragen, F. G., and Koeman, J. H.,** Mutagenicity of extracts of some vegetables commonly consumed in the Netherlands, *J. Agric. Food Chem.*, 31, 1020, 1983.

15. **Tamura, G., Gold, G., Feroluzzi, A., and Ames, B. N.,** Fecalase: a model for the activation of dietary glycosides to mutagens by intestinal flora, *Proc. Natl. Acad. Sci. U.S.A.*, 77, 4961, 1980.

16. **Kato, K., Mori, H., Fujii, M., Bunai, Y., Nishikawa, A., Shima, H., Takahashi, M., and Hirono, I.,** Lack of promotive effect of quercetin on methylazoxymethanol acetate carcinogenesis in rats, *J. Toxicol. Sci.*, 9, 319, 1984.

17. **Kato, K., Mori, H., Fujii, M., Bunai, Y., Nishikawa, A., Shima, H., Takahashi, M., and Hirono, I.,** Absence of initiating activity by quercetin in the rat liver, *Ecotoxicol. Environ. Saf.*, 10, 63, 1985.

18. **Hirose, M., Fukushima, S., Sakata, T., Inui, M., and Ito, N.,** Effect of quercetin on two-stage carcinogensis of the rat urinary bladder, *Cancer Lett.*, 21, 23, 1983.

19. **Grasso, P. and Goldberg, L.,** Problems confronted and lessons learnt in the safety evaluation of Brown FK, *Food Cosmet. Toxicol.*, 8, 539, 1968.

20. **Powell, R., Murray, M., Chen, C., and Lee, A.,** Survey of the manufacture, import and uses for benzidine, related substances and related dyes and pigments, *EPA Report* 560/13-79-005, Environmental Protection Agency, Washington, D.C., 1979.

21. **Cerniglia, C. E., Freeman, J. P., Franklin, W., and Pack, L. D.,** Metabolism of benzidine and benzidine-congener based dyes by human, monkey and rat intestinal bacteria, *Biochem. Biophys. Res. Commun.*, 107, 1224, 1982.

22. **Nony, C. R. and Bowman, M. C.,** Trace analysis of potentially carcinogenic metabolites of an azo dye and pigment in hamster and human urine as determined by two chromatographic procedures, *J. Chromatogr. Sci.*, 18, 64, 1980.

23. **Manning, B. W., Cerniglia, C. E., and Federle, T. W.,** Metabolism of the benzidine-based azo dye Direct Black 38 by human intestinal microbiota, *Appl. Environ. Microbiol.*, 50, 10, 1985.

24. **Hartter, D. R.,** Use and importance of nitroaromatic chemicals in the chemical industry, in *Toxicity of Nitroaromatic Compounds*, Rickert, D. E., Ed., Hemisphere Publishing Co., Washington, D.C., 1984, 1.

25. **Rosenkranz, H. S. and Mermelstein, R.,** The genotoxicity, metabolism and carcinogenicity of nitrated polycyclic aromatic hydrocarbons, *J. Environ. Sci. Health*, 3, 221, 1982.

26. **Lindmark, D. G. and Muller, M.,** Antitrichomonad action, mutagenicity, and reduction of metronidazole and other nitroimidazoles, *Antimicrob. Agents Chemother.*, 10, 476, 1976.

27. **Reddy, B. G., Pohl, L. R., and Krishna, G.,** The requirement of the gut flora in nitrobenzene-induced methemoglobinemia in rats, *Biochem. Pharmacol.*, 25, 1119, 1976.

28. **Levin, A. A. and Dent, J. G.,** Comparison of the metabolism of nitrobenzene by hepatic microsomes and cecal microflora from Fischer-344 rats *in vitro* and the relative importance of each *in vivo*, *Drug Metab. Dispos.*, 10, 450, 1982.

29. **Doolittle, D. J., Sherrill, J. M., and Butterworth, B. E.,** Influence of intestinal bacteria, sex of animal, and position of the nitro group on the hepatic genotoxicity of nitrotoluene isomers in vivo, *Cancer Res.*, 43, 2846, 1983.

30. **Mirsalis, J. C., Hamm, T. E., Sherrill, J. M., and Butterworth, B. E.,** Role of gut flora in genotoxicity of dinitrotoluene, *Nature*, 295, 322, 1982.

31. **El-Bayoumy, K., Sharma, C., Louis, Y. M., Reddy, B., and Hecht, S. S.,** The role of intestinal microflora in the metabolic reduction of 1-nitropyrene to 1-aminopyrene in conventional and germfree rats and in humans, *Cancer Lett.*, 19, 311, 1983.

32. **Manning, B. W., Federle, T. W., and Cerniglia, C. E.,** Use of a semicontinuous culture system as a model for determining the role of human intestinal microflora in the metabolism of xenobiotics, *J. Microbiol. Meth.*, 6, 81, 1987.

33. **Manning, B. W., Campbell, W. L., Franklin, W., Declos, K. B., and Cerniglia, C. E.,** Metabolism of 6-nitrochrysene by intestinal microflora, *Appl. Environ. Microbiol.*, 54, 197, 1988.

34. **Roe, F. J. C., Carter, R. L., and Adamthwaite, S.,** Induction of liver and lung tumours in mice by 6-aminochrysene administered during the first 3 days of life, *Nature*, 221, 1063, 1969.

35. **Felton, J. S., Knize, M. G., Shen, N. H., Andresen, B. D., Bjeldanes, L. F., and Hatch, F. T.,** Identification of the mutagens in cooked beef, *Environ. Health Perspect.*, 67, 17, 1986.

36. **Yamazoe, Y., Shimada, M., Kamataki, T., and Kato, R.,** Microsomal activation of 2-amino-3-methylimidazo[4,5-*f*]quinoline, a pyrolysate of sardine and beef extracts, to mutagenic intermediates, *Cancer Res.*, 43, 5768, 1983.

37. **Alldrick, A. J., Lake, B. G., Flynn, J., and Rowland, I. R.,** Metabolic conversion of IQ and MeIQ to bacterial mutagens, *Mutation Res.*, 163, 109, 1986.

38. **Ohgaki, H., Takayama, S., and Sugimura, T.,** Carcinogenicities of heterocyclic amines, *Mutation Res.*, 259, 399, 1991.

39. **Bashir, M., Kingston, D. G. I., Carman, R. J., Van Tassell, R. L., and Wilkins, T. D.,** Anaerobic metabolism of 2-amino-3-methyl-3H-imidazo[4,5-*f*]quinoline, (IQ) by human faecal flora, *Mutation Res.*, 190, 187, 1987.

40. **Carman, R. J., Van Tassell, R. L., Bashir, M., Kingston, D. G. I., and Wilkins, T. D.,** In vitro and in vivo metabolism of the dietary pyrolysis carcinogen, IQ, to a direct acting mutagen, 7-OHIQ, by human intestinal bacteria, in *Anaerobes Today*, Hardie, J. M. and Borriello, S. P., Eds., Wiley, Chichester, 1988, 224.

41. **Rumney, C. J., Rowland, I. R., and O'Neill, I. K.,** Conversion of IQ to 7-OHIQ by gut microflora, *Nutr. Cancer*, 19, 67, 1993.

42. **Carman, R. J., Van Tassell, R. L., Kingston, D. G. I., Bashir, M., and Wilkins, T. D.,** Conversion of IQ, a dietary pyrolysis carcinogen, to a direct-acting mutagen by normal intestinal bacteria of human, *Mutation Res.*, 206, 335, 1988.

43. **Rowland, I. R.,** The toxicology of N-nitrosocompounds, in *Nitrosamines Toxicology and Microbiology*, Hill, M. J., Ed., Ellis Horwood, Chichester, 1988, 117.

44. **Klubes, P., Cerna, I., Rabinowitz, A. D., and Jondorf, W. R.,** Factors affecting dimethylnitrosamine formation from simple precursors by rat intestinal bacteria, *Food Cosmet. Toxicol.*, 10, 757, 1971.

45. **Leach, S. A., Challis, B., Cook, A. R., Hill, M. J., and Thompson, M. H.,** Bacterial catalysis of the N-nitrosation of secondary amines, *Biochem. Soc. Trans.*, 13, 380, 1985.

46. **Massey, R. C., Key, P. E., Mallett, A. K., and Rowland, I. R.,** An investigation of the endogenous formation of apparent total N-nitroso compounds in conventional microflora and germ-free rats, *Food Chem. Toxicol.*, 26, 595, 1988.

47. **Walters, C. L., Downes, M. J., Edwards, M. W., and Smith, P. L. R.,** Determination of a non-volatile N-nitrosamine on a food matrix, *Analyst (London)*, 103, 1127, 1978.

48. **Leach, S.,** Mechanisms of endogenous N nitrosation, in *Nitrosamines Toxicology and Microbiology*, Hill, M. J., Ed., Ellis Horwoood, Chichester, 1988, 69.

49. **Rowland, I. R., Granli, T., Bockman, O. C., Key, P. E., and Massey, R. C.,** Endogenous N-nitrosation in man assessed by measurement of apparent total N-nitroso compound in faeces, *Carcinogenesis*, 12, 1395, 1991.

50. **Bruce, W. R., Baptista, J., Che, T., Furrer, R., Gingerich, J. S., Gupta, I., Krepinski, J. J., Grey, A. A., and Yates, P.,** General structure of 'fecapentaenes', the mutagenic substances in human feces, *Naturwissenschaften*, 69, 557, 1982.

51. **Hirai, N., Kingston, D. G. I., van Tassell, R. L., and Wilkins, T. D.,** Structure elucidation of a potent mutagen from human feces, *J. Am. Chem. Soc.*, 104, 6149, 1982.

52. **Baptista, J., Bruce, W. R., Gupta, I., Krepinski, J. J., van Tassell, R. L., and Wilkins, T. D.,** On the distribution of different fecapentaenes, the fecal mutagens, in the human population, *Cancer Lett.*, 22, 299, 1985.

53. **Schiffman, M. H., van Tassell, R. L., Robinson, A., Smith, L., Daniel, J., Hoover, R. N., Weil, R., Rosenthal, J., Nair, P. P., Schwarz, S., Pettigrew, H., Curale, S., Batist, G., Block, G., and Wilkins, T. D.,** Case control study of colorectal cancer and fecapentaene excretion, *Cancer Res.*, 49, 1322, 1989.

54. **de Kok, T. M. C. M., van Faasen, A., ten Hoor, F., and Kleinjans, J. C. S.,** Fecapentaene excretion and fecal mutagenicity in relation to nutrient intake and fecal parameters in humans on omnivorous and vegetarian diets, *Cancer Lett.*, 62, 11, 1992.

55. **de Kok, T. M. C. M., van Iersel, M. L. P. S., ten Hoor, F., and Kleinjans, J. C. S.,** In vitro study on the effects of fecal composition on fecapentaene kinetics in the large bowel, *Mutation Res.*, 302, 103, 1993.

56. **Wilkins, T. D., Lederman, M., van Tassell, R. L., Kingston, D. G. I., and Henion, J.,** Characterization of a mutagenic bacterial product in human feces, *Am J. Clin. Nutr.*, 33, 2513, 1980.
57. **van Tassell, R. L., MacDonald, D. K., and Wilkins, T. D.,** Production of a fecal mutagen by *Bacteroides* spp., *Infect. Immun.*, 37, 975, 1982.
58. **van Tassell, R. L., MacDonald, D. K., and Wilkins, T. D.,** Stimulation of mutagen production in human feces by bile and bile acids, *Mutation Res.*, 103, 233, 1982.
59. **Plummer, S. M., Grafstom, R. C., Yang, L. L., Curren, R. D., Linnainmaa, K., and Harris, C. C.,** Fecapentaene-12 causes DNA damage and mutations in human cells, *Carcinogenesis*, 7, 1607, 1986.
60. **Schmid, E., Bauchinger, M., Braselmann, H., Pfaendler, H. R., and Goggelmann, W.,** Dose-response relationship for chromosome aberrations induced by fecapentaene-12 in human lymphocytes, *Mutation Res.*, 191, 5, 1987.
61. **Curren, R. D., Putman, D. L., Yang, L. L., Hayworth, S. R., Lawler, T. E., Plummer, S. M., and Harris, C. C.,** Genotoxicity of fecapentaene-12 in bacterial and mammalian cell assay systems, *Carcinogenesis*, 8, 349, 1987.
62. **Hinzman, M. J., Novotny, C., Ullah, A., and Shamsuddin, A. M.,** Fecal mutagen fecapentaene-12 damages mammalian colon epithelial DNA, *Carcinogenesis*, 8, 1475, 1987.
63. **Vaughan, D. J., Furrer, R., Baptista, J., and Krepinsky, J. J.,** The effect of fecapentaenes on nuclear aberrations in murine colon epithelial cells, *Cancer Lett.*, 37, 199, 1987.
64. **Suzuki, K. and Bruce,W. R.,** Human faecal fractions can produce nuclear damage in the colonic epithelial cells of mice, *Mutation Res.*, 141, 35, 1984.
65. **Ward J. M., Anjo, T., Ohannesian, L., Keefer, L. K., Devor, D. E., Donovan, P. J., Smith, G. T., Henneman, J. R., Streeter, A. J., Konishi, N., Rehm, S., Reist, E. J., Bradford, W. W., and Rice, J. M.,** Inactivity of fecapentaene-12 as a rodent carcinogen or tumor initiator, *Cancer Lett.*, 42, 49, 1988.
66. **Weisburger, J. H., Jones, R. C., Wang, C. X., Backlund, J. Y. C., Williams, G. M., Kingston, D. G. I., Van Tassell, R. L., Keyes, R. F., Wilkins, T. D., De Wit, P. P., Van der Steeg, M., and Van der Gen, A.,** Carcinogenicity tests of fecapentaene-12 in mice and rats, *Cancer Lett.*, 49, 89, 1990.
67. **Shamsuddin, A. M., Ullah, A., Baten, A., and Hale, E.,** Stability of fecapentaene-12 and its carcinogenicity in F344 rats, *Carcinogenesis*, 12, 601, 1991.
68. **Zarkovic, M., Qin, X., Nakatsuru, Y., Oda, H., Nakamura, T., Shamsuddin, A. M., and Ishikawa, T.,** Tumor promotion by fecapentaene-12 in a rat colon carcinogenesis model, *Carcinogenesis*, 14, 1261, 1993.
69. **Hill, M. J.,** *Microbes and Human Carcinogenesis*, Edward Arnold, London, 1986.
70. **Hill, M. J.,** Gut flora and cancer in humans and laboratory animals, in *Role of the Gut Flora in Toxicity and Cancer*, Rowland, I. R., Ed., Academic Press, London, 1988, 461.
71. **Eyssen, H. and Caenepeel, P.,** Metabolism of fats, bile acids and steroids, in *Role of the Gut Flora in Toxicity and Cancer*, Rowland, I. R., Ed., Academic Press, London, 1988, 263.
72. **Rowland, I. R., Mallett, A. K., and Wise, A.,** The effect of diet on the mammalian gut flora and its metabolic activities, *CRC Crit. Rev. Toxicol.*, 16, 31, 1985.
73. **Clinton, S. K.,** Dietary protein and carcinogenesis, in *Nutrition, Toxicity, and Cancer*, Rowland, I. R., Ed., CRC Press, Boca Raton, FL, 1992, 455.
74. **MacDonald, I. A., Bokkenheuser, V. D., Winter, J., McLernon, A. M., and Mosbach, E. H.,** Degradation of steroids in the human gut, *J. Lipid Res.*, 24, 675, 1993.
75. **McMichael, A. J. and Potter, J. D.,** Host factors in carcinogenesis: certain bile-acid metabolic profiles that selectively increase the risk of proximal colon cancer, *J. Natl. Cancer Inst.*, 75, 185, 1985.
76. **Silverman, S. J. and Andrews, A. W.,** Bile acids: comutagenic activity in the *Salmonella*-mammalian microsome mutagenicity test: brief communication, *J. Natl. Cancer Inst.*, 59, 1557, 1977.
77. **Lupton, J. P. and Marchant, L. J.,** Independent effects of fiber and protein on colonic luminal ammonia concentration, *J. Nutr.*, 119, 235, 1989.
78. **Chipman, K.,** Bile as a source of reactive metabolites, *Toxicology*, 25, 99, 1982.
79. **Gregus, Z. and Klaasen, C. D.,** Enterohepatic circulation of toxicants, in *Gastrointestinal Toxicology*, Rozman, K. and Hanninen, O., Eds, Elsevier, Amsterdam, 1986, 57.
80. **Reddy, B. S., Weisburger, J. H., Narisawa, T., and Wynder, E. L.,** Colon carcinogenesis in germfree rats with 1,2-dimethylhydrazine and N-methyl-N-nitro-N-nitrosoguanidine, *Cancer Res.*, 74, 2368, 1974.
81. **Chipman, J. K., Millburn, P., and Brooks, T. M.,** Mutagenicity and *in vivo* disposition of bilairy metabolites of benzo(a)pyrene, *Toxicol. Lett.*, 17, 233, 1983.
82. **Nanno, M., Morotomi, M., and Takayama, H.,** Mutagenic activation of biliary metabolites of benzo(a)pyrene by β-glucuronidase-positive bacteria in human faeces, *J. Med. Microbiol.*, 22, 351, 1986.
83. **Kinoshita, N. and Gelboin, H. V.,** β-Glucuronidase catalysed hydrolysis of benzo(a)pyrene-3-glucuronide and binding to DNA, *Science*, 199, 307, 1978.
84. **Larsen, G. L.,** Deconjugation of biliary metabolites by microfloral β-glucuronidases, sulphatases and cysteine conjugate β-lyases and their subsequent enterohepatic circulation, in *Role of the Gut Flora in Toxicity and Cancer*, Rowland, I. R., Ed., Academic Press, London, 1988, 79.
85. **Thompson, R. Q., Stutevant, M., Bird, O. D., and Glazko, A. J.,** The effect of metabolites of chloramphenicol (chloromycetin) on the thyroid of the rat, *Endocrinology*, 55, 665, 1954.

86. **Berlin, M.,** Mercury, in *Handbook on Toxicology of Metals*, Friburg, L., Nordbery, G. F., and Vouk, V. B., Eds., Elsevier, Amsterdam, 1979, 503.

87. **Takeuchi, T.,** Biological reactions and pathological changes in human beings and animals caused by organic mercury contamination, in *Environmental Mercury Contamination*, Harting, R. and Dinman, B. D., Eds., Ann Arbor Science Publishers, Ann Arbor, MI, 1972.

88. **Miettinen, J. K.,** Absorption and elimination of dietary mercury (Hg2+) and methylation in man, in *Mercury, Mercurials and Mercaptans*, Miller, M. W. and Clarkson, T. W., Eds., Charles C. Thomas, Springfield, IL, 1971, 223.

89. **Clarkson, T. W.,** General principles underlying the toxic action of metals, in *Handbook on Toxicology of Metals*, Friburg, L., Nordbery, G. F., and Vouk, V. B., Eds., Elsevier, Amsterdam, 1979, 99.

90. **Rowland, I. R., Davies, M. J., and Evans, J. G.,** Tissue content of mercury in rats given methylmercuric chloride orally: Influence of intestinal flora, *Arch. Environ. Health*, 35, 155, 1980.

91. **Rowland, I. R., Robinson, R. D., Doherty, R. A., and Landry, T. D.,** Are developmental changes in methylmercury metabolism and excretion mediated by the intestinal microflora? in *Reproductive and Developmental Toxicity of Metals*, Clarkson, T. W., Nordberg, G. F., and Sager, P. R., Eds., Plenum Press, New York, 1983, 745.

92. **Klaasen, C. D.,** Biliary excretion of metals, *Drug Metab. Rev.*, 5, 165, 1976.

93. **Norseth, T.,** Biliary excretion and intestinal reabsorption of mercury in the rat after injection of methylmercuric chloride, *Acta Pharmacol. Toxicol.*, 33, 280, 1973.

94. **Rowland, I. R., Davies, M. J., and Grasso, P.,** Metabolism of methylmercuric chloride by the gastrointestinal flora of the rat, *Xenobiotica*, 8, 37, 1978.

95. **Nakamura, I., Hosokawa, K., Tamura, H., and Miura, T.,** Reduced mercury excretion with feces in germ-free mice after oral administration of methylmercury chloride, *Bull. Environ. Contam. Toxicol.*, 17, 528, 1977.

96. **Rowland, I. R., Robinson, R. D., and Doherty, R. A.,** Effects of diet on mercury metabolism and excretion in mice given methylmercury: role of gut flora, *Arch. Environ. Health*, 39, 401, 1984.

97. **Seko, Y., Miura, T., Takashi, M., and Koyama, T.,** Methylmercury decomposition in mice treated with antibiotics, *Acta Pharmacol. Toxicol.*, 49, 259, 1981.

98. **Doherty, R. A. and Gates, A. H.,** Epidemic methylmercury poisoning: Application of a mouse model., *Pediatr. Res.*, 7, 319, 1973.

99. **Ballatori, N. and Clarkson, T. W.,** Developmental changes in the biliary excretion of methylmercury and glutathione, *Science*, 216, 61, 1982.

100. **Schaedler, R. W.,** The relationship between the host and its intestinal microflora, *Proc. Nutr. Soc.*, 32, 41, 1973.

101. **Shahristani, H. and Shihbab, K. M.,** Variation of biological half-life of methylmercury in man, *Arch. Environ. Health*, 28, 324, 1974.

102. **Setchell, K. D. R. and Adlercreutz, H.,** Mammalian lignans and phyto-oestrogens. Recent studies on their formation, metabolism and biological role in health and disease, in *Role of the Gut Flora in Toxicity and Cancer*, Rowland, I. R., Ed., Academic Press, London, 1988, 315.

103. **Axelson, M., Sjovall, J., Gustafsson, B. E., and Setchell, K. D. R.,** Soya — a dietary source of the non-steroidal eostrogen equol in human and animals, *J. Endocrinol.*, 102, 49, 1984.

104. **Adlercreutz, H., Fotsis, T., Bannwart, C., Wahala, K., Brunow, G., and Hase, T.,** Determination of urinary lignans and phyto-oestrogens metabolites, potential antioestrogens and anticarcinogens, in urine of women on various habitual diets, *J. Steroid Biochem.*, 25, 791, 1986.

105. **Lee, H. P., Gourley, L., Duffy, S. W., Esteve, J., Lee, J., and Day, N. E.,** Dietary effect on breast cancer risk in Singapore, *Lancet*, 337, 1197, 1991.

106. **Serraino, M. and Thompson, L. U.,** The effect of flaxseed supplementation on early risk markers for mammary carcinogenesis, *Cancer Lett.*, 60, 135, 1991.

107. **Serraino, M. and Thompson, L. U.,** The effect of flaxseed supplementation on the initiation and promotional stages of mammary tumorigenesis, *Nutr. Cancer*, 17, 153, 1991.

108. **Serraino, M. and Thompson, L. U.,** Flaxseed supplementation and early markers of colon carcinogenesis, *Cancer Lett.*, 63 159, 1992.

109. **Alford, T. C., Do, H. M., Geelkhoed, G. W., Tsangaris, N. T., and Lippman, M. E.,** Steroid hormone receptors in human colon cancers, *Cancer*, 43, 980, 1979.

110. **Wattenberg, L. W. and Leong, J. L.,** Inhibition of the carcinogenic action of benzo(a)pyrene by flavones, *Cancer Res.*, 30, 1922, 1970.

111. **Nagase, S., Fujimake, C., and Isaka, H.,** Effect of administration of quercetin on the production of experimental liver cancer in rats fed 4-dimethylaminoazobenzene, *Proc. Jpn. Cancer Assoc.*, 26, 1964.

112. **Alldrick, A. J., Flynn, J., and Rowland, I. R.,** Effect of plant derived flavonoids and polyphenolic acids on the activity of mutagens from cooked food, *Mutation Res.*, 163, 225, 1986.

113. **Deschner, E. D., Ruperto, J., Wong, G., and Newmark, H. L.,** Quercetin and rutin as inhibitors of azoxymethanol-induced colonic neoplasia, *Carcinogenesis*, 12, 1193, 1991.

114. **Morotomi, M. and Mutai, M.,** In vitro binding of potent mutagenic pyrolyzates to intestinal bacteria, *J. Natl. Cancer Inst.*, 77, 195, 1986.

115. **Rowland, I. R., Mallett, A. K., Cole, C. B., and Fuller, R.,** Mutagen activation by hepatic fractions from conventional, germfree and monoassociated rats, *Arch. Toxicol.,* 11 (Suppl.), 261, 1987.

116. **Roe, F. J. C. and Grant, G. A.,** Inhibition by germ-free status of the development of liver and lung tumours in mice exposed neonatally to 7,12-dimethylbenzanthracene: implications in relation to tests for carcinogenicity, *Int. J. Cancer,* 6, 133, 1970.

117. **Mizutani, T. and Mitsuoka, T.,** Relationship between liver tumorigenesis and intestinal bacteria in gnotobiotic C3H/He male mice, in *Recent Advances in Germfree Research,* Sakai, S., Ed., Tokai University Press, Tokai, 1981, 639.

118. **Rumney, C. J., Rowland, I. R., Coutts, C. M., Randerath, K., Reddy, R., Shah, A. B., Ellul, A., and O'Neill, I. K.,** Effects of risk-associated human dietary macrocomponents on processes related to carcinogenesis in human-flora-associated (HFA) rats, *Carcinogenesis,* 14, 79, 1993.

119. **Ruddell, W. S. J., Bone, E. S., Hill, M. J., and Walters, C. L.,** Pathogenesis of gastric cancer in pernicious anaemia, *Lancet,* 1, 521, 1978.

120. **Venitt, S.,** Mutagens in human faeces, in *Role of the Gut Flora in Toxicity and Cancer,* Rowland, I. R., Ed., Academic Press, London, 1988, 399.

121. **Nishino, H., Iwashima, A., Fujiki, H., and Sugimura, T.,** Inhibition by quercetin of the promoting effect of teleocidin on skin papilloma formation in mice initiated with 7,12-dimethylbenz[a]anthracene, *Gann,* 75, 113, 1984.

122. **Das, M., Khan, W. A., Asokan, P., Bickers, D. R., and Mukhtar, H.,** Inhibition of polycyclic aromatic hydrocarbon-DNA adduct formation in epidermis and lungs of SENCAR mice by naturally occurring plant phenols, *Cancer Res.,* 47, 767, 1987.

123. **Huang, M. T., Wood, A. W., Newmark, H. L., Sayer, J. M., Yagi, H., Jerina, D. M., and Conney, A. H.,** Inhibition of the mutagenicity of bay-region diol-epoxides of polyaromatic hydrocarbons by phenolic plant flavonoids, *Carcinogenesis,* 4, 1631, 1983.

*Chapter 8*

# Structure, Function, and Metabolism of Host Mucus Glycoproteins

*Michael E. Quigley and Sean M. Kelly*

## CONTENTS

I. Introduction ........................................................................................................ 175
II. Factors Affecting the Integrity of the Mucus Gel Layer ....................................... 175
   A. Intracellular Organization and Biosynthesis ................................................... 175
   B. Structure and Function of Gastrointestinal Mucus Secretions ......................... 176
   C. Turnover ....................................................................................................... 177
      1. Regulation of Synthesis ........................................................................... 177
      2. Regulation of Secretion ........................................................................... 179
      3. Erosion ................................................................................................... 179
III. Analytical Techniques to Determine Bacterial Metabolism of Mucin ................... 179
   A. Constituent Sugars ........................................................................................ 180
   B. Sialic Acids ................................................................................................... 183
   C. Proteins and Amino Acids ............................................................................. 183
   D. Products of Fermentation ............................................................................... 184
IV. Degradation of Mucus Glycoproteins and Pathological Implications ................... 184
   A. Metabolism of Mucin .................................................................................... 184
      1. Penetration of the Mucus Gel Layer ......................................................... 184
      2. Breakdown of the Protective Polymeric Structure of Mucus ..................... 185
      3. Mucin Utilizing Bacteria ......................................................................... 185
   B. Mucus, Peptic Ulceration, and *Helicobacter pylori* ....................................... 186
   C. Mucus and Inflammatory Bowel Disease (IBD) .............................................. 188
V. Conclusions ........................................................................................................ 189
References ............................................................................................................... 190

## I. INTRODUCTION

Studies of the physical, chemical, and immunological changes that occur to mucus glycoproteins are important for understanding the possible etiology of certain pathological conditions. Bacterial metabolism of mucus glycoproteins may play a role both in health and disease which is apposite to mechanisms of how the depth and integrity of the mucus gel layer is maintained. The protective properties of mucus are discussed here. In order to assess the ability of particular species of bacteria to grow on mucus glycoproteins, accurate *in vitro* analytical determinations that can follow the fermentation processes are required. A number of new techniques are now available that offer attractive alternatives to previous methodologies in terms of specificity and simplicity. Mucus bacteriology and the etiology and pathology of gastrointestinal disorders such as gastric ulceration and inflammatory bowel disease are also included.

## II. FACTORS AFFECTING THE INTEGRITY OF THE MUCUS GEL LAYER

### A. INTRACELLULAR ORGANIZATION AND BIOSYNTHESIS

In many respects, the intracellular organization of gastrointestinal (GI) mucus cells is typical of all exocrine cell types that are concerned with the synthesis, processing, storage, and vectoral secretion of protein or glycoprotein products. Mucus glycoproteins (mucins) are produced and secreted from salivary glands, the esophagus, stomach, small and large intestine, as well as gall bladder and pancreatic ducts. Synthesis of the peptide moeity of the mucin molecule is carried out in the cisternae of the rough endoplasmic reticulum (RER) in which amino acid residues are incorporated into nascent mucin polypeptides.[1,2] The biosynthesis of mucin begins in the ribosomes with the production of peptides enriched in

0-8493-4524-3/95/$0.00+$.50
© 1995 by CRC Press, Inc.

serine, threonine, and proline (in domains that will eventually become glycosylated), and other areas more enriched in serine, hydrophobic, and polar amino acids (that will form "naked" domains or discrete "link" peptides). The peptides are then transported from the RER to the Golgi compartment via a membrane-vesicle shuttle system.[3] A number of studies[4-11] have confirmed that the Golgi complex is the major site of synthesis. Mucin production proceeds in the endoplasmic reticulum and/or Golgi complex via the formation of sugar-nucleotides and addition of N-acetylgalactosamine (GalNAc) to the oxygen atom of serine or threonine residues in acceptor peptides. This linkage is termed O-glycosidic and distinguishes mucin-type linkages from a second type of carbohydrate-protein linkage (N-glycosidic) in which N-acetyl glucosamine (GlcNAc) is joined to the nitrogen of asparagine residues.[12] Oligosaccharide core structures are formed in the Golgi which are then elongated, one residue at a time, to form a backbone with terminal sugars then added. The glycoprotein may then be modified, by adding sulfate groups and subunit polymerization, after which the mucins are then packaged in supranuclear secretory vesicles and subsequently migrate towards the cell apex, where they fuse with the plasma membrane intermittently disgorging their contents into the lumin. Under baseline conditions, in the absence of neural or pharmacological stimulation, this results in a steady, slow secretory rate.[13]

## B. STRUCTURE AND FUNCTION OF GASTROINTESTINAL MUCUS SECRETIONS

The gastrointestinal tract is composed of four principal layers: mucosal layer, submucosal layer, muscle layer, and serous layer with the different regions showing similar structural features.[14] The luminal surface is covered by a lining of columnar epithelial cells, goblet cells, and endocrine cells; the layer being one cell thick. In the small intestine, villi and crypts are present whereas in the large intestine only crypts are present.[15] These differences in structural features reflect a greater absorptive capacity in the small intestine as compared to the greater protective role of the colon, where the composition of the glycoproteins makes them more resistant to bacterial attack. The mucus layer consists of water (95%), glycoproteins, electrolytes, proteins, and nucleic acids.[16,17]

All mucus glycoproteins are characterized by their very high molecular weight and consist of a number of carbohydrate side-chains attached to a protein core.[16,17] A number of studies have investigated the oligosaccharide structure.[18-25] Both neutral and acidic species are present. The carbohydrate side-chains consist of five different sugars, all or only a few of which may be present. There is a core region containing GalNAc linking the oligosaccharide side-chain to the protein core; a backbone region, often branched, of repeating D-galactose and GlcNAc; and a peripheral region, at the nonreducing end where the terminal sugar is responsible for the antigenicity of the mucin. There are known to be five different types of core regions[26] consisting in its simplest form of just a single GalNAc residue linked to either serine or threonine of the protein core to more complex branched type structures such as those consisting of two GalNAc residues, one $\beta 1 \rightarrow 6$, the other $\beta 1 \rightarrow 3$ linked to GalNAc. The backbone regions vary in length and antigenicity. In linear sequences, oligosaccharides are joined by $1 \rightarrow 3$ linkages, whereas the branches result from the addition of $1 \rightarrow 6$ linkages. The best characterized linkages in the peripheral regions are the major blood group antigens H, A, B, Le[a], and Le[b] shown below:

$$
\begin{array}{ll}
\text{H} & \text{Gal}\beta 1 \rightarrow 3/4\text{GlcNAc} \ldots \\
& \uparrow 1,2 \\
& \text{Fuc}\alpha
\end{array}
$$

$$
\begin{array}{ll}
\text{A} \quad \text{GalNAc}\alpha 1 \rightarrow 3\text{Gal}\beta 1 & \rightarrow 3/4\text{GlcNAc} \ldots \\
& \uparrow 1,2 \\
& \text{Fuc}\alpha
\end{array}
$$

$$
\begin{array}{ll}
\text{B} \quad \text{Gal}\alpha 1 \rightarrow 3\text{Gal}\beta 1 & \rightarrow 3/4\text{GlcNAc} \ldots \\
& \uparrow 1,2 \\
& \text{Fuc}\alpha
\end{array}
$$

$$
\begin{array}{ll}
\text{Le}^{a} & \text{Gal}\beta 1 \rightarrow 3\text{GlcNAc} \ldots \\
& \uparrow 1,4 \\
& \text{Fuc}\alpha
\end{array}
$$

$$
\begin{array}{ll}
\text{Le}^{b} & \text{Gal}\beta 1 \rightarrow 3\text{GlcNAc} \ldots \\
& \uparrow 1,2 \qquad \uparrow 1,4 \\
& \text{Fuc}\alpha \qquad \text{Fuc}\alpha
\end{array}
$$

The wide range of different structures in the core, backbone, and peripheral regions results in considerable intra- and inter-species diversity in the composition of mucin side chains. Differences between healthy mucus and that in disease states, such as inflammatory bowel disease, and colon cancer also occur,[16,17,27-36] which results in a wide range of enzymes being necessary to allow complete breakdown of the side chains.

The main compositional changes from gastric to colonic mucus are outlined in Table 1. The presence of sialic acids and esters of sulfate confer a charged state to the mucus rendering it more viscous and less susceptible to bacterial attack. Post-mortem intestinal tissue has been fractionated to separate neutral and acidic mucins.[37] Major differences were found in the proportion of amino acid residues and carbohydrates in the different mucins. The acidic species (>10% sialic acids) contained higher proportions of sialic acids, GalNAc, proline, threonine, and glycine, whereas neutral species had elevated fucose, GlcNAc, galactose, serine, aspartate, and alanine. Increased proline in acidic mucins probably helps closer packing, which then assists the protective role of mucins lower in the GI tract.

Models proposed to describe the tertiary structure of mucins include the "windmill" model of gastric mucin,[16] a "beaded mucin" model for egg-white ß-ovomucin[38] and sputum mucins,[39] and a "flexible-thread" configuration for rat intestinal mucin.[40] The presence of cysteine residues in nonglycosylated regions allows the formation of disulfide bridges, thus linking glycoproteins into large aggregates. Carlstedt and Sheehan[41] showed that a plot of the log of the radius of gyration ($R_G$) vs. the log of relative molecular mass ($M_R$) showed a straight line which was close to that expected from a random coil. In addition, there was a similar relationship between $R_G$ and $M_R$ for whole mucins and their subunits which held true for gastric, cervical, and respiratory mucus, indicating that all three macromolecules consist of linear flexible chains.

The functional properties of mucus include its acting *in vivo* as a lubricant and as a barrier to diffusion.[42-49] It also binds to calcium ions,[50-52] mutagens,[53] antibiotics,[54,55] and bacteria.[56] The ability of mucus to carry out these functions depends on many factors, of which the concentration of the mucus glycoproteins plays a major role. At a concentration of about 20 mg/ml, glycoprotein molecules fill the entire solution volume. Any increases in the concentration of glycoprotein molecules causes overlap of their domains with an increase in noncovalent interactions with gel formation between 30 and 40 mg/ml.[16] These *in vitro* results can be compared to results found *in vivo*, where concentrations of mucins in the mucus gel layer varies throughout the intestinal tract, i.e., 47, 38, and 20 mg/ml in the stomach, duodenum, and colon, respectively. The rheological properties and solubility of mucus also depend strongly on interactions with other luminal constituents such as lipids, nucleic acids, inorganic acids and alkali, salts, serum proteins, and enzymes.[40] To investigate the ability of mucus to act as a barrier to diffusion, the amount of solute transferred versus time across a mucus layer was compared to that transferred across an aqueous layer using a diffusion chamber technique.[42] Species of low relative molecular mass were retarded by a factor of 2 to 5, but higher molecular mass compounds such as RNA and lysozyme were retarded by factors of 18 and 27, respectively. Increased turnover of the mucus gel layer due to the effects of bacterial metabolism will affect these normal protective functions, thus rendering the sensitive underlying epithelial cells more susceptible to pathological agents.

## C. TURNOVER

The ability of mucus *in vivo* to prevent chemical and mechanical damage to the underlying epithelial cells depends on its ability to form a gel of sufficient depth and integrity. A continuous dynamic balance exists *in vivo* between synthesis, secretion, and erosion.

### 1. Regulation of Synthesis

Factors affecting synthesis include competition of glycosyl transferases for a common acceptor, substrate specificity, and compartmentalization. After GalNAc has been attached to the protein core at serine or threonine residues, competition between at least three different glycosyl transferases exists, which results in different core structures and consequently in different degrees of branching and chain elongation. Substrate specificity is provided by the presence of mono-, di-, or trisaccharides, branch structures, or specific linkages that affect accessibility of the transferases. Compartmentalization is thought to be responsible for heterogeneous glycosylation either by concentrating different glycosidases in cis, medial, and trans Golgi cisternae or by the presence of functionally distinct Golgi stacks.[57-59] Malnutrition, fasting, and the presence of inhibitors[60-62] affect the glycosylation and synthesis of proteins which influences mucin biosynthesis. Methotrexate,[63] temporary ischemia,[64] radiation,[65] and inflammation[66]

**Table 1**  Amino Acid Composition (Mol %), Carbohydrate Composition and Distinguishing Features of Gastrointestinal Mucins

| | Gastric | Small intestine | Colon |
|---|---|---|---|
| **Amino acids** | | | |
| Threonine[a] | 13.9 | 27.9 | 19.0 |
| Serine[a] | 13.7 | 12.8 | 11.0 |
| Proline[b] | 13.0 | 13.4 | 11.4 |
| Aspartate | 7.6 | 5.4 | 7.3 |
| Glutamate | 10.6 | 6.2 | 9.9 |
| Glycine | 10.6 | 6.2 | 9.8 |
| Alanine | 5.6 | 4.1 | 6.4 |
| Valine | 6.0 | 4.3 | 5.4 |
| Cystine $\beta$ | 3.5 | 3.8 | ND |
| Others | 15.5 | 15.9 | 19.8 |
| | | | |
| **Monosaccharides** | | | |
| Fucose | 21.4 | 15.3 | 10.9 |
| Galactose | 35.4 | 25.9 | 30.1 |
| GalNAc[a] | 9.8 | 24.5 | 10.1 |
| GlcNAc | 9.8 | 24.5 | 10.1 |
| Sialic acids | 1.2 | 13.2 | 8.9 |
| Sulfate/GlcNAc | 0.3 | 0.47 | 0.3–0.5 |
| | | | |
| **Distinguishing features** | | | |
| Size of polymer ($M_r \times 10^{-6}$) | 2–44 | 2 | 1–15 |
| Size of subunit ($M_r \times 10^{-6}$) | 2–44 | 2 | 1–15 |
| Protein (% by weight) | 13–20 | 13–16 | 13–48 |
| Sulfomucin staining | + | ++ | ++++ |
| Sulfate (% by weight) | 1–3 | 2–3 | 3–5 |
| Sialic acid (mol %) | <2 | 8–16 | 10–20 |
| Mucus gel thickness (m) | 50–500 | ND | 16–150 |
| Oligosaccharide chain length | 6–15 | ND | 2–12 |
| Neutral/acidic chains (%) | 70/30 | ND | 13/18 |

*Note:*  The symbol $\beta$ denotes residues that form disulfide bridges that link one or more glycoproteins in the formation of high molecular weight mucin molecules. The remaining amino acid residues are present in higher amounts in the "naked" or nonglycosylated regions of mucus. ND, not determined; GalNAc, *N*-acetyl galactosamine; GlcNAc, *N*-acetyl glucosamine.

[a]  Denotes the residues that link the protein core to the carbohydrate side-chains.

[b]  High in the glycosylated regions.

From Neutra, R. N. and Forstner, J. F., *Physiology of the Gastrointestinal Tract*, 2nd ed., Johnson, L. R., Ed., Raven Press, New York, 1987, 1975.

affect the migration of epithelial cells. Maturation of precursor cells is affected by colcemid,[67] prostaglandins,[68] dietary factors,[69] and the aging process.[70] A number of antiinflammatory drugs such as salicylates and corticosteroids have been shown to decrease mucus production,[71-76] whereas the antiulcer drug carbenoxolone,[77,78] had a stimulatory effect. Certain bacterial toxins, such as both crude and pure cholera toxin and *Escherichia coli* enterotoxin result in increased cyclic AMP-independent mucus secretion.[79-83] Other factors affecting mucin biosynthesis include chemical irritants such as ethanol,[84] prostaglandins,[85] EGF,[86] and nonsteroidal antiinflammatory drugs such as indomethacin.[87]

## 2. Regulation of Secretion

It has been shown that the application of acetyl choline or its analog carbachol resulted in increased secretion. Many GI peptides and hormones such as CCK, gastrin, secretin, VIP, cerulein, substance P, neurotensin, and somatostatin fail to induce detectable loss of granules from intestinal goblet cells and have no effect on the crypt goblet cell response.[88-90] Serotonin, however, has been implicated in the induction of mucus secretion and may be responsible for the increased secretion observed in mucus secreting carcinoid tumors.[91] In addition, prostaglandins E and F have been shown to increase mucus production.[92]

## 3. Erosion

The naked nonglycosylated regions from gastric mucins in humans and pigs (MW $2 \times 10^6$) are susceptible to proteolysis by trypsin, pepsin, and pronase,[93-95] resulting in glycoprotein subunits (MW $5 \times 10^5$). Depolymerization is thought to occur quickly, since there is a rapid fall in viscosity within 30 min. After 2 h of exposure to pepsin, the gastric mucus gel virtually disappears. The result of depolymerization is that a range of soluble glycosylated glycoproteins arrive in the colon and are therefore available for fermentation by enteric bacteria. Estimates using intubation studies have indicated that 3 to 5 g/d of mucus glycoproteins arrive in the colon and are theoretically able to sustain a viable population of bacteria in the absence of any added dietary substrates.[96] Third world countries still experience seasonal variations in food availability and in intakes of specific nutrients, as did nearly all of our immediate ancestors.[97] During these periods, the only carbohydrate and protein available to bacteria are from endogenous secretions. So-called "yo-yo" diets would, in theory, be particularly harmful since during periods of starvation, bacterial erosion of the mucus gel layer could lead to increased permeability to inflammatory and carcinogenic agents that may be present in subsequent diets. Prolonged starvation is associated with an overgrowth of intestinal organisms[98] resulting in decreased mucins.[99] Malnutrition also results in enteric infection[100] that leads to the production of extracellular bacterial glycosidases and proteases which may deplete the mucin coat.[101] The link with bacterial fermentation is an important consideration in understanding the etiology of immunological bowel diseases, (e.g., due to malnutrition), where therapy using dietary changes relies, in part, on the establishment of a bacterial population beneficial to health.[102] Changing from a fiber-free diet to one high in guar gum, pectin, and carrageenan[103] results in decreased mucin degradation. Diets high in fruit, vegetables, and cereal products are recommended dietary guidelines.[104,105] These contain dietary sources of carbohydrate such as free sugars, resistant starch, fructooligosaccharides, and plant cell-wall material (pectins, cellulose, and other nonstarch polysaccharides) that collectively will be present in much higher quantities than endogenous glycoproteins. Those carbohydrates not metabolized by human intestinal enzyme secretions will reach the colon where they are then available for fermentation.[106-110] Food preparation methods can significantly increase the amount of starch that is resistant to human endogenous enzymes thus increasing the amount of carbohydrate that arrive in the colon and is therefore available for bacterial fermentation.[111] *In vitro* studies suggest that as a result of choosing a diet high in certain types of carbohydrates such as the fructose-containing inulins particular species of bacteria (e.g., bifidobacteria) may grow in preference to others.[112]

## III. ANALYTICAL TECHNIQUES TO DETERMINE BACTERIAL METABOLISM OF MUCIN

Various fermentation systems exist which attempt to study mucin utilization *in vitro*.[113,114] These systems include batch and continuous cultures with either mixed bacteria or monoculture inocula.

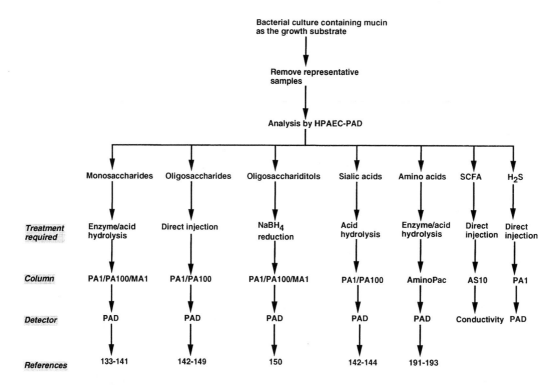

**Figure 1** Schematic diagram showing how high performance anion-exchange chromatography with pulsed amperometric detection (HPAEC-PAD) can be used to follow the disappearance of carbohydrate and protein, or the appearance of products of fermentation after bacterial metabolism of mucin in *in vitro* bacterial culture experiments. SCFA, short chain fatty acids.

## A. CONSTITUENT SUGARS

Monosaccharides, oligosaccharides, and oligosacchariditols may be quantified using a number of techniques including colorimetry,[115-120] gas-liquid chromatography,[121-129] or high-pressure liquid chromatography.[130-162] Colorimetric techniques can be used to measure individual sugars or groups of sugars. In bacterial fermentation studies, however, the presence of a number of interfering substances such as mineral ions and phenolic compounds may compromise the accuracy of values obtained. Thus, despite the simplicity of such techniques the more accurate chromatographic techniques will be expected to give more reliable information. A range of techniques are available for measuring individual mucin sugars by GLC. These techniques often allow simultaneous analysis of neutral sugars and hexosamines or neutral sugars and uronic acids. Accurate quantification is achieved, although the derivatizitation procedures are generally time consuming. The use of HPLC techniques has been extensively reviewed by Honda.[130] Until recently, HPLC techniques were not attractive alternatives to GLC. The use of reverse phase HPLC gives poor separation between individual monosaccharides and refractive index detectors result in considerable intereference in complex biological matrices. An important recent development in carbohydrate analysis has been the use of high-pressure anion-exchange chromatography in conjunction with pulsed amperometric detection (HPAEC-PAD).

The use of HPAEC-PAD is shown schematically in Figure 1 in which most of the applications require a single column. The technique is suitable for neutral sugars only or neutral sugars and hexosamines,[133-139] uronic acids,[140,141] sialic acids and sialylated oligosaccharides,[142-144] other oligosaccharides,[145-149] and oligosacchariditols.[150] The technique has been applied in the analysis of glycoprotein determinations,[145,147,149-157] in measurements of bacterial cell-wall polysaccharides,[136,158,159] and in the measurement of products of fermentation (see discussion later). In addition, a recently developed column (Carbopac MA-1) has been developed to analyze sugar alcohols. Preliminary studies have shown that this gives separation of hexosaminitols that are at least comparable to those achieved using cation exchange HPLC.[160] The separate measurement of GalNAc and *N*-acetyl galactosaminitol is important in quantifying how many carbohydrate side-chains are present on the glycoprotein molecule and therefore may be important in bacterial fermentation studies.

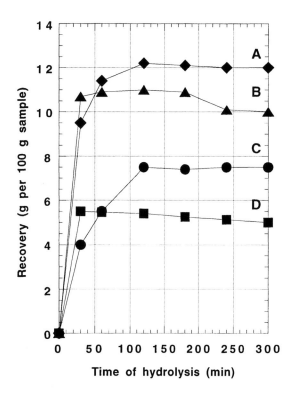

**Figure 2** Time course showing acid (2 mmol/l $H_2SO_4$) release of monosaccharides from pig gastric mucin. The sugars released are: A, N-acetylglucosamine; B, galactose; C, N-acetylgalactosamine; D, fucose.

The separation of neutral sugars and hexosamines for the specific measurement of mucus glycoproteins in the presence of exogenous carbohydrate has been previously described.[135] Figure 2 shows how constituent sugars may be released using $H_2SO_4$. The use of sulfuric acid as a hydrolyzing agent has the disadvantage that sulfate ions are strongly bound to the anion-exchange resin present in the separator column and may therefore selectively displace carbohydrates causing irreproducible chromatography. Sulfate ions were therefore removed during the chromatography using an anion guard column placed before the separator column as a bypass method around the analytical column. This technique obviates the need for the time consuming removal of sulfate ions before chromatography and is simpler to perform than GLC techniques since no derivatization procedures are required. The separation and quantification of chemically unmodified neutral sugars, hexosamines, and uronic acids in a single chromatographic run offers a simple and rapid technique for measurement of mucin sugars in the presence of exogenous sugars that is important in ileostomy and bacterial fermentation studies. This can be achieved using three columns placed in series as shown schematically in Figure 3. The three columns are required in order to (1) bypass sulfate ions around the analytical column (Dionex AG5 column), (2) separate uronic acids (Dionex AG10 column), and (3) separate neutral sugars and hexosamines (Dionex CarboPac PA-100 column). When a sample containing sulfate ions, uronic acids, and neutral and amino sugars is injected they follow the flow of the eluents until they reach column 1 (Figure 3a). Once here, the sulfate ions are selectively retained, whereas the carbohydrates pass through. On reaching the second column (AG10), uronic acids are retained whereas the nonacidic monosaccharides pass through only to be retained on the PA-100 column. This process occurs within 90 sec of injection of the sample. After 90 sec the flow of eluents through the columns is redirected by a valve activated under high pressure as shown in Figure 3b. When the sulfate ions are purged from the AG5 column, after 90 sec they pass straight to the detector and therefore never pass through the separator columns (AG10 and PA-100) where they would otherwise adversely affect the chromatography. During the next 30 min the neutral and amino sugars are separated on the PA-100 column using 2 mmol/l NaOH as eluent. Under these conditions, the uronic acids remain on the AG10 column. On increasing the sodium hydroxide concentration of eluent to 15 mmol/l the uronic acids are then eluted from the AG10 column. Reequilibration with the starting eluent (2 mmol/l NaOH) requires only 6 min. The use of three columns using only NaOH as eluent offers an attractive alternative to using NaOH/sodium acetate gradients[136,161] because the run times are kept to a minimum as no prolonged cleaning and reequilibration with the starting elution conditions is required. Also, no sample

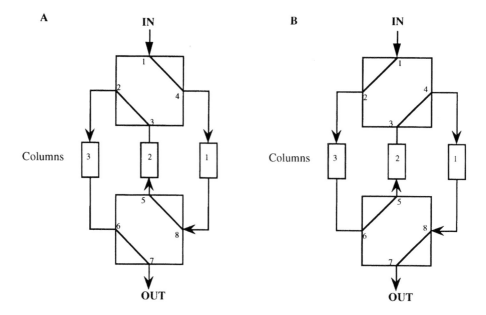

A IN    B IN

Columns    Columns

OUT    OUT

**Figure 3** Schematic diagram showing how three columns placed in series may be used in combination with a valve activated under high pressure to remove interfering sulfate ions and to separate neutral sugars, hexosamines, and uronic acids from mucin during a single chromatographic run. (A) The direction of flow of eluent (2 mmol/l NaOH) through three columns placed in series is in the "load" or inject position. Within 90 sec of injection of a sample, sulfate ions, present in the hydrolysate, are retained on column 1 (an AG5 column), the uronic acids are then retained on column 2 (an AG10 column) whereas the neutral monosaccharides and hexosamines reach column 3 (a CarboPac-PA100 column). (B) After 90 sec, the direction of flow of eluent through the columns is altered via a valve that is activated under high pressure. The sulfate ions are then purged from the AG5 column over the next 19 min and do not pass through columns 2 and 3 where they would otherwise adversely affect the chromatography. Within 30 min, neutral sugars and hexosamines are separated on column 3 whereas the uronic acids are retained indefinitely on column 2. By increasing the concentration of eluent to 15 mmol/l NaOH, the uronic acids are then released from column 2.

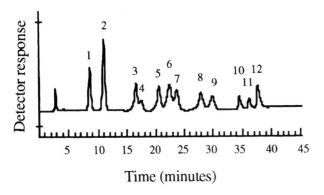

**Figure 4** High-pressure liquid chromatogram for a sugar standard mixture: (1) fucose; (2) deoxygalactose (internal standard 1); (3) arabinose; (4) galactosamine; (5) galactose; (6) glucosamine; (7) glucose; (8) xylose; (9) mannose; (10) galacturonic acid; (11) glucuronic acid; (12) mannuronic acid (internal standard 2).

preparation other than dilution of the sample is required. A typical chromatogram of a standard sugar mixture is shown in Figure 4.

The classical way to establish the chemical structure of polysaccharides consists of the following steps: (1) establish its sugar composition; (2) establish its glycosidic linkages by methylation analysis; (3) recognition of repeating units by specific chemical degradation procedures like partial hydrolysis and acetylation; and (4) NMR-spectrometric methods.[21-25] The isolation and characterization of different glycoproteins and oligosaccharides is important in some disease states and in following mucus oligosaccharide degradation. Simple and reproducible techniques to show individual variation and follow changes that occur during fermentation are desirable.

Two new techniques have been developed to isolate glycoproteins and their oligosaccharides. In the first, monosaccharides and oligosaccharides of glycoproteins are electrotransferred onto polyvinylidene fluoride (PVDF) membranes.[163] This technique involves (1) separation and purification of the glycoprotein by SDS-PAGE, (2) electrotransfer from the gel to the PVDF membrane,[164] (3) acid hydrolysis and monosaccharide analysis of the purified glycoprotein by HPAEC-PAD[135,137-139] or GLC and/or (4) glycosidase digestion and oligosaccharide mapping of the glycoprotein. In the second technique, the oligosaccharide side-chains are immobilized onto polystyrene microtiter wells.[165] The oligosaccharides are then deglycosylated or glycosylated with exoglycosidases or endoglycosidases or with glycosyltransferases and the modified glycan structures identified using lectin probes.[165] Complementary information may be obtained using HPAEC-PAD. For example, Hermentin et al.[155] have set up a database for mapping mono-, di-, tri-, and tetrasialylated and asialylated N-glycans by HPAEC to show differences in separation that are affected by adding or removing specific monosaccharides. Thus, changes in antigenicity of mucus by addition or removal of monosaccharides could readily be followed. Furthermore, Reddy and Bush[150] have separated a number of mucus oligosaccharides and oligosacchariditols which may be used as a simple technique to measure many oligosaccharides and oligosacchariditols in a single chromatographic run.

## B. SIALIC ACIDS

More than 25 different sialic acids have been reported in which N-acetyl neuraminic acid (Neu5Ac) is thought to be the precursor.[166] For example, the substitution of a H on the N-acetyl group with a hydroxyl group gives rise to N-glycolyl neuraminic acid (Neu5Gc), which is commonly found in mucus glycoproteins. Replacement of the amino group at the 5 position with a hydroxyl group (OH) gives rise to 2-keto-3-deoxynonulosonic acid (KDN), whereas most of the other sialic acids result from O-substitution of one or more of the OH groups of Neu5Ac, Neu5Gc, or KDN with acetyl, methyl, lactyl, or sulfate groups.

Sialic acids are released from glycoproteins by heating at 80°C for 1 h in 0.1 N $H_2SO_4$. The free sialic acids can be measured colorimetrically using orcinol,[167] diphenylamine,[168,169] direct Ehrlich,[169] tryptophan-perchloric acid,[170] hydrochloric acid,[171] and sulfuric-acetic acid procedures.[172] These techniques are both insensitive and nonspecific but the use of periodate in strong acid solution followed by extraction into cyclohexanone was found to increase the sensitivity with resorcinol by 12-fold and reduce the effects of interferences.[173] The released sialic acids may be derivatized and measured using GLC[174] or GLC-MS,[175] the latter requiring highly purified samples. In general, the use of HPLC has the advantage compared to GLC techniques of allowing rapid and direct quantification of underivatized samples, although the use of spectrophotometric detection renders some of the techniques nonspecific.[143,176-179] Wang et al.[142] separated five acidic monosaccharides including Neu5Ac and Neu5Gc using HPAEC-PAD, while Manzi et al.[143] used an identical system to separate 18 different sialic acids with varying degrees and positions of acetylation. The hydrolysis of sialic acids and measurement using the alkaline eluent as described by Wang et al. allowed measurement of total sialic acids since both the acid hydrolysis and alkaline eluent result in de-O-acetylation. For the measurement of individual sialic acids including different degrees and positions of O-acetylation then minimal migration or de-O-acetylation is essential during the depolymerization step. The use of 2 mol/l $CH_3COOH$ at 80°C for 3 to 5 h[180] followed by separation at neutral pH is preferential. The ability to measure individual sialic acids is important since the positions and degree of acetylation affects the ability of bacteria to degrade the mucin molecules.

## C. PROTEINS AND AMINO ACIDS

Total protein may be measured colorimetrically[181] or by the Kjeldahl technique after digestion at high temperature in the presence of catalysts. However, colorimetric techniques are subject to interferences and the accuracy of the Kjeldahl analysis relies on a factor (6.25) to calculate the amount of protein from the measured $N_2$. In fermentation studies, the measurement of total protein does not give detailed information on the preference, if any, of individual amino acids being metabolized. The analysis of amino acid residues requires optimized hydrolysis of proteins. Typically, 6N HCl has been used for this purpose but it can be problematical because variable recoveries are obtained for the individual amino acids, with tryptophan being particularly labile.[182] Changes of amino acids are reported to be catalyzed by the presence of carbohydrates, porphyrins, and lipids. Thus, removal of these compounds using enzymic hydrolysis and ethanolic extraction techniques before chemical hydrolysis could possibly lessen these problems. The side reactions are lessened by using sulfonic acid containing 0.2% tryptamine as a catalyst which is a strong acid but unlike HCl is a nonoxidizing acid.[183] D'Aniello et al.[184] have recently described a technique for hydrolyzing proteins using a combination of enzymic and acid hydrolyis where the duration of the acid hydrolysis is minimized.

Separation of the released amino acids using reversed phase HPLC has been described[185] but this technique requires extensive sample pretreatment, although excellent separation of 25 amino acids in less than 9 min has been reported.[186] As the amino acids are not naturally fluorescent and only phenylalanine, tyrosine, and tryptophan respond to UV detection, then various procedures for pre-column derivatization techniques have been developed.[187-189] The measurement of amino acids using electrochemical detection (HPAEC-PAD)[190-193] has also been employed because it offers the advantage that problems inherent in the use of precolumn labeling techniques are obviated. The use of anion-exchange chromatography results in good separation of more than 20 different amino acids in 40 min[193] and the technique is not subject to interferences from components which may be present in complex biological matrices.

## D. PRODUCTS OF FERMENTATION

The analysis of fermentation products is important since some short chain fatty acids (SCFA) such as butyrate can be beneficial to the host, whereas others such as phenols, $NH_3$, and amines are potentially harmful. GLC has been widely used in the analysis of fermentation products such as gases,[194] SCFA,[195] ethanol,[196] phenols,[197] amines,[198] and sulfide.[199] The main disadvantages of the above techniques are often in the need to extract the chemical of interest, (e.g., into chloroform), or the need for lengthy derivatization procedures. The direct measurement of these samples in feces after a single dilution step only minimizes the possibility of variable recoveries that occur during extraction procedures and significantly reduces the time taken to prepare the sample prior to analysis. Due to the minimal effect of interferences and picomolar sensitivity of electrochemical detection, the use of HPAEC-PAD offers a considerable advantage over the above techniques and may be used to give more reliable results. The technique has been applied to the measurement of sulfate in plant foods,[200] and aliphatic alcohols.[201] These techniques require developmental work for application to feces.

## IV. DEGRADATION OF MUCUS GLYCOPROTEINS AND PATHOLOGICAL IMPLICATIONS

### A. METABOLISM OF MUCIN

#### 1. Penetration of the Mucus Gel Layer

Bacteria may reside under or within the mucus gel layer or in the lumin of the gut. In order to penetrate mucus, various mechanisms have evolved in different bacteria. For example, *Helicobacter pylori* is a highly motile bacterium which is able to use an array of powerful flagellae to physically disrupt the mucus gel layer. This process may be aided by breakdown of the polymeric structure of mucus which has been attributed to the production of proteases[56] and later ammonia.[202,203] *H. pylori* has also been shown to be strongly bound to the sialylated ganglioside GM3, and so the removal or age-dependent decrease in the sialic acid content of mucins would hinder the binding ability of the bacteria. Other bacteria such as *E. coli* can also be found completely separated from luminal contents by a layer of mucin.[204] Experiments on rat bladders showed that these bacteria could penetrate the mucin layer within 2 h of infection, although the mode of entry was not described. A freeze-fixation technique was used to examine bacteria, mucin, and epithelial cells in the GI tract and bladders of adult germ-free rats. Scanning electron microscopy of the rat ileum demonstrated the presence of small cavities in the mucin layer in which bacteria could clearly be seen, suggesting that the barrier functions of the mucus gel may be compromised during bacterial invasion.[204] One strain of *E. coli* (R-DEC 1) that lacks flagellae was shown to possess pili that mediate adherence of the bacterium to partially purified brush borders of rabbit ileal mucosal epithelial cells.[205]

*Yersinia enterocolitis* is an enteroinvasive bacterium that causes gastroenteritis in humans. In areas of the intestine where the morphological impact of gastritis is at its most severe, there is an increased mucin content, synthesis, and secretion.[206] *Yersinia enterocolitis* can readily bind to rabbit small intestinal and colonic mucins at 37°C but not at 25°C.[207] This difference has been attributed to the production (at 37°C but not at 25°C) of plasmid-encoded proteins that could bind to the mucins. The expression of these proteins was also regulated by calcium ions. Upon secretion, these proteins bind to the outer membrane

**Table 2** Steps Involved in the Hydrolysis of Mucus by Normal Physical, Chemical, and Enzymatic Mechanisms and Examples of How the Mucosal Barrier may be Rendered More Permeable by Pathogenic Mechanisms

| Steps involved in mucin degradation | Normal mechanisms | Examples of pathogenicity |
| --- | --- | --- |
| Physical/chemical | Mechanical, diffusion, HCl | Ammonium carbonate (*H. pylori*) |
| Proteolytic cleavage of "naked" link peptides | Trypsin, pronase | Pepsin 1 linked to ulceration |
| Removal of terminal/antigenic glycosides, e.g., sialic acids and sulfate groups | Mucin oligosaccharide degrading (MOD) bacteria | Increased turnover by bacterial sialidases and sulfatases |
| Removal of carbohydrate side-chains | Enzymatic hydrolysis by enteric bacteria | Increased turnover in C-limited growth (e.g., malnutrition/starvation) |

of the organism and are referred to as Yersinia outer membrane proteins (Yops). Yops are thought not only to be involved in bacterial attachment to the intestinal wall but also to help penetration of the mucosa by altering the surface charge and hydrophobicity of the bacterium and thus promoting agglutination (see also Chapter 9).

## 2. Breakdown of the Protective Polymeric Structure of Mucus

Mucin degradation is a multi-step process, shown schematically in Table 2, which is thought to begin with proteolytic cleavage of the nonglycosylated "naked" regions of glycoprotein. Proteolysis occurs in all regions of the intestine due to the action of pepsin, pancreatin,[95] and fecal proteases.[208] This process results in a marked reduction in the gelation and viscosity of the mucin. Diffusion into the lumin of mucus that has been degraded by endogenous secretions results in an accumulation of subunits of 500 kDa or more that are highly glycosylated and resistant to further proteolytic attack. Ofusu et al.[209] infused radiolabeled mucin into rat intestines and found that the total precipitable mucin decreased from 80 to 5% in 3 h. It is thought that mucins undergo physical and/or chemical changes before bacterial glycosidic action. This hypothesis is supported by the observed degradative *in vitro* activity of trypsin and pronase on porcine gastric mucin which resulted in fragments that were one-quarter of the molecular weight of the original molecule.[210] The effect of degradation of the protein component of the naked region of mucin may have significant effects on the permeability of the mucus barrier. *Vibrio cholerae* produces a metalloproteinase that is secreted and digests nonglycosylated peptide regions of small intestinal mucin with a concomitant 57% drop in viscosity. This drop may compromise the mucus barrier and allow the organism and its products to gain access to the underlying mucosa.

## 3. Mucin Utilizing Bacteria

The next stage of mucin degradation is carried out by the action of glycosidases produced by enteric bacteria.[211] A range of different enzymes are required for complete degradation, however. This is due to the complexity of the oligosaccharide chains which consist of a variety of different carbohydrate linkages that vary in the size of chain, degree of branching, type of linkage, antigenicity, presence of sialic acids or sulfate groups, and varying degrees and position of substitution of *O*-acetyl groups. Thus, many bacteria cannot utilize mucin oligosaccharides, despite producing constitutive glycosidases, until the oligosaccharide has been at least partially degraded by other bacteria. A group of bacteria that are thought to facilitate the accessibility of other bacteria have been termed mucin oligosaccharide degrading bacteria (MOD).[212,213] Strains of *Ruminococcus torques*, *R. gnavus*, and certain *Bifidobacterium* species have all been included in the class of MOD. Of particular interest is that intergenus variability occurs with respect to enzyme production. For example, *R. torques* was found to contain a number of enzymes including α-1-3-galactosaminidase, α-1,2,3,4-fucosidase, sialidase, β-galactosidase, and β-galactosaminidase, but not α-1-3-galactosidase. Another strain, *R. gnavus*, contained all of these enzymes except α-1-3-galactosaminidase and β-galactosaminidase. This work highlighted the increased survival potential of different bacterial species to live in humans producing mucins of different antigenicity. In human A secretors, the terminal sugar is GalNAc which is α-1-3 linked to galactose so that the ability of *R. torques* to produce an enzyme that can hydrolyze this linkage gives this strain a greater survival potential. *Ruminococcus gnavus* has an important advantage in human B secretors because it is able to hydrolyze the terminal galactose residues. In contrast, bifidobacteria are able to hydrolyze terminal fucose residues, giving them a greater survival potential in human blood group H and Lewis secretors.[211] Although the above studies are of interest it is unclear yet how they relate to the *in vivo* scenario.

Although their impact on the colonic environment is likely to be relatively insignificant (due to their relatively low numbers), the ability of oral *Streptococcus* strains to utilize mucin as a source of carbohydrate has been studied in batch culture. Pure cultures of streptococci using saliva as a growth medium have indicated that salivary glycoproteins were degraded. While complex oral microflora consumed all available sugars, glycoproteins were only partially degraded in pure culture.[214] This situation suggests that, as in the colon, combinations of glycosidases from several sources are required for complete breakdown of mucins. It has been shown that some streptococci will grow on human saliva without added substrate,[215,216] although oral species could not grow on bovine submaxillary mucin where there is a high content of sialic acids. In the early stages of mucin degradation, terminal sugars must be initially hydrolyzed. This confers an advantage toward *Streptococcus mitis* which is able to hydrolyze terminal fucose sugars. This strain has been shown to be dominant in the early stages of plaque formation.[217]

Strains of *Bacteroides* may break down the mucin oligosaccharide backbone but not when terminal galactose (B secretors) or GalNAc (A secretors) residues are present. However, the oligosaccharide content of mucins are known to be heterogeneous so it is probable that *Bacteroides* spp. may metabolize part of the mucin oligosaccharides in the absence of MODs simply because terminal galactose or GalNAc are not always present. Thus, an advantage to bacteria living under the mucus gel layer could be to stimulate the production of high levels of immature mucins that are readily degradable without the need for a whole battery of enzymes. The secretion of toxins by some bacteria that stimulate the secretion of mucins may well serve this purpose.

A number of studies have been used to look at the effect on substrate utilization and fermentation products of mucin added to mixed cultures of bacteria.[218-222] In a three-stage continuous culture model of the colon it was shown that *Bacteroides fragilis* were the numerically predominant species in mucin and nonmucin containing cultures.[218] However, the addition of mucin to the system strongly stimulated bacteroides, bifidobacterial, and coliform populations. The effect of feeding mucin on glycosidase enzyme activity was generally to stimulate cell-associated production, whereas extracellular protease activity markedly increased. In these experiments, the effect of the continuous culture system gives a model of events in the proximal, medial, and distal regions of the colon. The system is shown in Figure 5. There was an increase in growth of coliforms and clostridia in vessel 3 (which simulates the relative effect of carbon-limited growth in the distal colon). Macfarlane and Gibson[219] showed that glycosidase formation by a strain of *B. fragilis* was inducible and strongly influenced by growth substrate. The formation of α-fucosidase and α-*N*-acetylgalactosaminidase was stimulated by fucose and repressed by galactose. With the exception of α-galactosidase, all other mucin degrading enzymes showed an inverse relationship to growth rate, with neuraminidase and fucosidase showing the largest changes. The addition of mucin to a mixed culture system containing fecal bacteria resulted in increased glycosidase and protease activity and an increase in SCFA secretion, which indicated that the glycoprotein was being actively broken down and fermented. The increase in protease activity was mainly by extracellular enzymes and, in particular, cysteine, serine, and metalloproteases.

The production of $H_2$ by many gut bacteria has led to different mechanisms of disposal, such as by excretion by breath or flatus, or by the production of $CH_4$ (methanogenic bacteria) or $H_2S$ (sulfate-reducing bacteria) (see Chapter 6). Utilization of chondroitin sulfate and mucin (both polymers are rich in sulfate) leads to much higher amounts of hydrogen sulfide production as compared to when starch is used as substrate in a mixed culture system.[222]

## B. MUCUS, PEPTIC ULCERATION, AND *HELICOBACTER PYLORI*

Although *H. pylori* is commonly recognized as inhabiting the stomach, considerations of its physiology and ecology may give an interesting insight into mucin pathology. The gastric mucosa is constantly

**Figure 5** A multiple-stage continuous culture model of the colon. The system consists of three vessels arranged in series with increasing operating volume and pH. A complex mixture of carbohydrates and proteins is fed to the system, with mucin also added as necessary. Vessel 1 (top) receives growth medium directly from the feed reservoir and sequentially feeds vessels 2 and 3 in turn and then out into a waste reservoir. Thus, vessel 1 approximates to the low pH, nutrient rich, fast growth rates found in the proximal colon. The other vessels are characterized by a progressively higher pH, slow growth rate, and nutrient depletion found in the transverse (vessel 2) and distal (vessel 3) colons. The system was operated at 37°C and under an atmosphere of oxygen-free nitrogen.

exposed to noxious agents such as acid and pepsin. Acid hypersecretors are recognized as a cause of duodenal ulceration.[223] However, there is considerable overlap between their acid output and that of normal subjects, indicating that acid quantity alone cannot explain ulceration. The amount of pepsin produced is not markedly different in ulcerated patients, but alterations in pepsin subgroups have been identified, namely increased secretion of pepsin I.[224] Pepsin I digests mucins more readily than other pepsin subgroups and therefore erodes the mucus barrier at a faster rate, which compromises the mucosal defense.

Defects in mucus production and structure have been demonstrated in ulcerated patients. Decreased production of the gel-forming polymeric glycoprotein component and a thinner layer of adherent mucus have both been described.[225,226] Indeed, *in vitro* studies have demonstrated impairment of the mucus pH gradient. The gradient across the mucus gel layer is not maintained in the face of an acid challenge in ulcer patients.[227]

An important new factor in the pathogenesis of peptic ulceration is *H. pylori*. There is now abundant evidence to link this bacterium to gastritis and peptic ulceration in humans.[228-230]

In the stomach, the organism resides close to the epithelial surface in the mucus layer. The bacterium is able to secrete ammonia as part of its normal metabolic function, and it has been postulated that this may assist in acid neutralization in a microniche. The exact mechanism by which *H. pylori* induces gastroduodenal ulceration is unknown, but it is likely that it exerts effects on both aggressive and defensive factors. Recent studies have demonstrated that hypergastrinemia observed in patients with duodenal ulceration may be due to *H. pylori* and so help explain the increased acid secretion demonstrated in these patients.[231] *H. pylori* also compromises the mucosal barrier allowing luminal factors access to the epithelial surface. Synthesis of enzymes involved in mucin degradation by *H. pylori* have been carried out. The organism seemingly carries out protease activity behavior typical of a metalloproteinase.[232,233] This protease may cause disintegration of the polymeric structure of gastric mucus and therefore loss of its gel forming properties, which are essential for the barrier forming function of mucus.

However, other groups have been unable to demonstrate any extracellular *H. pylori* protease activity and it has been suggested that the mucolytic activity of the organism may in fact be attributable to its high urease activity. The destructive effect of urea on the gastric mucosal barrier in the presence of a urease enzyme has been previously demonstrated. Sidebotham et al.[203] demonstrated that the break-up of the polymeric structure of mucus glycoproteins may result from the destabilizing effects of ammonium carbonate generated by the urease activity of *H. pylori*, and that this activity may therefore be responsible for its mucolytic activity. More recent work by Murakami et al.[202] has shown, using urea infusion studies in rats, that a quantitative relationship between ammonia production and the formation of histological lesions occurs.

Recently there is evidence that, in addition to its mucolytic activity, *H. pylori* may actually decrease mucus secretion by the mucosa.[234] Using a gastric cell line, *H. pylori* was found to compromise actual mucus production. Glycoprotein synthesis was not inhibited but secretion was which suggests a defect in exocytosis of glycoprotein by *H. pylori*. It is apparent therefore that *H. pylori* has a deleterious effect on the mucus component of mucosal defence and this may help explain, at least in part, the pathogenic mechanisms of the organism.

With regards to the large intestine, the biological role of *H. pylori* is unclear. The successful isolation of this organism from feces of Gambian and U.K. adults is a recent observation.[235,236] Thus, studies on the pathophysiology of *H. pylori* in the colon are in their infancy.

## C. MUCUS AND INFLAMMATORY BOWEL DISEASE (IBD)

Ulcerative colitis (UC) and Crohn's disease (CD) occur in regions of the intestine colonized by large numbers of bacteria and it has long been suspected that they have infectious etiologies. A possible bacterial link in these diseases is supported in CD which improves with bowel rest and alterations in the luminal environment using elemental diets and antibiotics.[237] Furthermore, UC can show similarities with infectious colitides such as cytomegalovirus colitis. The epithelium of the gut is continuously bathed by luminal contents containing large concentrations of bacteria. These bacteria contain several potent inflammatory products such as lipopolysaccharides, peptidoglycan polysaccharide complexes, and *N*-formyl methionyl oligopeptides such as FMLP.[238,239] However, conventional microbiological methods have thus far failed to identify a pathogenic agent.

In normal circumstances, the mucosal barrier protects the epithelium by excluding these products. However, in IBD the barrier may be defective which allows an inflammatory response in the mucosa. The intestine of patients with CD has been demonstrated to be more permeable to inert markers and have increased serum antibodies to commensal and pathogenic organisms. Hollander[240] recently demonstrated a permeability defect in first degree relatives of patients with CD. It has been suggested that this abnormal permeability may be the primary defect allowing luminal agents to excite an inflammatory response. Mucin depletion occurs more frequently in UC than in CD and has therefore been suggested as a diagnostic criterion by which to distinguish the diseases.[241,242]

In UC patients there is goblet cell depletion and alterations in the physicochemical properties of mucus, leading to defects in the mucosal barrier and therefore increased permeability. However, this defect is present only in inflamed mucosa and is not thought to be a primary effect.[243] Alterations in colonic glycoproteins in mucosa from patients with UC have been observed, in comparison to controls and patients with CD, although a more recent study using continuous gradient ion-exchange chromatography found no such alteration in UC mucus.[244]

To a considerable extent in the proximal colon, and universally in the distal colon, the terminal sugar in the mucin glycoprotein is replaced with a sialic acid (Neu5Ac) which prevents expression of the ABO antigens. Other blood group antigens such as Lewis, I, i, T, and Tn can be revealed by removing distal carbohydrate and sialic acid. Colonic mucins are present in higher concentrations and contain a greater sialylation and sulfation than in the proximal intestine which allows enhanced resistance to bacterial enzymes.[245] Many sialic acids have *O*-acetyl groups substituted at C4, 7, 8, or 9 positions which helps counteract bacterial attack especially at the C4 position. However, there is an age-related decrease in sialic acid content of mucus from >100 g/ml at age 25 to <20 g/ml at age 70 or older[246] which would help to explain an age-dependent increase of IBD.

It can be hypothesized that the presence of sialic acids may increase the rheological properties of mucus so that their removal would adversely affect the viscosity and hence protective properties of the mucus gel layer. However, changes in sialic acid content have been shown by several workers not to affect the rheological properties of gastric[247] or cervical mucus.[248] Increasing sialic acid content of mucins, therefore, probably arises to confer greater resistance to bacterial enzymic attack especially due to the presence of *O*-acetyl groups. Vatier et al.[249] suggested that sialic acids in gastric juice do not provide information on the protective properties of mucus but can be used to represent a reliable index of mucus glycoprotein erosion by pepsin. In order to distinguish between mucolytic and mucogenic effects of drugs, however, Guslandi[247] has suggested that the ratio of neutral to total mucoproteins, the so-called "mucoprotective index," is more useful. It is suggested that measurement of eroded glycoprotein fragments can be used to demonstrate disease.

Using combinations of stains, it has been shown that sulfomucins in the distal colon are found in deep crypts but that they mostly occur in upper crypts in the right colon.[245] In active UC a decrease in sulfomucins with a concomitant increase in sialomucins[241] has been observed, although Morita et al.[250] showed an increased sulfation of colonic mucin even in the uninflamed right colon. Alterations in the amounts of sulfomucins and sialomucins could have important effects on the susceptibility to bacterial attack. That bacterial erosion of intact mucus occurs is difficult to determine *in vivo*. However, a number of workers have found several sialidases and sulfatases[251-256] in feces and it is these enzymes that would need to be present before extensive degradation of mucin could occur in the heavily sialylated/sulfated mucins found in the colon. The production of large amounts of these enzymes that were able to degrade sufficient amounts of the mucus barrier could render the mucosal surface more permeable to bacteria and their products and thus be involved in IBD etiology. Recently, Corfield et al.[254] showed an increase in acylneuraminate lyase and protease activity in IBD but found that sialidase was not statistically different from normal subjects.

Roberton et al.[253] purified a glycosulfatase from a mucus glycopeptide-degrading *Prevotella* from the colon. The purified enzyme was highly active against GlcNAc, which predominates in the oligosaccharide chains of certain mucus glycoproteins such as rat gastric glycoprotein. This enzyme acted rapidly (19% desulfation in 10 min) but was incomplete (33% after 3 h). Crude periplasmic extracts removed 79% of the sulfate groups suggesting cooperation in mixed cultures. In the crude extract there are likely to be glycosidases which can remove sugars, thus rendering the sulfate sugars more accessible to desulfation. The removal of sulfate groups also facilitates hydrolysis of the remaining sugars. Thus, marked glycosidase activity was found after adding purified fecal sulfatase to an incubation mixture.[251] Tsai et al.[255] have isolated a sulfatase from feces and found that 2 bacteroides species (*B. fragilis* and *B. thetaiotaomicron*) were the organisms responsible. However, there are many sulfated substrates in the bowel including steroid sulfate, choline sulfate, glycosaminoglycans, sulfolipids, bilirubin sulfate, and sulfated glycoproteins so that a range of sulfatases will exist. It has been suggested[242] that the addition of synthetic sulfated or sialylated carbohydrates into the diet may be useful in therapeutic studies via competitive substrate inhibition. However, highly sulfated food additives such as native carrageenans and seaweeds both caused experimental UC in laboratory animals.[257,258] The causal factor is not certain but sulfate-reducing bacteria will convert sulfate to hydrogen sulfide[259] in the colon and hydrogen sulfide infused into laboratory animals has been shown to have a direct inflammatory response on colonocytes.[260]

## V. CONCLUSIONS

The action of glycosidases from particular species or strains of bacteria to remove sulfate and sialic acids may lead to glycoproteins that are easier to hydrolyze than intact mucins and result in a more permeable mucosa. Bacterial antigens such as FMLP and some bacterial products may then penetrate the mucosa, initiating an inflammatory response or causing further damage to the mucus barrier which may lead to ulceration. A postulated mechanism whereby disruption of the integrity of the mucosal layer may have pathological implications is shown in Figure 6.

Accurate analytical techniques are required in order to follow the mechanisms of these diseases. Colorimetric techniques are often used but nonspecific interferences that are present in fecal slurries may lead to unreliable data. New techniques have been discussed for the isolation and measurement of glycoproteins and oligosaccharides. A simple method for measuring mucin glycoproteins including neutral sugars, hexosamines, uronic acids, sialic acids (total and degree and position of acetylation), sulfate content of sugars (hexose and hexosamine sulfates), oligosaccharides (neutral and sialylated), and oligosacchariditols makes the use of HPAEC-PAD an exciting new technique for glycoprotein analysis. Optimization of this methodology is necessary to realistically determine the possible role of the colonic microflora in erosion of the mucin layer and potential pathological implications.

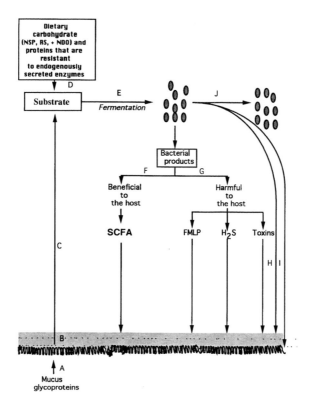

**Figure 6** Schematic diagram demonstrating postulated links between diet, bacteria, bacterial antigens and products of fermentation and their effect on the integrity of the mucosal layer. (A) The potentially sensitive epithelial cells are protected by a mucosal layer (B) consisting of glycoproteins, lipids, bile acids, electrolytes, proteins and nucleic acids. In normal individuals (C), the mucosal layer is shed into the lumin of the gut, where it may be metabolized by the colonic enteric bacteria, but will be replenished. (D) Dietary carbohydrates not hydrolyzed in the ileum arrive at the colon and will be available for fermentation (E). When dietary sources of carbohydrate such as nonstarch polysaccharides (NSP) and resistant starch are high, then products of fermentation (F) such as butyrate, that help maintain healthy colonocytes, will predominate. On diets that are low in these plant cell-wall materials, and chemically related compounds, as in malnutrition or diets that are high in cooked meat and fats, then slower intestinal transit times and more carbon limited growth of bacteria may occur. In this case, it is more likely that endogenous glycoproteins will be utilized; this may compromise mucosal defense. In addition, bacterial products and antigens that are harmful to the host are then more likely to cause an inflammatory response (G). Such a scenario may encourage the growth of pathogenic bacteria which may penetrate the mucosa thus perpetuating an inflammatory response (H, I), or the bacteria may move lower down the colon (J) where they may infect another portion of the mucosa.

## REFERENCES

1. **Kramer, M. F. and Geuze, J. J.,** Glycoprotein transport in the surface mucous cells of the rat stomach, *J. Cell Biol.,* 73, 533, 1977.
2. **Neutra, M. R. and Leblond, C. P.,** Radioautographic comparison of the uptake of galactose-H$^3$ and glucose-H$^3$ in the Golgi region of various cells secreting glycoproteins or mucopolysaccharides, *J. Cell Biol.,* 30, 137, 1966.
3. **Neutra, R. N. and Forstner, J. F.,** Gastrointestinal mucus: synthesis, secretion and function, in *Physiology of the Gastrointestinal Tract,* 2nd ed., Johnson, L. R., Ed., Raven Press, New York, 1987, 1975.
4. **Lev, R.,** The histochemistry of mucus producing cells in normal and diseased gastrointestinal mucosa, *Prog. Gastroenterol.,* 2, 13, 1970.
5. **Rambourg, A., Hernandez, W., and Leblond, C. P.,** Detection of complex carbohydrates in the Golgi apparatus of rat cells, *J. Cell Biol.,* 40, 395, 1969.
6. **Tartakoff, A. M.,** The combined function model of the Golgi complex: center for ordered processing of biosynthetic product of the rough endoplasmic reticulum, *Int. Rev. Cytol.,* 85, 221, 1983.

7. **Roth, J.,** Cytochemical localisation of terminal N-acetyl-galactosamine residues in cellular compartments of intestinal goblet cells: implications for the topology of O-localisation, *J. Cell Biol.*, 98, 399, 1984.

8. **Roth, J., Lucocq, J. M., Berger, E. G., Paulson, J. C., and Watkins, W. M.,** Terminal glycosylation is compartmentalized in the Golgi apparatus, *J. Cell Biol.*, 99, 229, 1984.

9. **Bennett, G., Leblond, C. P., and Haddad, A.,** Migration of glycoprotein from the Golgi apparatus to the surface of various cell types as shown by radioautography after labeled fucose injection into rats, *J. Cell Biol.*, 60, 258, 1974.

10. **Kramer, M. F., Geuze, J. J., and Strous, G. J. A. M.,** Site of synthesis, intracellular transport and secretion of glycoprotein in exocrine cells, *Ciba Found. Symp.*, 57, 25, 1978.

11. **Neutra, M. and Leblond, C. P.,** Synthesis of the carbohydrate of mucus in the Golgi complex as shown by electron radioautography of goblet cells from rats injected with glucose-H$^3$, *J. Cell Biol.*, 30, 119, 1966.

12. **Kornfeld, R. and Kornfeld, S.,** Structure of glycoproteins and their oligosaccharide units, in *The Biochemistry of Glycoproteins and Proteoglycans*, Lennarz, W. J., Ed., Plenum Press, New York, 1980, chap. 1.

13. **Specian, R. D. and Neutra, M. R.,** Mechanism of rapid mucus secretion in goblet cells stimulated by acetyl choline, *J. Cell Biol.*, 85, 626, 1980.

14. **Smith, P. L.,** Gastrointestinal physiology, in *Gastrointestinal Toxicology*, Rozman, K. and Hanninen, O., Eds., Elsevier, Amsterdam, 1986, 1.

15. **Van Hoogdalem, E. J., De Boer, A. G., and Breimer, D. D.,** Intestinal drug absorption enhancement: an overview, *Pharmac. Ther.*, 44, 407, 1989.

16. **Allen, A.,** Structure and function of gastrointestinal mucus, in *Physiology of the Gastrointestinal Tract*, Johnson, L. R., Ed., Raven Press, New York, 1981, chap. 21.

17. **Clamp, J. R., Allen, A., Gibbons, R. A., and Roberts, G. P.,** Chemical aspects of mucus, *Br. Med. Bull.*, 34, 25, 1978.

18. **Slomiany, B. L., Zolebska, E., and Slomiany, A.,** Structural characterization of neutral oligosaccharides of human H$^+$Le$^{b+}$ gastric mucin, *J. Biol. Chem.*, 14, 743, 1984.

19. **Slomiany, B. L., Zolebska, E., and Slomiany, A.,** Structures of the neutral oligosaccharides isolated from A-active human gastric mucin, *J. Biol. Chem.*, 259, 2863, 1984.

20. **Dekker, J., Aelmans, P. H., and Strous, G. J.,** The oligomeric structure of rat and human gastric mucins, *Biochem. J.*, 277, 423, 1991.

21. **Podolsky, D.,** Oligosaccharide structures of human colonic mucin, *J. Biol. Chem.*, 260, 8262, 1985.

22. **Podolsky, D. K.,** Oligosaccharide structures of isolated human colonic mucin species, *J. Biol. Chem.*, 15, 510, 1985.

23. **Wesley, A., Mantle, M., Man, D., Qureshi, R., Forstner, G., and Forstner, J.,** Neutral and acidic species of human intestinal mucin: evidence for different core peptides, *J. Biol. Chem.*, 260, 7955, 1985.

24. **Slomiany, B. L., Murty, V. L. N., and Slomiany, A.,** Isolation and characterization of oligosaccharides from rat colonic mucus glycoprotein, *J. Biol. Chem.*, 255, 9719, 1980.

25. **Carlson, D. M.,** Structures and immunochemical properties of oligosaccharides isolated from pig submaxillary mucins, *J. Biol. Chem.*, 243, 616, 1968.

26. **Feizi, T., Gooi, H. C., Childs, R. A., Picard, J. K., Uemara, K., Loomes, L. M., Thorpe, S. J., and Hounsell, E. F.,** in *Mucin Type Glycoproteins*, Feizi, T., Ed., Biochemical Society Transactions, 607th meeting, London, 1984, 591.

27. **Jabbal, I., Kells, D. I. C., Forstner, G., and Forstner, J.,** Human intestinal goblet cell mucin, *Can. J. Biochem.*, 54, 707, 1976.

28. **Smith, A. C. and Podolsky, D. K.,** Colonic mucin glycoproteins in health and disease, *Clin. Gastroenterol.*, 15, 1986.

29. **Rhodes, J. M.,** Colonic mucus and mucosal glycoproteins: the key to colitis and cancer?, *Gut*, 30, 1660, 1989.

30. **Isselbacher, K. J.,** Composition of human colonic mucin. Selective alteration in IBD, *J. Clin. Invest.*, 2, 142, 1983.

31. **Barresi, G., Tuccari, G., and Magazzu, G.,** Neutral mucins in coeliac disease, *Basic Appl. Histochem.*, 27, 55, 1983.

32. **Boland, C. R., Roberts, J. A., Siddiqui, B., Byrd, J., and Kim, Y. S.,** Cancer-associated colonic mucin in cultured human tumor cells and athymic (nude) mouse xenografts, *Cancer Res.*, 46, 5724, 1986.

33. **Kim, Y. S., Isaacs, R., and Perdomo, J. M.,** Alterations of membrane glycopeptides in human colonic adenocarcinoma, *Proc. Natl. Acad. Sci. U.S.A.*, 71, 4869, 1974.

34. **Lamont, J. T. and Isselbacher, K. J.,** Alterations in glycosyltransferase activity in human colon cancer, *J. Natl. Cancer Institute*, 54, 53, 1975.

35. **Boland, C. R., Montgomery, C. K., and Kim, Y. S.,** Alterations in human colonic mucin occuring with cellular differentiation and malignant transformation, *Proc. Natl. Acad. Sci. U.S.A.*, 79, 2051, 1982.

36. **Allen, A.,** Chemical aspects of mucus: Structure of gastrointestinal mucus glycoproteins and viscous and gel-forming properties of mucus, *Br. Med. Bull.*, 34, 28, 1978.

37. **Wesley, A. W., Forstner, J. F., and Forstner, G. G.,** Structure of intestinal-mucus glycoprotein from human postmortem or surgical tissue: inferences from correlation analyses of sugar and sulfate composition of individual mucins, *Carbohydr. Res.*, 115, 151, 1983.

38. **Robinson, D. S. and Monsey, J. B.,** The composition and proposed subunit structure of egg-white β-ovomucin, *Biochem. J.*, 147, 55, 1975.

39. **Roberts, G. P.,** Isolation and characterization of glycoproteins from sputum, *Arch. Biochem. Biophys.*, 173, 257, 1972.

40. **Forstner, J. F.,** Intestinal mucins in health and disease, *Digestion*, 17, 234, 1978.

41. **Carlstedt, I. and Sheehan, J. K.,** Is the macromolecular architecture of cervical, respiratory and gastric mucins the same?, *Biochem. Soc. Trans.*, 12, 615, 1984.

42. **Desai, M. A. and Vadgama, P.,** Estimation of effective diffusion coefficients of model solutes through gastric mucus: assessment of a diffusion chamber technique based on spectrophotometric analysis, *Analyst*, 116, 1113, 1991.

43. **Hannoun, B. J. M. and Siephenopoulos, G.,** Diffusion coefficients of glucose and ethanol in cell-free and cell-occupied calcium alginate membranes, *Biotechnol. Bioeng.*, 28, 829, 1986.

44. **Nimmerfall, F. and Rosenthaler, J.,** Significance of the goblet cell mucin layer, the outermost luminal barrier to passage through the gut wall, *Biochem. Biophys. Res. Commun.*, 94, 960, 1980.

45. **Smith, G. W., Wiggins, P. M., Lee, S. P., and Tasman-Jones, C.,** Diffusion of butyrate through pig colonic mucus *in vitro*, *Clin. Sci.*, 70, 271, 1986.

46. **Lucas, M. L.,** Estimation of sodium chloride diffusion coefficient in gastric mucin, *Dig. Dis. Sci.*, 29, 1984, 336.

47. **Smithson, K. W.,** Diffusion barrier in the small intestine, *Science*, 220, 221, 1983.

48. **Peppas, N. A., Hansen, P. J., and Buri, P. A.,** A theory of molecular diffusion in the intestinal mucin, *Int. J. Pharmac.*, 20, 107, 1984.

49. **Nicholas, C. V., Desai, M., Vagdama, P., McDonnell, M. B., and Lucas, S.,** pH dependence of hydrochloric acid diffusion through gastric mucus: correlation with diffusion through a water layer using a membrane-mounted glass pH electrode, *Analyst*, 116, 463, 1991.

50. **Forstner, J. F. and Forstner, G. G.,** Calcium binding to intestinal goblet cell mucin, *Biochem. Biophys. Acta*, 386, 283, 1975.

51. **Forstner, J. F. and Forstner, G. G.,** Effects of calcium on intestinal mucin: implications for cystic fibrosis, *Pediatr. Res.*, 10, 609, 1976.

52. **Forstner, J. F., Jabbal, I., Findlay, B. P., and Forstner, G. G.,** Interaction of mucins with calcium, H+ and albumin, *Mod. Probl. Pediatr.*, 19, 54, 1975.

53. **Timothy, C. K., Lynnette, N. G., Ferguson, R., Harris, P. J., Watson, M. E., and Robertson, A. M.,** *In vitro* adsorption of a hydrophobic mutagen to gastrointestinal mucus glycoprotein (mucin) and dietary fibre, *Chem. Biol. Interact.*, 82, 219, 1992.

54. **Niibuchi, J.-J., Aramaki, Y., and Tsuchiya, S.,** Binding of antibiotics to rat intestinal mucin, *Int. J. Pharmaceut.*, 30, 181, 1986.

55. **Kearney, P. and Marriot, C.,** The effects of mucus glycoproteins on the bioavailability of tetracycline. III. Everted gut studies, *Int. J. Pharmaceut.*, 38, 211, 1987.

56. **Slomiany, B. L. and Slomiany, A.,** Mechanism of *Helicobacter pylori* pathogenesis: focus on mucus, *J. Clin. Gastroenterology*, 14, S114, 1992.

57. **Balch, W. E., Dunphy, W. G., Braell, W. A., and Rothman, J. E.,** Reconstitution of the transport of protein between succesive compartments of the Golgi measured by the coupled incorporation of N-acetylglucosamine, *Cell*, 39, 405, 1984.

58. **Balch, W. E., Glich, B. S., and Rothman, J. E.,** Sequential intermediates in the pathway of intercompartmental transport in a cell-free system, *Cell*, 39, 525, 1984.

59. **Braell, W. A., Balch, W. E., Dobbertin, D. C., and Rothman, J. E.,** The glycoprotein that is transported between successive compartments of the Golgi in a cell-free system resides in stacks of cisternae, *Cell*, 39, 511, 1984.

60. **Allen, A. and Kent, P. W.,** Biosynthesis of intestinal mucins, *Biochem. J.*, 106, 301, 1968.

61. **O'Doherty, P. J. A. and Kuksis, A.,** Effect of puromycin *in vitro* on protein and glycerolipid biosynthesis in isolated epithelial cells of rat intestine, *Int. J. Biochem.*, 6, 435, 1975.

62. **Sherman, P., Forstner, J., Roomi, N., Khatri, I., and Forstner, G.,** Mucin depletion in the intestine of malnourished rats, *Am. J. Physiol.*, 248, G418, 1985.

63. **Jeynes, B. J. and Altman, G. G.,** Light and scanning electron microscopic observations of the effects of the sublethal doses of methotrexate on the rat small intestine, *Anat. Rec.*, 191, 1, 1978.

64. **Rijke, R. P. C., Hanson, W. R., Plaiser, H. M., and Osborne, J. W.,** The effect of ischemic villus cell damage on crypt cell proliferation in the small intestine, *Gastroenterology*, 71, 786, 1976.

65. **Shea-Donohue, T., Dorval, E. D., Montcalm, E., El-Bayer, H., Durakovic, A., Conklin, J. J., and Dubois, A.,** Alterations in gastric mucus secretion in rhesus monkeys after exposure to ionizing radiation, *Gastroenterology*, 88, 685, 1985.

66. **Eastwood, G. L. and Trier, J. S.,** Epithelial cell renewal in cultured rectal biopsies in ulcerative colitis, *Gastroenterology*, 64, 383, 1973.

67. **Clarke, R. M.,** Control of intestinal epithelial replacement: lack of evidence for a tissue specific blood-borne factor, *Cell Tissue Kinet.*, 7, 241, 1974.

68. **Cassidy, M. M., Lightfoot, F. G., and Vahouny, G. V.,** Structural-functional modulation of mucin secretory patterns in the gastrointestinal tract, *Prog. Clin. Biol. Res.*, 73, 97, 1981.

69. **Biol, M. C., Martin, A., Oehninger, C., Louisot, P., and Richard, M.,** Biosynthesis of glycoproteins in the gastrointestinal mucosa. II. Influence of diets, *Am. Nutr. Metab.*, 25, 269, 1981.

70. **Wattel, W., Van Huis, G. A., Kramer, M. F., and Geuze, J. J.,** Glycoprotein synthesis in the mucous cell of the vascularly perfused rat stomach, *Am. J. Anat.*, 156, 313, 1979.
71. **Glass, G. B. J. and Slomiany, B. L.,** Derangements of biosynthesis, production and secretion of mucus in gastrointestinal injury and disease, in *Mucus in Health and Disease*, Elstein, M. and Parke, D. V., Eds., Plenum Press, New York, 1977, 311.
72. **Menguy, R.,** Gastric mucus and the gastric mucous barrier, *Am. J. Surg.*, 117, 806, 1969.
73. **Parke, D. V.,** Pharmacology of mucus, *Br. Med. Bull.*, 34, 89, 1978.
74. **Menguy, R. and Masters, Y. F.,** The effects of aspirin on gastric mucus secretion, *Surg. Gynecol. Obstet.*, 92, 1, 1965.
75. **Menguy, R. and Masters, Y. F.,** Effect of cortisone on mucoprotein secretion by gastric antrum of dogs: pathogenesis of steroid ulcer, *Surgery*, 54, 19, 1963.
76. **Desbaillets, L. and Menguy, R.,** Inhibition of gastric mucous secretion by ACTH, *Am. J. Dig. Dis.*, 12, 583, 1967.
77. **Domschke, W., Domschke, S., Classen, M., and Demling, L.,** Some properties of mucus in patients with gastric ulcer: effect of treatment with carbenoxolone sodium, *Scand. J. Gastroenterol.*, 7, 647, 1972.
78. **Parke, D. V. and Symons, A. M.,** The biochemical pharmacology of mucus, in *Mucus in Health and Disease*, Elstein, M. and Parke, D. V., Eds., Plenum Press, New York, 1977, 423.
79. **Moon, H. W., Whipp, D. C., and Baetz, A. L.,** Comparative effects of enterotoxin from *Escherichia coli* and *Vibrio cholerae* on rabbit and swine small intestine, *Lab. Invest.*, 25, 133, 1971.
80. **Yardley, J. H., Bayless, T. M., Leubbers, E. H., Halstead, C. H., and Hendrix, T. R.,** Goblet cell mucus in the small intestine. Findings after net fluid production due to cholera toxin and hypertonic solutions, *Johns Hopkins Med. J.*, 131, 1, 1972.
81. **Forstner, J. F., Roomi, N. W., Fahim, R. E. F., and Forstner, G. G.,** Cholera toxin stimulates secretion of immunoreactive intestinal mucin, *Am. J. Physiol.*, 240, G10, 1981.
82. **Forstner, J. F., Roomi, N. W., Fahim, R., Gall, G., Perdue, M., and Forstner, G. G.,** Acute and chronic models for hypersecretion of intestinal mucin, in *Mucus and Mucosa*, Ciba Foundation Symposium 109, Pitman, London, 1984, 61.
83. **Donowitz, M., Charney, A. N., and Hynes, R.,** Propanalol prevention of cholera enterotoxin-induced intestinal secretion in the rat, *Gastroenterology*, 76, 482, 1979.
84. **Dinoso, V. P., Ming, S., and Meniff, J.,** Ultrastructural changes of the canine gastric mucosa after topical application of graded concentrations of ethanol, *Dig. Dis.*, 21, 626, 1976.
85. **Heim, H. K., Oestmann, A., and Sewing, K. Fr.,** Stimulation of glycoprotein and protein synthesis in isolated pig gastric mucosal cells by prostaglandins, *Gut*, 31, 412, 1990.
86. **Kelly, S. M. and Hunter, J. O.,** EGF stimulates synthesis and secretion of mucus glycoproteins in human gastric mucosa, *Clin. Sci.*, 79, 425, 1990.
87. **Rainsford, K. D.,** The effects of aspirin and other non-steroid anti-inflammatory/analgesic drugs on gastrointestinal mucus. Glycoprotein synthesis *in vivo*: relationship to ulcerogenic actions, *Biochem. Pharmacol.*, 27, 877, 1978.
88. **Menguy, R. and Thompson, A. E.,** Regulation of secretion of mucus from the gastric antrum, *Ann. N.Y. Acad. Sci.*, 140, 797, 1967.
89. **Neutra, M. R., O'Malley, L. J., and Specian, R. D.,** Regulation of intestinal goblet cell secretion, *Am. J. Physiol.*, 242, G380, 1982.
90. **Flemstrom, G., Heylings, J. R., and Garner, A.,** Gastric and duodenal $HCO_3$-transport in vitro: effects of hormones and local transmitters, *Am. J. Physiol.*, 242, G100, 1982.
91. **Abt, A. B. and Carter, S. L.,** Goblet cell carcinoid of the appendix, *Arch. Pathol.*, 100, 301, 1976.
92. **Lamont, J. T., Ventola, A. S., Maull, E. A., and Szabo, S.,** Cystamine and prostaglandin $F_{2b}$ stimulate rat gastric mucin release, *Gastroenterology*, 84, 306, 1983.
93. **Bell, A. E., Sellers, L. A., Allen, A., Cunliffe, W. J., Morris, E. R., and Ross-Murphy, S. B.,** Properties of gastric and duodenal mucus: effect of proteolysis, disulfide reduction, bile, acid, ethanol, and hypertonicity on mucus gel structure, *Gastroenterology*, 88, 269, 1985.
94. **Pearson, J. P., Allen, A., and Venables, C. W.,** Gastric mucus: isolation and polymeric structure of the undegraded glycoprotein — its breakdown by mucin, *Gastroenterology*, 78, 709, 1981.
95. **Scawen, M. and Allen, A.,** The action of proteolytic enzymes on the glycoprotein from pig gastric mucus, *Biochem. J.*, 163, 363, 1977.
96. **Salyers, A. A.,** Energy sources of major intestinal fermentative anaerobes, *Am. J. Clin. Nutr.*, 32, 158, 1979.
97. **Garn, S. M. and Leanard, W. R.,** What did our ancestors eat?, *Nutr. Rev.*, 47, 337, 1989.
98. **Mata, L. J. J., Jiminez, M., Cordon, R., Rosales, E., Prera, R., Schneider, E., and Vitera, F.,** Gastrointestinal flora of children with protein-calorie malnutrition, *Am. J. Clin. Nutr.*, 25, 1118, 1972.
99. **Sherman, P., Forstner, J., Roomi, N., Khatri, I., and Forstner, G.,** Mucin depletion in the intestine of malnourished rats, *Am. J. Physiol.*, 248, G418, 1985.
100. **Tomkins, A.,** Nutritional studies and severity of diarrhea among preschool children in rural Nigeria, *Lancet*, 1, 860, 1981.
101. **Prizont, R.,** Degradation of intestinal glycoproteins by pathogenic *Shigella flexneri*, *Infect. Immun.*, 36, 615, 1982.

194

102. **Seidman, E., Leloiko, N., Ament, M., Berman, W., Caplan, D., Evans, J., Kocoshis, S., Lake, A., Motil, K., Sutpen, J., and Thomas, D.,** Nutritional issues in paediatric inflammatory bowel disease, *J. Paed. Gastroenterol. Nutr.*, 12, 424, 1991.
103. **Shiau, S. Y. and Chang, G. W.,** Effects of dietary fibre on fecal mucinase and β-glucuronidase in rats, *J. Nutr.*, 113, 138, 1983.
104. **UK Department of Health,** Dietary reference values for food energy and nutrients for the United Kingdom, in *Report of the Panel on Dietary Reference Values of the Committee on Medical Aspects of Food Policy, — Report on Health and Social Subjects no. 41,* HMSO, London, 1992.
105. **U.S. Department of Agriculture and U.S. Department of Health and Human Services,** Nutrition and your health: dietary guidelines for Americans, in *Home and Garden Bull.*, 232, 3rd ed., U.S. Government Printing Office, Washington, D.C., 1990.
106. **Vercelotti, J. R., Salyers, A. A., and Wilkins, T. D.,** Complex carbohydrate breakdown in the human colon, *Am. J. Clin. Nutr.*, 31, 586, 1978.
107. **Bond, J. H. and Levitt, M. D.,** Fate of soluble carbohydrates in the colon of rats and man, *J. Clin. Invest.*, 57, 1158, 1976.
108. **Englyst, H. N. and Cummings, J. H.,** Digestion of the carbohydrates of banana *(Musaparadisica serpentum)* in the human small intestine, *Am. J. Clin. Nutr.*, 44, 42, 1986.
109. **Englyst, H. N. and Cummings, J. H.,** Digestion of the polysaccharides of some cereal foods in the human small intestine, *Am. J. Clin. Nutr.*, 44, 42, 1986.
110. **Fleming, S. E. and Rodriguez, M. A.,** Influence of dietary fibre in faecal excretion of volatile fatty acids by human adults, *J. Nutr.*, 113, 1613, 1983.
111. **Kingman, S. M. and Englyst, H. N.,** The influence of food preparation methods on the in-vitro digestibility of starch in potatoes, *Food Chem.*, 49, 181, 1994.
112. **Wang, X. and Gibson, G. R.,** Effects of the *in vitro* fermentation of oligofructose and inulin by bacteria growing in the human large intestine, *J. Appl. Bacteriol.*, 75, 373, 1993.
113. **Gibson, G. R., Cummings, J. H., and Macfarlane, G. T.,** Use of a three-stage continuous culture system to study the effect of mucin on dissimilatory sulfate reduction and methanogenesis by mixed populations of human gut bacteria, *Appl. Environ. Microbiol.*, 54, 2750, 1988.
114. **Macfarlane, G. T., Hay, S., Macfarlane, S., and Gibson, G. R.,** Effect of different carbohydrates on growth, polysaccharidase and glycosidase production by *Bacteroides ovatus,* in batch and continuous culture, *J. Appl. Bacteriol.*, 68, 179, 1990.
115. **Warren, L.,** The thiobarbituric acid assay of sialic acids, *J. Biol. Chem.*, 234, 1971, 1959.
116. **Englyst, H. N. and Hudson, G. J.,** Colorimetric method for routine measurement of dietary fibre as non-starch polysaccharides. A comparison with gas-liquid chromatography, *Food Chem.*, 24, 63, 1987.
117. **Englyst, H. N., Quigley, M. E., and Hudson, G. J.,** Determination of dietary fibre as non-starch polysaccharides with gas-liquid chromatogarphic, high-performance chromatographic or colorimetric measurement of constituent sugars, *Analyst*, 119, 1497, 1994.
119. **Scott, R. W.,** Colorimetric determination of hexuronic acids in plant materials, *Anal. Chem.*, 51, 936, 1979.
120. **Taylor, K. A. and Buchanan-Smith, J. G.,** A colorimetric method for the quantitation of uronic acids and a specific assay for galacturonic acid, *Anal. Biochem.*, 201, 190, 1992.
121. **Archer, M. A. Hamilton, J. K., and Smith, F.,** The reduction of sugars with sodium borohydride, *Am. Chem. Soc.*, 73, 4691, 1951.
122. **Barr, J. and Nordin, P.,** Microdetermination of neutral sugars found in glycoproteins, *Anal. Biochem.*, 108, 313, 1980.
123. **Henry, R. J., Blakeney, A. B., Harris, P. J., and Stone, B. A.,** Detection of neutral and amino sugars from glycoproteins and polysaccharides as their alditol acetates, *J. Chromatogr.*, 256, 419, 1983.
124. **Dierckxsens, G. C., De Meyer, L., and Tonino, J.,** Simultaneous determination of uronic acids, hexosamines and galactose of glycosaminoglycans by gas liquid chromatography, *Anal. Biochem.*, 130, 120, 1983.
125. **Chaplin, M. F.,** A rapid and sensitive method for the analysis of carbohydrate components in glycoproteins using gas liquid chromatography, *Anal. Biochem.*, 123, 336, 1982.
126. **Walters, J. S. and Hedges, J. I.,** Simultaneous determination of uronic acids and aldoses in plant tissues and sediment by capillary gas chromatography of *N*-hexylaldonamide and alditol acetates, *Am. Chem. Soc.*, 60, 988, 1988.
127. **Lehrfeld, J.,** Simultaneous gas liquid chromatography determination of aldonic acids and aldoses, *Anal. Chem.*, 57, 346, 1985.
128. **Lehrfeld, J.,** Gas liquid chromatography determination of aldonic acids as acetylated aldonamides, *Carbohydr. Res.*, 135, 179, 1985.
129. **Lehrfeld, J.,** Differential gas liquid chromatography method for determination of uronic acids in carbohydrate mixtures, *Anal. Biochem.*, 115, 410, 1981.
130. **Honda, S.,** High-performance liquid chromatography of mono- and oligosaccharides, *Anal. Biochem.*, 140, 1, 1984.
131. **Hughes, S. and Johnson, D. C.,** Amperometric detection of simple carbohydrates at platinum electrodes in alkaline solutions by application of a triple pulse potential waveform, *Anal. Chem. Acta,* 132, 11, 1981.

132. **Lee, Y. C.,** High performance anion-exchange chromatography for carbohydrate research, *Anal. Biochem.*, 189, 151, 1990.

133. **Martens, D. A. and Frankenberger, W. T., Jr.,** Determination of saccharides by high performance anion-exchange chromatography with pulsed amperometric detection, *Chromatographia*, 29, 7, 1990.

134. **Martens, D. A. and Frankenberger, W. T., Jr.,** Determination of saccharides in biological materials by high-performance anion-exchange chromatography with pulse amperometric detection, *J. Chrom.*, 546, 297, 1991.

135. **Quigley, M. E. and Englyst, H. N.,** Determination of neutral sugars and hexosamines by high performance liquid chromatography with pulsed amperometric detection, *Analyst*, 117, 1715, 1992.

136. **Clarke, A. C., Sarabin, V., Keenleyside, W., Maclachlan, P. R., and Whitfield, C.,** The compositional analysis of bacterial extracellular polysaccharides by high performance anion-exchange chromatography, *Anal. Biochem.*, 199, 68, 1991.

137. **Hardy, M. R., Townsend, R. R., and Lee, Y. C.,** Monosaccharide analysis of glycoconjugates by anion-exchange chromatography with pulsed amperometric detection, *Anal. Biochem.*, 170, 54, 1988.

138. **Hardy, M. R.,** Monosaccharide analysis of glycoconjugates by high-performance anion-exchange chromatography with pulsed amperometric detection, *Meth. Enzymol.*, 179, 76, 1990.

139. **Peelen, G. O. H., deJong, J. G. N., and Wevers, R. A.,** High-performance liquid chromatography of monosaccharides and oligosaccharides in a complex biological matrix, *Anal. Biochem.*, 198, 334, 1991.

140. **Martens, D. A. and Frankenberger, W. T., Jr.,** Determination of glycuronic acids by high-performance anion-exchange chromatography with pulsed amperometric detection, *Chromatographia*, 30, 651, 1990.

141. **Quigley, M. E. and Englyst, H. N.,** Determination of the uronic acid constituents of non-starch polysaccharides by high-performance anion-exchange chromatography with pulsed amperometric detection, *Analyst*, 119, 1151, 1994.

142. **Wang, W. T., Erlansson, K., Lindh, F., Lundgren, T., and Zopf, D.,** High-performance liquid chromatography of sialic acid-containing oligosaccharides and acidic monosaccharides, *Anal. Biochem.*, 190, 182, 1990.

143. **Manzi, A. E., Diaz, S., and Varki, A.,** High-pressure liquid chromatography on a pellicular resin anion-exchange column with pulsed amperometric detection: a comparison with six other systems, *Anal. Biochem.*, 188, 20, 1990.

144. **Townsend, R. R., Hardy, M. R., Cumming, D. A., Carver, J. P., and Bendiak, B.,** Separation of branched sialylated oligosaccharides using high pH anion-exchange chromatography with pulsed amperometric detection, *Anal. Biochem.*, 182, 1, 1989.

145. **Hardy, M. R. and Townsend, R. R.,** Separation of positional isomers of oligosaccharides and glycopeptides by high performance anion-exchange chromatography with pulsed amperometric detection, *Proc. Natl. Acad. Sci. U.S.A.*, 85, 3289, 1988.

146. **Hernandez, L. M., Ballou, L., and Ballou, C. E.,** Separation of yeast asparagine-linked oligosaccharides by high performance anion-exchange chromatography, *Carbohydr. Res.*, 203, 1, 1990.

147. **Basa, L. J. and Spellman, M. W.,** Analysis of glycoprotein derived oligosaccharides by high pH anion-exchange chromatography, *J. Chromatogr.*, 499, 205, 1990.

148. **Hotchkiss, A. T., Jr. and Hicks, K. B.,** Analysis of oligogalacturonic acids with 50 or fewer residues by high performance anion-exchange chromatography and pulsed amperometric detection, *Anal. Biochem.*, 184, 200, 1990.

149. **Pfeiffer, G., Geyer, H., Geyer, R., Kalsner, I., and Wendorf, P.,** Separation of glycoprotein-*N*-glycans by high-pH anion-exchange chromatography, *Biomed. Chrom.*, 4, 193, 1990.

150. **Reddy, G. P. and Bush, C. A.,** High performance anion-exchange chromatography of neutral milk oligosaccharides and oligosaccharide alditols derived from mucin glycoproteins, *Anal. Biochem.*, 198, 278, 1991.

151. **Willenbrock, F. W., Neville, C. A., Jacob, G. S., and Scudder, P.,** The use of HPLC-pulsed amperometry for the characterization and assay of glycosidases and glycosyltransferases, *Glycobiology*, 1, 223, 1991.

152. **Hardy, M. R.,** Liquid chromatography analysis of the glycoproteins, *LC-GC*, 7, 242, 1989.

153. **Shimamato, C., Desmukh, G. D., Rigot, W. L., and Boland, C. R.,** Analysis of cancer-associated colonic mucin by ion-exchange chromatography: evidence for a mucin species of lower molecular charge and weight in cancer, *Biochim. Biophys. Acta*, 991, 284, 1989.

154. **Yet, M. -G. and Wold, F.,** The distribution of glycan structures in individual *N*-glycosylation sites in animal and plant glycoproteins, *Arch. Biochem. Biophys.*, 278, 356, 1990.

155. **Hermentin, P., Witzel, R., Uliegenthert, J. F. G., Kamerling, J. P., Nimtz, M., and Conradt, H. S.,** A strategy for the mapping of *N*-glycans by high pH anion exchange chromatography with pulsed amperometric detection, *Anal. Biochem.*, 203, 281, 1992.

156. **Hermentin, P., Witzel, R., Doenges, R., Bauer, R., Haupt, H., Patel, T., Parekh, R. B., and Brazel, K.,** The mapping by high-pH anion-exchange chromatography with pulsed amperometric detection and capillary electrophoresis of the carbohydrate moieties of human plasma-acid glycoprotein, *Anal. Biochem.*, 206, 419, 1992.

157. **Townsend, R. R.,** Assessment of protein glycosylation using high-pH anion-exchange chromatography with pulsed electrochemical detection, in *Protein Glycosylation: Cellular, Biotechnical and Analytical Aspects*, Vol. 15, Conradt, H. S., Ed., V. C. H., Weinheim, Germany, 1990, 146.

158. **Yu Ip, C. C., Manam, R., Hepler, R., and Hennessey, J. P., Jr.,** Carbohydrate composition analysis of bacterial polysaccharides: optimized acid hydrolysis conditions for HPAEC-PAD analysis, *Anal. Biochem.*, 201, 343, 1992.

159. **Blake, D. A. and McClean, N. V.,** High-pressure, anion-exchange chromatography of proteoglycans, *Anal. Biochem.,* 190, 158, 1990.

160. **Hayase, T. and Chuan, Y.,** High performance cation-exchange chromatography of hexosaminitols, *Anal. Biochem.,* 208, 208, 1993.

161. **De Ruiter, G. A., Schols, H. A., Voragen, A. G. J., and Rombouts, F. M.,** Carbohydrate analysis of water-soluble uronic acid-containing polysaccharides with high-performance anion-exchange chromatography using methanolysis combined with TFA hydrolysis is superior to four other methods, *Anal. Biochem.,* 207, 176, 1992.

162. **Townsend, R. R. and Hardy, M. R.,** Analysis of glycoprotein oligosaccharides using high-pH anion-exchange chromatography, *Glycobiology,* 1, 139, 1991.

163. **Weitzhandler, M., Kadlecek, D., Avdalovic, N., Forte, J. G., Chow, D., and Townsend, R. R.,** Monosaccharide and oligosaccharide analysis of proteins transferred to polyvinyl fluoride membranes after sodium dodecyl sulfate-polyacrylamine gel electrophoresis, *J. Biol. Chem.,* 268, 5121, 1993.

164. **Laemmli, U. K.,** Cleavage of structural proteins during the assembly of the head of bacteriophage T4, *Nature (London),* 227, 680, 1970.

165. **Orberger, G., Gebner, R., Fuchs, H., Volz, B., Köttgen, E., and Tauber, R.,** Enzymatic modelling of the oligosaccharide chains of glycoproteins immobilized onto polystyrene surfaces, *Anal. Biochem.,* 214, 195, 1993.

166. **Schauer, R.,** Occurence of sialic acids, in *Sialic acids: Chemistry, Metabolism and Function,* Vol. 10, Schauer, R., Ed., Springer-Verlag, New York, 1982, 5.

167. **Werner, I. and Odin, L.,** On the presence of sialic acid in certain glycoproteins and in gangliosides, *Acta Soc. Med. Upsaliensis,* 57, 230, 1952.

168. **Ayala, W., Moore, L. V., and Hess, E. L.,** The purple color reaction given by diphenylamine reagent. I. With normal and rheumatic fever sera, *J. Clin. Invest.,* 30, 781, 1951.

169. **Pigman, P., Hawkins, W. L., Blair, M. G., and Holley, H. L.,** Sialic acid in normal and arthritic synovial fluids, *Arthritis Rheum.,* 1, 151, 1958.

170. **Seibert, F. B., Pfaff, M. L., and Siebert, M. V.,** A serum polysaccharide in tuberculosis and carcinoma, *Arch. Biochem. Biophys.,* 18, 279, 1948.

171. **Folch, J., Arsove, S., and Meath, J. A.,** Isolation of brain strandin, a new type of large molecule tissue component, *J. Biol. Chem.,* 191, 819, 1951.

172. **Hess, E. L., Coburn, A. F., Bates, R. C., and Murphy, P.,** A new method for measuring sialic acid levels in serum and its application to rheumatic fever, *J. Clin. Invest.,* 36, 449, 1957.

173. **Warren, L.,** The thiobarbituric acid of sialic acids, *J. Biol. Chem.,* 234, 1971, 1959.

174. **Schauer, R.,** Analysis of sialic acids, in *Methods in Enzymology,* Ginsburg, V., Ed., Academic Press, San Diego, 1987, 132.

175. **Reuter, G., Pfeil, R., Stoll, S., Schauer, R., Kamerling, J. P., Versluis, C., and Vliegenthart, J. F.,** Identification of new sialic acids derived from glycoprotein of bovine submandibular gland, *Eur. J. Biochem.,* 134, 139, 1983.

176. **Schauer, R.,** Isolation and purification of sialic acids, in *Sialic acids: Chemistry, Metabolism and Function,* Vol. 10, Schauer, R., Ed., Springer-Verlag, New York, 1982, 51.

177. **Schauer, R.,** Colorimetry and thin-layer chromatography of sialic acids, in *Sialic acids: Chemistry, Metabolism and Function,* Vol. 10, Schauer, R., Ed., Springer-Verlag, New York, 1982, 77.

178. **Schauer, R.,** Gas-liquid chromatography and mass spectrometry of sialic acids, in *Sialic acids: Chemistry, Metabolism and Function,* Vol. 10, Schauer, R., Ed., Springer-Verlag, New York, 1982, 95.

179. **Schauer, R.,** NMR spectroscopy of sialic acids, in *Sialic acids: Chemistry, Metabolism and Function,* Vol. 10, Schauer, R., Ed., Springer-Verlag, New York, 1982, 127.

180. **Varki, A. and Diaz, S.,** The release and purification of sialic acids from glycoconjugates: methods to minimize the loss and migration of *O*-acetyl groups, *Anal. Biochem.,* 137, 236, 1984.

181. **Lowry, O. H., Rosebrough, W. J., Farr, A. L., and Randall, R. J.,** Protein measurement with the folin phenol reagent, *J. Biol. Chem.,* 143, 265, 1951.

182. **Hunt, S.,** Degradation of amino acids accompanying *in vitro* protein hydrolysis, in *Chemistry and Biochemistry of the Amino Acids,* Barrett, G. C., Ed., Chapman and Hall, New York, 1985, 376.

183. **Simpson, R. J., Neuberger, M. R., and Liu, T. Y.,** Complete amino acid analysis of proteins from a single hydrolysate, *J. Biol. Chem.,* 251, 1936, 1976.

184. **D'Aniello, A., Petrucilli, L., Gardner, C., and Fisher, G.,** Improved method for hydrolyzing proteins without inducing racemization and for determining their true *D*-amino acid content, *Anal. Biochem.,* 213, 290, 1993.

185. **McClung, G. and Frankenberger, W. T., Jr.,** Comparison of reverse-phase high-performance liquid chromatographic methods for precolumn derivatized amino acids, *J. Liq. Chromatogr.,* 11, 613, 1988.

186. **Tarr, G. E.,** Rapid separation of amino acid phenylthiohydantoins by isocratic high performance liquid chromatography, *Anal. Biochem.,* 111, 27, 1981.

187. **Jones, B. N. and Gilligan, J. P.,** *O*-pthalaldehyde precolumn derivatization and reversed phase high-performance liquid chromatography of polypeptide hydrolysates and physiological fluids, *J. Chrom.,* 266, 471, 1983.

188. **Gardner, W. S. and Miller, W. H.,** Reverse-phase liquid chromatographic analysis of amino acids after reaction with *O*-pthaladehyde, *Anal. Biochem.,* 101, 471, 1983.

189. **Lindroth, P. and Mopper, K.,** High performance liquid chromatographic determination of subpicomole amounts of amino acids by pre-column fluorescence derivatization with *O*-pthalaldehyde, *Anal. Chem.*, 51, 1667, 1979.

190. **Hamilton, P. B.,** Ion exchange chromatography of amino acids. A single column, high resolving, fully automated procedure, *Anal. Chem.*, 35, 2055, 1963.

191. **Polta, J. A. and Johnson, D. C.,** The direct electrochemical detection of amino acids at a platinum electrode in an alkaline chromatographic effluent, *J. Liq. Chromatogr.*, 6, 1727, 1983.

192. **Pickering, M. V.,** Ion exchange chromatography of free amino acids, *LC-GC*, 7, 484, 1987.

193. **Martens, D. A. and Frankenberger, W. T., Jr.,** Pulsed amperometric detection of amino acids separated by anion-exchange chromatography, *J. Liq. Chromatogr.*, 15, 423, 1992.

194. **Allison, C. and Macfarlane, G. T.,** Effect of nitrate on methane production and fermentation by slurries of human faecal bacteria, *J. Gen. Microbiol.*, 134, 1397, 1988.

195. **Holdemann, L. V., Cato, E. P., and Moore, W. E. C., Eds.,** *Anaerobe Laboratory Manual*, 4th ed., Virginia Polytechnic Institute and State University, Blacksburg, 1977.

196. **Macfarlane, G. T., Gibson, G. R., and Cummings, J. H.,** Comparison of fermentation products in different regions of the human colon, *J. Appl. Bacteriol.*, 72, 57, 1992.

197. **Macfarlane, G. T., Cummings, J. H., Macfarlane, S., and Gibson, G. R.,** Influence of retention time on degradation of pancreatic enzymes by human colonic bacteria grown in a 3-stage continuous culture system, *J. Appl. Bacteriol.*, 67, 521, 1989.

198. **Pfundstein, B., Tricker, A. R., and Preussmann, R.,** Determination of primary and secondary amines in foodstuffs using gas chromatography and chemiluminescence detection with a modified thermal energy laser, *J. Chrom.*, 539, 141, 1991.

199. **Kage, S., Nagata, N., Kimura, K., and Kudo, K.,** Extractive alkylation and gas chromatographic analysis of sulfide, *J. Forensic Sci.*, 33, 217, 1988.

200. **Hafez, A. A., Goyal, S. S., and Rains, D. W.,** Quantitative determination of total sulfur in plant tissues using acid digestion and ion chromatography, *Agron. J.*, 83, 148, 1991.

201. **Lacourse, W. R., Johnson, D. C., Rey, M. A., and Slingsby, R. W.,** Pulsed amperometric detection of aliphatic alcohols in liquid chromatography, *Anal. Chem.*, 63, 134, 1991.

202. **Murakmi, M., Saita, H., Teramura, S., Dekigai, H., Asagoe, K., Kusaka, S., and Kita, T.,** Gastric ammonia has a potent ulcerogenic action on the rat stomach, *Gastroenterology*, 105, 1710, 1993.

203. **Sidebotham, R. L., Batten, J. J., Karim, Q. N., Spencer, J., and Baron, J. H.,** Breakdown of gastric mucus in the presence of *Helicobacter pylori*, *J. Clin. Pathol.*, 44, 52, 1991.

204. **Davis, C. P., Balish, E., and Uehling, D.,** Bacterial microenvironments: bacterial interaction with mucin layers in gastrointestinal tracts and bladders, *Scanning Electron Microsc.*, 2, 269, 1977.

205. **Cantley, J. R.,** The rabbit model of *Escherichia coli* (strain RDEC-1) diarrhea, in *Attachment of Organisms to the Gut Mucosa*, Vol. 1., Boedeker, E. C., Ed., CRC Press, Boca Raton, FL, 1984, 39.

206. **Mantle, M., Thakore, E., Hardin, J., and Gall, D. G.,** Effect of *Yersinia enterocolitica* on intestinal secretion, *Am. J. Physiol.*, 256, G319, 1989.

207. **Mantle, M. and Rombough, C.,** Growth in, and breakdown of, purified rabbit small intestinal mucin by *Yersinia enterocolitica*, *Infect. Immun.*, 61, 4131, 1993.

208. **Hutton, D. A., Pearson, J. P., Allen, A., and Foster, S. N. E.,** Mucolysis of the colonic mucus barrier by faecal proteinases: inhibition by interacting polyacrylate, *Clin. Sci.*, 78, 265, 1990.

209. **Ofusu, F., Forstner, J., and Forstner, G.,** Mucin degradation in the intestine, *Biochim. Biophys. Acta*, 543, 476, 1978.

210. **Allen, A. and Starkey, B. J.,** Neuraminidase in pig gastric mucus, *Biochim. Biophys. Acta*, 338, 364, 1974.

211. **Hoskins, L. C.,** Mucin degradation in the human gastrointestinal tract and its significance to enteric microbial ecology, *Eur. J. Gastroenterol. Hepatol.*, 5, 205, 1992.

212. **Hoskins, L. C. and Boulding, E. T.,** Mucin degradation in human colon ecosystems. Evidence for the existence and role of subpopulations producing glycosidases as extracellular enzymes, *J. Clin. Invest.*, 67, 163, 1981.

213. **Hoskins, L. C., Augustines, M., McKee, W. B., Boulding, E. T., Kriaris, M., and Niedermeyer, G.,** Mucin degradation in human colon ecosystems. Isolation and properties of fecal strains that degrade ABH blood group antigens and oligosaccharides from mucin glycoproteins, *J. Clin. Invest.*, 75, 944, 1985.

214. **De Jong, N. H. and Van der Hoeven, J. S.,** The growth of oral bacteria on saliva, *J. Dent. Res.*, 65, 85, 1986.

215. **Van der Hoeven, J. S., Van den Kieboom, C. W. A., and Camp, P. J. M.,** Utilization of mucin by oral *Streptococcus* species, *Antonie van Leeuwenhoek*, 57, 165, 1990.

216. **De Jong, N. H., Van der Hoeven, J. S., Van Os, J. H., and Olijve, H. J.,** Growth of oral *Streptococcus* and *Actinomyces viscous* in human saliva, *Appl. Environ. Microbiol.*, 47, 901, 1984.

217. **Killian, M. and Rolla, G.,** Initial colonization of teeth in monkeys as related to diet, *Infect. Immun.*, 14, 1022, 1976.

218. **Macfarlane, G. T., Hay, S., and Gibson, G. R.,** Influence of mucin glycosidase, protease and arylamidase activities of human gut bacteria grown in a 3-stage continuous culture system, *J. Appl. Bacteriol.*, 66, 407, 1989.

219. **Macfarlane, G. T. and Gibson, G. R.,** Formation of glycoprotein degrading enzymes by *Bacteroides fragilis*, *FEMS Microbiol. Lett.*, 77, 289, 1991.

220. **Vercellotti, J. R., Salyers, A. A., Bullard, W. S., and Wilkins, T. D.,** Breakdown of mucin and plant polysaccharides in the human colon, *Can. J. Biochem.*, 55, 1190, 1977.

221. **Macfarlane, G. T., Hay, S., Macfarlane, S., and Gibson, G. R.,** Effect of different carbohydrates on growth, polysaccharidase and glycosidase production by *Bacteroides ovatus,* in batch and continuous culture, *J. Appl. Bacteriol.*, 68, 179, 1990.

222. **Gibson, G. R., Cummings, J. H., and Macfarlane, G. T.,** Factors affecting uptake by bacteria growing in the human large intestine, in *Microbiology and Biochemistry of Strict Anaerobes Involved in Interspecies Hydrogen Transfer,* Belaich, J. P., Bruschi, M., and Garcia, J. L., Eds., Plenum Press, New York, 1990, 191.

223. **Baron, J. H.,** The clinical application of gastric secretion measurements, *Clin. Gastroenterol.*, 2, 293, 1973.

224. **Walker, V. and Taylor, W. H.,** Pepsin-1 secretion in chronic peptic ulceration., *Gut*, 21, 766, 1980.

225. **Younan, F., Pearson, J., Allen, A., and Venables, C.,** Changes in the structure of the mucus gel on the surface of the stomach in association with peptic ulcer disease, *Gastroenterology*, 82, 827, 1982.

226. **Allen, A., Ward, R., Cunliffe, W. J., Hutton, D. A., Pearson, J. P., and Venables, C. W.,** Changes in adherent mucus gel and pepsinolysis in peptic ulcer patients, *Dig. Dis. Sci.*, 30, 365, 1985.

227. **Quigley, E. M. M. and Turnberg, L. A.,** pH of the microclimate lining human gastric and duodenal mucosa *in vivo,* *Gastroenterology*, 87, 1876, 1987.

228. **Moss, S. and Calam, J.,** *Helicobacter pylori* and peptic ulceration: the present position, *Gut*, 33, 289, 1992.

229. **Rauws, E. A. and Tytgat, G. N. J.,** Cure of duodenal ulceration associated with eradication of *Helicobacter pylori,* *Lancet*, 335, 1233, 1990.

230. **Marshall, B. J., Barrett, J., Prakash, C., McCallum, R. W., and Guerrant, R. L.,** Urea protects *Helicobacter pylori* from the bactericidal effects of acid, *Gastroenterology*, 99, 697, 1990.

231. **Levi, S., Beardshall, K., Haddad, G., Playford, R., Ghsoh, P., and Calam, J.,** *Campylobacter pylori* and duodenal ulcer; the gastrin link, *Lancet*, 1, 1167, 1990.

232. **Slomiany, B. L., Bilski, J., Sarosiek, J., Murty, V. L. N., Dworkin, B., Vanthorn, K., Zielinski, J., and Slomiany, A.,** *Campylobacter pyloridis* degrades mucin and undermines gastric mucosal integrity, *Biochem. Biophys. Res. Commun.*, 144, 307, 1987.

233. **Sarosiek, J., Slomiany, A., and Slomiany, B. L.,** Evidence for weakening of gastric mucus integrity by *Campylobacter pylori,* *Scand. J. Gastroenterol.*, 23, 585, 1988.

234. **Micots, I., Augernon, C., Laboise, C. L., Muzeau, F., and Megraud, F.,** Mucin exocytosis; a major target for *Helicobacter pylori,* *J. Clin. Pathol.*, 46, 241, 1993.

235. **Kelly, S. M., Cummings, J. H., Macfarlane, G. T., Grimes, V., Macfarlane, S., and Gibson, G. R.,** Occurrence of *Helicobacter pylori* in fecal specimens from patients with dyspepsia in the UK, *Gastroenterology*, 104, A116, 1993.

236. **Gibson, G. R., Cummings, J. H., and Kelly, S. M.,** Isolation of *Helicobacter pylorii* from patients in the UK-implications for treatment, *Gastroenterology*, A106, A81, 1994.

237. **Saverymuttu, S., Hodgson, H. J. F., and Chadwick, V. S.,** Controlled trial comparing prednisilone with an elemental diet plus non-aborbable antibiotics in active Crohn's disease, *Gut*, 26, 994, 1985.

238. **Chadwick, V. S.,** Bacterial products as inflammatory modulators in inflammatory bowel disease, in *Inflammatory Bowel Disease,* Anagnostides, A. A., Hodgson, H. J. F., and Kirsner, J. B., Eds., Chapman and Hall, New York, 1991, 46.

239. **Chadwick, V. S., Mellor, D. M., Myers, D. B., Selden, A. C., Keshavarzian, A., Broom, M. F., and Hobson, C. H.,** Production of peptides inducing chemotaxis and lysosomal enzyme release in human neutrophils by intestinal bacteria *in vitro* and *in vivo, Scand. J. Gastroenterol.*, 23, 121, 1988.

240. **Hollander, D.,** Crohn's disease — a permeability disorder of the tight junction?, *Gut*, 29, 1821, 1988.

241. **Rhodes, J. M., Black, R. R., Gallimore, R., and Savage, A.,** Histochemical demonstration of desialation and desulphation of normal and inflammatory bowel disease rectal mucus by faecal extracts, *Gut*, 26, 1312, 1985.

242. **Rhodes, J. M.,** Colonic mucus and glycoproteins: the key to colitis and cancer?, *Gut*, 30, 1660, 1989.

243. **Almer, S., Franzen, L., Olaison, G., Smedh, K., and Strom, M.,** Increased absorption of polyethylene glycol 600 deposited in the colon in active ulcerative colitis, *Gut*, 34, 509, 1993.

244. **Raouf, A., Parker, N., Iddon, D., Ryder, S., Langdon-Brown, B., Milton, J. D., Walker, R., and Rhodes, J. M.,** Ion exchange chromatography of purified colonic mucus glycoproteins in inflammatory bowel disease: absence of a selective subclass defect, *Gut*, 32, 1139, 1991.

245. **Greco, V., Lauro, G., Fabrini, A., and Torsoli, A.,** Histochemistry of the colonic epithelial mucins in normal subjects and in patients with ulcerative colitis. A qualitative and histophotometric investigation, *Gut*, 8, 491, 1967.

246. **Corfield, A. P., Wagner, S. A., Safe, A., Mountford, R. A., Clamp, J. R., Kamerling, J. P., Vliegenthart, J. F. G., and Schauer, R.,** Sialic acids in human gastric aspirates: detection of 9-*O*-lactyl- and 9-*O*-acetylneuraminic acids and a decrease in total sialic acid concentration with age, *Clin. Sci.*, 84, 573, 1993.

247. **Guslandi, M.,** Sialic acid: ambiguous marker of pepsin-degraded mucus, *Dig. Dis. Sci.*, 34, 1477, 1989.

248. **Meyer, F. A., King, M., and Gelman, R. A.,** On role of sialic acid in the rheological properties of mucus, *Biochim. Biophys. Acta*, 392, 223, 1975.

249. **Vatier, J., Poitevin, B. S., and Mignon, M.,** Sialic acid content and proteolytic activity in gastric juice in humans: an approach for appreciating mucus glycoprotein erosion, *Dig. Dis. Sci.*, 33, 144, 1988.

250. **Morita, H., Kettlewell, M. G. W., Jewell, D. P., and Kent, P. W.,** Mucus production and its sulphation in the normal colonic mucosa and in patients with left-sided ulcerative colitis (UC), *Clin. Sci.*, 78, 10P, 1990.

251. **Tsai, H. H. and Rhodes, J. M.,** Purification and characterization of a novel mucin sulphatase from human faeces, *Med. Res. Soc.*, 79, 13P, 1990 (abstract).
252. **Rhodes, J. M., Gallimore, R., Elias, E., and Kennedy, J. F.,** Faecal sulphatase in health and in inflammatory bowel disease, *Gut*, 26, 466, 1985.
253. **Roberton, A. M., McKenzie, C. G., Sharpe, N., and Stubbs, L. B.,** A glycosulphatase that removes sulfate from mucus glycoprotein, *Biochemistry*, 293, 683, 1993.
254. **Corfield, A. P., Wagner, S. A., Clamp, J. R., Kriaris, M. S., and Hoskins, L. C.,** Mucin degradation in the human colon: production of sialidase, sialate, *O*-acetylesterase, *N*-acetyl neuraminate lyase, arylesterase, and glycosulphatase activities by strains of fecal bacteria, *Infect. Immun.*, 60, 3971, 1992.
255. **Tsai, H. H., Sunderland, D., Gibson, G. R., Hart, C. A., and Rhodes, J. M.,** A novel mucin sulphatase from human faeces: its identification, purification and characterization, *Clin. Sci.*, 82, 447, 1992.
256. **Yamagata, T., Saito, H., Habuchi, O., and Suzuki, S.,** Purification and properties of bacterial chondroitinases and chondrosulphatases, *J. Biol. Chem.*, 247, 1523, 1968.
257. **Marcus, R. and Watt, J.,** Seaweeds and ulcerative colitis in laboratory animals, *Lancet*, 30, 1969, 489.
258. **Marcus, S. N., Marcus, A. J., and Watt, J.,** Chronic ulcerative disease of the colon in rabbits fed native carrageenans, *Proc. Nutr. Soc.*, 42, 1551, 1983.
259. **Gibson, G. R.,** A review: physiology and ecology of the sulphate-reducing bacteria, *J. Appl. Bacteriol.*, 69, 769, 1990.
260. **Aslam, M., Batten, J. J., Florin, T. H. J., Sidebotham, R. L., and Baron, J. H.,** Hydrogen sulphide induced damage to the colonic mucosal barrier in the rat, *Gut*, 33 (Suppl. 2), F274, 1991.

# Chapter 9

# Bacterial Infections and Diarrhea

*George T. Macfarlane and Glenn R. Gibson*

## CONTENTS

I. Introduction ....................................................................................................................201
II. *Escherichia coli* ............................................................................................................ 203
    A. EPEC ....................................................................................................................203
    B. ETEC ....................................................................................................................204
    C. EIEC .....................................................................................................................206
    D. EHEC ...................................................................................................................206
    E. EAggEC ............................................................................................................... 206
III. *Salmonella* ....................................................................................................................206
IV. *Shigella* .........................................................................................................................207
V. *Vibrio* .............................................................................................................................209
VI. *Mycobacterium* ............................................................................................................ 209
VII. *Yersinia* ........................................................................................................................210
VIII. *Aeromonas* ..................................................................................................................211
IX. *Plesiomonas* ................................................................................................................ 212
X. *Campylobacter* ..............................................................................................................212
XI. *Bacteroides fragilis* ......................................................................................................213
XII. Clostridia ......................................................................................................................214
    A. *Clostridium septicum* ........................................................................................ 214
    B. *Clostridium perfringens* .....................................................................................214
    C. *Clostridium botulinum* ...................................................................................... 215
    D. *Clostridium difficile* ...........................................................................................215
References ...........................................................................................................................216

## I. INTRODUCTION

Acute enteritis or colitis can result from infection by viruses, fungi, protozoa, helminths, or bacteria. Although virally induced forms of enteric disease are most frequently diagnosed by physicians, colonic infection by bacteria is also a serious clinical problem. A number of established bacterial pathogens are of worldwide significance, while newer developments in laboratory diagnoses and antimicrobial therapy ensure that the list of infectious agents is constantly increasing.[1,2] Figure 1 shows the main colonic infections in humans, a number of which have no identifiable etiological agent but probably can be attributed to viruses. Acute enteric infection does not discriminate between age or economic status, and young and old in developing and industrialized communities may be affected at all times of the year. Enteric infections are of course especially prevalent in communities where standards of personal hygiene, food preparation, and water quality are low. An idea of the human misery that it causes may be gauged from the fact that in Third World countries about five million children die annually from diarrheal disease.

Diarrhea is probably the most common characteristic of enteric disease, but this symptom is not necessarily indicative of infection, since it can also result from ingestion of preformed toxins in contaminated foodstuffs. In general terms, the human colonic microflora may conveniently be divided into three broad categories:[3]

1. Autochthonous microorganisms: these are true inhabitants of the colon that have undergone evolution with the host.
2. Indigenous bacteria: organisms that at some stage have been able to colonize the individual.
3. Contaminants: bacteria which reside in gut contents on a short-term basis and do not permanently establish in the ecosystem.

0-8493-4524-3/95/$0.00+$.50

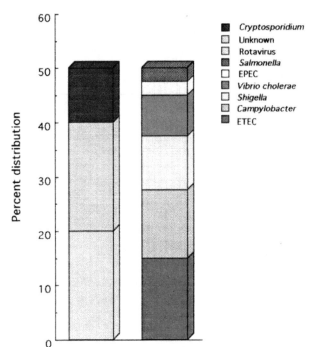

**Figure 1** Predominant colonic infections of humans. Left hand column shows the incidence of viruses or parasites The right hand column gives the major bacterial infections. Data are from the World Health Organization (1990).

The majority of infectious bacteria fall into the third group, although the normal flora may also contain a variety of pathogenic agents, such as certain species of *Bacteroides* and *Clostridium* (see discussion later). Occasionally, homeostasis may be perturbed by invasive microorganisms which are not part of the commensal microbiota. The result is often severe diarrheal disease, which is a manifestation of acute inflammation.

The principal human intestinal bacterial pathogens can generally be characterized according to virulence factors that enable them to overcome the host defense.[4,5] First, invasive microorganisms are able to multiply within enterocytes or colonocytes and ultimately cause cell death.[6] Examples include enteroinvasive *Escherichia coli*, salmonellae, shigellae, yersiniae, aeromonads, and certain vibrios and campylobacters. The second group include enteropathogenic and enterohemorrhagic strains of *E. coli*, as well as some shigellae. These are termed cytotoxic bacteria.[7] These organisms elaborate a variety of extracellular agents that directly cause cell injury. Clearly, many pathogens are ubiquitous, and are both invasive and toxigenic. Toxigenic microorganisms secrete enterotoxins which are polypeptides that adversely influence intestinal salt and water balance. Examples of these microorganisms include *Vibrio cholerae*, some shigellas, and enterotoxigenic *E. coli*.[8] The final group of bacteria that causes acute inflammation are adhesive (e.g., enteroaggregative *E. coli*) microorganisms that bind tightly to the colonic mucosa.[9]

The host, by necessity, has evolved a number of more or less effective defense mechanisms to exclude the establishment of these invading microorganisms. Similarly, potentially pathogenic species that are normally present in the colon also need to be suppressed. In pseudomembranous colitis, antibiotic therapy (primarily clindamycin, ampicillin, and cephalosporins) may allow proliferation of *Clostridium difficile* with subsequent secretion of an enterotoxin and cytotoxin[10] (see discussion later).

Sarker and Gyr[11] concluded that the major nonimmunological host defense mechanisms were gastric acid secretion, intestinal motility, the barrier effect of the normal gut microflora, lysozyme, pancreatic secretions, and bile. The first effective mechanism active against invaders is gastric acidity. The pH of stomach contents ranges from 1 to 3 which is lethal to most bacteria. However, for a few hours after a meal the acidity of gastric contents becomes reduced. Moreover, because food enters the stomach as a bolus, mixed with saliva, a certain degree of buffering occurs, which together with rapid gastric emptying, may facilitate the survival of pathogenic microorganisms. In patients with achlorhydria, or hypochlorhydria, bacteria are able to survive prolonged exposure to gastric secretions.[12,13]

The human small intestine is also a hostile environment for the establishment of bacteria. Rapid peristaltic movements do not permit effective colonization to occur. However, some microbial toxins are able to impair intestinal motility.[14] In some cases, pancreatic juices and bile secretions may also be lethal towards invading microorganisms. Lysozymes, which can be found in salivary glands, intestinal epithelial cells, and pancreatic fluid, cause rapid destruction of some bacteria.[15,16]

Mucins are constantly secreted by the gastrointestinal tract. These glycoproteins form an efficient physical barrier against invasion and colonization.[5] However, some gut species, e.g., bacteroides, bifidobacteria, and ruminococci are able to utilize mucin as a source of carbon and energy, thereby reducing its protective effects (see Chapter 8). As a result, there is a constant, though dynamic, balance between mucin secretion by the host and its utilization by the microflora. The outcome of this competition can have an important bearing on infection of the bowel.

A number of systemic and local immune systems operate in the colon in response to infection.[17-20] Cell-mediated immunity occurs when antigens penetrate the gut mucosa and stimulate Ag-sensitive cells in lymphoid follicles and draining mesenteric lymph nodes. However, secretory antibodies (especially sIgA) are especially important for mucosal defense.[21] The majority of plasma cells in the lamina propria are committed to IgA synthesis, followed by IgM, and to a considerably lesser degree, IgG and IgE. The protective effects of secretory IgA at the mucosal surface include neutralization of toxins and viruses, and inhibition of bacterial adherence. IgA may be passed to breast-fed infants through maternal milk and can survive passage through the gastrointestinal tract.[22] These antibodies have a certain affinity for mucins and consequently line the gut epithelium.[3] Eventually, however, large numbers of microorganisms overwhelm the antibody and, after weaning commences, the infant gut microbiota rapidly begins to resemble that of the adult.[23,24] Secretory IgA is not a totally effective barrier to infection, because a number of potential pathogens are able to produce proteases that specifically degrade these molecules.[25]

The final major defense against infection of the large bowel is afforded by the normal indigenous flora. This resistance to newly introduced bacteria is well known, and is responsible for the long-term stability of the gut microbiota. The barrier effect is classically seen in animal studies, where germ-free animals are more susceptible to orally delivered shigellae and salmonellae as compared to their conventional counterparts. Apart from competition for substrates and sites of colonization, metabolic waste products of the normal flora such as hydrogen sulfide, short chain fatty acids, and other organic acids inhibit the growth of many pathogens, most of which prefer a neutral or slightly alkaline environment for growth. In addition, many bacteria such as lactic acid species are able to exert direct antagonistic effects towards other microorganisms.[26-28]

Often, however, the host defense becomes compromised or overwhelmed with the result that colonic bacterial infection occurs. This situation may arise because of toxins elaborated by the pathogen, environmental factors, or associated underlying illness in the host. An overview of how bacteria cause infections of the intestinal tract is given in Figure 2. In this chapter, we will review some of the predominant bacterial infections of the human colon.

## II. *ESCHERICHIA COLI*

The family Enterobacteriaceae comprises a large group of Gram-negative rods that includes the genera *Escherichia*, *Shigella*, *Salmonella*, *Proteus*, *Klebsiella*, *Enterobacter*, *Serratia*, *Citrobacter*, and *Yersinia*.[29] Many species in this family are potentially or overtly pathogenic to humans. Forms of *E. coli* that cause diarrhea in humans can be differentiated from commensal species normally resident in the colon on the basis of O (somatic) antigens and H (flagella) antigens. The different serotypes of *E. coli* are able to cause characteristic diseases, with 5 variants: enteropathogenic (EPEC), enterotoxigenic (ETEC), enteroinvasive (EIEC), enterohemorrhagic (EHEC) and a more recently described group of enteroaggregative species (EAggEC).[9,30-32] Factors believed to contribute towards the virulent properties of each of these serotypes are summarized in Table 1.

The enormous research effort devoted towards the physiology, biochemistry, ecology, and molecular biology of this assemblage of microorganisms makes it likely that newer groups will also be described in the not too distant future.

## A. EPEC

Approximately 50 years ago, the isolation of distinct subgroups of Gram-negative bacteria from fecal samples of infants with diarrhea led to the serotyping of EPEC, and these organisms were among the first intestinal pathogens to be described.[1] They are most often associated with diarrheal disease in children. EPEC is able to adhere to intestinal enterocytes, thereby causing localized destruction of microvilli. In

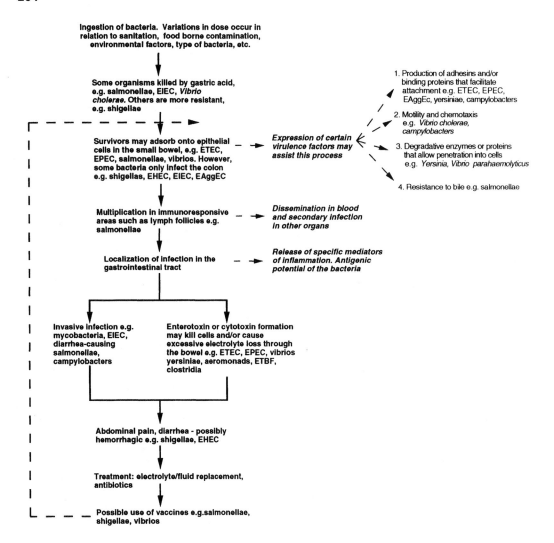

**Figure 2** Summary scheme detailing how bacteria may invade and cause disease in the large intestine.

addition, the organism is thought to produce a cytotoxin, thereby potentiating its pathological qualities.[33] Sertotyping of EPEC in stool specimens is indicative of infection, with the patient exhibiting symptoms of severe diarrhea and often vomiting.[34] Treatment is usually by oral rehydration and/or antibiotic therapy.

Many species of *E. coli* carry plasmids (R factors) that may confer increased resistance to the effects of antibiotics (see Chapter 2). Fortunately, these strains appear to be poor colonizers of the gut under normal circumstances.[35,36] However, during conditions of severe selective pressure (e.g., other forms of gut disease), this situation may change thereby enabling the organisms to proliferate.

## B. ETEC

ETEC mostly colonize the upper small intestine, and are noninvasive forms of *E. coli* that produce one or the most common forms of traveler's diarrhea in humans. Virulence determinants responsible for pathogenicity in these organisms are multifactorial. The bacteria are able to adhere to brush border microvilli in the intestine by the secretion of plasmid encoded adhesins.[37] Indeed, the classic adhesin studies of Smith were carried out with strain ETEC K88 (see Parry and Rooke[38]). Pili or fimbriae are described as filamentous appendages that are involved in the conjugative transfer of DNA.[37] Adhesins of human ETEC are composed of single subunits that polymerize into similar surface appendages. Their receptor specific function allows attachment and thus help evade the washing out effects of gut peristalsis.

**Table 1** Virulence Determinants Associated with *Escherichia coli* that Cause Infections in the Human Intestinal Tract

| Type of *E. coli* | Infection | Virulence factors | | | | | |
|---|---|---|---|---|---|---|---|
| | | Heat labile toxin (HL) | Heat stable toxin (SL) | Shiga-like toxin (SLA) | Fimbrial adhesins | Invasive protein | Binding protein |
| Enteropathogenic | Mainly infectious in children; importance in adults is less clear | – | – | + | – | – | + |
| Enterotoxigenic | Very common cause of traveler's diarrhea that is thought to be transmitted by contaminated water | + | + | – | + | – | + |
| Enteroinvasive | Associated with areas of poor hygiene, usually foodborne | – | – | – | ? | + | ? |
| Enterohemorrhagic | Causes a bloody diarrhea, spread by food and nonpasteurized milk | – | – | + | + | – | – |
| Enteroaggregative | Localized adhesion similar to EPEC, mainly affects children | – | + | – | + | – | – |

?, Uncertain whether virulence factor is produced.

Several antigens have been determined in ETEC that are thought to act as colonization factors,[39] the majority of which are encoded by genes contained in plasmids that are expressed at 37°C.[40-42]

Once established, ETEC elaborate one or both types of toxin. These are termed LT (heat labile) and ST (heat stable) forms. Both induce secretion rather than absorption of fluid in the gut. LT activates the adenylate cyclase-cyclic AMP system (a mechanism similar to the cholera toxin, but the effects are somewhat milder). The accumulation of c-AMP causes fluid secretion by villous cells and consequently severe diarrhea.[43] Toxin ST has a similar effect, but on guanylate cyclase-cGMP.[44,45] ETEC strains have been the subject of a considerable number of molecular studies such that their virulence mechanisms, i.e., invasion and toxin production now provide the standard bacterial descriptions of pathogenicity found in the majority of medical microbiology texts.

## C. EIEC

Enteroinvasive strains of *E. coli* exhibit distinctive pathological differences from ETEC. EIEC are often foodborne microorganisms and, unlike ETEC, where challenge by less than 100 bacteria can cause disease, a large infective dose is required[45] (usually above $10^8$). However, they do exert some degree of selectivity which is thought to be related to the presence of plasmid encoded outer membrane proteins of the organisms. The cure of these plasmids produces nonpathogenic forms of EIEC.[46]

EIEC illness resembles that of shigellae in that invasion of epithelial cells occurs, with proliferation of bacteria in the cytoplasm, production of intracellular microcolonies and spreading to adjacent cells.[30] Subsequent inhibition of protein synthesis by bacterial toxin production kills the host cells and may lead to the formation of ulcerated lesions due to initation of inflammatory events.

## D. EHEC

Hemorrhagic colitis is associated with EHEC (also known as verotoxin producing *E. coli* or VTEC), which is a noninvasive form of *E. coli*. Cytotoxin secreted by EHEC is similar to that of *Shigella*.[47] Similarities between this form of colitis and ulcerative colitis has led some investigators to postulate an etiological role for EHEC in the latter disease. Difficulties arise in interpretation of results, however, because *E. coli* is a common natural inhabitant of the infant and adult colon (see Chapter 1). Sloughing and necrosis of intestinal epithelial cells occurs during EHEC invasion, which predominantly affects the large intestine, and has mostly been attributed to serotype 0157:H7.[29]

## E. EAggEC

Tissue culture assays have led to the description of EAggEC, which evidence suggests is an invader of the colon but not of the small bowel.[48] Cell adhesion experiments have shown that the bacteria form localized aggregative adherence patterns ("stacked brick" appearance).[49] A strong epidemiological association between EAggEC and persistent infantile diarrhea exists in developing countries,[50,51] and a heat stable enterotoxin produced by EAggEC has been described.[51,52]

## III. *SALMONELLA*

According to the Kauffman-White scheme of differentiation, more than 2000 strains belong to the genus *Salmonella*.[53] Pathogenic salmonellae are able to cause gastroenteritis, typhoid fever, and septicemia in humans. Recently, however, these organisms have been reclassified to give (depending on the scheme) either one (*Salmonella enterica*) or three species (*Salmonella typhi*, *Salmonella paratyphi*, and *Salmonella enteriditis*), although many subgroups or serotypes may still be differentiated.[54] Epidemiologically, it is often convenient to use the original classification, which is based on the properties of various lipopolysaccharide and cell-wall antigens, since this allows specific outbreaks of the disease to be monitored. Recent serotypes and their percentage incidence reported from the U.S. are shown in Figure 3.

Until the emergence of campylobacter as an important gut pathogen, salmonellae were believed to be the predominant cause of foodborne diarrhea in industrialized countries.[55] Salmonellae cause diarrhea by adsorbing onto specific receptors in the intestine, and causing localized release of inflammatory mediators, e.g., prostaglandins. The result of this process is stimulation of cAMP production followed by diarrhea. These bacteria do not produce enterotoxins and therefore most resemble a true infection, whereby disease is directly attributable to cell invasion by microorganisms. The gastroenteritis may be mild and self limiting, but other systemic disorders can occur in susceptible patients, such as those with sickle cell anemia or

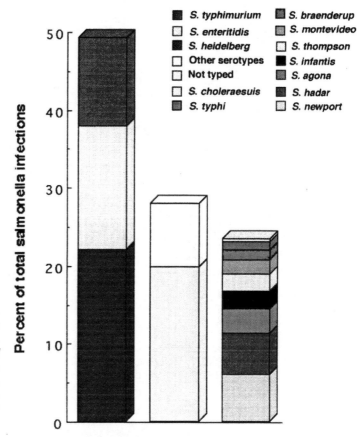

**Figure 3** Serotypes and percent incidence of *Salmonella* infections in the U.S. in 1989. From Farmer, J. J. and Kelly, M. T., *Manual of Clinical Microbiology*, 5th ed., Balows, A., Hausler, W. J., Herrmann, K. L., Isenberg, H. D., and Shadomy, H. J., Eds., ASM Publications, Washington, D.C., 1991, 360.

individuals who are immunocompromised. These complications may lead to septicemia, and perhaps meningitis, pneumonia, or osteomyelitis.[56,57] Virulence factors associated with salmonellae include synthesis of proteins that facilitate binding of the bacteria to intestinal mucus.[58-62] A number of molecular studies have shown that mutations in specific genetic loci reduce the capacity to penetrate intestinal cells.[63-65] Various serotypes carry plasmids that confer increased virulence in mice.[66-68]

Typhoid fever is a severe epidemic disease in underdeveloped countries, but may still occasionally occur in the developed world. *Salmonella typhi* is the most frequently implicated causative bacterium, and is exclusively a human pathogen. Paratyphoid fever is a milder form of typhoid caused by *S. paratyphi*. Typhoid causing salmonellae penetrate the gut mucosa through Peyer's patches in the small intestine and reach intestinal lymph nodes. Once here, they are able to survive and multiply within macrophages. The bacteria are then transported to mesenteric lymph nodes, and enter the bloodstream, from there they circulate in the blood, and possibly invade a number of organs, particularly those where reticuloendothelial cells are important, e.g., spleen, bone marrow, and liver. In the liver, multiplication in Kupffer cells occurs, with reinvasion of the blood and possibly the gall bladder[69,70] (the bacteria are relatively resistant to the effects of bile). Consequently, salmonellae reenter the intestine in higher numbers, often resulting in ulceration.

Treatment of typhoid fever is usually by administrating of antibiotics, e.g., ampicillin and chloramphenicol, while recent molecular studies to develop an oral vaccine have met with some success.[71-73]

## IV. *SHIGELLA*

Shigellosis or bacillary dysentery is endemic in developed and third world countries, where it contributes to about 10% of deaths from resulting from diarrheal-associated illness. This condition is typically characterized by production of low volume bloody diarrhea containing mucus, together with severe cramping and abdominal pain.[74,75] The etiological agents of shigellosis belong to the genus *Shigella*,

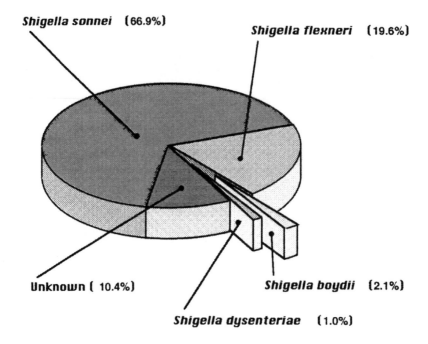

**Figure 4** Serotypes and incidence of *Shigella* infections in the U.S. in 1989. From Farmer, J. J. and Kelly, M. T., *Manual of Clinical Microbiology*, 5th ed., Balows, A., Hausler, W. J., Herrmann, K. L., Isenberg, H. D., and Shadomy, H. J., Eds., ASM Publications, Washington, D.C., 1991, 360.

members of which are Gram-negative, nonmotile facultative anaerobes. The major species involved in dysentery are *Shigella sonnei, Shigella flexneri, Shigella boydii*, and *Shigella dysenteriae*. The comparative incidence of these organisms is shown in Figure 4.

Shigellae are facultative intracellular parasites that exhibit strict host and tissue specificities for human colonic epithelial cells. Shigellosis characteristically results in death of the mucosal cells and their subsequent sloughing into the bowel lumin.[44] Although dissemination of the bacteria in tissues is usually self limiting in that the submucosa is not normally infected, the infection is characterized by general inflammation and ulceration of the colon. A number of complications can arise because of this invasive type of infection. These complications include rectal prolapse, distension of the colon (e.g., toxic megacolon), or other systemic effects including encephalopathy, leukamoid reactions, and hemolytic uremic syndrome.[76,77] In conjunction with severe infantile malnutrition, complications such as kwashiorkor, a protein deficiency syndrome, may also occur.[74]

The three main factors responsible for bacterial virulence in shigellae are:

1. The cell wall, which contains a highly antigenic lipopolysaccharide
2. Invasion of epithelial cells
3. Toxin production following invasion

*In vitro* studies have indicated that invasion into gut epithelial cells is a multi-stage process involving induced phagocytosis, escape from the phagocytic vacuole, multiplication and spread within the epithelial cell cytoplasm, passage into adjacent cell, and subsequent destruction of host tissues.[78-82]

As indicated earlier, certain strains of *E. coli* are able to produce a toxin similar to that of *S. dysenteriae*, which causes the most severe form of bacterial dysentery. The protein structure of the Shiga toxin consists of a single A subunit with multiple B subunits.[83,84] In *E. coli*, these biologically identical entities are termed Shiga-like toxins (Table 1).[85,86] The significance of this observation is that studies between shigellae and *E. coli* have been important in allowing the pathogenesis of infection to be more fully characterized (largely by genetic methods).[87-98] Evidence now indicates that virulence in *Shigella* species depends on the presence of a number of plasmids.[99-101]

A unique feature of shigellosis is that epidemiological studies show that a very low number of infective bacteria is required for symptoms to occur. For example, as few as ten infective organisms have been shown to cause dysentery in a volunteer group of healthy adults.[102] In comparison, the infective salmonella dose is between $10^5$ and $10^{10}$ bacteria,[103,104] while that of *Vibrio cholerae* is at least $10^8$.[105] As person-to-person transmission is the major route of *Shigella* infection,[98,106] investigators have looked at the relative acid resistance of these bacteria. *In vitro* data indicate that in comparison to other enteric infectious agents, *Shigella* spp. are comparatively acid tolerant, although this depends on the physiological condition of the bacteria at certain stages of the cell cycle, e.g., resistance was highest during the late exponential phase of growth.[107]

The range of symptomatic responses to *Shigella* infection is very variable and can be mild to severe, depending on the type and virulence of the pathogenic species involved. More importantly, widespread resistance to antibiotics is increasing, which makes treatment difficult and expensive.[108] Consequently, the development of a vaccine, using specific cell-wall polysaccharides, is being investigated.[109,110]

## V. *VIBRIO*

The most frequently encountered species of *Vibrio* pathogenic to humans are those that cause diarrhea, of which five strains are thought to be involved[111] (*Vibrio cholerae*,[112] *Vibrio parahaemolyticus*,[113] *Vibrio fluvialis*,[114] *Vibrio hollisae*,[115] and *Vibrio mimicus*[114,116]). Other species such as *Vibrio vulnificus* and *Vibrio alginolyticus* may be involved in wound infections.[5,117,118] By far the most heavily documented cases are those involving *V. cholerae* and *V. parahaemolyticus*.

Although *V. cholerae* is primarily a small intestinal pathogen, many of its symptomatic effects manifest in the colon. Cholera is still regarded as an endemic disease in parts of South America and Africa, as well as in some areas of the United States.[119] The bacterium is an inhabitant of fresh water that is spread by drinking water or contaminated food (mainly shellfish). A relatively large infective dose is required, since the organisms are sensitive to the effects of gastric acid. Consequently, hypochlorhydric patients are at greatest risk of infection.[120] The symptoms of cholera result from the production of a powerful enterotoxin secreted by the bacteria after attachment to the small intestinal mucosa. Vibrios with enhanced motility and chemotactic properties have a clear selective advantage.[119] Cholera toxin (CT) causes massive secretion of salt and water in the bowel, which cannot be attributed to the invasive properties of the bacterium.[121] The toxin has been well characterized[122] and consists of five B subunits that surround a single A subunit.[123] B subunits enable CT to bind to cell plasma membranes and then subunit A moves towards the cytoplasm of the cell.[124] Subsequent activation of adenylate cyclase occurs (by the activity of GTPase after ribosylation), causing cAMP dependent chloride ion excretion which is responsible for secretory diarrhea.[121] The most effective treatment is to rapidly administer replacement fluids with most of the bacteria being sensitive to the effects of tetracycline.[125] A killed whole cell vaccine, or one lacking toxin genes, is protective.[126]

The halophilic bacterium *V. parahaemolyticus* is an important gut pathogen that is mainly transmitted in contaminated fish or shellfish. Although *V. parahaemolyticus* infections have a worldwide distribution, they are most common in people living in Japan due to their predilection for consuming raw seafood. Clinical symptoms of infection include diarrhea (which is milder than cholera), nausea, and vomiting.[117,127,128] The organism is invasive, primarily infects the colon, and may produce a form of dysentery. A number of virulence factors have been associated with *V. parahaemolyticus*. Some of these factors may promote hemolysis and include phospholipase A,[129] a lyphophospholipase,[129] a heat resistant thermostable membrane damaging protein (TDH),[130-132] a related cytolytic heat labile hemolysin (TRH),[133] and a cytolysin that is genetically unrelated to either TDH or TRH.[134] TDH seems to play a major role in enterotoxicity of the organism, although the precise mechanism whereby membrane lysis occurs remains unclear.[135] The illness is usually self limiting but in a small number of cases may be fatal. Oral treatment to induce rehydration may be prescribed.[5]

## VI. *MYCOBACTERIUM*

Mycobacteria are readily recognizable because of their acid fastness and slow growth.[136] Mycobacterial species that cause disease in humans include *Mycobacterium tuberculosis* and *Mycobacterium leprae*, the respective etiological agents of tuberculosis and leprosy.[136-138] With regard to this chapter, *M. tuberculosis* only will be considered further, because this disease may affect the large intestine.

Tuberculosis is a chronic, contagious, severe (often fatal) disease that mainly occurs as granulomatous foci in the lung. Briefly, the ingestion of tubercle bacilli (*M. tuberculosis*) in droplets or aerosols occurs.

The bacteria are then ingested by macrophages of the pulmonary alveoli, where they multiply. The next stage in pathogenesis is migration of viable mycobacteria through the lymphatic system where a largely T cell-mediated immune response occurs. An inflammatory response throughout the sites of infection, including lymph nodes, is characteristic.[136] Some tubercles may disseminate through the body into other organs, including the colon. This secondary infection is known as gastrointestinal tuberculosis,[139] but it may also arise from swallowing mycobacteria in pulmonary secretions.[5]

Large intestinal tuberculosis primarily affects the cecum and ascending colon. Fever, abdominal pain, and weight loss are symptomatic, with a characteristic form of mucosal ulceration and chronic inflammation.[140] Because of certain similarities between this disorder and Crohn's disease (CD), mycobacteria have in the past been postulated as being etiologically important in CD. However, this hypothesis lacks definitive confirmation (see Chapter 10).

Current research on *M. tuberculosis* is directed towards the mechanism whereby the organism is able to survive and grow within macrophages (see References 141 to 143 for potential mechanisms whereby this may occur), as well as the development of a vaccine. It is possible that heat shock proteins which are elaborated by mycobacteria form the basis of an effective subunit vaccine.[144,145]

Mycobacterial infections have recently become a center of attention due to the observation that they predominate in patients suffering from AIDS.[146,147] The most frequently implicated organism is *Mycobacteria avium-intracellulare* (MAI). Interestingly, this type of opportunistic infection is detected in regional lymph nodes and the gastrointestinal tract.[148] Similarities between MAI intrusion in the bowel and Whipple's disease have also been made.[149]

## VII. *YERSINIA*

Yersiniae are small Gram-negative, occasionally pleomorphic, coccobacilli. These facultative anaerobes are fermentative and nutritionally undemanding. Three species are pathogenic to humans. *Yersinia pestis* is the etiological agent of bubonic plague, whereas both *Yersinia enterocolitica* and *Yersinia pseudotuberculosis* cause disease in the large bowel.[1,5,150,151] Carriage of yersiniae is widespread in wild animals and farm animals, and *Y. enterocolitica* is a common infectious agent in children, with contaminated foods, especially milk, being responsible for its dissemination.[152,153] Fifty-seven O antigens have been reported in *Y. enterocolitica* and 50 different serotypes, with bio-serotype 1/0:8 being most common in the U.S.[154] The bacteria are invasive and enterotoxigenic; species that cause enteric disease are intracellular parasites that are able to invade cells in the proximal colon, terminal ileum, Peyer's patches, macrophages and the appendix, although they may also translocate into the bloodstream causing sepsis.[151,155-157] The expression of virulence factors is strain dependent. For example, serogroups 03, 05, 08, 09, and 27 are enterotoxigenic. Chromosomal genes are involved in enterotoxin formation, although the invasion determinants are plasmid associated.[158]

Acute enteritis is uncommon in *Y. pseudotuberculosis* infections, but is more frequently found in patients infected with the more virulent *Y. enterocolitica*. Diarrhea, which is relatively long lasting and can resemble shigellosis, is the most common symptom of the infection, with leukocytes occurring in stools. This results from both the invasive and enterotoxin producing capabilities of the bacteria.[155] Other symptoms include fever, nausea, vomiting and abdominal pain, and complications such as arthritis, nephritis, and erythema can occur in adults.[159] After culture of bacteria from stool specimens, treatment may be carried out using specific antibodies.[160]

Pathogenic strains of *Yersinia* are characterized by the production of a virulence plasmid, pYV.[161,162] This 70 kb genetic element encodes a number of proteins that are responsible for pathogenicity. The proteins comprise a multifactorial mechanism active against host defense mechanisms, and include *Yersinia* outer membrane proteins (Yops). Two independent regulatory mechanisms control Yop gene expression and are subject to environmental stimuli (i.e., thermoregulation, such that Yop secretion only occurs at 37°C, and a low concentration of calcium ions).[163,164]

The function of a number of Yops have been determined. These functions and the biological activities of other virulence proteins secreted by *Yersinia* spp. are shown in Table 2. In summary, these proteins enable the bacteria to promote agglutination and hemagglutination, enhance adherence to cell membranes, collagen, and fibronectin, and confer increased resistance to phagocytosis and the bactericidal effect of serum.[165-169]

**Table 2** Pathological Characteristics of Proteins Secreted by
Gastrointestinal Species of *Yersinia*[a]

| Type of protein | Postulated function |
| --- | --- |
| Yad A | Forms multimeric fimbrillae on the bacterial surface that are associated with resistance to lysis by complement and phagocytosis, as well as assisting adherence to mammalian cells. |
| Invasin | Promotes bacterial attachment and entry into eukaryotic cells by binding to β1 integrins. |
| pH 6 antigen | Responsible for thermo-induced binding of the bacteria to epithelial cells. |
| V antigen | Putative antihost function that may be neutralized by antibody. |
| Yop E | Cytotoxin which becomes active after bacterial attachment to the host cell. |
| Yop D | Necessary for the entry of Yop D into the target cell. |
| Yop B | Hydrophobic protein that has two transmembrane regions. |
| Yop H | A phosphotyrosine phosphatase that inhibits phagocytosis. It may be an effective subversive mechanism towards the host immune response. |
| Yop M | Binds human thrombin and so inhibits aggregation of platelets and the onset of inflammation. |
| Yop O (YpkA) | Serine-threonine protein kinase inhibitor, essential for bacterial virulence. |
| Yop J | Not essential for virulence. Thought to have a role in the initial stages of invasion. |
| YopN/LcrE | Required for negative regulation of other Yops. |

*Note:* Yop, *Yersinia* outer membrane protein.

[a] Table is taken from References 163 to 167.

## VIII. *AEROMONAS*

Bacteria belonging to the genus *Aeromonas* are small, motile, Gram-negative, oxidase positive rods which are physiologically and ecologically similar to vibrios. Aeromonads may on occasion be pathogenic and are responsible for a variety of infections in cold- and warm-blooded animals, including humans. They are widely distributed in marine and fresh water aquatic environments, although the major reservoir of infection in humans appears to be sewage-contaminated water.

While there are many reports of aeromonads causing enteric disease in humans, this has been questioned by some authors.[170] One reason for this is that the bacteria can be isolated from the stools of about 3% of apparently healthy individuals.[171] However, studies show that *Aeromonas* isolates from feces of patients with diarrhea are phenotypically different from those found in healthy people, in that they are markedly more virulent.[172] Thus, only certain strains appear to be pathogenic. The weight of evidence tends to show that aeromonads are etiologic agents of human enteric disease, and some years ago in the U.S., these bacteria were isolated more frequently from diarrheal feces than either shigellae, salmonellae or campylobacters.[173]

The principal human pathogens are *Aeromonas hydrophila*, and to a lesser extent *Aeromonas sobria*. *Aeromonas hydrophila* is enterotoxic,[174] cytotoxic,[175] adherent,[176] and invasive,[177] although toxins are considered to be the principal virulence determinants. *Aeromonas hydrophila* produces two heat-labile hemolysins, one of which cross reacts with the heat-labile enterotoxins of *V. cholerae* and *E. coli*.[174,178] Compared to these other toxins, *A. hydrophila* cytonic enterotoxin induces more rapid accretion of fluid in the rabbit ileal loop test.[179] Two hemolysins (α and β) are produced by *A. hydrophila*, which are cytotoxic to HeLa cells and human fibroblasts,[180] and they give rise to hemorrhagic enteritis in the rat ileal loop test.[181] Beta hemolysin is apparently similar in a number of respects to the cardiotoxic thermostable hemolysin formed by *V. parahaemolyticus*, although the molecules are not related immunologically.[179] As well as colonizing the bowel, illness in humans may also be caused by exotoxin which is produced in some refrigerated foods.[182]

Invasion studies show that *A. hydrophila* can replicate in human HEp-2 cell monolayers.[177] This work has suggested that invasion is a chromosomally determined property, which is not dependent on the carriage and expression of virulence plasmids as is the case in *Salmonella typhimurium*,[183] *E. coli*,[184] and Shigellae.[100,184,185]

The organism infects the small and large bowels causing acute diarrhea in children and adults, who may exhibit cholera-like symptoms. The diarrhea is usually mild and watery, with an absence of blood or mucus in tropical or third world countries. However, in developed communities such as the U.S., a severe dysentery-like diarrhea is seen, with accompanying abdominal pain and vomiting.[186,187] *Aeromonas hydrophila* infection of large intestine has also been reported to progress to a more severe disease resulting in segmental colitis involving the cecum and ascending colon.[181]

## IX. *PLESIOMONAS*

The only species in the genus *Plesiomonas* is *Plesiomonas shigelloides*. The organism is a motile Gram-negative facultative anaerobe which is phenotypically similar to aeromonads and vibrios with which it usually associated. It is seldom found in feces of healthy individuals.[188] *Plesiomonas shigelloides* infections are normally waterborne, and localized infections normally occur in the body following trauma, or they may spread from the gut of patients with underlying disease.[189] Although like *Aeromonas*, there are conflicting reports concerning the ability of *P. shigelloides* to cause diarrheal disease, there is increasing evidence that some strains are intestinal pathogens in humans, with occasional fatalities being cited.[190]

*Plesiomonas shigelloides* is seldom an etiological agent of disease in western communities. It was first recognized as a causative organism in acute gastroenteritis in Japan.[191] Symptoms of *P. shigelloides* infection in the large bowel usually include production of watery diarrhea. Treatment is by electrolyte and fluid replacement therapy, while trimethoprim-sulfa and tetracycline appear to be effective in eliminating the organism from the bowel.[192,193]

While few data are available concerning the mechanisms whereby *P. shigelloides* causes disease in the large intestine, a virulence plasmid has been identified.[194] It is also known that some strains are able to invade HeLa cells,[195] and there are reports of the bacterium producing heat-labile and heat-stable enterotoxins.[196,197] Like aeromonads, some, but not all, strains of *P. shigelloides* appear to be invasive and toxigenic.

## X. *CAMPYLOBACTER*

Campylobacters are Gram-negative, oxidase positive, motile, microaerophilic-curved rods or helical bacteria. They are nutritionally demanding organisms whose normal habitats are the intestinal tracts of domestic and wild animals. Over the last two decades, these organisms have increasingly been recognized as major causes of diarrheal disease in humans.[198,199] Consequently, they have been the subject of a number of reviews covering the epidemiology and pathophysiology of their infections.[200-202] In parts of the U.K., reported campylobacter infections have long exceeded those attributed to salmonellae and *Shigella sonnei* (Figure 5), and a similar situation exists in the U.S.[203] Campylobacter enteritis may be a consequence of infection by one of several species (Table 3), although *Campylobacter jejuni* and to a lesser degree *Campylobacter coli* are the most important.[204,205] With the exception of *Campylobacter fennelliae* and *Campylobacter cinaedi*, which seem to be transmitted by venereal contact,[206,207] inter-individual campylobacter infection is probably very rare. The main reservoirs of infection are contaminated water, and variety of foods, particularly raw milk and raw or undercooked poultry or red meat.[208,211]

Campylobacters infect the small intestine and the large bowel where they cause an enterocolitis. The infectious dose of campylobacters is extremely variable,[212] although human volunteer studies show that 500 bacteria are sufficient to cause diarrhea.[213,214] The symptoms of campylobacter infection are particularly severe and debilitating, with fever, headache, abdominal pain, and, in many cases, bloody diarrheal feces containing mucus and leukocytes. However, the disease is usually self limiting and death is unusual.[215,216] Nevertheless, complications may occur as sequelae to campylobacter infections including bacteremia,[217] appendicitis,[218] pseudomembranous colitis,[219] toxic megacolon,[220] and cholecystitis.[221] Many clinicians and researchers now hold the view that the majority of camplylobacter infections go unreported, such that the number of recorded cases grossly underestimates the incidence of the infection.[222]

Large numbers of campylobacters are excreted in the stools of infected persons and their culture and identification is therefore relatively straightforward.[223,224] However, recovery of the organisms from water or contaminated foods often poses problems, particularly when they are present in low numbers, or are viable but nonculturable.[225,226] It has been suggested that the polymerase chain reaction (PCR) may be useful for detection and typing of the bacteria under these circumstances.[226]

Motility is believed to be an important factor that enables campylobacters to penetrate the mucus linings of the bowel, and subsequently colonize the intestinal epithelium.[227-229] The bacteria also manifest

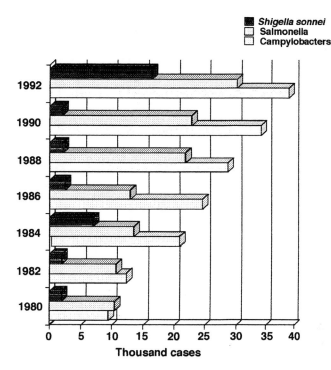

Shigella sonnei
Salmonella
Campylobacters

**Figure 5** Laboratory reports of bacterial infections of the intestinal tract in England and Wales between 1982 and 1992. From *Interim Report on Campylobacter*, HMSO, London, 1993.

**Table 3** Pathogenic or Potentially Pathogenic Campylobacters in Humans

*C. jejuni*
*C. coli*
*C. fetus* ssp. *fetus*
*C. hyointestinalis*
*C. upsaliensis*
*C. laridis*
*C. fennelliae*
*C. cinaedi*

a number of other putative pathogenicity features: for example, *C. jejuni* is enteroinvasive, and is able to survive within epithelial cells,[230] as well as being enterotoxigenic. The organism produces a variety of invasins,[231,232] enterotoxins,[233,234] and cytotoxins.[235,236] Flagellae are believed to be involved in adhesion and invasion.[237] Campylobacter heat-labile enterotoxin genes may be located on plasmids in some strains, but there is evidence in other isolates for chromosomal involvement.[238] The characteristics of this toxin are similar to heat-labile cholera and *E. coli* enterotoxins.[233,234]

Although the relative importance of these mechanisms in causing disease is uncertain due to the absence of a suitable animal model for studying human campylobacteriosis, a recent report indicates that the newborn piglet may be a suitable for this purpose.[239] It has been suggested that the invasive properties of campylobacters are the most important expressions of virulence.[239] There is a degree of experimental support for this contention, because some studies have shown that some campylobacter isolates from patients with enteritis do not produce enterotoxins or cytotoxins.[240]

## XI. *BACTEROIDES FRAGILIS*

*Bacteroides fragilis* and related species are strictly anaerobic, nonmotile Gram-negative rods that occur in high numbers as normal commensals in the healthy large intestine[241] (see Chapter 1). They are also the most clinically important anaerobic pathogens in humans, where the majority of infections that occur in diffuse sites of the body usually originate from the large bowel.[242,243]

**Table 4** Toxins and Possible Spreading Factors Formed by *Clostridium septicum*

| Toxin or spreading factor | Activities |
|---|---|
| α-Toxin | Lethal and hemolytic |
| β-Toxin | DNase |
| γ-Toxin | Hyaluronidase |
| δ-Toxin | Thiol-activated cytolysin and hemolysin |
| H-Toxin | Hemolysin inhibited by long chain saturated fatty acids |
| Gelatinase | Breakdown of tissue proteins |
| Neuraminidase | Mucin destruction in large intestine ? |
| N-Acetyl glucosaminidase | Mucin destruction ? |
| β-Galactosidase | Mucin destruction ? |

However, these bacteria may also be overtly pathogenic in the large intestine, with some *B. fragilis* strains recently emerging as etiological agents of diarrhea in humans and other animals.[244-246] The organisms do not seem to be adherent or invasive, and disease appears to occur as a result of enterotoxin formation.[244] These enterotoxigenic *B. fragilis* (ETBF) can be isolated from the feces of children and adults with watery diarrhea, and studies indicate that the enterotoxin may possibly be a heat-labile low molecular mass protein of about 19.5 kDa.[247] The enterotoxin reportedly induces rapid structural changes in human colonic epithelial cells with crypt hyperplasia and sloughing of the epilthelium.[247]

As yet, few studies have been made on ETBF infections in humans. Consequently, little is known of the incidence and medical significance of diarrheal disease caused by these organisms.

## XII. CLOSTRIDIA

### A. *CLOSTRIDIUM SEPTICUM*

This bacterium was an important cause of gas gangrene in the First World War,[248] but since then the nature of human infection has changed markedly. Although *Clostridium septicum* is not a member of the normal gut flora, the majority of *C. septicum* associated diseases now seem to be opportunistic atraumatic infections, originating from the large bowel.[249-251]

*Clostridium septicum* is the primary etiological agent in neutropenic enterocolitis as well as a number of necrotizing infections of humans and other animals.[252,253] The disease typically manifests as a rapidly spreading hemorrhagic infection of the ileocecal region of the large bowel, which mainly affects debilitated patients suffering from acute leukemia and other malignancies, coronary heart disease, enterocolitis, and diabetes.[254-259] Pronounced edema and wall thickening of the gut are also characteristic anatomic characteristics of the infection. Although *C. septicum* infections are not especially common, they are dramatic and have unusually high morbidity and mortality rates, with the patients chances of survival estimated to range from between 21 and 35%.[258-260] Part of the explanantion for this is that in the early stages of the infection, the symptoms are similar to those of appendicitis, and unless surgical and antibiotic treatment is prompt, death can occur within hours.[256]

Tissue invasion by *C. septicum* is probably facilitated by secretion of a variety toxins and spreading factors (Table 4), although with the exception of α-toxin, their role in pathogenicity is unclear. α-Toxin is lethal, necrotizing, and hemolytic.[261] This protein does not exhibit phospholipase C activity and is unrelated to the α-toxin of *C. perfringens*.[262] Two other hemolysins (δ-toxin and H-toxin) are also produced,[263,264] but virtually nothing is known of the nutritional, physiological, or genetic factors that affect the expression of virulence determinants in *C. septicum*.

### B. *CLOSTRIDIUM PERFRINGENS*

*Clostridium perfringens* is a member of the normal colonic microflora in humans and animals. This Gram-positive, nonmotile, anaerobic rod is the major clostridial pathogen in humans and animals.[248] Sixteen putative virulence determinants have been identified, of which 12 are toxins.[265] Four major lethal toxins have been characterized (alpha, beta, epsilon, iota),[266] and their formation by individual strains is used as a basis for differentiating *C. perfringens* into five types (A–E) as shown in Table 5. Type A is the predominant human pathogen, causing gas gangrene and necrotic enteritis;[248,249] type B is responsible

**Table 5** Toxin Formation and Typing
of *Clostridium perfringens*

| | Type | | | | |
|---|---|---|---|---|---|
| **Toxin** | **A** | **B** | **C** | **D** | **E** |
| Alpha | + | + | + | + | + |
| Beta | − | + | + | − | − |
| Epsilon | − | + | − | + | − |
| Iota | − | − | − | − | + |

for enteritis in sheep, as well as lamb dysentery; type C is the etiological agent of enteritis necroticans in humans and enteritis in animals;[267] type D produces enterotoxemia in animals; and type E causes enteritis in sheep and cattle.

With respect to diseases of the intestinal tract in humans, type A *C. perfringens* are enterotoxigenic and are important food poisoning agents.[268] They are a major cause of food-related illness in both the U.K.[269] and the U.S.[270] This results from the ingestion of food, particularly meats and gravies, that are heavily contaminated (>$10^6$ bacteria per g) with the organism. In the small bowel, the vegetative cells sporulate and upon lysis, release an enterotoxin, which has pore forming activities in epithelial cells, and is cytotoxic.[265] The enterotoxin is one of the most extensively studied clostridial toxins, and is not a component of the spore coat, but appears to be an intracellular product.[265,271] Evidence suggests that modification of the enterotoxin by pancreatic proteases potentiates its effects on the brush border.[271,272] The symptoms of clostridial intoxication appear within 8 to 24 h and manifest as short-term diarrhea with abdominal pain and nausea.[268,269]

*Clostridium perfringens* also causes a nonfood poisoning diarrhea which seems to occur mainly in elderly people and may be antibiotic associated.[273-275] This disease is more severe than that resulting from typical *C. perfringens* food poisoning.[276]

The other major intestinal disease caused by *C. perfringens* is enteritis necroticans, which is also known as Pig-Bel or Darmbrand.[267,268] This disease is a necrotizing hemorrhagic jejunitis, although the ileum is sometimes involved. The etiologic agents are type C strains.[277] Beta toxin, which is lethal and necrotizing, is responsible for destruction of villi in the small bowel.[267,277] It is a 28 kDa protein that may be associated with a plasmid.[278,279] Symptoms of enteritis necroticans include vomiting, severe cramps, nausea, shock, small bowel obstruction, dehydration, and bloody diarrhea, possibly followed by constipation.[267,268] The small intestine becomes inflamed with edematous and necrotic patches that turn gangrenous.

## C. *CLOSTRIDIUM BOTULINUM*

Botulism is an often fatal condition which may result from ingestion of preformed botulinum toxin (types A, B, E, and F) in foods,[280] or very rarely, by contamination of wounds with *Clostridium botulinum*.[281] The third, and now most common form of the disease, is infant botulism,[282,283] so-called because children under five months of age are almost exclusively affected.[284] Infant botulism results from the germination of *C. botulinum* spores in the large intestine.[284-286] In some countries, honey has been implicated as a reservoir of infection.[284,287] Colonization of the bowel by *C. botulinum* seems to be influenced by the species composition of the infant gut, with a more suppressive microflora being found in breast-fed children.[282,288] The bacteria variously elaborate neurotoxins (types A, B, and F) in the colon, which can be detected in feces.[284,289-291] Because the organism is not a member of the normal gut flora, its isolation from stools is indicative of infection.[291]

Symptoms of infant botulism include muscular paralysis, lack of appetite, lethargy and general weakness. The disease is not diarrheal, and often times, the patients may be constipated.[284] The majority of cases require hospitalization,[292] where treatment is by general supportive therapy, with attention being paid to the child's respiratory needs. Antitoxin therapy is not used because of undesirable side effects.[284]

## D. *CLOSTRIDIUM DIFFICILE*

This anaerobic bacterium is the principal etiologic agent of pseudomembranous or antibiotic-associated colitis. Under normal circumstances, the gut microflora provides an effective barrier to the ingress of *Clostridium difficile*, however, the infection may become established in patients treated with certain antibiotics, particularly ampicillin, clindamycin, and cephalosporins.[293] Bacteroides, bifidobacteria, and lactobacilli are thought to be anatogonistic towards *C. difficile.*[294]

The disease is relatively uncommon and is associated with formation of ulcer-like lesions in the large bowel, leading to characteristic pseudomembranes, which consist of fibrin, mucin, leukocytes, and other debris.[295,296]

Two toxins are elaborated by *C. difficile* (designated A and B).[297] They are formed simultaneously and have estimated moleculer weights of 300 kDa.[298] Toxin B is cytotoxic, however toxin A causes fluid accumulation in the rabbit ileal loop test and is strongly enterotoxigenic.[299,300] Toxin B is not active in the intestine on its own which leads some investigators to postulate that toxin A is primarily responsible for the clinical manifestations of *C. difficile* infections.[301] Although almost nothing is known about the structure of toxin B, toxin A has been well characterized as a multivalent protein with repeating amino acid sequences at the C-terminus.[302] Receptors for toxin A are thought to exist on the mucosal surface, where binding to epithelial cell membranes occurs. Evidence shows that this results in extensive tissue damage, with the secretion of a hemorrhagic fluid into the bowel. An inflammatory response contributes towards the severity of the disease.[303] *Clostridium difficile* toxin may be assayed in stools on the basis of cytotoxicity in tissue culture.[304,305]

# REFERENCES

1. **Sawyer, M. K. and Gehlbach, S. H.,** Bacterial diseases of the colon, *Prim. Care,* 15, 125, 1988.
2. **Griffiths, J. K.,** Colonic infections, *Curr. Opin. Gastroenterol.,* 9, 83, 1993.
3. **Drasar, B. S. and Roberts, A. K.,** Control of the large bowel microflora, in *Human Microbial Ecology,* Hill, M. J. and Marsh, P. D., Eds., CRC Press, Boca Raton, FL, 1989, 87.
4. **Cohen, M. B.,** Etiology and mechanisms of acute infectious diarrhea in infants in the United States, *J. Pediatr.,* 118, 534, 1991.
5. **Cohen, M. B. and Giannella, R. A.,** Bacterial infections: pathophysiology, clinical features and treatment, in *The Large Intestine: Physiology, Pathophysiology and Disease,* Phillips, S. F., Pemberton, J. H., and Shorter, R. G., Eds., Raven Press, New York, 1991, 395.
6. **Giannella, R. A., Broitman, S. A., and Zamcheck, N.,** The influence of gastric acidity on bacterial and parasitic infection, *Ann. Intern. Med.,* 78, 271, 1973.
7. **O'Brien, A. D. and Holmes, R. A.,** Shiga and shiga-like toxins, *Microbiol. Rev.,* 51, 775, 1987.
8. **Carpenter, C. C.,** Cholera and other enterotoxin-related diarrheal disease, *Rev. J. Infect. Dis.,* 126, 551, 1972.
9. **Levine, M. M.,** *Escherichia coli* cause diarrhea: enterotoxigenic, enteropathogenic, enteroinvasive, enterohemorrhagic and enteroadherent, *J. Infect. Dis.,* 155, 377, 1987.
10. **Bartlett, J. G.,** Antibiotic-associated pseudomembranous colitis, *Rev. Infect. Dis.,* 1, 530, 1979.
11. **Sarker, S. A. and Gyr, K.,** Non-immunological defence mechanisms of the gut, *Gut,* 33, 987, 1992.
12. **Bartle, H. J. and Harkins, M. J.,** The gastric secretion: its bacterial value to man, *Am. J. Med. Sci.,* 169, 373, 1925.
13. **Giannella, R. A., Broitman, S. A., and Zamcheck, N.,** Gastric acid barrier to ingested microorganisms in man: studies *in vivo* and *in vitro, Gut,* 13, 251, 1972.
14. **Justus, P. G., Martin, J. L., and Goldberg, D. A.,** Myoelectric effects of *Clostridium difficile:* motility-altering factors distinct from its cyclotoxin and enterotoxin in rabbits, *Gastroenterology,* 83, 836, 1982.
15. **Rubinstein, E., Mark, Z., and Haspel, J.,** Antibacterial activity of pancreatic fluid, *Gastroenterology,* 88, 926, 1985.
16. **Walker, W. A.,** Host defense mechanisms in gastrointestinal tract, *Pediatrics,* 57, 901, 1976.
17. **Brown, W. R.,** The gut and regulation of immune responses, *View. Dig. Dis.,* 15, 5, 1983.
18. **Stossell, T. P.,** Phagocytosis, *N. Engl. J. Med.,* 290, 717, 1979.
19. **Root, R. K.,** Humoral immunity and complement, in *Principles and Practice of Infectious Diseases,* Mandell, G. L., Douglas, R. G., and Bennett, J. E., Eds., John Wiley & Sons, New York, 1979, 21.
20. **Barrowman, J. A.,** *Physiology of the Gastro-intestinal Lymphatic System,* Cambridge University Press, Cambridge, 1978.
21. **Frank, M. M.,** The complement system in host defense and inflammation, *Rev. Infect. Dis.,* 1, 483, 1979.
22. **Wing, E. J. and Remington, J. S.,** Lymphocytes and macrophages in cell mediated immunity, in *Principles and Practice of Infectious Diseases,* Mandell, G. L., Douglas, R. G., and Bennett, J. E., Eds., John Wiley & Sons, New York, 1979, 83.
23. **Cooperstock, M. S. and Zedd, A. A.,** Intestinal flora of infants, in *Human Intestinal Microflora in Health and Disease,* Hentges, D. J., Ed., Academic Press, New York, 1983, 79.
24. **Benno, Y., Sawada, K., and Mitsuoka, T.,** The intestinal microflora of infants: composition of faecal flora in breast-fed and bottle-fed infants, *Microbiol. Immunol.,* 28, 975, 1984.
25. **Kornfeld, S. J. and Plaut, A. G.,** Secretory immunity and the bacterial IgA proteases, *Rev. Infect. Dis.,* 3, 521, 1981.

26. **Dodd, H. M. and Gasson, M. J.,** Bacteriocins of lactic acid bacteria, in *Genetics and Biotechnology of Lactic Acid Bacteria,* Gasson, M. J. and de Vos, W. M., Eds., Blackie Academic & Professional, Glasgow, England, 1994, 211.

27. **Klaenhammer, T. R., Ahn, C., Fremaux, C., and Milton, K.,** Molecular properties of *Lactobacillus* bacteriocins, in *Bacteriocins, Microcins and Lantibiotics,* James, R., Ladzunski, C., and Plattus, F., Eds., Springer-Verlag, Berlin, 1992, 37.

28. **Fowler, G. G. and Gasson, M. J.,** Antibiotics — nisin, in *Food Preservatives,* Russel, N. J. and Gould, G. W., Eds., Blackie, London, 1991, 135.

29. **Brenner, D. J.,** Introduction to the family Enterobacteriaceae, in *The Prokaryotes, A Handbook on the Biology of Bacteria: Ecophysiology, Isolation, Identification, Application,* 2nd ed., Balows, A., Trüper, H. G., Dworkin, M., Harder, W., and Schleifer, K. H., Eds., Springer-Verlag, Berlin, 1992, 2673.

30. **Keusch, G.,** The enteric bacteria: diarrhea and dysentery, in *Mechanisms of Microbial Disease,* Schaechter, M., Medoff, G., and Schlessinger, D., Eds., Williams & Wilkins, Baltimore, MD, 1989, 256.

31. **Farmer, J. J. and Kelly, M. T.,** *Enterobacteriaceae,* in *Manual of Clinical Microbiology,* 5th ed., Balows, A., Hausler, W. J., Herrmann, K. L., Isenberg, H. D., and Shadomy, H. J., Eds., American Society for Microbiology Publications, Washington, D.C., 1991, 360.

32. **Knutton, S., Shaw, R. K., Bhan, M. K., Smith, H. R., McConnell, M. M., Cheasty, T., Williams, P. H., and Baldwin, T. J.,** Ability of enteroaggregative *Escherichia coli* strains to adhere in vitro to human intestinal mucosa, *Infect. Immun.,* 60, 2083, 1992.

33. **Guerrant, R. L., Lohr, J. A., and Williams, E. K.,** Acute infectious diarrhea. I. Epidemiology, etiology and pathogenesis, *Pediatr. Infect. Dis.,* 5, 353, 1986.

34. **Clausen, C. R. and Christie, D. L.,** Chronic diarrhea in infants caused by adherent enteropathogenic *Escherichia coli,* *J. Pediatr.,* 100, 358, 1982.

35. **Mason, T. G. and Richardson, G.,** A review: *Escherichia coli* and the human gut: some ecological considerations, *J. Appl. Bacteriol.,* 1, 1, 1981.

36. **Linton, A. H., Howe, K., Bennett, P. M., and Richmond, M. H.,** The colonization of the human gut by antibiotic resistant *Escherichia coli* from chickens, *J. Appl. Bacteriol.,* 43, 465, 1977.

37. **Wadstrom, T.,** Adherence traits and mechanisms of microbial adhesion in the gut, *Clin. Trop. Med. Comm. Dis.,* 3, 3, 1988.

38. **Parry, S. H. and Rooke, D. M.,** Adhesins and colonization factors of *Escherichia coli,* in *The Virulence of Escherichia coli Reviews and Methods,* Sussman, M., Ed., Academic Press, London, 1985, 79.

39. **Vibould, G. I., Binsztein, N., and Svennerholm, A. M.,** A new fimbrial putative colonization factor, PCFO20, in human enterotoxigenic *Escherichia coli, Infect. Immun.,* 61, 5190, 1993.

40. **Evans, D. G., Silver, R. P., Evans, D. J., Chase, D. G., and Gorbach, S. L.,** Plasmid-controlled colonization factor associated with virulence in *Escherichia coli* enterotoxigenic for humans, *Infect. Immun.,* 12, 656, 1975.

41. **Honda, T., Arita, M., and Miwatani, T.,** Characterization of new hydrophobic pili of human enterotoxigenic *Escherichia coli:* a possible new colonization factor, *Infect. Immun.,* 43, 959, 1984.

42. **Smyth, C. J.,** Two mannose-resistant haemagglutinins on enterotoxigenic *Escherichia coli* of serotype O6. K15. K16 or H⁻ isolated from travellers and infantile diarrhoea, *J. Gen. Microbiol.,* 128, 2081, 1982.

43. **Guerrant, R. L., Ganguly, U., and Casper, A. G. T.,** Effect of *Escherichia coli* on fluid transport across canine small bowel: mechanism and time course with enterotoxin and whole bacterial cells, *J. Clin. Invest.,* 52, 1707, 1973.

44. **Rennels, M. B. and Levine, M. M.,** Classical bacterial diarrhea: perspectives and update: *Salmonella, Shigella, Escherichia coli, Aeromonas* and *Plesiomonas, Pediatr. Infect. Dis.,* 5, S91, 1986.

45. **Gross, R. J.,** The pathogenesis of *Escherichia coli* diarrhoea, *Rev. Med. Microbiol.,* 2, 37, 1991.

46. **Schaeter, M.,** Normal microbial flora, in *Mechanisms of Microbial Disease,* Schaechter, M., Medoff, G., and Schlessinger, D., Eds., Williams & Wilkins, Baltimore, MD, 1989, 177.

47. **O'Brien, A. D., Lively, T. A., and Chen, M. E.,** *Escherichia coli* 157: H7 strains associated with hemorrhagic colitis in the United States produce a *Shigella dysenteriae* 1 (Shiga)-like cytotoxin, *Lancet,* 1, 1235, 1983.

48. **Vial, P. A., Robins-Browne, R., Lior, H., Prado, V., Kaper, J. B., Nataro, J. P., Maneval, D., Elsayed, A., and Levine, M. M.,** Characterization of enteroadherent-aggregative *Escherichia coli,* a putative agent of diarrheal disease, *J. Infect. Dis.,* 158, 70, 1988.

49. **Nataro, J. P., Kaper, J. B., Robins-Browne, R., Prado, V., Vial, P., and Levine, M. M.,** Patterns of adherence of diarrhoeagenic *Escherichia coli* to HEp-2 cells, *Pediatr. Infect. Dis.,* 6, 829, 1987.

50. **Bhan, M. K., Raj, P., Levine, M. M., Kaper, J. B., Bhandari, N., Srivastava, R., Kumar, R., and Sazawi, S.,** Enteroaggregative *Escherichia coli* associated with persistent diarrhea in a cohort of rural children in India, *J. Infect. Dis.,* 159, 1061, 1989.

51. **Cravioto, A., Tello, A., Navarro, A., Ruiz, J., Villafan, H., Uribe, F., and Eslava, C.,** Association of *Escherichia coli* HEp-2 adherence patterns with type and duration of diarrhoea, *Lancet,* 337, 262, 1991.

52. **Savarino, S. J., Fasano, A., Robertson, A. C., and Levine, M. M.,** Enteroaggregative *Escherichia coli* elaborate a heat-stable enterotoxin demonstrable in an '*in vitro*' rabbit intestinal model, *J. Clin. Invest.,* 87, 1450, 1991.

53. **World Health Organization,** Center for Reference and Research on *Salmonella,* Antigenic formulae of the *Salmonella,* WHO International *Salmonella* Center, Insititut Pasteur, Paris, 1980.

54. **Minor, L. L.,** The genus *Salmonella,* in *The Prokaryotes, A Handbook on the Biology of Bacteria: Ecophysiology, Isolation, Identification, Application,* 2nd ed., Balows, A., Truper, H. G., Dworkin, M., Harder, W., and Schleifer, K. H., Eds., Springer-Verlag, Berlin, 1992, 2760.

55. **Rubin, R. H. and Weinstein, L.,** *Salmonellosis: Microbiological, Pathological and Clinical Features,* Stratton Book Corp., New York, 1983.

56. **Cohen, J. I. and Bartlett, J. A.,** Extraintestinal manifestation of *Salmonella* infections, *Medicine,* 66, 349, 1987.

57. **Hornick, R. B.,** Typhoid fever and other *Salmonella* infections, in *Tropical and Geographical Medicine,* Warren, K. S. and Mahmoud, A. A. F., Eds., McGraw-Hill, New York, 1984, 710.

58. **Ensgraber, M. and Loos, M.,** A 66-kilodalton heat shock protein of *Salmonella typhimurium* is responsible for binding of the bacterium to intestinal mucus, *Infect. Immun.,* 60, 3072, 1992.

59. **Kusters, J. G., Mulders-Kremers, G. A. W. M., van Doornik, C. E. M., and van der Zeijst, B. A. M.,** Effects of multiplicity of infection, bacterial protein synthesis, and invasion of human cell lines by *Salmonella typhimurium,* *Infect. Immun.,* 61, 5013, 1993.

60. **Altermeyer, R. M., McKern, J. K., Bossio, J. C., Rosenshine, I., Finlay, B. B., and Galan, J. E.,** Cloning and molecular characterization of a gene involved in *Salmonella* adherence and invasion of cultured epithelial cells, *Mol. Microbiol.,* 7, 89, 1993.

61. **Stone, B. J., Garcia, C. M., Badjer, J. L., Hassett, T., Smith, R. I. F., and Miller, V. L.,** Identification of novel loci affecting entry of *Salmonella enteritidis* into eukaryotic cells, *J. Bacteriol.,* 174, 3945, 1992.

62. **Galen, J. E., Ginocchio, C. J., and Costeas, P.,** Molecular and functional characterization of the *Salmonella* invasion gene *invA:* homology of InvA to members of a new protein family, *J. Bacteriol.,* 174, 4338, 1992.

63. **Galan, J. E. and Curtis, R.,** Cloning and molecular characterization of genes whose products allow *Salmonella typhimurium* to penetrate tissue culture cells, *Proc. Natl. Acad. Sci. U.S.A.,* 86, 6383, 1989.

64. **Ginocchio, C. J., Pace, J., and Galan, J. E.,** Identification and molecular characterization of a *Salmonella typhimurium* gene involved in triggering the internalization of salmonellae into cultured cells, *Proc. Natl. Acad. Sci. U.S.A.,* 89, 5976, 1992.

65. **Leung, K. Y. and Finlay, B. J.,** Intracellular replication is essential for the virulence of *Salmonella typhimurium,* *Proc. Natl. Acad. Sci. U.S.A.,* 88, 11470, 1991.

66. **Fierer, J., Eckmann, L., Fang, F., Pfeifer, C., Finlay, B. B., and Guiney, D.,** Expression of the salmonella virulence plasmid gene *spvB* in cultured macrophages and nonphagocytic cells, *Infect. Immun.,* 61, 5231, 1993.

67. **Chikami, G. K., Fierer, J., and Guiney, D. G.,** Plasmid-mediated virulence in *Salmonella dublin* demonstrated by the use of a Tn*5-oriT* construct, *Infect. Immun.,* 50, 420, 1985.

68. **Heffernan, E. J., Fierer, J., Chikami, G., and Guiney, D.,** Natural history of oral *Salmonella dublin* infection in BALB/c mice: effect of an 80-kilobase-pair plasmid on virulence, *J. Infect. Dis.,* 155, 1254, 1987.

69. **Keusch, G. M. and Thea, D. M.,** The salmonellae: typhoid fever and gastroenteritis, in *Mechanisms of Microbial Disease,* Schaechter, M., Medoff, G., and Schlessinger, D., Eds., Williams & Wilkins, Baltimore, MD, 1989, 266.

70. **Taylor, D. N., Pollard, R. A., and Blake, P. A.,** Typhoid in the United States and the risk to the international traveler, *J. Infect. Dis.,* 148, 599, 1983.

71. **Chatfield, S. N., Fairweather, N. F., Charles, I. G., Pickard, D., Levine, M., Hone, D., Posada, M., Strugnell, R. A., and Dougan, G.,** Construction of a genetically defined *Salmonella typhi* Ty2 *aroA, aroC* mutant for the engineering of a candidate oral typhoid-tetanus vaccine, *Vaccine,* 10, 53, 1992.

72. **Chatfield, S. N., Charles, I. G., Makoff, A. J., Oxer, M. D., Dougan, G., Pickard, D., Slater, D., and Fairweather, N. F.,** Use of the *nirB* promoter to direct the stable expression of heterologous antigens in *Salmonella* oral vaccine strains: development of a single dose oral tetanus vaccine, *Bio. Technol.,* 10, 888, 1992.

73. **Dougan, G.,** The molecular basis for the virulence of bacterial pathogens: implications for oral vaccine development, *J. Gen. Microbiol.,* 140, 215, 1994.

74. **Barrett-Connor, E. and Connor, J. D.,** Extraintestinal manifestations of shigellosis, *Am. J. Gastroenterol.,* 53, 234, 1970.

75. **Christie, A. B.,** *Infectious Diseases: Epidemiology and Clinical Practice,* 2nd ed., Churchill Livingstone, London, 1974.

76. **Formal, S. B., Hale, T. L., and Sansonetti, P.,** Invasive enteric pathogens, *Rev. Infect. Dis.,* 14, S702, 1983.

77. **Thea, D. M. and Keusch, G.,** Digestive system, in *Mechanisms of Microbial Disease,* Schaechter, M., Medoff, G., and Schlessinger, D., Eds., Williams & Wilkins, Baltimore, MD, 1989, 628.

78. **Ogawa, H., Nakamura, A., and Nakaya, R.,** Cinemicrographic study of tissue cell cultures infected with *Shigella flexneri,* *J. Med. Sci. Biol.,* 21, 259, 1968.

79. **Bernardini, M. L., Mounier, J., d'Hauteville, H., Coquis-Rondon, M., and Sansonetti, P. J.,** Identification of *icsA,* a plasmid locus of *Shigella flexneri* that governs bacterial intra- and intercellular spread through interaction with F-actin, *Proc. Natl. Acad. Sci. U.S.A.,* 86, 3867, 1989.

80. **Keusch, G. T., Grady, G. F., and Mata, L. J.,** The pathogenesis of *Shigella* diarrhea. I. Enterotoxin production by *Shigella dysenteriae,* *J. Clin. Invest.,* 51, 1212, 1972.

81. **Vicari, G., Olitzki, A. L., and Olitzki, Z.,** The action of thermolabile toxin of *Shigella dysenteriae* on cell culture in vitro, *Br. J. Exp. Pathol.,* 41, 179, 1960.

82. **LaBrec, E. H., Schneider, H., Magnani, T. J., and Formal, S. B.,** Epithelial cell penetration as an essential step in the pathogenesis of bacillary dysentary, *J. Bacteriol.,* 88, 1503, 1964.

83. **Acheson, D. W. K., Calderwood, S. B., Boyko, S. A., Lincicombe, L. L., Kane, A. V., Donohue-Rolfe, A., and Keusch, G. T.,** Comparison of shiga-like toxin I B-subunit expression and localization in *Escherichia coli* and *Vibrio cholerae* by using *trc* or iron-regulated promoter systems, *Infect. Immun.*, 61, 1098. 1993.

84. **Donohue-Rolfe, A., Keusch, G. T., Edson, C., Thorley-Lawson, D., and Jacewicz, M.,** Pathogenesis of *Shigella* diarrhea. IX. Simplified high yield purification of *Shigella* toxin and characterization of subunit composition and function by the use of subunit-specific monoclonal and polyclonal antibodies, *J. Exp. Med.*, 160, 1767, 1984.

85. **O'Brien, A. D. and LaVeck, G. D.,** Purification and characterization of a *Shigella dysenteriae* 1-like toxin produced by *Escherichia coli*, *Infect. Immun.*, 40, 675, 1983.

86. **Seidah, N. G., Donohue-Rolfe, A., Lazure, C., Auclair, F., Keusch, G. T., and Chretien, M.,** Complete amino acid sequence of *Shigella* toxin B-chain. A novel polypeptide containing 69 amino acids and one disulfide bridge, *J. Biol. Chem.*, 261, 13928, 1986.

87. **O'Brien, A. D., LaVeck, G. D., Thompson, M. R., and Formal, S. B.,** Production of *Shigella dysenteriae* type 1-like cytotoxin by *Escherichia coli*, *J. Infect. Dis.*, 146, 763, 1982.

88. **Makino, S., Sasakawa, C., and Yoshikawa, M.,** Genetic relatedness of the basic replicon of the virulence plasmid in shigellae and enterovasive *Escherichia coli*, *Microb. Pathol.*, 5, 267, 1988.

89. **Luria, S. E. and Burrous, J. K.,** Hybridization between *Escherichia coli* and *Shigella*, *J. Bacteriol.*, 74, 461, 1957.

90. **Hale, T. L., Guerry, P., Seid, R. C., Kapfer, C., Wingfield, M. E., Reaves, C. B., Baron, L. S., and Formal, S. B.,** Expression of lipopolysaccharide O antigen in *Escherichia coli* K-12 hybrids containing plasmid and chromosomal genes from *Shigella dysenteriae* 1, *Infect. Immun.*, 46, 470, 1984.

91. **Hale, T. L., Oaks, E. V., and Formal, S. B.,** Identification and antigenic characterization of virulence-associated plasmid-coded proteins of *Shigella* spp. and enteroinvasive *Escherichia coli*, *Infect. Immun.*, 50, 620, 1985.

92. **Hale, T. L., Sansonetti, P. J., Scad, P. A., Austin, S., and Formal, S. B.,** Characterization of virulence plasmids and plasmid-associated outer membrane proteins in *Shigella flexneri*, *Shigella sonnei*, and *Escherichia coli*, *Infect. Immun.*, 40, 340, 1983.

93. **Endo, Y., Tsurugi, K., Yutsudo, T., Takeda, Y., Ogasawara, T., and Igarashi, K.,** Site of action of a vero toxin (VT2) from *Escherichia coli* O157:H7 and of Shiga toxin on eukaryotic ribosomes. RNA N-glycosidase activity of the toxins, *Eur. J. Biochem.*, 171, 45, 1988.

94. **Falkow, S., Schneider, H., Baron, L. S., and Formal, S. B.,** Virulence of *Escherichia-Shigella* genetic hybrids in the guinea pig, *J. Bacteriol.*, 86, 1251, 1963.

95. **Formal, S. B., Gemski, P., Baron, L. S., and LaBrec, E. H.,** Genetic transfer of *Shigella flexneri* antigens to *Escherichia coli* K-12, *Infect. Immun.*, 1, 279, 1970.

96. **Formal, S. B., Labrec, E. H., Kent, T. H., and Falkow, S. B.,** Abortive intestinal infection with an *Escherichia coli-Shigella flexneri* hybrid strain, *J. Bacteriol.*, 89, 1374, 1965.

97. **Gemski, P., Sheanan, D. G., Washington, O., and Formal, S. B.,** Virulence of *Shigella flexneri* hybrids expressing *Escherichia coli* somatic antigens, *Infect. Immun.*, 6, 104, 1972.

98. **Hale, T. L.,** Genetic basis of virulence in *Shigella* species, *Microbiol. Rev.*, 55, 206, 1991.

99. **Sansonetti, P. J., d'Hauteville, H., Ecobichon, C., and Pourcel, C.,** Molecular comparison of virulence plasmids in *Shigella* and enteroinvasive *Escherichia coli*, *Microbiol.*, 134A, 295, 1983.

100. **Sansonetti, P. J., Kopecko, D. J., and Formal, S. P.,** *Shigella sonnei* plasmids: evidence that a large plasmid is neccesary for virulence, *Infect. Immun.*, 34, 75, 1981.

101. **Sansonetti, P. J., Kopecko, D. J., and Formal, S. P.,** Involvement of a plasmid in the invasive ability of *Shigella flexneri*, *Infect. Immun.*, 35, 852, 1982.

102. **Dupont, H. L., Levine, M. M., Hornick, R. B., and Formal, S. B.,** Inoculum size in shigellosis and implications for expected mode of transmission, *J. Infect. Dis.*, 159, 1126, 1989.

103. **Blaser, M. J. and Newman, L. J.,** A review of human salmonellosis. I. Infective dose, *Rev. Infect. Dis.*, 4, 1096, 1982.

104. **Hornick, R. B., Greisman, S. E., Woodward, T. E., Dupont, H. L., Dawkins, A. T., and Snyder, M. J.,** Typhoid fever: pathogenesis and immunological control, *N. Engl. J. Med.*, 283, 739, 1970.

105. **Cash, R. A., Music, S. I., Libonati, J. P., Snyder, M. J., Wenzel, R. P., and Hornick, R. B.,** Response of man to infection with *Vibrio cholerae*. I. Clinical, serologic, and bacteriologic responses to a known inoculum, *J. Infect. Dis.*, 129, 45, 1974.

106. **Keusch, G. T. and Bennish, M. L.,** Shigellosis: recent progress, persisting problems and research issues, *Pediatr. Infect. Dis.*, 8, 713, 1989.

107. **Gorden, J. and Small, P. L. C.,** Acid resistance in enteric bacteria, *Infect. Immun.*, 61, 364, 1993.

108. **Robbins, J. B., Chu, C. Y., and Schneerson, R.,** Hypothesis for vaccine development: serum IgG LPS antibodies confer protective immunity to non-typhoidal salmonellae and shigellae, *Clin. Infect. Dis.*, 15, 346, 1992.

109. **Hale, T. L. and Formal, S. B.,** Oral shigella vaccines, *Curr. Top. Microbiol. Immunol.*, 146, 205, 1989.

110. **Taylor, D. N., Trofa, A. C., Sadoff, J., Chu, C., Bryla, D., Shiloach, J., Cohen, D., Ashkenazi, S., Lerman, Y., Egan, W., Schneerson, R., and Robbins, J. B.,** Synthesis, characterization, and clinical evaluation of conjugate vaccines composed of the O-specific polysaccharides of *Shigella dysenteriae* Type 1, *Shigella flexneri* Type 2a, and *Shigella sonnei* (*Plesiomonas shigelloides*) bound to bacterial toxins, *Infect. Immun.*, 61, 3678, 1993.

111. **Farmer, J. J. and Hickman-Brenner, F. W.,** The genera *Vibrio* and *Photobacterium,* in *The Prokaryotes, A Handbook on the Biology of Bacteria: Ecophysiology, Isolation, Identification, Application,* 2nd ed., Balows, A., Truper, H. G., Dworkin, M., Harder, W., and Schleifer, K. H., Eds., Springer-Verlag, Berlin, 1992, 2952.

112. **Barua, D. and Burrows, W., Eds.,** *Cholera,* W. B. Saunders, Philadelphia, 1974.

113. **Fujino, T., Sakaguchi, E., Sakasaki, R., and Takeda, Y.,** *International symposium on Vibrio parahaemolyticus,* Saikon Publishing Company, Tokyo, 1974.

114. **Huq, M. I., Alam, A. K. M. J., Brenner. D. J., and Morris, G. K.,** Isolation of *Vibrio*-like group, EF-6, from patients with diarrhea, *J. Clin. Microbiol.,* 11, 621, 1980.

115. **Hickman, F. W., Farmer, J. J., Hollis, D. G., Fanning, G. R., Stiegerwalt, A. G., Weaver, R. E., and Brenner, D. J.,** Identification of *Vibrio hollisae* sp. nov. from patients with diarrhea, *J. Clin. Microbiol.,* 15, 395, 1982.

116. **Brenner, D. J., Hickman-Brenner, F. W., Lee, J. V., Steigerwalt, A. G., Fanning, G. R., Hollis, D. G., Farmer, J. J., Weaver, R. E., and Seidler, R. J.,** *Vibrio furnissiii,* (formerly aerogenic biogroup of *Vibrio fluvialis*), a new species isolated from human feces and the environment, *J. Clin. Microbiol.,* 18, 816, 1983.

117. **Blake, P. A., Weaver, R. E., and Hollis, D. G.,** Diseases of humans (other than cholera) caused by vibrios, *Ann. Rev. Microbiol.,* 34, 341, 1980.

118. **Hughes, J. M., Hollis, D. B., Gangarosa, E. J., and Weaver, R. E.,** Non-cholera vibrios infections in the United States. Clinical, epidemiologic and laboratory features, *Ann. Intern. Med.,* 88, 602, 1978.

119. **Kelly, M. T.,** Cholera: a worldwide perspective, *Pediatr. Infect. Dis.,* 5, S101, 1986.

120. **Levine, M. M.,** Cholera in Louisiana: old problem, new light, *N. Engl. J. Med.,* 302, 345, 1980.

121. **Donowitz, M. and Welsh, M. J.,** Regulation of mammalian small intestinal electrolyte secretion, in *Physiology of the Gastrointestinal Tract,* Johnson, L. R., Ed., Raven Press, New York, 1987, 1351.

122. **Dirata, V. J., Peterson, K. M., and Mekalanos, J. J.,** Regulation of cholera toxin synthesis, in *Molecular Basis of Bacterial Pathogenesis,* Iglewski, B. H. and Clark, V. L., Eds., Academic Press, San Diego, 1990, 355.

123. **Apter, F. M., Lencer, W. I., Finkelstein, R. A., Mekalanos, J., and Neutra, M. R.,** Monoclonal immunoglobulin A antibodies directed against cholera toxin prevent the toxin-induced chloride secretory response and block toxin binding to intestinal epithelial cells in vitro, *Infect. Immun.,* 61, 5271, 1993.

124. **Holmgren, J.,** Actions of cholera toxin and the prevention and treatment of cholera, *Nature,* 292, 413, 1981.

125. **Pickering, L. K.,** Antimicrobial therapy of gastrointestinal infections, *Pediatr. Clin. North Am.,* 30, 373, 1983.

126. **Levine, M. M., Kaper, J. B., Herrington, D., Losonsky, G., Morris, J. G., Clements, M. L., Black, R. E., Tall, B., and Hall, R.,** Volunteer studies of deletion mutants of *Vibrio cholerae* O1 prepared by recombinant techniques, *Infect. Immun.,* 56, 161, 1988.

127. **Bolen, J. L., Zaminska, S. A., and Greenough, W. B.,** Clinical features in enteritis due to *Vibrio parahaemolyticus,* *Am. J. Med.,* 57, 638, 1974.

128. **Joseph, S. W., Colwell, R. R., and Kaper, J. B.,** *Vibrio parahaemolyticus* and related halophilic vibrios, *Crit. Rev. Microbiol.,* 10, 77, 1982.

129. **Yanagase, Y., Inoue, K., Ozaki, M., Ochi, T., Amano, T., and Chazono, M.,** Hemolysins and related enzymes of *Vibrio parahaemolyticus,* *Biken J.,* 13, 77, 1970.

130. **Douet, J. P., Castroviejo, M., Dodin, A., and BeBear, C.,** Purification and characterisation of Kanagawa haemolysin from *Vibrio parahaemolyticus,* *Res. Microbiol.,* 143, 569, 1992.

131. **Sakurai, J., Matsuzaki, A., and Miwatani, T.,** Purification and characterization of thermostable direct hemolysin of *Vibrio parahaemolyticus,* *Infect. Immun.,* 8, 775, 1973.

132. **Sakurai, J., Bahavar, M. A., Jinguji, Y., and Miwatani, T.,** Interaction of thermostable direct hemolysin of *Vibrio parahaemolyticus* with human erythrocytes, *Infect. Immun.,* 8, 775, 1973.

133. **Shirai, H., Ito, H., Hirayama, T., Nakamoto, Y., Nakabayashi, N., Kumagai, K., Takeda, Y., and Nishibuchi, M.,** Molecular epidemiologic evidence for association of thermostable direct hemolysin (TDH) and TDH-related hemolysin of *Vibrio parahaemolyticus* with gastroenteritis, *Infect. Immun.,* 58, 3568, 1990.

134. **Taniguchi, A., Kubomura, S., Hirano, H., Mizue, K., Ogawa, M., and Mizuguchi, Y.,** Cloning and characterization of a gene encoding a new thermostable hemolysin from *Vibrio parahaemolyticus,* *FEMS Microbiol. Letts.,* 67, 339, 1990.

135. **Bernheimer, A. W. and Rudy, B.,** Interactions between membranes and cytolytic peptides, *Biochim. Biophys. Acta,* 864, 123, 1988.

136. **Dannenberg, A. M.,** Pathogenesis of tuberculosis, in *Pulmonary Diseases and Disorders,* Fishman, A. P., Ed., McGraw-Hill, New York, 1980, 1264.

137. **Kaplan, G. and Cohen, Z. A.,** The immunology of leprosy, *Int. Rev. Exp. Pathol.,* 28, 45, 1986.

138. **Good, R. C.,** The genus *Mycobacterium* — medical, in *The Prokaryotes, A Handbook on the Biology of Bacteria: Ecophysiology, Isolation, Identification, Application, 2nd ed.,* Balows, A., Trüper, H. G., Dworkin, M., Harder, W., and Schleifer, K. H., Eds., Springer-Verlag, Berlin, 1992, 1238.

139. **Bentley, G. and Webster, J. H. H.,** Intestinal tuberculosis: a 10 year review, *Br. J. Surg.,* 54, 90, 1967.

140. **Tabrisky, J., Lindstrom, R. B., Peters, R., and Lachman, R. S.,** Tuberculosis enteritis: review of a protean disease, *Am. J. Gastroenterol.,* 63, 49, 1975.

141. **McDonough, K. A., Kress, Y., and Bloom, B. R.,** Pathogenesis of tuberculosis: interaction of *Mycobacterium tuberculosis* with macrophages, *Infect. Immun.*, 61, 2763, 1993.

142. **Armstrong, J. A. and Hart, P. D.,** Response of cultured macrophages to *Mycobacterium tuberculosis*, with observations on fusion of lysosomes with phagosomes, *J. Exp. Med.*, 134, 713, 1971.

143. **Schlesinger, L. S., Bellinger-Kawahara, C. G., Payne, N. R., and Horwitz, M. A.,** Phagocytosis of *Mycobacterium tuberculosis* is mediated by human monocyte complement component C3, *J. Immunol.*, 144, 2771, 1990.

144. **Mustafa, A. S., Lundin, K. E. A., and Oftung, F.,** Human T cells recognize mycobacterial heat shock proteins in the context of multiple HLA-DR molecules: studies with healthy subjects vaccinated with *Mycobacterium bovis* BCG and *Mycobacterium leprae*, *Infect. Immun.*, 61, 5294, 1993.

145. **Tantimavanich, S., Nagai, S., Nomaguchi, H., Kinomoto, M., Ohara, N., and Yamada, T.,** Immunological properties of ribosomal proteins from *Mycobacterium bovis* BCG, *Infect. Immun.*, 61, 4005, 1993.

146. **Barnes, P. F., Bloch, A. B., Davidson, P. T., and Snider, D. E.,** Tuberculosis in patients with human immunodeficiency virus infection, *N. Engl. J. Med.*, 324, 1644, 1991.

147. **Horsburgh, C. R.,** *Mycobacterium avium* complex infection in the acquired immunodeficiency syndrome, *N. Engl. J. Med.*, 324, 1332, 1991.

148. **Greene, J. B., Sidhu, G. S., and Lewin, S.,** *Mycobacterium avium-intracellulare:* a cause of disseminated life-threatening infection in homosexuals and drug abusers, *Ann. Intern. Med.*, 105, 184, 1986.

149. **Strom, R. L. and Gruminger, R. P.,** AIDS with *Mycobacterium avium-intracellulare* lesions resembling those of Whipple's disease, *N. Engl. J. Med.*, 309, 1323, 1983.

150. **Bottone, E. J.,** *Yersinia enterocolitica:* a panoramic view of a charismatic microorganism, *Crit. Rev. Microbiol.*, 5, 211, 1977.

151. **Kohl, S.,** *Yersinia enterocolitica* infections in children, *Pediatr. Clin. North Am.*, 26, 433, 1979.

152. **Black, R. E., Jackson, R. J., and Tsai, T.,** Epidemic *Yersinia enterocolitica* infection due to contaminated chocolate milk, *N. Engl. J. Med.*, 298, 76, 1978.

153. **Schieven, B. C. and Randall, C.,** Enteritis due to *Yersinia enterocolitica*, *J. Pediatr.*, 84, 402, 1974.

154. **Gilmour, A. and Walker, S. J.,** Isolation and identification of *Yersinia enterocolitica* and the *Yersinia enterocolitica*-like bacteria, in *Journal of Applied Bacteriology Symposium Supplement*, Lund, B. M., Sussman, M., Jones, D., and Stringer, M. F., Eds., Blackwell Scientific Publications, Oxford, 1988, 213S.

155. **Marks, M. I., Pai, C. H., Lafleur, L., Lackman, L., and Hammerberg, O.,** *Yersinia enterocolitica* gastroenteritis: a propective study of clinical, bacteriologic, and epidemiologic features, *J. Pediatr.*, 96, 26, 1980.

156. **Vantrappen, G., Ponett, E., and Geboes, K.,** *Yersinia* enteritis and enterocolitis: gastroenterologic aspects, *Gastroenterology*, 72, 220, 1977.

157. **Spira, T. J. and Kabins, S. A.,** *Yersinia enterocolitica* septicemia with septic arthritis, *Arch. Intern. Med.*, 136, 1305, 1976.

158. **Scotland, S. M.,** Toxins, in *Journal of Applied Bacteriology Symposium Supplement*, Lund, B. M., Sussman, M., Jones, D., and Stringer, M. F., Eds., Blackwell Scientific Publications, Oxford, 1988, 109S.

159. **Tertti, R., Gransfors, K., and Lehtononen, O. P.,** An outbreak of *Yersinia pseudotuberculosis* infection, *J. Infect. Dis.*, 149, 245, 1984.

160. **Williams, E. K., Lohr, J. A., and Guerrant, R. L.,** Acute infectious diarrhea. II. Diagnosis, treatment and prevention, *Pediatr. Infect. Dis.*, 5, 458, 1986.

161. **Portnoy, D. A., Moseley, S. L., and Falkow, S.,** Characterization of plasmids and plasmid-associated determinants of *Yersinia enterocolitica* pathogenesis, *Infect. Immun.*, 31, 775, 1981.

162. **Portnoy, D. A., Wolf-Watz, H., Bolin, I., Beeder, A. B., and Falkow, S.,** Characterization of common virulence plasmids in *Yersinia* species and their role in the expression of outer membrane proteins, *Infect. Immun.*, 43, 108, 1984.

163. **Gemski, P., Lazere, J. R., and Casey, T.,** Plasmid associated with pathogenicity and calcium dependency of *Yersinia enterocolitica*, *Infect. Immun.*, 27, 682, 1980.

164. **Skurnik, M.,** Expression of antigens encoded by the virulence plasmid of *Yersinia enterocolitica* under different growth conditions, *Infect. Immun.*, 47, 183, 1985.

165. **Heeseman, J. and Gruter, L.,** Genetic evidence that the outer membrane protein YOP1 of *Yersinia enterocolitica* mediates adherence and phagocytosis of resistance to human epithelial cells, *FEMS Microbiol. Lett.*, 40, 37, 1987.

166. **Lian, C. J., Hwang, W. S., and Pai, C. H.,** Plasmid-mediated resistance to phagocytosis in *Yersinia enterocolitica*, *Infect. Immun.*, 55, 1176, 1987.

167. **Straley, S. C., Skrzypak, E., Plano, G. Y., and Bliska, J. B.,** Yops of *Yersinia* spp. pathogenic for humans, *Infect. Immun.*, 61, 3105, 1993.

168. **Schulze-Koops, H., Burkhardt, H., Heesman, J., von der Mark, K., and Emmrich, F.,** Plasmid-encoded outer membrane protein YadA mediates specific binding of enteropathogenic yersiniae to various types of collagen, *Infect. Immun.*, 60, 2153, 1992.

169. **Pai, C. H. and DeStephano, L.,** Serum resistance associated with virulence in *Yersinia enterocolitica*, *Infect. Immun.*, 35, 605, 1982.

170. **Morgan, D. R., Johnson, P. C., DuPont, H. L., Satterwhite, T. K., and Wood, L. V.,** Lack of correlation between known virulence properties of *Aeromonas hydrophila* and enteropathogenicity for humans, *Infect. Immun.*, 50, 62, 1985.

171. **Von Gravenitz, A. and Zinterhofer, L.,** The detection of *Aeromonas hydrophila* in stool specimens, *Health Lab. Sci.*, 7, 124, 1970.

172. **Atkinson, H. M., Adams, D., and Trust, T. J.,** *Aeromonas* adhesion antigens, *Experentia*, 43, 372, 1987.

173. **George, W. L., Nakata, M. M., Thompson, J., and White, M. L.,** *Aeromonas*-related diarrhea in adults, *Arch. Intern. Med.*, 145, 2207, 1985.

174. **Rose, J. M., Houston, C. W., Coppenhaver, D. H., Dixon, J. D., and Kurovsky, A.,** Purification and chemical characterization of a cholera toxin-cross reactive cytolytic enterotoxin produced by a human strain of *Aeromonas hydrophila*, *Infect. Immun.*, 57, 1165, 1989.

175. **Johnson, W. M. and Lior, H.,** Cytotoxicity and suckling mouse reactivity of *Aeromonas hydrophila* isolated from human sources, *Can. J. Microbiol.*, 27, 1019, 1981.

176. **Atkinson, H. M. and Trust, T. J.,** Hemagglutination properties and adherence ability of *Aeromonas hydrophila*, *Infect. Immun.*, 27, 938, 1980.

177. **Lawson, M. A., Burke, V., and Chang, B. J.,** Invasion of HEp-2 cells by fecal isolates of *Aeromonas hydrophila*, *Infect Immun.*, 47, 680, 1985.

178. **Rose, J. M., Houston, C. W., and Kurosky, A.,** Bioactivity and immunological characterization of a cholera toxin-cross reactive cytolytic enterotoxin from *Aeromonas hydrophila*, *Infect. Immun.*, 57, 1170, 1989.

179. **Wadstrom, T. and Ljungh, A.,** Correlation between toxin formation and diarrhoea in patients infected with aeromonas, *J. Diarrhoeal Dis. Res.*, 6, 113, 1988.

180. **Wretlind, B., Mollby, S., and Wadstrom, T.,** Separation of two hemolysins from *Aeromonas hydrophila* by isoelectric focusing, *Infect. Immun.*, 4, 503, 1971.

181. **Farraye, F. A., Peppercorn, M. A., Ciano, P. S., and Kavesh, W. N.,** Segmental colitis associated with *Aeromonas hydrophila*, *Am. J. Gastroenterol.*, 84, 436, 1989.

182. **Kirov, S. M. and Brodribb, F.,** Exotoxin production by *Aeromonas* spp. in foods, *Lett. Appl. Microbiol.*, 17, 208, 1993.

183. **Jones, G. W., Robert, D. K., Svinarich, D. M., and Whitfield, F. J.,** Association of adhesive, invasive and virulent phenotypes of *Salmonella typhimurium* with autonomous 60-megadalton plasmids, *Infect. Immun.*, 38, 476, 1987.

184. **Sansonetti, P. J., d'Hauteville, H., Formal, S. B., and Toucas, M.,** Plasmid-mediated invasiveness of "Shigella-like" *Escherichia coli*, *Ann. Microbiol.*, 133A, 351, 1982.

185. **Kopecko, D. J., Baron, L. S., and Buysse, J.,** Genetic determinants of virulence in *Shigella* and dysenteric strains of *Escherichia coli*: their involvement in the pathogenesis of dysentery, *Curr. Top. Microbiol. Immunol.*, 118, 71, 1985.

186. **Rahman, A. F. M. S. and Willoughby, J. M. T.,** Dysentery-like syndrome associated with *Aeromonas hydrophila*, *Br. Med. J.*, 281, 976, 1980.

187. **Gracey, M., Burke, V., and Robinson, J.,** *Aeromonas*-associated gastroenteritis, *Lancet*, ii, 1304, 1982.

188. **Arai, T., Ikejima, N., Itoh, T., Sakai, S., Shimada, T., and Sakazaki, R.,** A survey of *Plesiomonas shigelloides* from aquatic environments, domestic animals, pets and humans, *J. Hyg.*, 84, 203, 1980.

189. **Holmberg, S. D., Schell, W. L., Fanning, G. R., Wachsmuth, I. K., Hickman-Brenner, F. W., and Farmer, J. J.,** *Aeromonas* intestinal infections in the United States, *Ann. Intern. Med.*, 105, 683, 1986.

190. **Sawel, G. V., Das, B. C., Acland, P. R., and Heath, D. A.,** Fatal infection with *Aeromonas sobria* and *Plesiomonas shigelloides*, *Br. Med. J.*, 292, 525, 1986.

191. **Sakazaki, R., Nakaya, R., and Fukumi, H.,** Studies on so-called Paracolon SC 27 (Ferguson), *Jpn. Med. Sci. Biol.*, 12, 355, 1959.

192. **Rolston, K. V. I. and Hopfer, R. L.,** Diarrhea due to *Plesiomonas shigelloides* in cancer patients, *J. Clin. Microbiol.*, 20, 597, 1984.

193. **Remhardt, J. F. and George, W. L.,** Comparative in vitro activities of selected antimicrobial agents against *Aeromonas* species and *Plesiomonas shigelloides*, *Antimicrob. Agents Chemother.*, 27, 643, 1985.

194. **Herrington, D. A., Tzipori, S., Robins-Browne, R. M., Tall, B. D., and Levine, M. M.,** In vitro and in vivo pathogenicity of *Plesiomonas shigelloides*, *Infect. Immun.*, 55, 979, 1987.

195. **Binns, M. M., Vaughn, S., Sanyal, S. C., and Timmis, K. N.,** Invasionsfahigkeit von *Plesiomonas shigelloides*, *Zbl. Bakt. Hyg. A*, 257, 343, 1984.

196. **Foster, B. G. and Rao, V. B.,** Isolation and characterization of *Aeromonas shigelloides* endotoxin, *Tex. J. Sci.*, 27, 367, 1976.

197. **Manorama, T. V., Agarwal, A. K., and Sanyal, S. C.,** Enterotoxins of *Plesiomonas shigelloides*. Partial purification and characterization, *Toxicology*, 22, 269, 1983.

198. **Newell, D. G.,** Ed., *Campylobacter*, MTP Press Ltd., Lancaster, England, 1982.

199. *Interim Report on Campylobacter*, HMSO, London, 1993.

200. **Blaser, M. J., Taylor, D. N., and Feldman, R. A.,** Epidemiology of *Campylobacter jejuni* infections, *Epidem. Rev.*, 5, 157, 1983.

201. **Walker, R. I., Caldwell, M. B., Lee, E. C., Guerry, P., Trust, T. J., and Ruis-Palacios, G. M.,** Pathophysiology of *Campylobacter* enteritis, *Microbiol. Rev.*, 50, 81, 1986.

202. **Griffiths, P. L. and Park, R. W. A.,** Campylobacters associated with human diarrhoeal disease, *J. Appl. Bacteriol.*, 69, 281, 1990.

203. **Blaser, M. J., Wells, J. G., Feldman, R. A., Pollard, R. A., and Allen, J. R.,** The collaborative campylobacter diarrheal disease study group. *Campylobacter* enteritis in the United States: a multi-center study, *Ann. Intern. Med.*, 98, 360, 1983.

204. **Skirrow, M. B. and Benjamin, J.,** 1001 Campylobacters: cultural characteristics of intestinal campylobacters from man and animals, *J. Hyg.*, 85, 427, 1980.

205. **Karmali, M. A., Penner, J. L., Fleming, P. C., Williams, A., and Hennessy, J. N.,** The serotype and biotype distribution of clinical isolates of *Campylobacter jejuni/coli* over a three year period, *J. Infect. Dis.*, 147, 243, 1983.

206. **Totten, P. A., Fennel, C. L., Tenover, F. C., Wezenberg, J. M., Perine, P. L., Stamm, W. E., and Holmes, K. K.,** *Campylobacter cinaedi* (sp. nov.) and *Campylobacter fennelliae* (sp. nov.): two new *Campylobacter* species associated with enteric disease in homosexual men, *J. Infect. Dis.*, 151, 131, 1985.

207. **Totten, P. A., Patton, C. M., Tenover, F. C., Barrett, T. J., Stamm, W. E., Steigerwalt, A. G., Lin, J. Y., Holmes, K. K., and Brenner, D. J.,** Prevalence and characterization of hippurate negative *Camylobacter jejuni* in King County, Washington, *J. Clin. Microbiol.*, 25, 1747, 1987.

208. **Stehr-Green, J., Mitchell, P., Nicholls, C., McEwan, S., and Payne, A.,** *Campylobacter* enteritis-New Zealand, *MMWR Morb. Mortal. Wkly. Rep.*, 40, 116, 1991.

209. **Humphrey, T. J. and Beckett, P.,** *Campylobacter jejuni* in dairy cows and raw milk, *Epidemiol. Infect.*, 98, 263, 1987.

210. **Harris, N. V., Weiss, N. S., and Nolan, C. M.,** The role of poultry and meats in the etiology of *Camylobacter jejuni/coli* enteritis, *Am. J. Pub. Health*, 76, 407, 1986.

211. **Walker, R. I., Caldwell, M. B., Lee, E. C., Guerry, P., Trust, T. J., and Ruiz-Palacios, G. M.,** Pathophysiology of *Campylobacter* enteritis, *Microbiol. Rev.*, 50, 81, 1986.

212. **Turnbull, P. C. B. and Rose, P.,** *Campylobacter jejuni* and salmonella in raw red meats. A Public Health Laboratory Service Survey, *J. Hyg.*, 88, 29, 1982.

213. **Robinson, D. A.,** Infective dose of *Campylobacter jejuni* in milk, *Br. Med. J.*, 282, 1584, 1981.

214. **Black, R. E., Levine, M. M., Clements, M. L., Hughs, T. P., and Blaser, M. J.,** Experimental *Campylobacter jejuni* infection in humans, *J. Infect. Dis.*, 157, 472, 1988.

215. **Blaser, M. J., Berkowitz, I. D., Laforce, F. M., Cravens, J., Reller, L. B., and Wang, W.-L.,** Campylobacter enteritis: clinical and epidemiological features, *Ann. Intern. Med.*, 91, 179, 1979.

216. **Drake, A. A., Gilchrist, M. J. R., Washington, J. A., Huizinga, K. A., and Van Scoy, R. E.,** Diarrhea due to *Campylobacter fetus* subspecies jejuni: a clinical review of 73 cases, *Mayo Clin. Proc.*, 56, 414, 1981.

217. **Tauxe, R. V., Hargrett-Bean, N., Patton, C. M., and Wachsmuth, I. K.,** Campylobacter isolates in the United States, 1982–1986, *Morbid. Mortal. Wkly. Rep.*, 37, 1, 1988.

218. **Anders, B. J., Lauer, B. A., and Paisley, J. W.,** Campylobacter gastroenteritis in neonates, *Am. J. Dis. Child.*, 135, 900, 1981.

219. **Michalak, D. M., Perrault, J., Gilchrist, M. J., Dozois, R. R., Carney, J. A., and Sheedy, P. F.,** *Campylobacter fetus* ss. *jejuni*: a cause of massive lower gastrointestinal hemorrhage, *Gastroenterology*, 79, 742, 1980.

220. **McKinley, M. J., Taylor, M., and Sangree, M. H.,** Toxic megacolon with *Campylobacter* colitis, *Com. Med.*, 44, 496, 1980.

221. **Mertens, A. and DeSmet, M.,** Campylobacter cholecystitis, *Lancet*, 1, 1092, 1980.

222. **Kendal, E. J. C. and Tanner, E. I.,** Campylobacter enteritis in general practice, *J. Hyg.*, 88, 155, 1982.

223. **Skirrow, M. B. and Benjamin, J.,** Is enrichment culture necessary for the isolation of *Campylobacter jejuni* from faeces?, *J. Clin. Pathol.*, 37, 478, 1984.

224. **Doyle, M. P.,** *Campylobacter* in foods, in *Campylobacter Infection in Man and Animals*, Butzler, J. P., Ed., CRC Press, Boca Raton, FL, 1983, 163.

225. **Heisick, J.,** Comparison of enrichment broths for isolation of *Campylobacter jejuni*, *Appl. Environ. Microbiol.*, 50, 1313, 1985.

226. **Birkenhead, D., Hawkey, P. M., Heritage, J., Gascoyne-Binzi, D. M., and Kite, P.,** PCR for the detection and typing of campylobacters, *Lett. Appl. Microbiol.*, 17, 235, 1993.

227. **Caldwell, M. B., Guerry, P., Lee, E. C., Burans, J. P., and Walker, R. I.,** Reversible expression of flagella in *Campylobacter jejuni*, *Infect. Immun.*, 50, 941, 1985.

228. **Newell, D. G., McBride, H., and Dolby, J.,** Investigations on the role of flagella in the colonisation of infant mice with *Campylobacter jejuni* and attachment of *Campylobacter jejuni* to human epithelial cell lines, *J. Hyg.*, 95, 217, 1985.

229. **Morooka, T., Umeda, A., and Amako, K.,** Motility as an intestinal colonisation factor for *Campylobacter jejuni*, *J. Gen. Microbiol.*, 131, 1973, 1985.

230. **Konkel, M. E., Hayes, S. F., Joens, L. A., and Gieplak, W. J.,** Characteristics of the internalisation and intracellular survival of *Campylobacter jejuni* in human epithelial cell cultures, *Microb. Pathol.*, 13, 357, 1992.

231. **Konkel, M. E. and Joens, L. A.,** Adhesion to and invasion of Hep cells by *Campylobacter* species, *Infect. Immun.*, 57, 2984, 1989.

232. **Babakhani, F. K. and Joens, L. A.,** Primary swine intestinal cells as a model for studying *Campylobacter jejuni* invasiveness, *Infect. Immun.*, 61, 2723, 1993.

233. **Klipstein, E. A. and Engert, R. E.,** Properties of crude *Campylobacter jejuni* heat labile enterotoxin, *Infect. Immun.*, 45, 314, 1984.

234. **Johnson, W. M. and Lior, H.,** A new heat-labile cytolethal distending toxin (CLDT) produced by *Campylobacter* species, *Microbiol. Pathol.*, 4, 115, 1988.

235. **Klipstein, E. A., Engert, R. E., Short, H., and Schenk, E. A.,** Pathogenic properties of *Campylobacter jejuni*: assay and correlation with clinical manifestations, *Infect. Immun.*, 50, 43, 1985.

236. **Johnson, W. M. and Lior, H.,** Cytotoxic and cytotonic factors produced by *Campylobacter jejuni*, *Campylobacter coli* and *Campylobacter laridis*, *J. Clin. Microbiol.*, 24, 275, 1986.

237. **Wassenaar, T. M., Bleumink-Pluym, N., and Van der Zeijst, B. A. M.,** Inactivation of *Campylobacter jejuni* flagellin genes by homologous recombination demonstrates the *fla*A but not *fla*B is required for invasion, *EMBO J.*, 10, 2055, 1991.

238. **Lee, E., McCardell, B., and Guerry, P.,** Characterisation of a plasmid-encoded enterotoxin in *Campylobacter jejuni*, in *Campylobacter III*, Pearson, A. D., Skirrow, M. B., Lior, H., and Rowe, B., Eds., PHLS, London, 1985, 1.

239. **Babakhani, F. K., Bradley, G. A., and Joens, L. A.,** Newborn piglet model for campylobacteriosis, *Infect. Immun.*, 61, 3466, 1993.

240. **Fricker, C. R. and Park, R. W. A.,** A two-year study of the distribution of "thermophilic" campylobacters in human, environmental and food sample from the Reading area with particular reference to toxin production and heat-stable serotype, *J. Appl. Bacteriol.*, 66, 477, 1989.

241. **Salyers, A. A.,** *Bacteroides* of the lower gastrointestinal tract, *Annu. Rev. Microbiol.*, 38, 293, 1984.

242. **Brooke, I. A.,** A 12 year study of aerobic and anaerobic bacteria in intra-abdominal and postsurgical abdominal wound infections, *Surg. Gynecol. Obstet.*, 169, 387, 1989.

243. **Guzman, C. A., Plate, M., and Pruzzo, C.,** Role of neuraminidase-dependent adherence in *Bacteroides fragilis* attachment to human epithelial cells, *FEMS Microbiol. Letts.*, 71, 187, 1990.

244. **Myers, L. L., Shoop, D. S., Stackhouse, L. L., Newman, F. S., Flaherty, R. J., Letson, G. W., and Sack, R. B.,** Isolation of enterotoxigenic *Bacteroides fragilis* from humans with diarrhea, *J. Clin. Microbiol.*, 25, 2330, 1987.

245. **Myers, L. L., Collins, J. E., and Shoop, D. S.,** Ultrastructural lesions of enterotoxigenic *Bacteroides fragilis* in rabbits, *Vet. Pathol.*, 28, 336, 1991.

246. **Duimstra, J. R., Myers, L. L., Collins, J. E., Benfield, D. A., Shoop, D. S., and Bradbury, W. C.,** Enterovirulence of enterotoxigenic *Bacteroides fragilis* in gnotobiotic pigs, *Vet. Pathol.*, 28, 514, 1991.

247. **Myers, L. L. and Weikel, C. S.,** Enterotoxin as a virulence factor in *Bacteroides fragilis*-associated diarrhoeal disease, in *Medical and Environmental Aspects of Anaerobes*, Duerden, B. I., Brazier, B. J. S., Seddon, S. V., and Wade, W. G., Eds., Wrightson Biomedical Publishing Ltd., Petersfield, Hampshire, England, 1992, 90.

248. **MacLennan, J. D.,** The histotoxic clostridial infections of man, *Bact. Rev.*, 26, 177, 1962.

249. **Kolbeinsson, M. E., Holder, W. D., and Aziz, S.,** Recognition, management, and prevention of *Clostridium septicum* abscess in immunosuppressed patients, *Arch. Surg.*, 126, 642, 1991.

250. **Sjolin, S. U. and Hansen, A. K.,** *Clostridium septicum* gas gangrene and an intestinal malignant lesion, *J. Bone Joint Surg.*, 73A, 772, 1991.

251. **Randall, J. M., Hall, K., and Coulthard, M. G.,** Diffuse pneumocephalus due to *Clostridium septicum* cerebritis in haemolytic uraemic syndrome: CT demonstration, *Neuroradiology*, 35, 218, 1993.

252. **Borriello, S. P. and Carman, R. J.,** Clostridial diseases of the gastrointestinal tract in animals, in *Clostridia in Gastrointestinal Disease*, Borriello, S. P., Ed., CRC Press, Boca Raton, FL, 1985, 195.

253. **Borriello, S. P.,** Newly described clostridial diseases of the gastrointestinal tract: *Clostridium perfringens* enterotoxin-associated diarrhoea and neutropenic enterocolitis due to *Clostridium septicum*, in *Clostridia in Gastrointestinal Disease*, Borriello, S. P., Ed., CRC Press, Boca Raton, FL, 1985, 224.

254. **Alpern, R. J. and Dowell, V. R.,** *Clostridium septicum* infections and malignancy, *JAMA*, 209, 385, 1969.

255. **Nirgiotis, J. G., Cox, C. S., and Feldtman, R. W.,** Atraumatic *Clostridium septicum* infections, *Ther. Chall.*, December, 21, 1991.

256. **Kudsk, K. A.,** Occult gastrointestinal malignancies producing metastatic *Clostridium septicum* infections in diabetic patients, *Surgery*, 4, 765, 1992.

257. **Hurley, L. and Howe, K.,** Mycotic aortic aneurysm infected by *Clostridium septicum*-a case history, *Angiol. J. Vas. Dis.*, July, 585, 1991.

258. **Kornbluth, A. A., Danzig, J. B., and Bernstein, L. H.,** *Clostridium septicum* infection and associated malignancy: report of 2 cases and review of the literature, *Medicine*, 68, 30, 1989.

259. **Koransky, J. R., Stargel, M. D., and Dowell, V. R.,** *Clostridium septicum* bacteremia — its clinical significance, *Am. J. Med.*, 66, 63, 1979.

260. **Pelfrey, T. M., Turk, R. T., Peoples, J. B., and Elliot, D. W.,** Surgical aspects of *Clostridium septicum* septicemia, *Arch. Surg.*, 119, 546, 1984.

261. **Ballard, J., Bryant, A., Stevens, D., and Tweten, R. K.,** Purification and characterization of the lethal toxin (alpha-toxin) of *Clostridium septicum, Infect. Immun.*, 60, 784, 1992.

262. **Smith, L. D. S.,** *The Pathogenic Anaerobic Bacteria,* Charles C. Thomas, Springfield, IL, 1968.

263. **Willis, A. T.,** *Clostridia of Wound Infections,* Butterworth and Co., London, 1979.

264. **Takano, S., Masatoshi, N., and Kato, I.,** Activation of phospholipase $A_2$ in rabbit erythrocyte membranes by a novel hemolytic toxin (H-toxin) of *Clostridium septicum, FEMS Microbiol. Letts.*, 68, 319, 1990.

265. **Rood, J. I. and Cole, S. T.,** Molecular genetics and pathogenesis of *Clostridium perfringens, Microbiol. Rev.*, 55, 621, 1991.

266. **McDonel, M. L.,** Toxins of *Clostridium perfringens* types A,B,C,D and E, in *Pharmacology of Bacterial Toxins,* Dorner, F. and Drewy, H., Eds., Pergamon, New York, 1986, 447.

267. **Walker, P. D.,** Pig-Bel, in *Clostridia in Gastrointestinal Disease,* Borriello, S. P., Ed., CRC Press, Boca Raton, FL, 1985, 93.

268. **Hobbs, B. C.,** *Clostridium perfringens* and *Bacillus cereus* infections, in *Food-Borne Infections and Intoxications,* Deiman, H., Ed., Academic Press, New York, 1969, 131.

269. **Anon.,** Foodborne disease surveillance in England and Wales 1984, *Commun. Dis. Rep.*, 34, 3, 1986.

270. **Shandera, W. X., Tacket, C. O., and Blake, P. A.,** Food poisoning due to *Clostridium perfringens* in the United States, *J. Infect. Dis.*, 147, 169, 1983.

271. **Granum, P. E., Whitaker, J. R., and Skelkvaale, R.,** Trypsin activation of enterotoxin from *Clostridium perfringens* type A. Fragmentation and some physicochemical properties, *Biochim. Biophys. Acta*, 668, 325, 1981.

272. **Granum, P. E. and Richardson, M.,** Chymotrypsin treatment increases the activity of *Clostridium perfringens* enterotoxin, *Toxicology*, 29, 898, 1991.

273. **Borriello, S. P., Larson, H. E., Welch, A. R., Barclay, F., Stringer, M. F., and Bartholomew, B. A.,** Enterotoxigenic *Clostridium perfringens:* a possible cause of antibiotic-associated diarrhoea, *Lancet*, 1, 305, 1984.

274. **Borriello, S. P., Larson, H. E., Barclay, F. E., and Welch, A. R.,** *Clostridium perfringens* enterotoxin-associated diarrhea, in *Recent Advances in Anaerobic Bacteriology,* Borriello, S. P. and Hardie, J. M., Eds., Martinius Nijhoff, Boston, 1987, 33.

275. **Jackson, S. G., Yip-chuck, D. A., Clark, J. B., and Brodsky, M. H.,** Diagnostic importance of *Clostridium perfringens* enterotoxin analysis in recurring enteritis among elderly, chronic care psychiatric patients, *J. Clin. Microbiol.*, 23, 748, 1986.

276. **Larson, H. E. and Borriello, S. P.,** Infectious diarrhea due to *Clostridium perfringens, J. Infect. Dis.*, 157, 390, 1988.

277. **Granum, P. E.,** *Clostridium perfringens* toxins involved in food poisoning, *Int. J. Food Microbiol.*, 10, 101, 1990.

278. **Duncan, C. L., Rokos, E. A., Christenson, C. M., and Rood, J. I.,** Multiple plasmids in different toxigenic types of *Clostridium perfringens:* possible control of beta toxin production, in *Microbiology-1978,* Schlessinger, S., Ed., American Society for Microbiology, Washington, D.C., 1978, 246.

279. **Jolivet-Reynaud, C., Popoff, M. R., Vinit, M. A., Ravisse, P., Moreau, H., and Alouf, J. E.,** Enteropathogenicity of *Clostridium perfringens* β-toxin and other clostridial toxins, *ZBL Bacteriol. Mikrobiol. Hyg. Abt. 1*, 15, 145, 1986.

280. **Gilbert, R. J., Rodhouse, J. C., and Haugh, C. A.,** Anaerobes and food poisoning, in *Clinical and Molecular Aspects of Anaerobes,* Borriello, S. P., Ed., Wrightson Biomedical Publishing Ltd., Petersfield, Hampshire, England, 1990, 85.

281. **MacDonald, K. L., Rutherford, G. W., Friedman, S. M., Dietz, J. R., Kaye, B. R., McKinley, G. F. Tenney, J. F., and Cohen, M. L.,** Botulism and botulism-like illness in chronic drug abusers, *Ann. Intern. Med.*, 102, 616, 1985.

282. **Arnon, S. S.,** Infant botulism: anticipating the second decade, *J. Infect. Dis.*, 154, 201, 1986.

283. **Anon,** Infant botulism (editorial), *Lancet*, 2, 1256, 1986.

284. **Arnon, S. S.,** Infant botulism, in *Clinical and Molecular Aspects of Anaerobes,* Borriello, S. P., Ed., Wrightson Biomedical Publishing, Petersfield, Hampshire, England, 1990, 41.

285. **Sugiyama, H.,** *Clostridium botulinum* neurotoxins, *Microbiol. Rev.*, 44, 419, 1980.

286. **Arnon, S. S.,** Infant botulism, pathogenesis, clinical aspects and relation to crib death, in *Biomedical Aspects of Botulism,* Lewis, G. E., Ed., Academic Press, New York, 1981, 331.

287. **Sugiyama, H., Mills, D., and Kuo, L. J. C.,** Number of *Clostridium botulinum* spores in honey, *J. Food Protect.*, 41, 848, 1978.

288. **Sullivan, N. M., Mills, D. C., Riemann, H. P., and Arnon, S. S.,** Inhibition of growth of *Clostridium botulinum* by intestinal microflora from healthy infants, *Microbiol. Ecol. Health Dis.*, 1, 179, 1988.

289. **Smith, G. E., Hinde, F., Westmoreland, D., Berry, P. R., and Gilbert, R. J.,** Infantile botulism, *Arch. Dis. Child.*, 64, 871, 1989.

290. **Turner, H. O., Brett, E. M., Gilbert, R. G., and Ghosh, A. C.,** Infant botulism in England, *Lancet*, 2, 1277, 1978.

291. **Midura, T. F. and Arnon, S. S.,** Infant botulism, identification of *Clostridium botulinum* and its toxin in faeces, *Lancet*, 2, 16, 1976.

292. **Sonnabend, O. A. R., Sonnabend, W. F. F., Krech, U., Molz, G., and Sigvist, T.,** Continuous microbiological and pathological study of 70 sudden and unexpected infant deaths: toxigenic intestinal *Clostridium botulinum* in 9 cases of sudden infant death syndrome, *Lancet*, 1, 1273, 1978.

293. **Bartlett, J. G.,** Pseudomembranous colitis, in *Human Intestinal Microflora in Health and Disease,* Hentges, D. J., Ed., Academic Press, London, 1983, 448.

294. **Rolfe, R. D., Helebian, S., and Finegold, S. M.,** Bacterial interference between *Clostridium difficile* and the normal fecal flora, *J. Infect. Dis.,* 143, 470, 1981.

295. **Tedesco, F. J.,** Clindamycin-associated colitis. Review of the clinical spectrum of 47 cases, *Dig. Dis.,* 21, 26, 1976.

296. **Tedesco, F. J., Barton, R. W., and Alpers, D. H.,** Clindamycin-associated colitis: a prospective study, *Ann. Intern. Med.,* 81, 429, 1974.

297. **Lyerly, D. M., Johnson, J. L., and Wilkins, T. D.,** The toxins of *Clostridium difficile,* in *Clinical and Molecular Aspects of Anaerobes,* Borriello, S. P., Ed., Wrightson Biomedical Publishing Ltd., Guildford, Surrey, England, 1990, 137.

298. **Lyerly, D. M., Phelps, C. J., Toth, J., and Wilkins, T. D.,** Characterization of toxins A and B of *Clostridium difficile* with monoclonal antibodies, *Infect. Immun.,* 54, 70, 1986.

299. **Lima, A. A. M., Lyerly, D. M., Wilkins, T. D., Innes, D. J., and Guerrant, R. L.,** Effects of *Clostridium difficile* toxins A and B in rabbit small and large intestine *in vivo* and on cultured cells *in vitro, Infect. Immun.,* 56, 582, 1988.

300. **Mitchell, T. J., Ketley, J. M., Burdon, D. W., Candy, D. C. A., and Stephen, J.,** Biological mode of action of *Clostridium difficile* Toxin A: a novel enterotoxin, *J. Med. Microbiol.,* 23, 211, 1987.

301. **Lyerly, D. M., Saum, K. E., MacDonald, D., and Wilkins, T. D.,** Effect of toxins A and B given intragastrically to animals, *Infect. Immun.,* 47, 349, 1985.

302. **Price, S. B., Phelps, C. J., Wilkins, T. D., and Johnson, J. L.,** Cloning of the carbohydrate-binding portion of the toxin A gene of *Clostridium difficile, Curr. Microbiol.,* 16, 55, 1987.

303. **Triadafilopoulos, G., Pothoulakis, C., O'Brien, M. J., and LaMont, J. T.,** Differential effects of *Clostridium difficile* toxins A and B on rabbit ileum, *Gastroenterology,* 93, 273, 1987.

304. **Rietra, P. J., Slaterus, K. W., Zanen, H. C., and Meuwissen, S. G. M.,** Clostridial toxins in faeces of healthy infants, *Lancet,* 2, 319, 1978.

305. **Donta, S. T. and Myers, M. G.,** *Clostridium difficile* toxin in asymptomatic neonates, *J. Pediatr.,* 100, 431, 1982.

*Chapter 10*

# The Role of Intestinal Bacteria in Etiology and Maintenance of Inflammatory Bowel Disease

*Vinton S. Chadwick and Robert P. Anderson*

## CONTENTS

    I. Definitions and Clinical Overview ..................................................................... 227
   II. IBD, Genetics, and Intestinal Bacteria ............................................................. 228
  III. IBD, Environmental Factors, and the Intestinal Microflora ................................ 230
  IV. Evidence Implicating Intestinal Luminal Bacteria in IBD ................................. 230
       A. Luminal Factors, Disease Distribution, and Site of Recurrence After Surgery ................. 230
       B. Quantitative and Qualitative Changes in Intestinal Luminal Bacteria in IBD .................. 232
       C. Changes in Intestinal Microflora with Surgical Diversion
          or Ileostomy Reservoir Procedures ...................................................... 234
    V. Intestinal Mucosal Bacteria and IBD ................................................................ 236
       A. Conventional Mucosal Bacteria ......................................................... 236
       B. L-forms of Mucosal Bacteria in IBD ................................................. 238
       C. Mycobacteria and IBD ..................................................................... 238
   VI. Translocation of Bacteria Across the Intestine in IBD ...................................... 239
  VII. Immunological Evidence for Involvement of Bacteria in IBD ............................ 240
       A. Immunological Evidence for Involvement of Mycobacteria in IBD ................... 241
       B. Specificity of Intestinal B and T Cells in IBD ...................................... 242
 VIII. Metabolic Activity of Commensal Bacteria in IBD ........................................... 242
   IX. Animal Models of IBD and Intestinal Bacteria ................................................ 242
    X. Inflammatory Products of Intestinal Bacteria and IBD ..................................... 243
       A. Endotoxin ....................................................................................... 243
       B. Peptidoglycan-Polysaccharide Complexes ........................................... 243
       C. Muramyl Peptides ............................................................................ 244
       D. Formyl Peptides .............................................................................. 245
   XI. Conclusions ................................................................................................... 247
References ............................................................................................................. 248

## I. DEFINITIONS AND CLINICAL OVERVIEW

Inflammatory bowel disease (IBD) is a term used by clinicians, pathologists, and researchers to describe Crohn's disease (CD), ulcerative colitis (UC), and nonspecific colitis. The latter may be an intermediate form of the others and is common in children. In this chapter, however, the focus will be on the classical forms of CD and UC, which have distinctive pathophysiological characteristics while sharing some genetic, clinical, and therapeutic features. Both CD and UC are chronic relapsing inflammatory disorders of the intestine and in this respect differ from self-limited inflammations which are caused by infections or toxins. These inflammations and those caused by ischemia or therapeutic radiation will not be discussed, although intestinal bacteria may play an important role in such conditions.

UC only affects the large intestine, whereas CD can affect any site from mouth to anus, although there is a predilection for distal small and proximal large intestine. Differentiating CD from UC can be problematic in patients with disease confined to the colon. The pathological and clinical features of these conditions are summarized in Table 1.

Although CD and UC are often confined to the intestine, and in that sense are not multi-system diseases, several extraintestinal conditions are associated with IBD. These complications (Table 2), which may occasionally antedate the bowel disease, may implicate the intestinal flora in disease pathogenesis.

IBD occurs worldwide but is most common in North America and Northern Europe and among caucasians in comparable latitudes in the southern hemisphere. This disease affects up to two million

**Table 1**  Characteristic Clinical and Pathological Features
of Inflammatory Bowel Disease

|  | Ulcerative colitis | Crohn's disease |
|---|---|---|
| *Clinical features* | | |
| Gradual onset | usual | usual |
| Diarrhea | usually blood and mucus | usually watery |
| Abdominal pain | infrequent | frequent |
| Fever | only if severe | frequent |
| *Endoscopic features* | | |
| Rectal involvement | invariable | frequently absent |
| Distribution | continuous and symmetrical | segmental and asymmetrical |
| Ulceration | diffuse with friability | focal aphthoid or linear |
| Cobblestoning | absent | frequent submucosal nodules |
| Fistulas | absent | quite common |
| *Pathological features* | | |
| Inflammation | mucosal | mucosal and submucosal |
| Goblet cell mucus | depleted | retained |
| Crypt abscesses | frequent | less frequent |
| Crypt architecture | distorted | often normal |
| Granulomas | absent | frequent (>60%) |

*Note:*  Anemia, hypoalbuminemia, raised sedimentation rate (ESR) and C-reactive protein
(CRP) are common features in both diseases.

**Table 2**  Extraintestinal Complications
of Inflammatory Bowel Disease

| Eyes | uveitis |
|---|---|
| | episcleritis |
| Mouth | aphthous ulceration |
| Skin | erythema nodosum |
| | pyoderma gangrenosum |
| Joints | pauciarticular peripheral arthritis |
| | ankylosing spondylitis |
| Liver/biliary | primary sclerosing cholangitis |
| | peri-cholangitis |
| | gallstones[a] |
| Renal tract | hyperoxaluria and nephrolithiasis[a] |
| | hydronephrosis[a] |

[a]  Associated with ileal resection or ileal Crohn's
disease.

people, and most of those are between the ages of 16 and 40. These diseases are a major cause of morbidity in this young population. It has been estimated in Scotland (population five million) that by the end of the century, CD alone will account for 35,300 hospital bed-days annually.[1] Fifteen percent of patients require surgery in any ten-year period. Hospital outpatient workloads are enormous. While some patients enjoy good health and long periods of remission, others suffer chronic ill health, major disruption of lifestyle, and a small number lose so much intestinal function that they require intravenous nutrition, sometimes for long periods. Table 3 summarizes the considerable advances in medical and surgical therapy in recent years. While therapeutic advances are encouraging, the major need is for further understanding of the genetic and environmental causes of these diseases.

## II. IBD, GENETICS, AND INTESTINAL BACTERIA

IBD is definitely a genetic disorder. Confirmatory evidence for this statement includes observations of familial aggregation, with high relative risks of disease in first degree relatives, concordance in twins,

**Table 3** Established Advances in
Medical and Surgical Therapy of IBD

| | |
|---|---|
| Antiinflammatory agents | 5-Aminosalicylates (5-ASA derivatives) |
| | delayed release oral preparations |
| | stable enema preparations |
| | Corticosteroids |
| | conventional systemic or local |
| | nonabsorbable or nonsystemic preparations |
| Antibiotics | Metronidazole for colonic CD |
| Immunosuppressants | Azathioprine/6-mercaptopurine |
| | Methotrexate |
| | Cyclosporin A |
| Operations for UC | panproctocolectomy/ileostomy |
| | ileo-anal pouch |
| Operations for CD | segmental resections |
| | strictureplasty |
| Dietary therapy | elemental and defined formula diets in CD |

*Note:* Selective antiinflammatory agents such as leukotriene B5 antagonists
and selective immunosuppressives (i.e., monoclonal antibodies, etc.)
are currently under clinical trial.

association with genetic syndromes, linkage with genetic markers, the presence of subclinical markers, and the racial predisposition in Jews.[2]

If a proband has UC, then first degree relatives have an eightfold relative risk of developing UC and a twofold risk of developing CD. If the proband has CD then there is a tenfold risk of first degree relatives developing CD and a fourfold risk of UC.[3] Concordance for CD in identical twins is 44% compared with 4% for dizygotic twins. In UC, concordance in monozygotic twins is 6% and there is an absence of concordance in dizygotic twins.[4] Interestingly, no case where one monozygotic twin has developed CD and the other UC has been reported, suggesting that genotype determines phenotype and yet CD and UC do co-exist within families, including first degree relatives. Environmental factors do not explain the familial incidence and systematic evaluation of spouses[5] of patients with IBD and the rare reported "clusters"[6,7] of cases show that the frequency in these cases is no greater than expected by chance.

The best model to explain genetic influences in etiology is "genetic heterogeneity."[2] According to this model, several different physiological defects resulting from several different disordered genes could result in predisposition to IBD. Indeed, multiple hits may be necessary, for example a defect in intestinal permeability, plus a particular type of immune response gene. Either of these abnormalities, in isolation, may not be sufficient to predispose to IBD and in other family or racial groups, other genes and therefore other pathophysiological mechanisms may be more important. Thus, IBD is associated with Turner's syndrome and the Hermansky-Pudlak syndrome, both genetic disorders, as well as with HLA DR4 in Japanese with CD and with certain C3 complement phenotypes (F and FS) in Danes with small bowel but not large bowel CD. These observations confirm genetic linkage but suggest multiple genetic influences. Complications of IBD such as primary sclerosing cholangitis and ankylosing spondylitis show linkage with other immune response genes, HLA B8 and HLA B27, respectively. Family studies indicate that first degree relatives of patients with IBD may exhibit certain disease markers such as autoantibodies to gut epithelial cell associated antigens (ECAC), antineutrophil cytoplasmic antibodies (ANCA), abnormal colonic mucins (species 4 deficiency, found in UC) and may have an underlying disorder of intestinal permeability (reviewed by Yang et al.[2]).

Unlikely as it may seem at first, the composition of the intestinal flora may be determined by the genetic constitution of the host and the genes involved may be relevant to disease pathogenesis.[8] The nasal[9] and fecal flora[8] are much more similar in monozygotic than in dizygotic twins. Patients with CD have an obligately anaerobic fecal flora composed of greater numbers of Gram-positive bacteria (*Eubacterium*, *Peptostreptococcus*, *Coprococcus*) and Gram-negative rods (*Bacteroides* and *Fusobacterium*) than is the equivalent flora of healthy subjects.[10] Approximately one-third of relatives of CD patients also have this "characteristic" flora, some of whom have developed CD subsequently.

## III. IBD, ENVIRONMENTAL FACTORS, AND THE INTESTINAL MICROFLORA

The geographical distribution, rising and falling incidence over short time scales, evidence for birth cohort effects, different prevalence of IBD in identical racial groups in different locations and the association with well established epidemiological risk factors confirm that IBD is an environmental disorder.[11]

In the northern hemisphere, there is an unexplained north-south gradient in incidence of CD. The incidence of CD has risen and fallen in some countries, while the incidence of UC increased in Scandinavia from <7 to 12 per $10^5$, between 1965 and 1983, mostly due to an increase in ulcerative proctitis, and the incidence of CD stayed constant at 6 per $10^5$.[12] A birth cohort effect has been shown in Sweden, where the incidence of extensive UC and CD has been falling in the age group 20 to 29 and rising in the age group 30 to 39 years.[12] The prevalence of IBD in Jews born and living in Israel is 220 per $10^5$, but in those domiciled in Israel but born in Asian and African countries it is 140, and for those born in America only 78.[13] Adverse perinatal events were 4.4 times more frequent in Swedish patients with CD,[14] and perinatal diarrheal illness had an odds ratio of 3.2 times in Canadian patients with UC.[15] Environmental influences are clearly prominent in IBD etiology[16] considering the well established association of nonsmoking and UC and smoking and CD.

A common theme in the above recent epidemiological studies is the probable importance of geographical location in early life, probably including the neonatal period and the likely role of adverse perinatal events as initiators of disordered immune reponses manifest later in life as IBD. Acquisition of appropriate immune tolerance to dietary antigens and appropriate but nonaggressive responses to ubiquitous bacterial antigens is occurring in this period. Immune responses to enteral antigen encountered in this "vulnerable period" have been implicated in food allergic disorders of infancy. Here, the apparent inability to develop appropriate tolerance mechanisms can recur following gastrointestinal infection in infants which produces transient and sometimes prolonged allergic disorders (i.e., milk protein, egg protein, soy protein intolerance, etc.).[17] However, although clinically important, these childhood disorders do not progress to CD or UC, either because this sort of mechanism is irrelevant to pathogenesis of IBD, or because IBD occurs by a similar mechanism but in individuals with different immune response genes.

The circumstantial evidence implicating normal intestinal luminal constituents (including ubiquitous dietary and bacterial antigens) in the pathogenesis of inflammation in IBD is now overwhelming, and reviewed below. The precise nature of the early environmental factors in IBD etiology remains obscure. However, an intestinal infection with a pathogenic virus or other agent at a crucial time point, which disturbs development of appropriate genetic, microbial, and dietary interactions in the neonatal gut would be consistent with current evidence. Thereafter, IBD would manifest as a problem of maintaining or re-establishing immunological tolerance whenever the intestinal mucosal barrier is transiently damaged.

## IV. EVIDENCE IMPLICATING INTESTINAL LUMINAL BACTERIA IN IBD

The evidence linking the inflammatory process in IBD with the intestinal commensal flora is compelling but circumstantial. Evidence comes initially from clinical and pathological observations on disease distribution, recurrence after surgery, and response to medical therapy, including antibiotics. Second, there is more formal microbiological evidence based on both microscopy and culture techniques. Third, there is evidence of permeation of inflammatory bacterial products across the intestinal mucosal barrier in IBD. Fourth, there is evidence of host immunity against commensal bacteria and their products. Finally, there is evidence from the role of bacteria in experimental animal models of IBD utilizing specific pathogen or germ-free lines, or especially susceptible strains with spontaneous disease or models created by transgenic or gene knockout manipulations.

### A. LUMINAL FACTORS, DISEASE DISTRIBUTION, AND SITE OF RECURRENCE AFTER SURGERY

From considerations of the distribution of UC (Table 1), one would suspect that the normal intestinal flora represented the source of the putative luminal factors, or ubiquitous antigens, which might be the target of host inflammatory responses. Indeed, while UC is confined to the colon and rectum, it is the more distal regions of the large intestine which are most likely to be involved in those with less than total colitis. These are not sites of highest bacterial concentrations but more likely sites of increasing bacterial death due to decreasing availability of nutrients and, at least in the rectum, reduced anaerobicity. Left hemicolectomy for

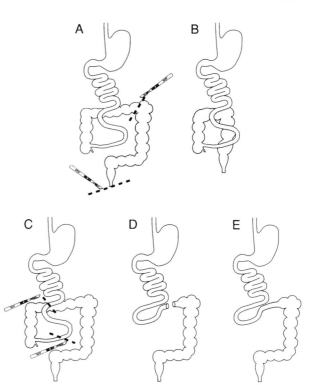

**Figure 1** Recurrence after surgical operations for ulcerative colitis and Crohn's disease. (A) Ulcerative colitis involving the L hemi-colon only and where L hemicolectomy was performed; (B) ulcerative colitis with inevitable recurrence of severe disease in residual colon after operation as described in A; (C) Crohn's disease involving distal ileum and proximal colon and where ileal resection with R hemicolectomy was performed; (D) Crohn's disease showing freedom from disease recurrence after operation described in C when defunctioning ileostomy as shown; (E) Crohn's disease showing recurrence of disease after operation described in C when primary or secondary anastomosis is performed. Recurrence involves the neo-terminal ileum.

localized left sided UC was abandoned, since it was invariably followed by severe colitis in the neo-left colon, even when resection margins were histologically free of disease (Figure 1).[18] It is difficult to implicate factors other than luminal bacteria in this situation. After total colectomy, intestinal inflammation does not recur in those patients with free draining ileostomies, but pouchitis occurs in a substantial proportion of patients with continent ileostomies[19] or ileo-anal pouch[20] operations (Figure 2). These pouches are distal intestinal reservoirs colonized by a bacterial flora intermediate between the normal ileal and colonic flora (see discussion later). Pouchitis occurs only in patients with previous UC, and not in those with previous polyposis coli who have colectomies for the prevention of colon cancer.

These important clinical observations suggest that intestinal inflammation in patients with UC is not strictly dependent on the presence of colonic epithelium (since the pouches are lined by ileal mucosa) but is directed against luminal factors present in reservoirs containing a large bowel type of flora and only occurs in those genetically predisposed or previously sensitized. The precise targets of the host response have not been determined but are presumably factors at highest concentrations in the most distal region of the bowel, whether that be rectum, colon, or ileal reservoir, the most likely factors being bacteria or their metabolic products.

The importance of luminal factors in the pathogenesis of CD has received rather more attention by clinical researchers than is the case for UC. In contrast to the situation described for UC, the distribution of CD suggests that both dietary and bacterial factors may be implicated. It is important to recognize that patients with CD may have a much more extensive intestinal abnormality than is evident from macroscopically visible lesions and extends into relatively sterile parts of the intestinal tract. If commensal bacteria or their products are involved, then their role may be to amplify the inflammatory response, thus determining the site of the major inflammatory lesions and the predilection for terminal ileum and colon. The evidence implicating luminal factors in pathogenesis is particularly compelling. Recent work on endoscopically defined anastomotic recurrence after right hemicolectomy for ileocolonic CD[21,22] showed that disease recurrence was essentially inevitable with rates as high as 80% at 6 months. The only patients protected from recurrence were those with temporary proximal ileostomy, none of who developed recurrence (Figure 1).[22] Following ileostomy closure, however, these patients were as susceptible to recurrence as those with immediate closure, which suggests that the normal fecal stream over the anastomotic site was critical in determining recurrence. This recent evidence adds support to previous work showing that diversion of the fecal stream may lead to improvement in CD and that reinfusion of

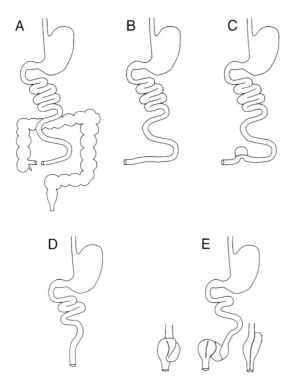

**Figure 2** Surgical operations used in treatment of either ulcerative colitis or Crohn's disease. (A) Exclusion of the colo-rectum; (B) pan-proctocolectomy with Brooke type free draining ileostomy; (C) pan-proctocolectomy with Kock type of continent ileal pouch; (D) pan-proctocolectomy with ileo-anal anastomosis; (E) pan-proctocolectomy with ileo-anal pouch (three variations shown).

ileostomy effluent into excluded colon may trigger events associated with disease relapse.[23] Factors in ileostomy effluent provoking symptomatic responses were apparently greater than 0.22 μm and there was little response to ultrafiltrate challenge. These studies and the apparent benefits of antibiotics, especially metronidazole, in colonic CD[24] may implicate intestinal bacteria in CD but does not exclude a major role for dietary factors.

Elemental diet (ED) therapy, as the only source of nutrition, induces rapid remission over 4 weeks in 70 to 80% of patients with CD, and is effective in both large and small bowel disease.[25] However, relapse is frequent on resumption of normal food and occurs most rapidly in those patients with colonic disease. Interpretation of the mechanism by which ED reduces disease activity is difficult because beneficial effects have also been reported with defined formula diets (containing polypeptides)[26] and even total parenteral nutrition, in combination with oral feeding.[27] Specific food intolerance (especially to milk and wheat proteins) has been identified by sequential challenge in some patients following induction of remission with ED. Prolonged remissions have been claimed for patients maintained on "patient tailored exclusion diets."[28] The effects of these dietary manipulations on the normal intestinal bacterial flora represents one possible mechanism for their actions, the other being removal or reduction of exposure to dietary antigens. The combination of ED and broad spectrum nonabsorbable antibiotics in CD was as effective as corticosteroids in one short-term trial,[29] which reinforced the importance of luminal factors in the inflammatory process.

It is of course not surprising that patients with inflamed and sometimes ulcerated intestines should show untoward reactions to intestinal luminal contents, either via direct mechanisms, priming or stimulation of immune responses normally downregulated in health. The evidence presented above could all be consistent with a secondary or permissive role for luminal factors in UC or CD. However, they may still be the dominant factors in perpetuating the inflammatory response, and whatever initiated the original disturbance may no longer be present and therefore inaccessible to intervention. Similarly, factors triggering relapse in those patients in remission may be multiple, and bear no relation to the initiating process and thus also be inaccessible to intervention.

## B. QUANTITATIVE AND QUALITATIVE CHANGES IN INTESTINAL LUMINAL BACTERIA IN IBD

Comparative studies of fecal bacteria in IBD are difficult because of differences in methodology, including elapsed time between collection and culture, number and variety of selective media, the nature of the sample (whether solid, liquid, blood or mucus), and whether counts are expressed per gram wet or dry weight. The appropriateness of the statistical analyses and precision of the methods may be criticized in some studies. It is likely that methods have progressively improved, so that more recent studies may more closely reflect the true scenario.

The earliest report of fecal bacterial counts in UC was in 1950 when Seneca and Henderson[30] applied aerobic culture techniques to samples from 17 patients. They found an increase in total bacterial concentration and coliform counts in UC. In contrast, Cooke[31] could find no significant difference in numbers or types of organisms in 20 patients with UC, though reported species were confined to *Escherichia coli, Enterococcus faecalis*, lactobacilli, bacteroides, clostridia, anaerobic streptococci, and proteus. The patients were described as being in acute relapse and were hospitalized suggesting moderate, to severe, activity of disease.

Gorbach and others[32] took a more sophisticated approach by studying the ecology of the fecal flora in untreated and treated patients. They set out to test the hypothesis that an imbalance or "dysbiosis" of the intestinal microflora might occur in IBD and chose typical cases of UC (n = 17) and CD (n = 8), with a range of extent and severity. None of the patients had been treated with steroids or antimicrobials for 6 weeks. The post-treatment study group comprised 12 patients with UC on salazopyrine and 8 UC and 2 CD patients treated with prednisone. Controls were 30 normal individuals (120 stool samples), 5 of whom took salazopyrine and 5 placebo to simulate a control-treated group. Stools were refrigerated for up to 24 h and 1 g of stool homogenized in sterile saline. Nine different media were used and concentrations were determined for total aerobes, total anaerobes, coliforms, streptococci, staphylococci, and micrococci. The authors' first conclusion was that the microflora of patients with mild to moderately active UC, whether distal or extensive, was not significantly different to that of normal controls. In severe extensive UC, concentrations of coliforms increased to 8.6 ($log_{10}$ per ml), compared with 6.4 in normals and 7.1 in distal colitis ($p < 0.01$). Significant increases in coliforms were also observed in both moderate and severe CD, 7.7 compared with 6.4 in normals ($p < 0.05$) and there was also a significant reduction in lactobacilli in CD (6.1 vs. 7.2, $p = 0.05$). There were no statistically significant changes with therapy in patients and only a trend for an increase in Gram-positive organisms in normal subjects on salazopyrine. The authors commented that the finding of increased fecal coliform concentrations with active disease and decrease with improvement could be a result rather than a cause of the diarrhea. They emphasized that the methods used yielded only gross estimates.

There have been several studies of the fecal flora of patients with CD and their first degree relatives.[33,34] A characterisitic anaerobic flora with more Gram-positive coccobacilli and Gram-negative rods has been reported. A major study involved 80 subjects in all, with 15 CD patients, 45 first degree relatives, and 20 unrelated healthy subjects.[10] Distribution of disease and previous operations were documented and an elapsed time of only 2 h between collection and processing of samples permitted. Total anaerobes, Gram-negative anaerobes, Gram-positive anaerobes (including eubacteria, bifidobacteria, cocci, coccobacilli and others), and total aerobes were quantified. Concentrations of Gram-positive coccobacilli were 15-fold, and of Gram-negative rods 2.5-fold, greater in CD than in healthy subjects (HS). This observation was not a consequence of operative resection. In CD, some Gram-negative rods (*Bacteroides, Fusobacterium*) increased at the expense of others (*Eubacterium, Bifidobacterium*). No effect of treatment with salazopyrine on fecal bacteria was noted. In a five to seven year follow up, three of the asymptomatic children who were relatives of patients with CD, and who had an abnormal flora, developed symptoms suggestive of CD, indicating that an abnormal flora antedates clinically apparent disease.

The above study and previous ones by the same group, which show genetic influences over the colonic microflora by comparison of monozygotic and dyzygotic twins[8] suggest that the "abnormal" flora in CD patients may be determined by genetic factors. Thus, the "characteristic flora" may be indigenous to subjects predisposed to CD. Others have shown that the composition of the colonic microflora is stable with time,[35] so that it is possible that the "abnormal" flora is present from a very early age. The suggestion of genetic control of the gut flora is consistent with observations made previously on the nasal flora.[9] The genes predisposing to CD may be simply be in linkage disequilibrium with those controlling the gut flora or the "abnormal" flora may be a causative factor in the pathogenic process of CD. One suggestion has been that *Coprococcus comes* strain Me46, a coccobacillus, may be proinflammatory under certain conditions by virtue of its ability to activate complement and its resistance to opsonization by specific IgG antibodies.[10,36] Other examples of possible involvement of specific organisms in CD and UC are considered later. In summary then, a genetic predispostion to CD may be accompanied by a genetic predisposition to permit intestinal colonization by an abnormal flora.

The anaerobic fecal flora in CD showing the increase in Gram-positive coccobacilli and in total Gram-negative rods has been confirmed by others.[34] Among the Gram-negative species, *Bacteroides vulgatus* accounted for 40% of the total in patients with CD but only 6% of the flora in healthy subjects. In a subsequent study,[37] higher percentages of Gram-negative rods were again found in patients with CD and coccoid rods were found in all patients with CD, but in none of the healthy subjects. Bacterial concentrations were expressed per gram dry weight of feces in these studies.

More recently, a semi-quantitative method for determination of fecal bacteria has been described[38] and applied to patients with both UC (37 patients), CD (42 patients), and healthy controls (21). Among the CD patients, one-half had clinically active disease and distribution, and the operative history and activity were recorded. Among the UC patients, 18 had active disease and overall, 11 patients had disease confined to the rectum. In essence, a standard loop was used to initiate a set of overlapping streaks to cross 5 regions of a primary isolation plate. Eight different media were used and growth scored as 1 to 5+. This method has previously been applied to urine and neonatal fecal samples. Validation was against the surface viable count method, but on only ten stool samples. Comparison between the tests was confined to simple correlations and no true test of agreement was applied. A 3+ score corresponded to values ranging over 3 logs in the surface count method, though in other categories the range was 1 to 1.5 logs. The advantages of this method, include its simplicity, which enables greater numbers of patients to be studied than is possible using standard quantitative methods.

The results of the study showed a significant increase in total numbers of aerobes in patients with active CD, compared to quiescent CD, UC, or controls, and in *E. coli* scores between active and inactive CD. Total anaerobic bacteria and bacteroides scores were similar in CD, UC, and controls. Lactobacilli and bifidobacteria scores were lower in CD. No differences in scores for *Bacteroides vulgatus* or *B. fragilis* were observed. There were no differences between UC and controls for any parameter. These results for aerobes and coliforms in CD are consistent with those of Gorbach et al.[32] however they are markedly discrepant from those of Wensinck,[33] van de Merwe and Mol,[39] van de Merwe et al.,[10] Ruseler-van Embden and Both-Patoir,[34] and Ruseler-van Embden and van Lieshout,[37] all of whom found increased numbers of coccobacilli and Gram-negative rods in patients with CD. The earlier study of West et al.[40] found no changes in the fecal flora in UC or CD, but only five patients with CD were studied. Keighley et al.[41] found increased coliforms, *B. fragilis*, and lactobacilli in operative ileo-colonic aspirates in patients with CD, but patients with UC had flora indistinguishable from controls irrespective of disease activity, but there was no increase in Gram-positve anaerobic coccobacilli.

It is impossible to come to a definitive conclusion concerning the fecal microflora in IBD, since there is no consensus. There is general agreement that diarrhea *per se* can lead to an increase in fecal aerobes and coliforms, so when these have been reported, they may be secondary and nonspecific. In UC, most authors have found no other distinctive change in number or composition of the flora irrespective of disease activity. The real question therefore is how we evaluate the so-called characteristic fecal flora of CD? Do patients with CD have a tenfold increase in anaerobic Gram-positive coccobacilli such as *Peptostreptococcus*, *Coprococcus*, and *Eubacterium* or not? Furthermore, do they have an increase in the proportion or absolute number of Gram-negative rods, and is the proportion of *B. vulgatus* increased as high as 40%? Is the positive finding of such changes more likely to be correct than a negative finding? These questions definitely need resolving.

## C. CHANGES IN INTESTINAL MICROFLORA WITH SURGICAL DIVERSION OR ILEOSTOMY RESERVOIR PROCEDURES

Surgical diversion of the fecal stream has been reported to be beneficial to the inflammatory process in patients with CD (Figure 2).[23,42] This is to some extent surprising, since fecal diversion may have deleterious effects when performed for other disorders. In particular, diversion colitis is a well recognized entity. The pathogenesis of this is incompletely understood, but the beneficial effects of the normal microflora in colonization resistance and in the formation of short chain fatty acids (SCFA), especially butyrate,[43] may be reduced when nutrients are diverted from the microbial environment. The condition responds to SCFA enemas.[44] It is therefore particularly relevant to consider microbial changes in the excluded colo-rectum after diversion, since in spite of the potentially deleterious effects, overall, they appear to be beneficial to those with CD.

In a study of 16 patients,[45] 10 with IBD, 8 of these with CD, the flora from the excluded colo-rectum was compared with feces of 16 healthy controls. There was a tenfold reduction in total anaerobes in the patients and these reductions were mainly in eubacteria or bifidobacteria. Interestingly, higher isolation rates for anaerobes, particularly peptostreptococci and bacteroides were found in patients with IBD, compared to those with other disorders such as diverticular disease. Again, we see a possible proclivity for patients with IBD presumably those with CD, who predominated in this study, to harbor an increased anaerobic flora including Gram-positive coccobacilli and Gram-negative rods. Among aerobes, enterobacteria were increased, and probably reflected loss of colonization resistance. Species such as *Proteus*, *Providencia*, and *Morganella* were found in patients but not controls. If these changes are beneficial to patients with CD, then a reduction in anaerobic flora is possibly the most likely change. It has to be remembered that diversion also reduces exposure to food mitogens and endogenous secretions, including bile salts. Changes in bacterial populations are not the only factors to consider.

In UC, there is no clinical improvement following diversion of the fecal stream.[42] The role of luminal bacteria in the inflammatory response, may however be inferred from the complication of pouchitis, which occurs after continent ileostomy or ileo-anal pouch operations for UC (Figure 2), but not after similar operations for polyposis coli. Conventional (Brooke) ileostomy is regarded as a curative operation in UC.[46,47] Inflammation in the terminal ileum is very rare and even then very mild. The ileostomy flora is not the same as normal ileal flora, but major changes are observed when the flora of conventional ileostomies are compared with those from an ileal or ileo-anal reservoir.[19,48,49] Several studies have been reported. In one,[50] there were 15 patients in each of three groups with conventional Brooke ileostomies, continent Kock pouch ileostomies, or pelvic ileal pouches. Two patients in the Kock group and four of the pelvic pouch group had acute or sub-acute pouchitis, which in every case subsequently responded to a course of metronidazole. Anaerobic bacterial counts were usually greater than 7 logs in patients with pouches and less than 3 logs in patients with free draining ileostomies. Concentrations of aerobes were similar among groups. The commonest anaerobe was *Bacteroides fragilis* ($10^5$ to $10^{11}$) in both pouch groups. Among aerobes, *E.coli*, *Proteus*, and *Klebsiella* predominated. Mucosal histology was either normal or showed mild chronic inflammation except in those with pouchitis, five of whom had acute inflammatory changes. Other studies[51,52] failed to demonstrate specific changes in the fecal flora of patients with pouchitis compared to those with similar pouches but without symptoms or acute mucosal inflammation. Nevertheless, a role for anaerobic flora is indicated due to the response to metronidazole and the association of inflammation with pouches containing high counts of anaerobes and not in free draining ileostomies with much lower anaerobe counts. Whether inflammation occurs clearly depends on the aggressiveness of the host response and does not require quantitative or qualitative changes in bacteria above those already present as a consequence of the presence of the pouch. In this respect, findings are consistent with those in UC where there are no consistent changes in quantity or quality of the fecal flora when remission is compared with relapse.

Failure to find microbiological correlates of acute pouchitis has not been universal. Santavirta et al.[53] compared mucosal morphology and fecal bacteriology in 30 patients with ileo-anal J-pouches and 10 patients with conventional ileostomies. They confirmed the increase in total bacterial counts and anaerobic counts in patients with pouches. The ratio of anaerobes to aerobes was, on average, 5000 with a pouch, and only 1 with conventional ileostomy. Some degree of mucosal inflammation was present in up to 70% of patients with pouches, whereas mucosal inflammation was graded as 0 in 9 out of 10, and 1 in only 1 out of 10 patients with conventional ileostomies. The authors claim a correlation between numbers of aerobes and the grade of acute inflammation, but this is markedly influenced by results in only two patients. Similarly, the correlation between grade of chronic inflammation and number of anaerobes was significant, but weak. Five of the nine patients with chronic inflammation had anaerobic counts within the range of those without any inflammation. This study therefore confirms that pouchitis can occur without significant change in fecal bacteria, when compared to those with pouches and no inflammation, while suggesting that the total load of anaerobes, and possibly aerobes, may be on the high side in some of those with inflammation.

In contrast to the situation in respect of the pouch itself, bacterial overgrowth proximally has been related to symptoms of diarrhea and evidence of malabsorption of nutrients in some patients after Kock ileostomies.[51] Jejunal aspirates showed an increase in anaerobic bacterial counts (range $10^3$ to $10^8$ per g aspirate). The symptoms, the malabsorption, and the number of jejunal bacteria decreased after treatment with metronidazole. Presumably, similar problems complicate ileo-anal anastomosis and are best viewed as metabolic and pathophysiological consequences of small bowel bacterial overgrowth distinct from the inflammatory processes of pouchitis. Jejuno-ileal motility is not much altered following creation of an ileo-anal pouch.[54] Transit is, however, slower[55] than after conventional ileostomy and considerably slower than normal orocecal transit. It is possible that this adaptation to colectomy in combination with loss of the ileocecal valve predisposes to bacterial overgrowth in more proximal regions of the intestine.

## V. INTESTINAL MUCOSAL BACTERIA AND IBD

Examination of the fecal flora is an appropriate way of assessing the composition of the colonic luminal flora, but almost certainly does not provide insights into the specific mucosally associated flora,[56] which has been demonstrated in experimental animals and probably exists in humans.

### A. CONVENTIONAL MUCOSAL BACTERIA

Since intramural bacteria had been visualized in tisssue from patients with CD, and a response to metronidazole and other antibiotics reported, the possibility that an abnormal mucosal flora might exist in CD has been investigated,[57] using full thickness operative samples from histologically abnormal and normal tissue from CD specimens, and control tissue from resections for colon cancer. Tissue for bacteriological study was snap frozen in glycerol transport broth and stored at $-20°C$ for up to one month before thawing and processing in an anaerobic chamber. A mucosal flora, or at least a positive culture, was found with all the large bowel and three quarters of the small bowel samples. Greater numbers of bacteria were associated with colonic tissue than jejunal tissue ($10^7$ to $10^8$ compared with $10^3$ to $10^4$) and ileal tissue was quantitatively intermediate. However, no increase in numbers of bacteria were found in CD tissue either affected or unaffected compared with controls. No bacteria were isolated from five mesenteric lymph nodes from five patients with CD. In controls, roughly one-half of the isolates were aerobic and half anaerobic, whereas in CD three quarters were aerobic isolates. Thus, in contrast to the luminal flora where anaerobes outnumber facultative aerobes by 100:1, the mucosal flora are equally or predominantly facultative, at least in terms of what can be cultured. In contrast with animal studies where certain strains of bacteria are isolated from different sites in the gut, no highly selective distribution was evident in this study. Many of the aerobic isolates were Gram-negative rods, bacteroides being the most common anaerobe isolated.

The advent of fiber-optic colonoscopy provided the opportunity to obtain multiple mucosal biopsies under direct vision. After standardized preparation, biopsies were taken using a sterile instrument introduced via a protective tube down its biopsy channel.[58] The forceps were pushed through a paraffin seal at the end of the tube prior to taking the biopsy from the mucosa. Tissue from each part of the colon was examined with diseased mucosa and adjacent normal appearing mucosa being identified. Strictly anaerobic culture techniques were used. Effects of sampling artifacts were examined by studying paired biopsies from closely adjacent regions, but no differences were observed. There were no differences in numbers of organisms isolated from ascending, transverse, or descending colon (6.3, 5.6, and 5.3 logs, respectively) while mean counts in the sigmoid, though significantly lower at 4.7 logs per mg, were attributed to enema preparation prior to colonoscopy. Mean values for numbers of anaerobes isolated were higher in polyps, cancers, and IBD affected mucosa, as compared to their respective control adjacent tissues, but only the difference for polyps reached significance. No difference in total numbers of organisms and recovery of genera were seen between disease states. Recovery of genera in all cases was *Bacteroides > Fusobacterium > Clostridium > Eubacterium > Peptostreptococcus*. This study has limited relevance to IBD since only five patients unspecified in respect of disease type or activity were included. Based on disease distribution, however, it is likely that at least three of the five patients had CD. While total numbers of anaerobes and genera showed no differences in diseased versus unaffected adjacent tissue in IBD, a greater number of species were isolated from the inflamed tissue.

In a study of the frequency of isolation of selected obligate or facultative enteropathogens from mucosal biopsies in patients with "chronic colitis," there were 29 patients with UC and 14 with CD.[59] In UC, microorganisms were isolated from 14 (48%) patients' biopsies and CD from 7 (50%). In the 20 healthy controls who had a rectal rather than a colonic biopsy, none of the organisms specifically searched for were grown. In patients with "nonspecific colitis or proctitis," 21 and 35% had positive isolates. Obligate enteropathogens (*Salmonella*, *Shigella*, *Yersinia*, *Campylobacter*, and *Cryptosporidium*) were not found in the UC or CD cases. Of the facultative pathogens, *Klebsiella* and *Pseudomonas* were isolated in 31 and 24%, respectively of patients with UC and 21 and 7% of patients with CD. These organisms were also found in "nonspecific colitis." In all colitis groups, *Chlamydia* were grown from approximately

20% of biopsies. Finally, *Aeromonas hydrophila* was found in 3.4% of UC biopsies and one or other obligate pathogen in 8% of patients with "nonspecific colitis."

Interpretation of these findings is difficult. In the case of IBD, the facultative pathogens may be acting via superinfection of mucosal sites or as commensals occupying a mucosal niche. The failure to find *Klebsiella* or *Pseudomonas* in normal rectal mucosa may be related to the fact that although these organisms can both be found among the normal gut flora they occur only in small numbers. Changes in IBD which may favor colonization of the mucosa with these genera is worthy of study. It still seems likely that the healthy mucosa is colonized with other Gram-negative rods, since they were found in a high percentage of control "normal" colonic tissue in the studies described previously. The role of *Chlamydia* which have been implicated in venereal proctitis but not previously in colitis is equally obscure though interesting.

Superinfection with pathogenic bacteria such as *Salmonella*, *Campylobacter*, and *Clostridium difficile* can be associated with IBD relapses or even at first presentation of IBD,[60] but this is the exception rather than the rule. If in addition we must consider superinfection with facultative pathogens particularly in terms of juxta-mucosal ecology, then a substantial proportion of presentations and relapses might be explained by "superinfection events" of one kind or another.

To shed some light on the rather controversial role of superinfection in IBD relapse, a prospective study was undertaken involving 64 patients with IBD, 49 with CD, and 15 with UC. Patients entered into the study at the time of relapse and colonoscopy was performed to obtain two or three biopsies from each region of the colon and the terminal ileum. Samples were transported in ringer solution and plated onto selective media. Stool specimens were also examined on the same media. Among the obligate enteropathogens, *C. difficile* was found in five patients with CD and one with UC. *Campylobacter jejuni* was found in one patient (CD) and *Salmonella typhimurium* in one patient (UC). Enteropathogenic *E. coli* were isolated from three patients (CD). The total number of positives was 11, or 17%. The most common pathogen was *C. difficile* as previously found in a similar study,[61] but its role in relapse is uncertain.[62] A carrier state of 15%[63,64] has been reported in asymptomatic adults and in patients with IBD,[61] *C. difficile* disappeared during treatment with steroids or salazopyrine. The rare occurrence of *Campylobacter jejuni*[65,66] and *Salmonella typhimurium*[67] confirms previous studies. *Yersinia enterocolitica* was not found in any patient, though positive serology was observed in four patients and attributed to nonspecific immune stimulation. *Aeromonas*, *Shigella*, and *Chlamydia* were not detected in this study and no evidence of superinfection with protozoa including *Giardia lamblia* was found, in contrast to other reports.[68] The authors conclude that enteropathogenic organisms play only a minor role in etiopathogenesis of IBD, and if found, require careful observation but not specific intervention with drugs. The same paper provides interesting data on culture of other organisms from tissue specimens. They isolated 15 bacterial and 6 fungal species, which were described as not generally related with enteropathogenicity. *Klebsiella pneumoniae* and *Pseudomonas aeruginosa* were found in 11 and 8%, respectively. *E. coli* was the most common organism being found in 84%, other genera included *Proteus*, *Enterobacter*, *Citrobacter*, and *Hafnia*. Among the fungi, *Candida albicans* was found in 39% with a particularly high frequency in CD (45%).

The question of whether it is "normal" to have such a rich flora associated with the intestinal mucosa and whether some of these organisms are facultative pathogens or innocent commensals cannot be answered from existing data. We lack microbiological studies on serial biopsies from healthy subjects to act as controls for the samples from patients with IBD. Suitable controls should be available for study, in ethically approved studies, since colonoscopy is performed frequently to investigate a possible large bowel cause for anemia. In many such patients, the large intestine proves to be completely normal. Until these studies are performed, it remains possible that the mucosal microflora is altered in IBD. The role of facultative pathogens is still questionable while the role of obligate pathogens in causation and relapse is almost certainly a minor one.

Studies of mucosal flora to date suggest that it is composed of organisms that are also culturable from feces, though the ratio of aerobes to anaerobes may differ markedly. There may also be qualitative differences in organisms in association with IBD. One such difference is in adhesivity of *E. coli* isolated from the stools of patients with IBD, both UC and CD, compared with isolates from healthy controls, or those with infectious diarrhea due to *C. jejuni*.[69] In these studies, an adhesion index is calculated after incubation of subcultured *E. coli* with buccal epithelial cells from a single donor, in the presence of D-mannose, and determination of the proportion of cells with more than 50 adherent Gram-negative rods. Adherent and nonadherent control organisms were included in the assay. Taking an adhesion index of

25% as the upper limit in control "nonadhesive" strains, 86% of isolates from patients with IBD were adhesive, compared with none in normal controls and 27% in infectious diarrhea. Adhesive strains were found with equal frequency in UC, in remission or relapse, and there was no difference between UC and CD. The adhesive property appears not to be a simple consequence of inflammation, is persistent in culture, is mannose resistant and therefore not due to type 1 fimbria. It may, by analogy with other virulence factors, be plasmid encoded. The authors suggest that transfer of plasmids from enteropathogenic intestinal bacteria to commensal *E. coli* may account for this phenomenon. Others have suggested that *E. coli* obtained from patients with ulcerative colitis can degrade mucins[70] and produce necrotoxins and hemolysins to a greater extent than those from control subjects. IgA$_1$ protease activity of the colonic flora has also been implicated as a possible pathogenic factor, but there is little evidence to associate it specifically with IBD.[71]

## B. L-FORMS OF MUCOSAL BACTERIA IN IBD

Another qualitative change in mucosal flora in IBD is the frequency of isolation of bacterial L-forms, otherwise known as cell wall deficient (CWD) organisms or spheroplasts. Two studies[72,73] have demonstrated a high recovery of bacterial L-forms from gut tissue in both CD and UC in up to 40 to 50% of samples, compared with tissue from control subjects where recovery was 1 to 6%. L-forms were recovered from both involved and uninvolved tissue and in 13% of uninvolved lymph nodes in CD. Isolation was independent of *in vivo* therapy with antibiotics, which are known to induce L-formation *in vitro*. Attempts to produce reversion to parental forms met with little success and revealed commensal organisms such as *Pseudomonas* species, *E. coli*, and *Enterococcus faecalis,* but not mycobacteria. However, other groups specifically aiming to culture tissue for mycobacteria, and also recovering CWD (spheroplasts)[74,75] which were variably acid fast, have succeeded in demonstrating *Myeobacterium paratuberculosis* in one or more of the many CWD isolates, using ribosomal RNA or DNA probes. This was the exception rather than the rule, however (see discussion later).

Theoretically, where the putative microbial etiological agent in a disease may be a cell wall-defective variant and therefore difficult to isolate, serological evidence of infection may be helpful. Shafii et al.[76] showed that 22 of 25 sera from patients with CD reacted in an immunofluorescence assay against CWD revertant strains of *Pseudomonas*-like organisms isolated from patients with CD. Weak positive fluorescence in UC sera and control sera was readily preabsorbed with a mixture of *P. aeruginosa, E. coli,* and *B. thetaiotaomicron.* This procedure did not diminish responses with CD sera. Fluorescence intensity with CD sera correlated positively with disease activity. *Pseudomonas maltophilia*[77] has been suggested as a possible CWD organism in CD, but serological evidence does not support this.[78] Serological evidence for and against involvement of chlamydia in CD has been published. Recent studies failed to find either an increased frequency of chlamydial antibodies (*C. trachomatis*) in CD or chlamydial DNA sequences in CD tissue using PCR.[79] Serological studies using mycobacterial antigens and cell mediated responses to different bacterial antigens are considered later.

Noncultivatable CWD organisms described as mollicute like have been shown to cause human chronic ocular inflammatory disease. These organisms, which parasitize vitreous leukocytes, have been found in the eyes of patients with UC[80] and CD[81] with idiopathic uveitis, leading to the suggestion that similar infections of the gut may underly the intestinal diseases.

If L-forms do play a pathogenic role in IBD, they could explain past failures to visualize organisms from tissues. They are resistant to host defenses, probably because they lack cell wall antigens, and are thus less antigenic which enables them to survive intracellularly. It is also possible that the high recovery of L-forms from both UC and CD tissue reflects secondary invasion or translocation of bacteria which then undergo L-transformation under the influence of hostile host factors such as leukocyte lysozyme which can damage bacterial cell walls and promote spheroplast formation.

## C. MYCOBACTERIA AND IBD

Mycobacteria are ubiquitous in the environment, so that isolation of mycobacteria from patients with CD does not necessarily imply involvement in etiopathogenesis. As with other facultative pathogens, *M. avium, M. paratuberculosis,* and *M. kansasii* can only be implicated in the pathogenesis of inflammation, if they are repeatedly isolated from normally sterile body sites, or are found repeatedly in association with a particular disease and not in appropriate controls. Finally, evidence of humoral or cellular immunity would be expected, since it is hard to envisage an inflammatory response to an organism in the absence of an immune response.

After much study, *M. paratuberculosis* remains a possible etiological agent in CD.[82] It is a member of the *M. avium* group, but can be distinguished from the other members of the group by use of DNA probes which recognize the multiple copies of a DNA insertion element known as IS900 in the *M. paratuberculosis* genome.[83] This organism has been isolated from only a small proportion of CD tissues examined to date (3 to 4%),[84-87] but never apparently from control or UC tissue. By implication then, if it causes CD, the bacilli must be extremely sparsely distributed in tissue (pauci-bacillary) or present as spheroplasts. *M. paratuberculosis* causes Johne's disease in ruminants[88] when it is multi-bacillary and usually easy to isolate, even though the organism has fastidious growth requirements. Johne's disease is characterized by an ileitis which is similar to that of CD. The difference in bacillary load in the two disorders is said to reflect the nature of the immune response with depressed cell mediated immunity (CMI) being found in Johne's and well developed CMI being found in patients with CD which accounts for the tuberculoid type of histopatholgy with granulomata and a low density of bacilli. Thus, it is argued that isolation rates of 3 to 4% in CD are still consistent with an etiological role for *M. paratuberculosis*.[82]

*Mycobacterium avium* infection is common and multi-bacillary in humans with AIDS because CMI is deficient.[89] Although *M. avium* has been isolated from CD, UC, and control tissues (1 to 2%) it is a distinct species and not the same as the one found in AIDS patients. This so-called Wood Pigeon strain can also cause Johne's disease in ruminants.[90] The situation is therefore far from clear. The argument that mycobacteria may be found as L-forms is also difficult to accept, since so few of the spheroplasts isolated from mucosal tissue reverted to mycobacteria, as referred to above.[72]

It might be hoped that recombinant DNA techniques would shed light on the situation. Using DNA solution hybridization,[91] mycobacterial DNA was reported in CD tissues, but other studies using Southern blotting of extracted DNA were negative. These studies had a reported sensitivity of one mycobacterial genome per 100 human cells.[92]

With the advent of PCR, and subsequently "nested PCR" techniques, sensitivity could be increased 1000-fold. Primers based on the IS900 insertion sequence, and a probe hybridizing to this sequence, were used to detect *M. paratuberculosis* genomes with a sensitivity of one bacterial genome per $10^4$ human cells. It is argued that even with this sensitivity a negative PCR may not exclude significant infection.[82] A positive PCR does not prove that the organism is causing the disease either, it may be a contaminant *in vivo* and cross-contamination *in vitro* is always a risk. Nevertheless, the pattern of results in patient groups should be subject to analysis and possibly interpretation.

Using DNA extracts of full thickness, samples of intestine removed at surgery from 40 patients with CD, 23 with UC, and 40 controls, *M. paratuberculosis* genomes were detected in 65% of CD (small intestine and colon), 4.3% of UC, and 12.5% of control tissues.[93] The authors concluded that positive results in controls were consistent with previously unsuspected alimentary prevalence in humans, but the overall results were consistent with an etiological role for *M. paratuberculosis* in CD. In another study,[94] 6 out of 18 cultures of CD tissues yielded positive results in PCR assays for IS900 DNA (only 1 out of 6 non-IBD controls was positive), but were described as being of low signal intensity. *M. avium* (IS902 PCR positive) was found in two of the CD, two of UC, and two of the non-IBD controls. This study was blinded and the PCR results bore no relation to visible bacillary forms of mycobacteria or spheroplasts in culture, which suggests that *M. paratuberculosis*, if present in the original tissue, did not replicate in culture.

The above studies are consistent, but not overwhelmingly so, with a role for *M. paratuberculosis* in CD, though this role may be one of passive association rather than active pathogenicity. Others, using PCR with a reported sensitivity of one bacterial genome per million human cells have failed to find even an association.[95]

## VI. TRANSLOCATION OF BACTERIA ACROSS THE INTESTINE IN IBD

It is now recognized that normal intestinal bacteria can pass from the intestinal lumen to extra-intestinal sites under a variety of clinically important circumstances, including immunosuppression, surgical stress, and trauma.[96] There is extensive experimental evidence indicating that anaerobic bacteria in the gut lumen, which do not translocate readily, may play a pivotal role in limiting translocation of facultative anaerobes such as *E. coli*, other enterobacteria, *Pseudomonas* species, and *Enterococcus* species.[97] High risk patients have been treated with selective antimicrobials to reduce translocation, and preserve the anaerobic flora of the gut, thereby reducing systemic infections.[98,99] Mechanisms controlling bacterial translocation are obscure, but both host and microbial factors are clearly important. Bacteria which can survive within macrophages, such as *Salmonella* or *Listeria* can translocate readily, and commensal bacteria may reach mesenteric lymph nodes within host phagocytes.[100] Although anaerobes do not easily translocate, they may do so in conjunction with aerobes where there is gross intestinal damage, for example caused by irradiation or mesenteric ischemia. There is also compelling evidence that aerobes such as *Ent. faecalis* can translocate across an intact intestinal mucosa, possibly by being taken up by enterocytes and then by macrophages.[101] This process also occurs with particles such as yeast and starch, which may be part of the normal antigen sampling mechanism of the gut.[96] As far as bacteria are concerned, translocation is increased as the concentration of a particular species increases in the lumin, and there is experimental evidence to show that elimination of the mucus layer results in an increase in populations of bacteria directly adherent to the enterocyte surface and increased translocation to extraintestinal sites. It is likely that IgA prevents adhesion and since it is not an opsonin and would not facilitate phagocytosis, which is a key translocation step. Conversely, IgM and IgG could promote translocation by this mechanism. The ability or otherwise of phagocytes to kill intracellular bacteria ultimately will determine the outcome of translocation.[96]

The possible relevance of the translocation phenomena, discussed above, to IBD relates first to direct microscopic observations of bacteria in the intestinal tissue of patients with CD[102] and UC.[103] A second line of evidence comes from bacteriological studies performed at the time of surgery, which show increased recovery of bacteria from serosal surfaces and mesenteric lymph nodes in IBD.[104,105] These studies support the idea that translocation is increased in IBD, but we are uncertain of the extent of this phenomenon in milder degrees of inflammatory activity, and do not know how frequently bacteria can be visualized or cultured in intestinal samples of healthy patients. When the intestinal mucosa is ulcerated, one might well expect translocation, but investigators also report detection of bacteria in nonulcerated mucosa.[103] Without further clarification, it is impossible to judge the significance of these findings in relation to the inflammatory response in IBD. Inflammatory products of intestinal bacteria certainly cross the epithelial barrier in active IBD, so the significance of small numbers of bacteria in mucosal tissues, as opposed to the adjacent lumin is arguable. In terms of the systemic consequences of translocation in IBD, there is a 44-fold increased representation of IBD among patients with native valve endocarditis[106] which may be related to the disease itself, or diagnostic manipulations, or to immunosuppresssive therapy.

## VII. IMMUNOLOGICAL EVIDENCE FOR INVOLVEMENT OF BACTERIA IN IBD

A major difficulty in defining a role for a particular bacterium in IBD is that isolation of the organism from feces or tissue does not necessarily implicate it in the disease process. Demonstration of a specific serological response would at least indicate host recognition of the agent. This determination could then be used as an argument for involvement of an organism if this response was greater in degree than responses to other bacteria and found in a large proportion of the patients with the disease, but not in appropriate disease controls, or healthy subjects. Unfortunately, this logic has substantial limitations in practice.

Notwithstanding the paucity of direct evidence implicating obligate enteropathogens in IBD, serological responses to the major pathogens have been studied. Blaser et al.[107] examined reponses to *Campylobacter*, *Yersinia*, *Listeria*, and *Brucella* in 40 patients with active CD (CDAI 154-509), who were well characterized and compared with age- and sex-matched healthy controls. Test antigens were well characterized prior to use. In a complement fixation assay, reciprocal geometric mean titer of the sera from patients with CD were higher to each antigen than was the titer of the controls, but this was only statistically significant in the case of *Yersinia pseudotuberculosis*. Ranges of serological responses were widespread in both patients and controls. The authors concluded that during periods of active bowel inflammation, polyclonal B cell stimulation may result in increased serum levels of cross-reactive antibodies directed at numerous species. Although significant in this investigation, *Yersinia* antibody titer had been negative in another study of 15 patients with recently diagnosed CD and it seemed unlikely this particular group of patients would have been exposed, on epidemiological grounds, to *Y. pseudotuberculosis*. Antibody titer did not correlate with any measure of disease activity, distribution, or duration. As part of the same study, an ELISA was used to measure serological responses to the common mycobacterial antigen arabinomannan. Mean and range of antibody levels were similar in patients and controls.

It is not surprising that serum antibody titer to commensal organisms including the facultative enteropathogens may be raised in IBD. Brown and Lee[108] found elevated titers to *B. fragilis* and

*Ent. faecalis* in patients with CD and UC, which correlated with disease severity rather than fecal concentrations. Tabaqchali et al.[109] found increased anti-*E. coli* antibodies in sera from patients with CD and UC, while Matthews et al.[110] found elevated antibody responses to *Peptostreptococcus* and *Eubacterium*, two of the major fecal anaerobes. The Rotterdam group[111] who described the characteristic fecal flora of CD described previously, also found increased agglutinating antibody titer against strains of *Eubacterium*, *Peptostreptococcus*, and *Coprococcus* species in patients with CD compared with appropriate disease controls and healthy subjects. Subsequently, the worldwide occurrence of agglutinating antibodies against four particularly selected coccoid anaerobes, *Eubacterium contortum*, (Me44 and Me47), *Peptostreptococcus productus* (C18) and *Coprococcus comes* (Me46), was assessed in 937 sera from patients with CD, UC and control groups, from 19 centers in 17 countries.[112] Results were expressed as positive or negative and tests read blind. Positive agglutination tests were found in 59% of CD, 29% of UC patients, and in 8% of healthy or diseased controls. Relative risks for CD patients of a positive test were 16.5 compared with healthy controls, and 5 for patients with CD confined to the colon compared with UC patients. In general, with the exception of the South African sera, all centers showed the same pattern of results, although the frequency of positive tests in CD varied from 35 to 83%. In South African sera, positive tests were equal in CD and UC (35%). Subsequently, purified antigens from *Coprococcus comes* have been used in an ELISA with the objective of providing a screening diagnostic test for CD.[113]

The above data are consistent with the proposition that the immune system is more frequently exposed to antigenic challenge with anaerobic coccobacilli, together with other intestinal anaerobes and aerobes in patients with IBD, than in healthy controls. Within IBD patient groups, those with CD may have a greater exposure than those with UC, either because of differential changes in bacterial concentrations within the lumin, or juxta-mucosally in the two diseases. Furthermore, in CD, inflammation usually involves a greater depth of intestinal wall and may more often be associated with bacteria within the bowel wall or adjacent mesenteric lymph nodes. Another study found elevated antibody titer to *Y. enterocolitica* and *Klebsiella pneumoniae* in patients with CD and UC, with highest levels in CD and no relation to disease activity, confirming the pattern referred to above.[114]

## A. IMMUNOLOGICAL EVIDENCE FOR INVOLVEMENT OF MYCOBACTERIA IN IBD

Studies of humoral immunity to mycobacterial antigens have given both positive and negative results in CD. Using a panel of three mycobacterial antigens, which were heterogeneous and not species specific, positivity to all three antigens was found in 18% of patients with CD and to at least one antigen in 84%.[115] Using an ELISA and eight mycobacterial sonicates,[116] a large proportion of patients with CD had antibodies that bound most antigens but there were no statisitical differences from responses with control sera, or sera from patients with UC, including responses to *M. paratuberculosis*. The findings were interpreted as indicating widespread contact with environmental mycobacteria, and not supporting a role for *M. paratuberculosis* in CD. Other studies of humoral immunity did not support a role for *M. paratuberculosis* either.[117,118]

It has been argued that the pauci-bacillary (tuberculoid) granulomatous type of mycobacterial infection characteristic of tuberculoid leprosy and the most likely model for *M. paratuberculosis* and CD is associated with strong T cell immunity. Studies of cell mediated immunity may be more relevant than antibody responses. An investigation of proliferative responses of both peripheral blood and mesenteric mononuclear cells (PBMC and MLMNC) to a range of mycobacterial and nonmycobacterial antigens revealed no specific sensitization to mycobacteria in association with IBD.[119] In particular, proliferative responses to *M. paratuberculosis* and *M. avium* were low in all groups. Increased CMI responses to nonmycobacterial antigens such as *Y. enterocolitica* and other enteric organisms (but not *Chlamydia trachomatis*) was observed in CD and UC MLMNC compared with controls.

Previous studies by Fiocchi et al.[120] demonstrated that mucosal T cells showed enhanced proliferation to bacteroides and staphylococcal antigens. Therefore, mucosal lymphocytes would be expected to be sensitized to bacterial antigens in the vicinity. Since *Y. enterocolitica*, like *M. paratuberculosis*, is an intracellular organism provoking a granulomatous response with demonstrable CMI, it is difficult to explain the absence of CMI to *M. paratuberculosis*, if it is the causative agent of CD.

One study showed that PBMNCs from patients with CD demonstrated little proliferative response to *M. paratuberculosis* antigen, but developed marked suppressor activity.[121] The possibility that suppression of CMI, or tolerance to this organism, is present in those with CD may still be considered, even though the association of tolerance with simultaneous inflammatory responses may be mechanistically impossible to explain. In contrast, reduced suppressor activity in response to mycobacterial antigens was observed by others,[122] in PBMNCs and lamina peripheral mononuclear cells (LPMNCs), similar to the nonspecific antigen suppression of PBMNCs previously reported.[123]

## B. SPECIFICITY OF INTESTINAL B AND T CELLS IN IBD

Hybridomas have been prepared from active B cells in lymphoid tissue draining lesions of CD and UC, by fusion of MLN suspensions with murine myeloma cells.[124] Supernatants were screened against pooled mycobacterial antigens, *B. fragilis* and *B. vulgatus*, two long-term mycobacterial isolates from patients with CD, and a pool of other organisms. Thirty percent of IgG secreting hybridomas from CD lymphocytes reacted with pooled mycobacteria and a similar proportion with pooled antigens from other intestinal organisms compared with about 15% in each case from UC derived hybridomas. Mesenteric lymphocytes are thus sensitized to some mycobacterial or common antigen, but there is no obvious clonality, indicating that one particular genus is most important in etiology. Similar conclusions were reached in a study that performed a molecular genetic analysis of the arrangement of immunoglobulin and antigen specific T cell receptor genes of isolated lamina propria lymphocytes from resected gut from patients with CD, UC, and other gastrointestinal disorders.[125] The sensitivity of the technique was sufficient to detect a monoclonal population of as little as 1% clonal expansion in a mixed cell population. In essence, the B and T cells were polyclonal, consistent with a broad immune response to a multiplicity of antigens, thus providing no insight into antigens which may be important in etiology of inflammation.

## VIII. METABOLIC ACTIVITY OF COMMENSAL BACTERIA IN IBD

Changes in the composition, morphology, and properties of the commensal flora in IBD outlined earlier are difficult to evaluate and depend largely on culture techniques that convey little information about bacterial turnover and metabolism. Determination of concentrations of short chain fatty acids (SCFA) and other organic acids as markers for anaerobic bacterial metabolism has shown increased succinate concentrations in stool samples from patients with UC,[126] presumably reflecting higher numbers or metabolic activity of certain groups of anaerobes. Other studies found increased concentrations of SCFA, especially N-butyrate, but related this to deficient absorption/utilization by the colonic mucosa, rather than increased production by luminal bacteria.[127]

## IX. ANIMAL MODELS OF IBD AND INTESTINAL BACTERIA

In experimental carrageenan colitis in guinea pigs, there is compelling evidence to show that without a commensal flora (germ-free state) there is no colitis.[128] Furthermore, a specific role for obligate anaerobes (metronidazole-sensitive), in particular *B. vulgatus*, was shown in the initiation of colitis. Established colitis was not responsive to metronidazole which implies that other commensal organisms may maintain the inflammatory response once established.[129] Further compelling evidence exists for a role of the commensal flora in nonsteroidal antiinflammatory drug (NSAID)-induced ileitis,[130] radiation-induced colitis, graft-vs.-host disease of the intestine, and ischemic colitis in experimental animals as the germ-free state markedly attenuates or abolishes the inflammatory response in each case.

No current animal model exactly resembles IBD. Some models, such as acetic acid colitis,[131] are merely applications of toxic substances to the intestine of conventionally caged rodents. These are models of intestinal inflammation, and perhaps only give an insight into the general processes of mucosal damage, inflammation, and restitution. The differential susceptibility of inbred strains of rats and mice, for example Buffalo (resistant) vs. Lewis (susceptible) after intramural injection of peptidoglycan,[132] may also assist in implicating colitis-related genes. Transgenic technology has further advanced this line of research with the development of the HLA-B27/β-microglobulin transgenic rat.[133] One line of rats, but not mice, expressing these genes spontaneously develops colitis and small intestinal inflammation, destructive peripheral and axial arthropathy, skin and nail dystrophy, orchitis, myocarditis, and keratitis. These manifestations resemble those seen in human HLA-B27-associated ankylosing spondyloarthropathy, perhaps with the exception that enteritis is widely recognized in human disease. (However, the recent study of De Vos et al.[134] documents asymptomatic ileitis and colitis in as many as 71% of HLA-B27 arthropaths.) Although this transgenic model is not strictly of IBD, it does offer great opportunities for studies of pathogenesis, and it is the first demonstration of a restricted DNA sequence affecting intestinal inflammation.

## X. INFLAMMATORY PRODUCTS OF INTESTINAL BACTERIA AND IBD

The inflammatory response in IBD may be directed toward factors arising from the gut lumin, rather than the intestinal mucosa itself.[135-137] Furthermore, radiolabeled granulocyte studies of Saverymuttu et al.[138] have shown that the proportion of the injected radiolabel entering the lumin and finally collected in stool is directly related to the severity of IBD. Though this luminal migration of neutrophils is not specific to IBD, it suggests that proinflammatory, chemotactic factors arising in the lumin may maintain mucosal inflammation. Indeed, it is now well established that soluble products of bacteria can produce all the histopathologic and immunopathologic responses attributable to intact viable organisms.

The commensal intestinal flora secrete, or contain within them, a number of potent inflammatory products such as lipopolysaccharide (LPS or endotoxin), peptidoglycan-polysaccharide complexes (PG-PS), muramyl peptides (MDPs), and N-formylmethionyl oligopeptides such as N-formylmethionyl-leucyl-phenylalanine (FMLP). It is not known why these particular products stimulate such a wide range of inflammatory responses while other products do not. There are substantial interactions among these substances; the priming, immunoadjuvant, and direct effects of these products could account for most of the intestinal inflammatory and immunological activity observed in IBD.

### A. ENDOTOXIN

Endotoxin (LPS) is a major component of the outer cell wall of Gram-negative bacteria. These bacteria do not secrete endotoxin but rather shed this outer cell wall component during normal growth and cell division, and at the time of cell lysis or death. As these processes are occurring continuously in the heavily colonized gut, free endotoxin is always present in gut contents. Probably 1 to 3 mg of LPS is present per gram of feces.[139]

The healthy intestine is an efficient barrier to all except trace amounts of endotoxin taken up by pinocytosis.[140] A striking increase in endotoxin permeation is produced experimentally by altering intestinal permeability, probably by opening up paracellular pathways.[141] Systemic endotoxemia occurs following saturation of hepatic clearance mechanisms and with lymphatic transport.[142] Indeed, in experimental peritonitis, about one-half of the absorbed endotoxin occurs via lymph, and the remainder by portal venous blood.[143] In humans, transient endotoxinemia may follow colonoscopy as a result of either bowel instrumentation alone or mucosal biopsy. The incidence quoted varies from 9 to 65%.[144] Although this may have no clinical significance, it supports the view that injury to the mucosal barrier of the gut results in an impressive transmucosal leak of luminal endotoxin.

In IBD, endotoxemia is present in all patients with very active disease, but in only about 10 to 12% of regular outpatients. In one study, plasma endotoxin levels correlated with the Van Hees index of disease activity and endotoxin levels fell progressively during therapy with total parenteral nutrition and steroids.[145] The addition of total gut irrigation to the aforementioned therapy resulted in a substantially more rapid decline in endotoxin levels and quicker onset of remission.

Endotoxin activation of inflammatory cells is thought to be mediated by a specific plasma membrane receptor.[146-148] Cell activation by endotoxin is accompanied by an increase in intracellular calcium concentration.[149] Much attention has been focused on macrophage/monocyte responses to endotoxin and their role in septic shock.[150] Endotoxin activation of macrophages/monocytes results in release of tumor necrosis factor (TNF), interleukins 1 and 6 (IL-1, IL-6), platelet activating factor (PAF), granulocyte-monocyte colony stimulating factor (GMCSF), and arachidonic acid metabolites (LTB4, LTD4, TXA2, PGE2, and PGI2).[151] Endotoxin also acts in a permissive manner *in vivo* and *in vitro* to heighten neutrophil and monocyte responsiveness to later challenge by stimulants such as bacterial formyl peptides.[149,152] *In vivo* this neutrophil priming effect may also be enhanced by endotoxin-stimulated release of macrophage derived GMCSF, PAF, and TNF. T and B lymphocyte activation is also part of the spectrum of endotoxin effects; LPS is mitogenic and a direct B cell stimulant, which induces immunoglobulin production.[150]

### B. PEPTIDOGLYCAN-POLYSACCHARIDE COMPLEXES

Peptidoglycan-polysaccharide (PG-PS) polymers are structural components of the cell walls of both Gram-positive and Gram-negative bacteria.[153,154] PG-PS polymers induce arthritis, uveitis, pancarditis, and hepatic granulomata after enteral injection. The arthritis has an acute phase (complement dependent) and a chronic phase (probably T cell dependent) and histologically resembles rheumatoid arthritis.[155] Intramural injection of streptococcal PG-PS into the cecum in mice induces chronic granulomatous inflammation.[156] Peptidoglycan was demonstrated in the granulomata by immunofluorescence. The inflammatory response to PG-PS was described as resembling CD but not showing mucosal ulceration. A phenomenon analogous to flare-ups in IBD could be induced in these colitic animals by intravenous injection of endotoxin.[157]

Absorption from the gut of radiolabeled PG-PS was demonstrated by the appearance of immunoprecipitable radioactivity in several organs.[156,158] The commonly described hepatobiliary complications of IBD prompted the investigation of PG-PS in bile. Biliary excretion of PG-PS delivered intraluminally or intravenously was less than 1%.[159] It was not clear whether significant metabolism of PG-PS occurred prior to excretion in bile.

Different strains of rats show variable susceptibility to the inflammatory effects of PG-PS. Lewis rats are high responders; Wistar rats, intermediate; and Buffalo rats, least susceptible.[160,161] Furthermore, in high-responder rats, bacterial overgrowth produced by the creation of self-filling jejunal blind loops produces liver dysfunction and intrahepatic biliary abnormalities resembling sclerosing cholangitis. These responses were abrogated by metronidazole, which implicated the anaerobic flora[162] presumably via absorption of bacterial products, since viable bacteria were not recovered from the liver or blood. Increased titers of anti-PG antibodies (IgA, IgM, and IgG in plasma, and secretory IgA in intestinal contents) were present in rats with blind loops. Other bacterial products could also be implicated in this model of sclerosing cholangitis as both endotoxin[163] and FMLP[164] cause release of peptidoleukotrienes from macrophages, which are then excreted in bile,[165] and FMLP itself undergoes efficient biliary excretion.[166,167]

Although PG-PS produces different responses in different strains of rats, which suggests genetic variation, PG-PS from different species of bacteria vary considerably in inflammatory potential. Some, such as Group A streptococci, enterobacteria, and mycobacteria produce protracted granulomatous inflammation whereas others such as *Peptosteptococcus* produce only transient inflammation. Clearly, both the pattern and chronicity of inflammation to PG-PS are critically dependent on the intrinsic properties of the microorganisms involved and the host responses.

Peptidoglycan-polysaccharide complexes have a broad spectrum of proinflammatory effects.[168-170] Like LPS, they activate macrophages, with secretion of broad-spectrum cytokines including TNF, interleukin-1, interleukin-6, prostaglandins, thromboxanes, and leukotrienes. They also activate complement and kinin systems. They activate both T and B lymphocytes. The complexity of these actions and the potential for prolonged effects may be due to the pharmacokinetics of PG-PS, which is phagocytosed and poorly biodegradable. A prolonged leukocytosis and monocytosis, lasting up to 20 weeks, follows PG-PS injection into susceptible Lewis rats. It has been suggested that PG-PS causes persistent stimulation of GMCSF.[155] Consequently, tolerance mechanisms may not operate for PG-PS as readily as they do for endotoxin.

## C. MURAMYL PEPTIDES

Discovery of the potent immunoadjuvant properties of Freund's complete adjuvant (FCA), an emulsion of mycobacterial cell wall in oil, stimulated a search for the cell wall substituent conveying bioactivity. Eventually, the minimal structure capable of replacing mycobacterial cell wall in FCA was shown to be the low molecular weight glycopeptide N-acetyl muramyl-L-alanyl-D-isoglutamine (MDP).[171,172] Many bacteria, both Gram-positive and Gram-negative, possess glycopeptides similar to MDP in structure and bioactivity.[154] These peptides are subunits of the peptidoglycan polymers that endow the bacterial cell wall with structural rigidity. It is to these subunits that the antibiotic vancomycin binds.[173,174] At the time of death, peptidoglycan can be degraded by microbial and mammalian enzymes to release free muramyl peptide. Muramyl peptide is also released from macrophages following phagocytosis of bacteria.[175] As the gut is a reservoir of MDP-like molecules, it seems surprising that no attempts have been made to involve these substances in the pathogenesis of inflammatory or infectious diseases of the bowel.

Muramyl peptides have diverse biological effects. There is accumulating evidence that these activities are mediated via a specific receptor on inflammatory cells.[176,177] They can produce uveitis and polyarthritis in mice, are pyrogenic and proinflammatory, are somnogenic (sleep-inducing), activate macrophages to release monokines, prime both macrophages and neutrophils so that responses to FMLP are enhanced,[178] are mitogenic for B cells, and enhance T-helper cell and T-suppressor cell function.[179]

No study has systematically investigated absorption of muramyl peptides from the gut; however, there is indirect evidence for absorption, because orally administered muramyl peptides retain immunoadjuvant activity, albeit at doses 20 times higher than those required by systemic administration.[178] Absorbed muramyl peptide appears to be excreted almost completely in urine.[180-183] Indeed, endogenous muramyl peptides have been isolated from urine[184,185] as well as measured by immunoassay[186-188] following penicillin treatment in humans.[189]

## D. FORMYL PEPTIDES

The observation that bacterial secretions were chemotactic for mammalian neutrophils was made more than 20 years ago.[190,191] Schiffman et al.[192] later demonstrated a directed rabbit neutrophil response (chemotaxis) to low molecular weight (150 to 1500 $M_r$) relatively heat stable factors from filtrates of *E. coli* culture medium. Chemotactic activity was resistant to aminopeptidase but not carboxypeptidase digestion, prompting the suggestion that the factor(s) may be N-terminal-blocked peptide(s). Further characterization of this chemotactic activity was limited by the low levels of material available. Schiffman and colleagues subsequently reported that synthetic formyl-methionine and formyl-methionyl peptides, but not the same molecules without the N-formyl group, possessed leukocyte chemotactic activity and that these molecules could be similar to the bacterial chemotactic factors.[193]

Formylmethionine is unique to prokaryotes[194] and mitochondria[195] as the first amino acid assembled at the ribosome during synthesis of all proteins. During posttranscriptional processing, formylmethionine and the subsequent 20 or so amino acids direct the newly synthesized protein to its site of action, e.g., cell membrane or extracellular medium if secreted.[196] This amino acid sequence is called the "signal sequence" and is usually clipped off once the nascent polypeptide arrives at its site of action.[197] Hence, Schiffman et al.[193] reasoned that bacteria are a ready source of formylmethionyl peptides fitting the general requirements for chemotactic activity defined by earlier studies.

Detailed structure-activity and genetic studies have now defined the DNA sequence and requirements for ligand potency of the plasma membrane receptor that mediates the inflammatory responses of formyl peptides.[198] Optimal activity is achieved with a ligand possessing an N-terminal blocked by a formyl group, then followed by methionine and at least two further hydrophobic amino acids. There is no strict requirement for the carboxyl-terminal to be free or blocked.[199-201] Bioactivity falls 10- to 100-fold with substitution of methionine by its oxidized form, methionine sulfoxide.[202] The only known endogenous mammalian ligand for the formyl peptide receptor is substance P.[203] However, substance P has an affinity for the formyl peptide receptor 100- to 1000-fold less than the index peptide formyl-methionyl-leucyl-phenylalanine (FMLP).

Improved chromatography and the development of immunoassays has allowed the purification of authentic FMLP from *E. coli* culture supernatants.[204,205] Several other N-blocked methionyl peptides were sequenced from bioactive fractions of HPLC-fractionated *E. coli* supernatant, but FMLP accounted for most of the bioactivity present.[204] FMLP-like immunoreactivity was also demonstrated; co-eluting with FMLP on HPLC from the culture supernatants of various other gut commensals: *Proteus vulgaris*, *Klebsiella pneumoniae*, *Enterococcus faecalis* and *Bacteroides fragilis*.[206] Authentic FMLP has also been isolated and sequenced from *Helicobacter pylori* culture supernatants in our laboratory.[207,208] Others have isolated a potent proinflammatory formyl tetrapeptide (F-Met-Leu-Phe-Ile) from *Staphylococcus aureus* culture medium.[209,210] Recently, another class of nonformylated bacterial oligopeptide, the *Ent. faecalis* sex pheromones has also been described as having mild formyl peptide receptor agonist activity (approximately 100-fold less potent than FMLP).[211]

It appears that the production of formyl peptide receptor agonists is relatively ubiquitous among both aerobic and anaerobic bacteria. FMLP is one of a group of naturally occurring formyl peptides activating acute inflammatory cells. Other naturally occurring nonformylated peptides, including substance P and bacterial pheromones, also have mild formyl peptide receptor agonist properties, but their significance *in vivo* is unclear.

Library searches for the *E. coli* gene encoding an amino terminal met-leu-phe sequence revealed only one polypeptide including this sequence, the UmuD protein,[212] which is part of the SOS operon whose function is to repair damaged DNA.[213] The met-leu-phe sequence occurs immediately at the amino-terminal of UmuD, implying the methionine residue is N-formylated. The SOS operon is "de-repressed" by single-strand DNA that is activated following exposure of genomic DNA to ultraviolet light, hydrogen peroxide, or other mutagens. A genetically engineered *E. coli* strain (MC 1000) that contains a plasmid (pSB13) encoding a single copy of the UmuD gene with a heat-sensitive repressor[214] was used to show a precursor product relationship.[215] When de-repressed at 42°C, the MC 1000 strain produced substantial

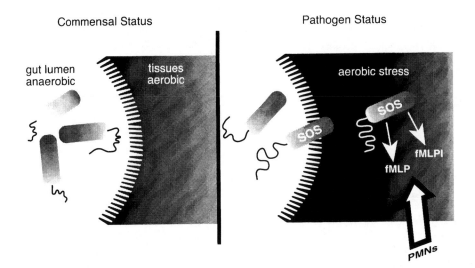

**Figure 3** Activation of "SOS" genes in *E. coli* exposed to aerobic stress associated with high host tissue $pO_2$. The inflammatory bacterial peptide fMet-Leu-Phe (fMLP) is produced from the N-terminus of a SOS protein called UmuD, which is an enzyme involved in bacterial DNA repair. Increased fMLP production may indicate to the host that the status of the bacteria has changed from commensal to pathogen.

amounts of a cell-associated, 12 kD polypeptide binding a specific anti-FMLP antiserum. The filtered supernatant from this MC 1000 culture medium contained two immunoreactive oligopeptides. Amino acid sequencing identified these secreted peptides as FMLP, and F-met-leu-phe-ile. (The fourth amino acid coded by the UmuD gene is isoleucine.) Exposure of conventional bacteria, aerobic *E. coli* K12, and anaerobic *B. vulgatus* to mutagens (hydrogen peroxide or UV light) was followed by a fivefold increase in secretion of FMLP-like immunoreactivity. This increase was blocked by pretreatment with chloramphenicol, implying a requirement for synthesis of new protein.[215]

The inducible UmuD precursor protein would appear to be a source of FMLP and F-met-leu-phe-ile in *E. coli*. However, the magnitude of contribution from this protein's synthesis to total *E. coli* production of FMLP remains unclear. There may be as yet unsequenced genes that also code for FMLP. Other *E. coli* formyl peptides whose sequences are known, do not correspond to the signal peptides of SOS operon products.[212]

Mammalian inflammatory cells have evolved a receptor recognizing a defined group of secreted bacterial peptides. Immunologically naive inflammatory and immune-competent cells do not recognize bacterial cells *per se*. Presumably, this detection system has evolved to allow an immediate directed inflammatory response to invasive bacteria, independent of delayed immune mechanisms. Understanding the possible bacterial origins of FMLP has suggested that the mammalian FMLP receptor may mediate a dynamic signaling system. The presence of bacteria in an adverse niche with increased oxidant stress, related to host production of free radicals, or simply elevated oxygen partial pressure, is marked by increased secretion of FMLP. Furthermore, FMLP itself stimulates neutrophils to release oxygen free radicals so potentially forming a positive feedback system. Hence, the formyl peptide receptor may have evolved to detect this inducible peptide as a marker of tissue invasion by bacteria (Figure 3).

Formyl peptide receptors have been described on human, nonhuman primate, rat, rabbit, murine, porcine, and equine granulocytes.[216] In humans, the FMLP receptor is inducible and the specific granule represents a mobilizable reservoir of FMLP receptors that are shifted to the plasma membrane with degranulation.[217-219] The "quiescent" neutrophil generally expresses 5 to $15 \times 10^3$ plasma membrane FMLP receptors, although this figure varies with the method of cell preparation and assay (fluorescent vs. radiolabeled ligand).[220-222] Neutrophils activated by exposure to cytokine,[223] growth factors,[224] endotoxin,[152,222] or over-vigorous preparation[220,225] express up to tenfold more surface receptors. Commensurate with this increase, whether cause or effect, neutrophil secretory and oxidative responses to equivalent doses of FMLP are dramatically enhanced *in vitro*[152,226,227] and *in vivo*.[178,228,229] This pretreatment with increased responsiveness has been termed "priming."[230] Indeed, in experimental animals, FMLP can substitute for the second injection of endotoxin in the Schwartzman phenomenon[231] and endotoxin pretreatment markedly increases inflammatory responses in the lung following FMLP challenge.[232] It is noteworthy that circulating neutrophils from patients with active CD and UC have increased numbers of FMLP receptors and enhanced responsiveness to FMLP.[233]

The biological actions of FMLP are very similar to those of endogenous chemoattractants, leukotriene $B_4$,[234] complement fragment 5a,[235] and interleukin 8.[236,237] At "low" levels, $10^{-10}$ to $10^{-8}$ $M$, exposure of neutrophils to FMLP concentration gradients stimulates cell polarization and directional movement up the gradient.[238] At higher concentrations of FMLP, $10^{-8}$ to $10^{-5}$ $M$, cell adhesiveness increases, motility ceases, and degranulation occurs with release of proteases to the interstitium.[239] The neutrophil is also stimulated to synthesize and release prostaglandins, endogenous chemoattractants ($LTB_4$), and oxygen free radicals. Not only neutrophils but monocytes, eosinophils, basophils, macrophages, endothelium, mast cells, activated B cells, and certain tumor cell lines possess specific receptors for FMLP.[239]

In view of the potent biological effects of FMLP, it is not surprising that a pronounced inflammatory response follows the introduction of FMLP across epithelial barriers such as the lungs, and skin.[240,241] In the intestine, colonic infusions in mice and rats[242] and rectal administration in rabbits resulted in experimental colitis,[243] although the concentrations used in these studies were at least 1000-fold greater than those estimated by bioassay of intestinal contents.[244] Changes in vascular permeability, blood flow, and mucosal permeability were produced with FMLP in the rat small intestine.[245] These effects were not found in animals rendered neutropenic. Infusion of FMLP into mesenteric arteries resulted in the appearance of leukotrienes ($LTB_4$) in mesenteric veins which suggests the presence of resident cells in the gut which respond to formyl peptides by eicosanoid synthesis.[246] Furthermore, FMLP is spasmogenic for intestinal smooth muscle, an effect mediated by enteric nerves and M1 cholinergic pathways and dependent on the integrity of the mucosal layer of the gut.[247]

Studies in the rat with both tritiated FMLP and radioiodinated formyl-methionyl-leucyl-tyrosine demonstrated that the intestinal mucosa is relatively impermeable to these peptides, whereas substantial enzymatic degradation of peptides occurs intraluminally.[248] Nevertheless, in both the ileum and the colon, the mucosal barrier is not 100% efficient, and intact peptide (1%) was absorbed. Absorbed intact peptide was excreted in bile and underwent enterohepatic circulation.[166] In rats with experimental colitis, colonic absorption and enterohepatic circulation of intact peptides increased eightfold[247] and, when colonic loops were pretreated with dithiothreitol to increase mucosal permeability, absorption and biliary recovery increased 50-fold.[249] There was a good correlation between changes in colonic mucosal permeability measured by [$^{51}$Cr]-EDTA absorption and biliary recovery of formylmethionyl peptide.[249]

The formyl peptides, especially FMLP, may be important mediators of intestinal inflammation in IBD and could contribute to some of the extra-intestinal effects, particularly in the liver and biliary system. Direct experimental evidence for involvement of formyl peptides in the pathophysiology of IBD is still awaited, however.

## XI. CONCLUSIONS

The development of satisfactory treatments for IBD and its extraintestinal complications is hampered by our ignorance of the etiology and pathogenesis of these conditions. There is strong evidence implicating luminal factors including both bacteria and their inflammatory products in the maintenance and possibly initiation of host inflammatory responses. The intestine is a vast reservoir of commensal microorganisms and their products. The intestinal mucosal barrier is not 100% effective and probably frequently breached. Immune responses directed against ubiquitous microorganisms and products could explain chronic relapsing inflammation on the basis of either an abnormally permeable mucosal barrier, changes in the composition or number of the luminal flora, or an over-aggressive host response to normal levels of exposure. IBD is certainly both a genetic and environmental disorder; however, the role of intestinal bacteria, commensal, facultative, or obligate pathogen remains enigmatic.

# REFERENCES

1. **Sedgwick, D., Drummond, J., Clarke, J., and Ferguson, A.,** Workload implications of the relentless increase in incidence of Crohn's disease, *Gut,* 31, A1205, 1990.
2. **Yang, H., Shohat, T., and Rotter, J. I.,** The genetics of inflammatory bowel disease, in *Current Topics in Gastroenterology. Inflammatory Bowel Disease,* MacDermott, R. P. and Stenson, W. F., Eds., Elsevier Science Publishing Co., New York, 1992, 17.
3. **Orholm, M., Munkholm, P., Langholz, E., Haaen Nielsen, O., Sorensen, T. I. S., and Binder, V.,** Familial occurrence of inflammatory bowel disease, *N. Engl. J. Med.,* 324, 84, 1991.
4. **Tysk, C., Lindberg, E., Jarnerot, G., and Floderus-Myrhed, B.,** Ulcerative colitis and Crohn's disease in an unselected population of monozygotic and dizygotic twins. A study of heritability and the influence of smoking, *Gut,* 29, 990, 1988.
5. **Weterman, I. T. and Pena, A. S.,** Familial incidence of Crohn's disease in The Netherlands and a review of the literature, *Gastroenterology,* 86, 449, 1984.
6. **Allan, R. N., Pease, P., and Ibbotson, J. P.,** Clustering of Crohn's disease in a Cotswold village, *Q. J. Med.,* 59, 473, 1986.
7. **Reilly, R. P. and Robinson, T. J.,** Crohn's disease — Is there a long latent period?, *Postgrad. Med. J.,* 62, 353, 1986.
8. **van de Merwe, J. P., Stegeman, J. H., and Hazenberg, M. P.,** The resident faecal flora is determined by genetic characteristics of the host. Implications for Crohn's disease?, *Antonie van Leeuwenhoek,* 49, 119, 1983.
9. **Hoeksma, A. and Winkler, K. C.,** The normal flora of the nose in twins, *Acta Leiden.,* 32, 123, 1963.
10. **van de Merwe, J. P., Schroder, A. M., Wensinck, F., and Hazenberg, M. P.,** The obligate anaerobic faecal flora of patients with Crohn's disease and their first-degree relatives, *Scand. J. Gastroenterol.,* 23, 1125, 1988.
11. **Ekbom, A. and Adami, H. -O.,** The epidemiology of inflammatory bowel disease, in *Current Topics in Gastroenterology. Inflammatory Bowel Disease,* MacDermott, R. P. and Stenson, W. F., Eds., Elsevier Science Publishing Co., New York, 1992, 1.
12. **Ekbom, A., Helmick, C., Zack, M., and Adami, H. O.,** The epidemiology of inflammatory bowel disease: a large population-based study in Sweden, *Gastroenterology,* 100, 350, 1991.
13. **Niv, Y. and Abukasis, G.,** Prevalence of ulcerative colitis in the Israeli kibbutz population, *J. Clin. Gastroenterol.,* 13, 98, 1991.
14. **Ekbom, A., Adami, H. O., Helmick, C. G., Jonzon, A., and Zack, M. M.,** Perinatal risk factors for inflammatory bowel disease: a case-control study, *Am. J. Epidemiol.,* 132, 1111, 1990.
15. **Koletzko, S., Griffiths, A., Corey, M., Smith, C., and Sherman, P.,** Infant feeding practices and ulcerative colitis in childhood, *Br. Med. J.,* 302, 1580, 1991.
16. **Wakefield, A. J., Sawyer, A. M., Hudson, M., Dhillon, A. P., and Pounder, R. E.,** Smoking, the oral contraceptive pill, and Crohn's disease, *Dig. Dis. Sci.,* 36, 1147, 1991.
17. **Walker-Smith, J. A.,** Cows milk protein intolerance of infancy, *Arch. Dis. Child.,* 50, 347, 1975.
18. **Schwarz, R. J. and Pezim, M. E.,** Failure of right-sided coloanal anastomosis for treatment of left-sided ulcerative colitis. Report of a case, *Dis. Colon Rectum,* 34, 618, 1991.
19. **Phillips, S. F.,** Biological effects of a reservoir at the end of the small bowel, *World J. Surg.,* 11, 763, 1987.
20. **Madden, M. V., Farthing, M. J. G., and Nicholls, R. J.,** Inflammation in ileal reservoirs: "pouchitis," *Gut,* 31, 247, 1990.
21. **Rutgeerts, P., Geboes, K., Vantrappen, G., Beyls, J., Kerremans, R., and Hiele, M.,** Predictability of the postoperative course of Crohn's disease, *Gastroenterology,* 99, 956, 1990.
22. **Rutgeerts, P., Geboes, K., Peeters, M., Hiele, M., Penninckx, F., Aerts, R., Kerremans, R., and Vantrappen, G.,** Effect of faecal stream diversion on recurrence of Crohn's disease in the neoterminal ileum, *Lancet,* 338, 771, 1991.
23. **Harper, P. H., Lee, E. C. G., Kettlewell, M. G. W., Bennett, M. K., and Jewell, D. P.,** Role of the faecal stream in the maintenance of Crohn's colitis, *Gut,* 26, 279, 1985.
24. **Blichfeldt, P., Blomhoff, J. P., Myhre, E., and Gjone, E.,** Metronidazole in Crohn's disease. A double blind cross-over clinical trial, *Scand. J. Gastroenterol.,* 13, 123, 1978.
25. **Teahon, K., Bjarnason, I., Pearson, M., and Levi, A. J.,** Ten years' experience with an elemental diet in the management of Crohn's disease, *Gut,* 31, 1133, 1990.
26. **Giaffer, M. H., North, G., and Holdsworth, C. D.,** Controlled trial of polymeric versus elemental diet in treatment of active Crohn's disease, *Lancet,* 335, 816, 1990.
27. **Greenberg, G. R., Fleming, C. R., Jeejeebhoy, K. N., Rosenberg, I. H., Sales, D., and Tremaine, W. J.,** Controlled trial of bowel rest and nutritional support in the management of Crohn's disease, *Gut,* 29, 1309, 1988.
28. **Alun Jones, V.,** Comparison of total parenteral nutrition and elemental diet in induction of remission of Crohn's disease. Long-term maintenance of remission by personalized food exclusions diets, *Dig. Dis. Sci.,* 32, 100S, 1987.
29. **Saverymuttu, S., Hodgson, H. J. F., and Chadwick, V. S.,** Controlled trial comparing prednisolone with an elemental diet plus non-absorbable antibiotics in active Crohn's disease, *Gut,* 26, 994, 1985.
30. **Seneca, H. and Henderson, E.,** Normal intestinal bacteria in ulcerative colitis, *Gastroenterology,* 15, 34, 1950.

31. **Cooke, E. M.,** A quantitative comparison of the faecal flora of patients with ulcerative colitis and that of normal persons, *J. Pathol. Bacteriol.*, 9, 439, 1967.

32. **Gorbach, S. L., Nahas, L., Plaut, A. G., Weinstein, L., Patterson, J. F., and Levitan, R.,** Studies of intestinal microflora. V. Fecal microbial ecology in ulcerative colitis and regional enteritis: relationship to severity of disease and chemotherapy, *Gastroenterology*, 54, 575, 1968.

33. **Wensinck, F.,** The faecal flora of patients with Crohn's disease, *Antonie van Leeuwenhoek*, 41, 214, 1975.

34. **Ruseler-van Embden, J. G. H. and Both-Patoir, H. C.,** Anaerobic gram-negative faecal flora in patients with Crohn's disease and healthy subjects, *Antonie van Leeuwenhoek*, 49, 125, 1983.

35. **Holdeman, L. V., Good, I. J., and Moore, W. E. C.,** Human fecal flora: variation in bacterial composition within individuals and a possible effect of emotional stress, *Appl. Environ. Microbiol.*, 31, 359, 1976.

36. **van de Merwe, J. P. and Stegeman, J. H.,** Binding *Coprococcus comes* to the Fc portion of IgG. A possible role in the pathogenesis of Crohn's disease?, *Eur. J. Immunol.*, 15, 860, 1985.

37. **Ruseler-van Embden, J. G. H. and van Lieshout, L. M. C.,** Increased faecal glycosidases in patients with Crohn's disease, *Digestion*, 37, 43, 1987.

38. **Giaffer, M. H., Holdsworth, C. D., and Duerden, B. I.,** The assessment of faecal flora in patients with inflammatory bowel disease by a simplified bacteriological technique, *J. Med. Microbiol.*, 35, 238, 1991.

39. **van de Merwe, J. P. and Mol, G. J. J.,** A possible role of *Eubacterium* and *Peptostreptococcus* species in the aetiology of Crohn's disease, *Antonie van Leeuwenhoek*, 46, 587, 1980.

40. **West, B., Lendrum, R., Hill, M. J., and Walker, G.,** Effects of sulphasalazine (Salazopyrin) on faecal flora in patients with inflammatory bowel disease, *Gut*, 15, 960, 1974.

41. **Keighley, M. R. B., Arabi, Y., Dimock, F., Burdon, D. W., Allan, R. N., and Alexander-Williams, J.,** Influence of inflammatory bowel disease on intestinal microflora, *Gut*, 19, 1099, 1978.

42. **Harper, P. H., Truelove, S. C., Lee, E. C. G., Kettlewell, M. G. W., and Jewell, D. P.,** Split ileostomy and ileocolostomy for Crohn's disease of colon and ulcerative colitis: a 20 year survey, *Gut*, 24, 106, 1983.

43. **Roediger, W. E. W.,** Role of anaerobic bacteria in the metabolic welfare of the colonic mucosa in man, *Gut*, 21, 793, 1980.

44. **Guillemot, F., Colombel, J. F., Neut, C., Verplanck, N., Lecomte, M., Romond, C., Paris, J. C., and Cortot, A.,** Treatment of diversion colitis by short chain fatty acids: prospective and double-blind study, *Dis. Colon Rectum*, 34, 861, 1991.

45. **Neut, C., Colombel, J. F., Guillemot, F., Cortot, A., Gower, P., Quandalle, P., Ribet, M., Romond, C., and Paris, J. C.,** Impaired bacterial flora in human excluded colon, *Gut*, 30, 1094, 1989.

46. **Brooke, B. N.,** Outcome of surgery for ulcerative colitis, *Lancet*, 2, 532, 1956.

47. **Rhodes, J. B. and Kirsner, J. B.,** The early and late course of patients with ulcerative colitis after ileostomy and colectomy, *Surg. Gynecol. Obstet.*, 121, 1303, 1965.

48. **Metcalf, A. M. and Phillips, S. F.,** Ileostomy diarrhoea, *Clin. Gastroenterol.*, 15, 705, 1986.

49. **Philipson, B., Brandberg, A., Jagenburg, R., Kock, N. G., Lager, I., and Ahren, C.,** Mucosal morphology, bacteriology, and absorption in intra-abdominal ileostomy reservoir, *Scand. J. Gastroenterol.*, 10, 145, 1975.

50. **Luukkonen, P., Valtonen, V., Sivonen, A., Sipponen, P., and Jarvinen, H.,** Fecal bacteriology and reservoir ileitis in patients operated on for ulcerative colitis, *Dis. Colon Rectum*, 31, 864, 1988.

51. **Kelly, D. G., Phillips, S. F., Kelly, K. A., Weinstein, W. M., and Gilchrist, M. J. R.,** Dysfunction of the continent ileostomy: clinical features and bacteriology, *Gut*, 24, 193, 1983.

52. **O'Connell, P. R., Rankin, D. R., Weiland, L. H., and Kelly, K. A.,** Enteric bacteriology, absorption, morphology and emptying after ileal pouch-anal anastomosis, *Br. J. Surg.*, 73, 909, 1986.

53. **Santavirta, J., Mattila, J., Kokki, M., and Matikainen, M.,** Mucosal morphology and faecal bacteriology after ileoanal anastomosis, *Int. J. Colorectal Dis.*, 6, 38, 1991.

54. **Stryker, S. J., Borody, T. J., Phillips, S. F., Kelly, K. A., Dozois, R. R., and Beart, R. W., Jr.,** Motility of the small intestine after proctocolectomy and ileal pouch-anal anastomosis, *Ann. Surg.*, 201, 351, 1985.

55. **Soper, N. J., Orkin, B. A., Kelly, K. A., Phillips, S. F., and Brown, M. L.,** Gastrointestinal transit after proctocolectomy with ileal pouch-anal anastomosis or ileostomy, *J. Surg. Res.*, 46, 300, 1989.

56. **Savage, D. C.,** Associations of indigenous microorganisms with gastrointestinal mucosal epithelia, *Am. J. Clin. Nutr.*, 23, 1495, 1970.

57. **Peach, S., Lock, M. R., Katz, D., Todd, I. P., and Tabaqchali, S.,** Mucosal-associated bacterial flora of the intestine in patients with Crohn's disease and in a control group, *Gut*, 19, 1034, 1978.

58. **Edmiston, C. E., Jr., Avant, G. R., and Wilson, F. A.,** Anaerobic bacterial populations on normal and diseased human biopsy tissue obtained at colonoscopy, *Appl. Environ. Microbiol.*, 43, 1173, 1982.

59. **Horing, E., Gopfert, D., Schroter, G., and von Gaisberg, U.,** Frequency and spectrum of microorganisms isolated from biopsy specimens in chronic colitis, *Endoscopy*, 23, 325, 1991.

60. **Schumacher, G., Sandstedt, B., Mollby, R., and Kollberg, B.,** Clinical and histologic features differentiating non-relapsing colitis from first attacks of inflammatory bowel disease, *Scand. J. Gastroenterol.*, 26, 151, 1991.

61. **Greenfield, C., Aguilar Ramirez, J. R., Pounder, R. E., Williams, T., Danvers, M., Marper, S. R., and Noone, P.,** *Clostridium difficile* and inflammatory bowel disease, *Gut*, 24, 713, 1983.

62. **Weber, P., Koch, M., Heizmann, W. R., Scheurlen, M., Jenss, H., and Hartmann, F.,** Microbic superinfection in relapse of inflammatory bowel disease, *J. Clin. Gastroenterol.,* 14, 302, 1992.
63. **Nakamura, S., Mikawa, M., Nakashio, S., Takabatake, M., Okado, I., Yamakawa, K., Serikawa, T., Okumura, S., and Nishida, S.,** Isolation of *Clostridium difficile* from feces and the antibody sera of young and elderly adults, *Microbiol. Immunol.,* 25, 345, 1981.
64. **Gilligan, P. H., McCarthy, L. R., and Genta, V. W.,** Relative frequency of *Clostridium difficile* in patients with diarrheal disease, *J. Clin. Microbiol.,* 14, 26, 1981.
65. **Goodman, M. J., Pearson, K. W., McGhie, D., Dutt, S., and Deodhar, S. G.,** *Campylobacter* and *Giardia lamblia* causing exacerbation of inflammatory bowel disease, *Lancet,* 2, 1247, 1980.
66. **Blaser, M. J., Hoverson, D., Ely, I. G., Duncan, D. J., and Wang, W. L.,** Studies of *Campylobacter jejuni* in patients with inflammatory bowel disease, *Gastroenterology,* 86, 33, 1984.
67. **Thaylor-Robison, S., Miles, R., Whitehead, A., and Dickinson, R. J.,** Salmonella infection and ulcerative colitis, *Lancet,* 1, 1145, 1989.
68. **Scheurlen, C., Kruis, W., Spengler, U., Weinzierl, M., Paumgartner, G., and Lamina, J.,** Crohn's disease is frequently complicated by Giardiasis, *Scand. J. Gastroenterol.,* 23, 833, 1988.
69. **Burke, D. A. and Axon, A. T. R.,** Adhesive *Escherichia coli* in inflammatory bowel disease and infective diarrhoea, *Br. Med. J.,* 297, 102, 1988.
70. **Cooke, E. M., Ewins, S. P., Gywel-Jones, J., and Lennard-Jones, J. E.,** Properties of strains of *Escherichia coli* carried in different phases of ulcerative colitis, *Gut,* 15, 143, 1974.
71. **Barr, G. D., Hudson, M. J., Priddle, J. D., and Jewell, D. P.,** Colonic bacterial proteases to $IgA_1$ and sIgA in patients with ulcerative colitis, *Gut,* 28, 186, 1987.
72. **Belsheim, M. R., Darwish, R. Z., Watson, W. C., and Schieven, B.,** Bacterial L-form isolation from inflammatory bowel disease patients, *Gastroenterology,* 85, 364, 1983.
73. **Ibbotson, J. P., Pease, P. E., and Allan, R. N.,** Cell-wall deficient bacteria in inflammatory bowel disease, *Eur. J. Clin. Microbiol.,* 6, 429, 1987.
74. **Chiodini, R. J., Van-Kruiningen, H. J., Thayer, W. R., and Coutu, J. A.,** Spheroplastic phase of mycobacteria isolated from patients with Crohn's disease, *J. Clin. Microbiol.,* 24, 357, 1986.
75. **Markesich, D. C., Graham, D. Y., and Yoshimura, H. H.,** Progress in culture and subculture of spheroplasts and fastidious acid-fast bacilli isolated from intestinal tissues, *J. Clin. Microbiol.,* 26, 1600, 1988.
76. **Shafii, A., Soher, S., Lev, M., and Das, K. M.,** An antibody against revertant forms of cell-wall-deficient bacterial variant in sera from patients with Crohn's disease, *Lancet,* 2, 332, 1981.
77. **Parent, K. and Mitchell, P. D.,** Cell wall-deficient variants of pseudomonas-like (Group Va) bacteria in Crohn's disease, *Gastroenterology,* 75, 368, 1978.
78. **Ibbotson, J. P., Pease, P. E., and Allan, R. N.,** Serological studies in Crohn's disease, *Eur. J. Clin. Microbiol.,* 6, 286, 1987.
79. **McGarity, B. H., Robertson, D. A. F., Clarke, I. N., and Wright, R.,** Deoxyribonucleic acid amplification and hybridisation in Crohn's disease using a chlamydial plasmid probe, *Gut,* 32, 1011, 1991.
80. **Wirostko, E., Johnson, L., and Wirostko, B.,** Ulcerative colitis associated chronic uveitis. Parasitization of intraocular leucocytes by mollicute-like organisms, *J. Submicrosc. Cytol. Pathol.,* 22, 231, 1990.
81. **Johnson, L. A., Wirostko, E., and Wirostko, W. J.,** Crohn's disease uveitis. Parasitization of vitreous leukocytes by mollicute-like organisms, *Am. J. Clin. Pathol.,* 91, 259, 1989.
82. **McFadden, J. J. and Seechurn, P.,** Mycobacteria and Crohn's disease. Molecular approaches, in *Current Topics in Gastroenterology. Inflammatory Bowel Disease,* MacDermott, R. P. and Stenson, W. F., Eds., Elsevier Science Publishing Co., New York, 1992, 259.
83. **Green, E., Tizard, M., Thompson, J., Winterbourne, D., McFadden, J., and Hermon-Taylor, J.,** Sequence and characteristics of IS900, an insertion sequence identified in a human Crohn's disease isolate of *Mycobacterium paratuberculosis, Nucleic Acid Res.,* 17, 9063, 1989.
84. **Chiodini, R. J., Van Kruiningen, H. J., Thayer, W. R., Merkal, R. S., and Coutu, J. A.,** Possible role of mycobacteria in inflammatory bowel disease. I. An unclassified *Mycobacterium* species isolated from patients with Crohn's disease, *Dig. Dis. Sci.,* 29, 1073, 1984.
85. **McFadden, J. J., Butcher, P. D., Chiodini, R., and Hermon-Taylor, J.,** Crohn's disease-isolated mycobacteria are identical to *Mycobacterium paratuberculosis,* as determined by DNA probes that distinguish between mycobacterial species, *J. Clin. Microbiol.,* 25, 796, 1987.
86. **Gitnick, G., Collins, J., Beaman, B., Brooks, D., Arthur, M., Imaeda, T., and Palieschesky, M.,** Preliminary report on isolation of mycobacteria from patients with Crohn's disease, *Dig. Dis. Sci.,* 34, 925, 1989.
87. **Graham, D. Y., Markesich, D. C., and Yoshimura, H. H.,** Mycobacteria and inflammatory bowel disease. Results of culture, *Gastroenterology,* 92, 436, 1987.
88. **Chiodini, R. J., van Kruiningen, H. J., and Merkal, R. S.,** Ruminant paratuberculosis (Johne's disease). The current status and future prospects, *Cornell Vet.,* 74, 218, 1984.
89. **Collins, F.,** *Mycobacterium avium*-complex infections and development of acquired immunodeficiency syndrome: Causal opportunist or causal co-factor, *Int. J. Leprosy,* 54, 458, 1986.

90. **McFadden, J., Collins, J., Beaman, B., Arthur, M., and Gitnick, G.,** Mycobacteria in Crohn's disease: DNA probes identify the wood pigeon strain of *Mycobacterium avium* and *Mycobacterium paratuberculosis* from human tissue, *J. Clin. Microbiol.,* 30, 3070, 1992.

91. **Yoshimura, H. H., Graham, D. Y., Estes, M. K., and Merkal, R. S.,** Investigation of association of mycobacteria with inflammatory bowel disease by nucleic acid hybridization, *J. Clin. Microbiol.,* 25, 45, 1987.

92. **Butcher, P. D., McFadden, J. J., and Hermon-Taylor, J.,** Investigation of mycobacteria in Crohn's disease tissue by Southern blotting and DNA hybridisation with cloned mycobacterial genomic DNA probes from a Crohn's disease isolated mycobacteria, *Gut,* 29, 1222, 1988.

93. **Sanderson, J. D., Moss, M. T., Tizard, M. L. V., and Hermon-Taylor, J.,** *Mycobacterium paratuberculosis* DNA in Crohn's disease tissue, *Gut,* 33, 890, 1992.

94. **Moss, M. T., Sanderson, J. D., Tizard, M. L. V., Hermon-Taylor, J., El-Zaatari, F. A. K., Markesich, D. C., and Graham, D. Y.,** Polymerase chain reaction detection of *Mycobacterium paratuberculosis* and *Mycobacterium avium* subsp. *silvaticum* in long term cultures from Crohn's disease and control tissues, *Gut,* 33, 1209, 1992.

95. **Bell, J. I. and Jewell, D. P.,** *Mycobacterium paratuberculosis* DNA cannot be detected in Crohn's disease tissues, *Gastroenterology,* 100, A611, 1991.

96. **Wells, C. L., Maddaus, M. A., and Simmons, R. L.,** Proposed mechanisms for the translocation of intestinal bacteria, *Rev. Infect. Dis.,* 10, 958, 1988.

97. **Wells, C. L., Maddaus, M. A., Jechorek, R. P., and Simmons, R. L.,** Role of intestinal anaerobic bacteria in colonization resistance, *Eur. J. Clin. Microbiol. Infect. Dis.,* 7, 107, 1988.

98. **Guiot, H. F. L., van den Broek, P. J., van der Meer, J. W. M., and van Furth, R.,** Selective antimicrobial modulation of the intestinal flora of patients with acute nonlymphocytic leukemia: a double-blind, placebo-controlled study, *J. Infect. Dis.,* 147, 615, 1983.

99. **Wade, J. C., de Jongh, C. A., Newman, K. A., Crowley, J., Wiernik, P. H., and Schimpff, S. C.,** Selective antimicrobial modulation as prophylaxis against infection during granulocytopenia: trimethoprim-sulfamethoxazole vs. nalidixic acid, *J. Infect. Dis.,* 147, 624, 1983.

100. **Wells, C. L., Maddaus, M. A., and Simmons, R. L.,** Role of the macrophage in the translocation of intestinal bacteria, *Arch. Surg.,* 122, 48, 1987.

101. **Wells, C. L.,** Relationship between intestinal microecology and the translocation of intestinal bacteria, *Antonie van Leeuwenhoek,* 58, 87, 1990.

102. **Aluwihare, A. P. R.,** Electron microscopy in Crohn's disease, *Gut,* 12, 509, 1971.

103. **Ohkusa, T., Okayasu, I., Tokoi, S., and Ozaki, Y.,** Bacterial invasion into the colonic mucosa in ulcerative colitis, *J. Gastroenterol. Hepatol.,* 8, 116, 1993.

104. **Ambrose, N. S., Johnson, M., Burdon, D. W., and Keighley, M. R.,** Incidence of pathogenic bacteria from mesenteric lymph nodes and ileal serosa during Crohn's disease surgery, *Br. J. Surg.,* 71, 623, 1984.

105. **Laffineur, G., Lescut, D., Vincent, P., Quandalle, P., Wurtz, A., and Colombel, J. -F.,** Translocation bacterienne dans la maladie de Crohn, *Gastroenterol. Clin. Biol.,* 16, 777, 1992.

106. **Kreuzpaintner, G., Horstkotte, D., Heyll, A., Losse, B., and Strohmeyer, G.,** Increased risk of bacterial endocarditis in inflammatory bowel disease, *Am. J. Med.,* 92, 391, 1992.

107. **Blaser, M. J., Miller, R. A., Lacher, J., and Singleton, J. W.,** Patients with active Crohn's disease have elevated serum antibodies to antigens of seven enteric bacterial pathogens, *Gastroenterology,* 87, 888, 1984.

108. **Brown, W. R. and Lee, E.,** Radioimmunological measurements of bacterial antibodies. Human serum antibodies reactive with *Bacteroides fragilis* and enterococcus in gastrointestinal and immunological disorders, *Gastroenterology,* 66, 1145, 1974.

109. **Tabaqchali, S., O'Donoghue, D. P., and Bettelheim, K. A.,** *Escherichia coli* antibodies in patients with inflammatory bowel disease, *Gut,* 19, 108, 1978.

110. **Matthews, N., Mayberry, J. F., Rhodes, J., Neale, L., Munro, J., Wensinck, F., Lawson, G. H. K., Rowland, A. C., Berkhoff, G. A., and Barthold, S. W.,** Agglutinins to bacteria in Crohn's disease, *Gut,* 21, 376, 1980.

111. **Wensinck, F. and van de Merwe, J. P.,** Serum agglutinins to *Eubacterium* and *Peptostreptococcus* species in Crohn's and other diseases, *J. Hyg.,* 87, 13, 1981.

112. **Wensinck, F., van de Merwe, J. P., and Mayberry, J. F.,** An international study of agglutinins to *Eubacterium, Peptostreptococcus* and *Coprococcus* species in Crohn's disease, ulcerative colitis and control subjects, *Digestion,* 27, 63, 1983.

113. **Hazenberg, M. P., van de Merwe, J. P., Pena, A. S., Pennock-Schroder, A. M., and van Lieshout, L. M. C.,** Antibodies to *Coprococcus* comes in sera of patients with Crohn's disease. Isolation and purification of the agglutinating antigen tested with an ELISA technique, *J. Clin. Lab. Immunol.,* 23, 143, 1987.

114. **Ibbotson, J. P., Pease, P. E., and Allan, R. N.,** Serological studies in Crohn's disease, *Eur. J. Clin. Microbiol.,* 6, 286, 1987.

115. **Elsaghier, A., Prantera, C., Moreno, C., and Ivanyi, J.,** Antibodies to *Mycobacterium paratuberculosis*-specific protein antigens in Crohn's disease, *Clin. Exp. Immunol.,* 90, 503, 1992.

116. **Stainsby, K. J., Lowes, J. R., Allan, R. N., and Ibbotson, J. P.,** Antibodies to *Mycobacterium paratuberculosis* and nine species of environmental mycobacteria in Crohn's disease and control subjects, *Gut,* 34, 371, 1993.

117. **Tanaka, K., Wilks, M., Coates, P. J., Farthing, M. J. G., Walker-Smith, J. A., and Tabaqchali, S.,** *Mycobacterium paratuberculosis* and Crohn's disease, *Gut,* 32, 43, 1991.

118. **Brunello, F., Pera, A., Martini, S., Marino, L., Astegiano, M., Barletti, C., Gastaldi, P., Verme, G., and Emanuelli, G.,** Antibodies to *Mycobacterium paratuberculosis* in patients with Crohn's disease, *Dig. Dis. Sci.,* 36, 1741, 1991.

119. **Ibbotson, J. P., Lowes, J. R., Chahal, H., Gaston, J. S. H., Life, P., Kumararatne, D. S., Sharif, H., Alexander-Williams, J., and Allan, R. N.,** Mucosal cell-mediated immunity to mycobacterial, enterobacterial and other microbial antigens in inflammatory bowel disease, *Clin. Exp. Immunol.,* 87, 224, 1992.

120. **Fiocchi, C., Battisto, J. R., and Farmer, R. G.,** Studies on isolated gut mucosal lymphocytes in inflammatory bowel disease. Detection of activated T-cells and enhanced proliferation to *Staphylococcus aureus* and lipopolysaccharides, *Dig. Dis. Sci.,* 26, 728, 1981.

121. **Ebert, E. C., Bhatt, B. D., Liu, S., and Das, K. M.,** Induction of suppressor cells by *M. paratuberculosis* antigen in inflammatory bowel disease, *Clin. Exp. Immunol.,* 83, 320, 1991.

122. **Dalton, H. R., Hoang, P., and Jewell, D. P.,** Antigen induced suppression in peripheral blood and lamina propria mononuclear cells in inflammatory bowel disease, *Gut,* 33, 324, 1992.

123. **Hodgson, H. J. F., Wands, J. R., and Isselbacher, K. J.,** Decreased suppressor cell activity in inflammatory bowel disease, *Clin. Exp. Immunol.,* 32, 451, 1978.

124. **Chao, L. P., Steele, J., Rodriques, C., Lennard-Jones, J., Stanford, J. L., Spiliadis, C., and Rook, G. A. W.,** Specificity of antibodies secreted by hybridomas generated from activated B cells in the mesenteric lymph nodes of patients with inflammatory bowel disease, *Gut,* 29, 35, 1988.

125. **Kaulfersch, W., Fiocchi, C., and Waldmann, T. A.,** Polyclonal nature of the intestinal mucosal lymphocyte populations in inflammatory bowel disease. A molecular genetic evaluation of the immunoglobulin and T-cell antigen receptors, *Gastroenterology,* 95, 364, 1988.

126. **Onderdonk, A. B. and Bartlett, J. G.,** Bacteriological studies of experimental ulcerative colitis, *Am. J. Nutr.,* 32, 258, 1979.

127. **Roediger, W. E. W., Heyworth, M., Willoughby, P., Piris, J., Moore, A., and Truelove, S. C.,** Luminal ions and short chain fatty acids as markers of functional activity of the mucosa in ulcerative colitis, *J. Clin. Pathol.,* 35, 323, 1982.

128. **Onderdonk, A. B., Franklin, M. L., and Cisneros, R. L.,** Production of experimental ulcerative colitis in gnotobiotic guinea pigs with simplified microflora, *Infect. Immun.,* 32, 225, 1981.

129. **Onderdonk, A. B.,** Experimental models for ulcerative colitis, *Dig. Dis. Sci.,* 30, 40S, 1985.

130. **Robert, A. and Asano, T.,** Resistance of germ free rats to indomethacin-induced intestinal lesions, *Prostaglandins,* 14, 331, 1977.

131. **MacPherson, B. R. and Pfeiffer, C. J.,** Experimental production of diffuse colitis in rats, *Digestion,* 17, 135, 1978.

132. **Sartor, R. B., Green, K. D., and Anderle, S. K.,** Spontaneously reactivating granulomatous enterocolitis with extraintestinal inflammation induced by bacterial cell wall polymers in rats, *Gastroenterology,* 96, A442, 1989.

133. **Hammer, R. E., Maika, S. D., Richardson, J. A., Tang, J. P., and Taurog, J. D.,** Spontaneous inflammatory disease in transgenic rats expressing HLA-B27 and human $\beta_2$m: an animal model of HLA-B27-associated human disorders, *Cell,* 63, 1099, 1990.

134. **De Vos, M., Cuvelier, C., Mielants, H., Veys, E., Barbier, F., and Elewaut, A.,** Ileocolonoscopy in seronegative spondyloarthropathy, *Gastroenterology,* 96, 339, 1989.

135. **Saverymuttu, S. H., Peters, A. M., Lavender, J. P., Pepys, M. B., Hodgson, H. J. F., and Chadwick, V. S.,** Quantitative fecal indium-111 labeled leucocyte excretion in the assessment of disease activity in Crohn's disease, *Gastroenterology,* 85, 1333, 1983.

136. **Nixon, J. B. and Riddell, R. M.,** Histopathology of ulcerative colitis, in *Inflammatory Bowel Disease,* Allan, R., Keighley, M., Alexander-Williams, J., and Hawkins, C., Eds., Churchill-Livingstone, Edinburgh, U.K., 1983, 194.

137. **Saverymuttu, S. H., Chadwick, V. S., and Hodgson, H. J. F.,** Granulocyte migration in ulcerative colitis, *Eur. J. Clin. Invest.,* 15, 60, 1985.

138. **Saverymuttu, S. H., Chadwick, V. S., and Hodgson, H. J. F.,** Comparison of fecal granulocyte excretion in ulcerative colitis and Crohn's colitis, *Dig. Dis. Sci.,* 29, 1000, 1984.

139. **Rogers, M. J., Moore, R., and Cohen, J.,** The relationship between faecal endotoxin and faecal microflora of the C57BL, *J. Hyg.,* 95, 397, 1985.

140. **Nolan, J. P., Hare, D. K., McDevitt, J. J., and Ali, M. V.,** *In vitro* studies of intestinal endotoxin absorption. I. Kinetics of absorption in the isolated everted gut sac, *Gastroenterology,* 72, 434, 1977.

141. **Gans, H. and Matsumoto, K.,** The escape of endotoxin from the intestine, *Surg. Gynecol. Obstet.,* 139, 395, 1974.

142. **Fink, P. C., Suin de Boutemard, C., and Haeckel, R.,** Endotoxaemia in patients with Crohn's disease: a longitudinal study of elastase/alpha$_1$-proteinase inhibitor and limulus-amoebocyte-lysate reactivity, *J. Clin. Chem. Clin. Biochem.,* 26, 117, 1988.

143. **Olafsson, P., Nylander, G., and Olsson, P.,** Endotoxin: route of transport in experimental peritonitis, *Am. J. Surg.,* 151, 443, 1986.

144. **Kelley, C. J., Ingoldby, C. J. H., Blenkharn, J. I., and Wood, C. B.,** Colonoscopy related endotoxemia, *Surg. Gynecol. Obstet.,* 161, 332, 1985.

145. **Wellmann, W., Fink, P. C., Benner, F., and Schmidt, F. W.,** Endotoxaemia in Crohn's disease. Treatment with whole gut irrigation and 5-aminosalicilylic acid, *Gut,* 27, 814, 1986.

146. **Lei, M. -G. and Morrison, D. C.,** Specific lipopolysaccharide-binding proteins on murine splenocytes. I. Detection of lipopolysaccharide-binding sites on splenocytes and splenocyte subpopulations, *J. Immunol.,* 141, 996, 1988.

147. **Fox, E. S., Thomas, P., and Broitman, S. A.,** Uptake and modification of $^{125}$I-lipopolysaccharide by isolated rat Kupffer cells, *Hepatology,* 8, 1550, 1988.

148. **Parent, J. B.,** Membrane receptors on rat hepatocytes for the inner core region bacterial lipopolysaccharides, *J. Biol. Chem.,* 265, 3455, 1990.

149. **Forehand, J. R., Pabst, M. J., Phillips, W. A., and Johnston, R. B., Jr.,** Lipopolysaccharide priming human neutrophils for an enhanced respiratory burst. Role of intracellular free calcium, *J. Clin. Invest.,* 83, 74, 1989.

150. **Morrison, D. C. and Ryan, J. L.,** Endotoxins and disease mechanisms, *Ann. Rev. Med.,* 38, 417, 1987.

151. **Chadwick, V. S., Schlup, M. M. T., Cooper, B. T., and Broom, M. F.,** Enzymes degrading bacterial chemotactic F-met peptides in human ileal and colonic mucosa. *J. Gastroenterol. Hepatol.,* 5, 375, 1990.

152. **Olsen, U. B. and Bille-Hansen, V.,** Endotoxin pretreatment enhances neutrophil FMLP-receptor binding and activity in guinea pigs, *Agents Actions,* 21, 177, 1987.

153. **Schleifer, K. H. and Krause, R. M.,** The immunochemistry of peptidoglycan, *J. Biol. Chem.,* 264, 986, 1971.

154. **Kotani, S., Tsujimoto, M., Koga, T., Nagao, S., Tanaka, A., and Kawata, S.,** Chemical structure and biological activity relationship of bacterial cell walls and muramyl peptides, *Fed. Proc.,* 45, 2534, 1986.

155. **Wells, A. F., Hightower, J. A., Parks, C., Kufoy, E., and Fox, A.,** Systemic injection of group A streptococcal peptidoglycan-polysaccharide complexes elicits persistent neutrophilia and monocytosis associated with polyarthritis in rats, *Infect. Immun.,* 57, 351, 1989.

156. **Sartor, R. B., Bond, T. M., Compton, K. Y., and Cleland, D. R.,** Intestinal absorption of bacterial cell wall polymers in rats, *Adv. Exp. Med. Biol.,* 216A, 835, 1986.

157. **Green, K. D. and Sartor, R. B.,** Systemic lipopolysaccharide reactivates peptidoglycan-polysaccharide-induced intestinal inflammation in rats, *Gastroenterology,* 94, A154, 1988.

158. **Sartor, R. B., Bond, T. M., and Schwab, J. H.,** Systemic uptake and inflammatory effects of luminal bacterial cell wall polymers in rats with acute colonic injury, *Infect. Immun.,* 56, 2101, 1988.

159. **Lichtman, S. N., Sartor, R. B., and Schwab, J. H.,** Elimination of PG-PS in rat bile, *Gastroenterology,* 94, A262, 1988.

160. **Wilder, R. L., Calandra, G. B., Garvin, A. J., Wright, K. D., and Hamsen, C. T.,** Strain and sex variation in the susceptibility of streptococcal cell wall induced polyarthritis in rats, *Arthr. Rheum.,* 25, 1064, 1982.

161. **Sartor, R. B.,** Animal models of intestinal inflammation, in *Inflammatory Bowel Disease,* MacDermott, R. P. and Stenson, W. F., Eds., Elsevier, New York, 1992, 337.

162. **Lichtman, S. N., Sartor, R. B., Keku, J., and Schwab, J. H.,** Hepatic inflammation in rats with experimental small bowel bacterial overgrowth, *Gastroenterology,* 98, 414, 1990.

163. **Keppler, D., Hagmann, W., and Rapp, S.,** Role of leukotrienes in endotoxin action in vivo, *Rev. Infect. Dis.,* 9, S580, 1987.

164. **Marchiseppi, I., Postiglione, M., Di Carlo, E., and Valentino, M.,** LTB$_4$ production in human polymorphonuclear leukocytes incubated with nFMLP, *Boll. Soc. Ital. Biol. Sper.,* 64, 501, 1988.

165. **Wettstein, M., Gerok, W., and Haussinger, D.,** Metabolism of cysteinyl leukotrienes in non-recirculating rat liver perfusion: hepatocyte heterogeneity in uptake and biliary excretion, *Eur. J. Biochem.,* 181, 115, 1989.

166. **Anderson, R. P., Woodhouse, A. F., Myers, D. B., Broom, M. F., Hobson, C. H., and Chadwick, V. S.,** Hepatobiliary excretion and enterohepatic circulation of bacterial chemotactic peptide (FMLP) in the rat, *J. Gastroenterol. Hepatol.,* 2, 45, 1987.

167. **Anderson, R. P., Butt, T. J., and Chadwick, V. S.,** Hepatobiliary excretion of bacterial formyl-methionyl peptides in rat. Structure activity studies, *Dig. Dis. Sci.,* 37, 248, 1992.

168. **Sartor, R. B.,** Role of intestinal microflora in initiation and perpetuation of inflammatory bowel disease, *Can. J. Gastroenterol.,* 4, 271, 1990.

169. **Chetty, C. and Schwab, J. H.,** Endotoxin-like products of gram-positive bacteria, in *Handbook of Endotoxin,* Vol. 1, Rietschel, E.T., Ed., Elsevier, Amsterdam, 1984, 376.

170. **Stimpson, S. A., Schwab, J. H., and Janusz, M. J.,** Acute and chronic inflammation induced by peptidoglycan structures and polysaccharide complexes, in *Biological Properties of Peptidoglycan,* Seidl, H. P. and Schleifer, K. H., Eds., Walter deGruyter, Berlin, 1986, 273.

171. **Ellouz, F., Adam, A., Ciorbaru, R., and Lederer, E.,** Minimal structural requirements for adjuvant activity of bacterial peptidoglycan derivatives, *Biochem. Biophys. Res. Comm.,* 59, 1317, 1974.

172. **Adam, A., Ciorbaru, R., Ellouz, F., Petit, J. F., and Lederer, E.,** Adjuvant activity of monomeric bacterial cell wall peptidoglycans, *Biochem. Biophys. Res. Comm.,* 56, 561, 1974.

173. **De Pedro, M. A. and Schwarz, U.,** Affinity chromatography of murein precursors on vancomycin-sepharose, *FEMS Microbiol. Lett.,* 9, 215, 1980.

174. **Perkins, H. R. and Nieto, M.,** The preparation of iodinated vancomycin and its distribution in bacteria treated with the antibiotic, *Biochem. J.*, 116, 83, 1970.

175. **Vermeullen, M. W. and Gray, G. R.,** Processing of *Bacillus subtilis* petidoglycan by a mouse macrophage cell line, *Infect. Immun.*, 46, 476, 1984.

176. **Silverman, D. H. S., Krueger, J. M., and Karnovsky, M. L.,** Specific binding sites for muramyl peptides on murine macrophages, *J. Immunol.*, 136, 2195, 1986.

177. **Tenu, J. P., Adam, A., Souvannavong, V., Yapo, A., Petit, J. F., and Douglas, K.,** Photoaffinity labeling of macrophages and B-lymphocytes using [125]I-labeled aryl-azide derivatives of muramyldipeptide, *Int. J. Immunopharmacol.*, 11, 653, 1989.

178. **Helmberg, A., Bock, G., Wolf, H., and Wick, G.,** An orally administered bacterial immunomodulator primes rabbit neutrophils for increased oxidative burst in response to opsonized zymosan, *Infect. Immun.*, 57, 3576, 1989.

179. **Bahr, G. M. and Chedid, L.,** Immunological activities of muramyl peptides, *Fed. Proc.*, 45, 2541, 1986.

180. **Tomasic, J., Ladesic, B., Valinger, Z., and Hrsak, I.,** The metabolic fate of [14]C-labeled peptidoglycan monomer in mice. I. Identification of the monomer and the corresponding pentapeptide in urine, *Biochimica et Biophysica Acta*, 629, 77, 1980.

181. **Parant, M., Parant, F., Chedid, L., Yapo, A., Petit, J. F., and Lederer, E.,** Fate of the synthetic immunoadjuvant, muramyl dipeptide ([14]C-labeled) in the mouse, *Int. J. Immunopharmacol.*, 1, 35, 1979.

182. **Ladesic, B., Tomasic, J., Kveder, S., and Hrsak, I.,** The metabolic fate of [14]C-labelled immunoadjuvant peptidoglycan monomer. II. In vitro studies, *Biochimica et Biophysica Acta*, 678, 12, 1981.

183. **Ambler, L. and Hudson, A. M.,** Pharmacokinetics and metabolism of muramyl dipeptide and nor-muramyl dipeptide [[3]H-labeled] in the mouse, *Int. J. Immunopharmacol.*, 6, 133, 1984.

184. **Krueger, J. M., Karnovsky, M. L., Martin, S. A., Pappenheimer, J. R., Walter, J., and Biemann, K.,** Peptidoglycans as promoters of slow-wave sleep. II. Somnogenic and pyrogenic activities of some naturally occurring muramyl peptides; correlates with mass spectrometric structure determination, *J. Biol. Chem.*, 259, 12659, 1984.

185. **Krueger, J. M., Pappenheimer, J. R., and Karnovsky, M. L.,** The composition of sleep-promoting factor isolated from human urine, *J. Biol. Chem.*, 257, 1664, 1982.

186. **Masayasu, H., Ono, K., and Takegoshi, T.,** Radioimmunoassay for $N_a$-(N-acetylmuramyl-alanyl-D-isoglutaminyl)-$N_e$-stearyl-lysine, *Chem. Pharmacol. Bull.*, 32, 4124, 1984.

187. **Franken, N., Seidl, P. H., Zauner, E., Kolb, H. J., Schleifer, K. H., and Weiss, L.,** Quantitative determination of human IgG antibodies to the peptide subunit determinant of peptidoglycan by an enzyme-linked immunosorbent assay, *Mol. Immunol.*, 22, 573, 1985.

188. **Bahr, G. M., Eshhar, Z., Ben-Yitzhak, R., Modabber, F. Z., Arnon, R., Sela, M., and Chedid, L.,** Monoclonal antibodies to the synthetic adjuvant muramyl dipeptide: characterization of the specificity, *Mol. Immunol.*, 20, 745, 1983.

189. **Park, H., Zeiger, A. R., and Schumacher, H. R.,** Detection of soluble peptidoglycan in urine after penicillin administration, *Infect. Immun.*, 43, 139, 1984.

190. **Keller, H. U. and Sorkin, E.,** Studies on chemotaxis. V. On the chemotactic effect of bacteria, *Int. Arch. Allergy*, 31, 505, 1967.

191. **Ward, P. A., Lepow, I. H., and Newman, L. J.,** Bacterial factors chemotactic for polymorphonuclear leukocytes, *Am. J. Pathol.*, 52, 725, 1968.

192. **Schiffman, E., Corcoran, B. A., and Wahl, S. M.,** N-Formylmethionyl peptides as chemoattractants for leucocytes, *Proc. Natl. Acad. Sci. U.S.A.*, 72, 1059, 1975.

193. **Schiffman, E., Corcoran, B. A., Ward, P. A., Smith, E., and Becker, E. L.,** The isolation and partial purification of neutrophil chemotactic factors from *Escherichia coli*, *J. Immunol.*, 114, 1831, 1975.

194. **Inouye, M. and Halegova, S.,** Secretion and membrane localization of proteins in *Escherichia coli*, *CRC Crit. Rev. Biol.*, 7, 339, 1980.

195. **Miura, A., Amaya, Y., and Mori, M.,** A metalloprotease involved in the processing of mitochondrial precursor proteins, *Biochem. Biophys. Res. Commun.*, 134, 1151, 1986.

196. **Briggs, M. S., Cornell, D. G., Dulhy, R. A., and Geirasch, L. M.,** Conformations of signal peptides induced by lipids suggests initial steps in protein export, *Science*, 233, 206, 1986.

197. **Novak, P., Ray, P. H., and Dev, I. K.,** Localization and purification of two enzymes from *Escherichia coli* capable of hydrolyzing a signal peptide, *J. Biol. Chem.*, 261, 420, 1986.

198. **Williams, L. T., Snyderman, R., Pike, M. C., and Lefkowitz, R. J.,** Specific receptor sites for chemotactic peptides on human polymorphonuclear leukocytes, *Proc. Natl. Acad. Sci. U.S.A.*, 74, 1204, 1977.

199. **Showell, H. J., Freer, R. J., Zigmond, S. H., Schiffman, E., Aswanikumar, S., Corcoran, B., and Becker, E. L.,** The structure activity relationships of synthetic peptides as chemotactic factors and inducers of lysosomal enzyme secreters for neutrophils, *J. Exp. Med.*, 143, 1154, 1976.

200. **Freer, R. J., Day, A. R., Radding, J. A., Schiffman, E., Aswanikumar, S., Showell, H. J., and Becker, E. L.,** Further studies on the structural requirements for synthetic peptide chemoattractants, *Biochemistry*, 19, 2404, 1980.

201. **Freer, R. J., Day, A. R., Muthukumararaswamy, N., Pinon, D., Wu, A., Showell, H. J., and Becker, E. L.,** Formyl peptide chemoattractants: a model of the receptor on rabbit neutrophils, *Biochemistry*, 21, 257, 1982.

202. **Clark, R. A. and Klebanoff, S. J.,** Chemotactic factor inactivation by the myeloperoxidase-hydrogen peroxide-halide system, *J. Clin. Invest.,* 64, 913, 1979.

203. **Marasco, W. A., Showell, H. A., and Becker, E. L.,** Substance P binds to the formylpeptide chemotaxis receptor on the rabbit neutrophil, *Biochem. Biophys. Res. Commun.,* 99, 1065, 1981.

204. **Broom, M. F., Sherriff, R. M., Tate, W. P., Collings, J., and Chadwick, V. S.,** Partial purification and characterisation of a formyl-methionine deformylase from rat small intestine, *Biochem. J.,* 257, 51, 1989.

205. **Marasco, W. A., Phan, S. H., Krutzsch, H., Showell, H. J., Feltner, D. E., Nairn, R., Becker, E. L., and Ward, P. A.,** Purification and identification of formyl-methionyl-leucyl-phenylalanine as the major peptide neutrophil chemotactic factor produced by *Escherichia coli, J. Biol. Chem.,* 259, 5430, 1984.

206. **Hobson, C. H., Roberts, E. C., Broom, M. F., Mellor, D. M., Sherriff, R. M., and Chadwick, V. S.,** Radioimmunoassay for formyl methionyl leucyl phenylalanine. I. Development and application to assessment of chemotactic peptide production by enteric bacteria, *J. Gastroenterol. Hepatol.,* 5, 32, 1990.

207. **Mooney, C., Keenan, J., Munster, D., Wilson, I., Allardyce, R., Bagshaw, P., Chapman, B., and Chadwick, V. S.,** Neutrophil activation by *Helicobacter pylori, Gut,* 32, 853, 1991.

208. **Broom, M. F., Sherriff, R. M., Munster, D., and Chadwick, V. S.,** Identification of formyl Met-Leu-Phe in culture filtrates of *Helicobacter pylori, Microbios,* 72, 239, 1992.

209. **Rot, A., Henderson, L. E., Sowder, R., and Leonard, E. J.,** *Staphylococcus aureus* tetrapeptide with high chemotactic potency and efficacy for human leukocytes, *J. Leukoc. Biol.,* 45, 114, 1989.

210. **Rot, A., Henderson, L. E., and Leonard, E. J.,** *Staphylococcus aureus*-derived chemoattractant activity for human monocytes, *J. Leukoc. Biol.,* 40, 43, 1986.

211. **Sannomiya, P., Craig, R. A., Clewell, D. B., Suzuki, A., Fujino, M., Till, G. O., and Marasco, W. A.,** Characterization of a class of nonformylated *Enterococcus faecalis*-derived neutrophil chemotactic peptides: the sex pheromones, *Proc. Natl. Acad. Sci. U.S.A.,* 87, 66, 1990.

212. **Perry, K. L., Elledge, S. T., Mitchell, B. B., Marsh, L., and Walker, G. C.,** UmuDC and mucAB operons whose products are required for UV light- and chemical-induced mutagenesis: UmuD, MucA, and LexA proteins share homology, *Proc. Natl. Acad. Sci. U.S.A.,* 82, 4331, 1985.

213. **Battista, J. L. R., Donnelly, C. E., Ohta, T., and Walker, G. C.,** The SOS response and induced mutagenesis, in *Mutation and the Environment Part A,* Mendelsohn, M. L. and Albertini, R. J., Eds., Wiley-Liss, New York, 1990, 169.

214. **Burckhardt, S. E., Woodgate, R., Scheuermann, R. H., and Echols, H.,** UmuD mutagenesis protein of *Escherichia coli*: overproduction, purification, and cleavage by RecA, *Proc. Natl. Acad. Sci. U.S.A.,* 85, 1811, 1988.

215. **Broom, M. F., Sherriff, R. M., Ferry, D. M., and Chadwick, V. S.,** fMet-Leu-Phe and the SOS operon in E. coli: a model of host bacterial interactions, *Biochem. J.,* 291, 895, 1993.

216. **Styrt, B.,** Species variation in neutrophil biochemistry and function, *J. Leukoc. Biol.,* 46, 63, 1989.

217. **Fletcher, M. P. and Gallin, J. I.,** Human neutrophils contain an intracellular pool of putative receptors for the chemoattractant N-formyl-methionyl-leucyl-phenylalanine, *Blood,* 62, 792, 1983.

218. **Gallin, J. I. and Seligmann, B. E.,** Mobilization and adaptation of human neutrophil chemoattractant fMet-Leu-Phe receptors, *Fed. Proc.,* 43, 2732, 1984.

219. **Gardner, J. P., Melnick, D. A., and Malech, H. L.,** Characterization of the formyl-peptide chemotactic receptor appearing at the phagocytic cell surface after exposure to phorbol myristate acetate, *J. Immunol.,* 136, 1400, 1986.

220. **Tennenberg, S. D., Zemlan, F. P., and Solomkin, J. S.,** Characterization of N-formyl-methionyl-leucyl-phenylalanine receptors on human neutrophils, *J. Immunol.,* 141, 3937, 1988.

221. **Leonard, E. J., Noer, K., and Skeel, A.,** Analysis of human monocyte chemoattractant binding by flow cytometry, *J. Leukoc. Biol.,* 38, 403, 1985.

222. **Allen, C. A., Broom, M. F., and Chadwick, V. S.,** Flow cytometery analysis of the expression of neutrophil FMLP receptors, *J. Immunol. Meth.,* 149, 159, 1992.

223. **Atkinson, Y. H., Marasco, W. A., Lopez, A. F., and Vadas, M. A.,** Recombinant human tumor necrosis factor-a. Regulation of N-formylmethionylleucylphenylalanine receptor affinity and function on human neutrophils, *J. Clin. Invest.,* 81, 759, 1988.

224. **Weisbart, R. H., Golde, D. W., and Gasson, J. C.,** Biosynthetic human GM-CSF modulates the number and affinity of FMLP receptors, *J. Immunol.,* 137, 3584, 1987.

225. **Haslett, C., Guthrie, L. A., Kopaniak, M. M., Johnston, R. B., and Henson, P. M.,** Modulation of multiple neutrophil functions by preparative methods or trace concentrations of bacterial lipopolysaccharides, *Am. J. Pathol.,* 119, 101, 1985.

226. **Fittschen, C., Sandhaus, R. A., Worthen, G. S., and Henson, P. M.,** Bacterial lipopolysaccharide enhances chemoattractant-induced elastase secretion by human neutrophils, *J. Leukoc. Biol.,* 43, 547, 1988.

227. **Wirthmueller, U., De Weck, A. L., and Dahinden, C. A.,** Platelet-activating factor production in human neutrophils by sequential stimulation with granulocyte-macrophage colony-stimulating factor and the chemotactic factors C5A or formyl-methionyl-leucyl-phenylalanine, *J. Immunol.,* 142, 3213, 1989.

228. **Tennenberg, S. D. and Solomkin, J. S.,** Neutrophil activation in sepsis. The relationship between fMetLeuPhe receptor mobilization and oxidative activity, *Arch. Surg.,* 123, 171, 1987.

229. **Zimmerli, W., Seligmann, B., and Gallin, J. I.,** Exudation primes human and guinea pig neutrophils for subsequent responsiveness to chemotactic peptide N-formyl-methionylleucylphenylalanine, *J. Clin. Invest.*, 77, 925, 1986.

230. **Guthrie, L. A., McPhail, L. C., Henson, P. M., and Johnston, R. B., Jr.,** Priming of neutrophils for enhanced release of oxygen metabolites by bacterial lipopolysaccharide, *J. Exp. Med.*, 160, 1656, 1984.

231. **Fehr, J., Dahinden, C., and Russi, R.,** Formylated chemotactic peptides can mimic the secondary provoking endotoxin injection in the generalized Schwartzman reaction, *J. Infect. Dis.*, 150, 160, 1984.

232. **Worthen, G. S.,** Lipid mediators, neutrophils, and endothelial damage, *Am. Rev. Respir. Dis.*, 136, 455, 1987.

233. **Anton, P. A., Targan, S. R., and Shanahan, F.,** Increased neutrophil receptors for and response to proinflammatory bacterial peptide formyl-methionyl-leucyl-phenylalanine in Crohn's disease, *Gastroenterology*, 97, 20, 1989.

234. **Ford-Hutchinson, A. W.,** Leukotriene $B_4$ and neutrophil function: a review, *J. R. Soc. Med.*, 74, 831, 1981.

235. **Van Epps, D. E., Bender, J. G., Simpson, S. J., and Chenowith, D. E.,** Relationship of chemotactic receptors for formyl peptide and C5a to CR1, CR3, Fc receptors on human neutrophils, *J. Leukoc. Biol.*, 47, 519, 1990.

236. **Baggiolini, M., Walz, A., and Kunkel, S. L.,** Neutrophil-activating peptide-1/interleukin 8, a novel cytokine that activates neutrophils, *J. Clin. Invest.*, 84, 1045, 1989.

237. **Westwick, J., Li, S. W., and Camp, R. D.,** Novel neutrophil-stimulating peptides, *Immunol. Today*, 10, 146, 1989.

238. **Marasco, W. A., Becker, E. L., and Oliver, J. M.,** The ionic basis of chemotaxis. Separate cation requirements for neutrophil orientation and locomotion in a gradient of chemotactic formylpeptide, *Am. J. Pathol.*, 98, 749, 1980.

239. **Becker, E. L.,** The formylpeptide receptor of the neutrophil. A search and conserve operation, *Am. J. Pathol.*, 129, 16, 1987.

240. **Desai, U., Kreutzer, D. L., Showell, H., Arroyave, C. V., and Ward, P. A.,** Acute inflammatory pulmonary reactions induced by chemotactic factors, *Am. J. Pathol.*, 96, 71, 1979.

241. **Mellor, D. M., Myers, D. B., and Chadwick, V. S.,** The cored sponge model of *in vivo* leucocyte chemotaxis, *Agents Actions*, 18, 550, 1986.

242. **Chester, J. F., Ross, J. S., Malt, R. A., and Weitzman, S. A.,** Acute colitis produced by chemotactic peptides in rats and mice, *Am. J. Pathol.*, 121, 284, 1985.

243. **LeDuc, L. E. and Nast, C. C.,** Chemotactic peptide-induced acute colitis in rabbits, *Gastroenterology*, 98, 989, 1990.

244. **Chadwick, V. S., Mellor, D. M., Myers, D. B., Selden, A. C., Keshavarzian, A., Broom, M. F., and Hobson, C. H.,** Production of peptides inducing chemotaxis and lysosomal enzyme release in human neutrophils by intestinal bacteria *in vitro* and *in vivo*, *Scand. J. Gastroenterol.*, 23, 121, 1988.

245. **von Ritter, C., Sekizuka, E., Grisham, M. B., and Granger, D. N.,** The chemotactic peptide N-formyl methionyl-leucyl-phenylalanine increases mucosal permeability in the distal ileum of the rat, *Gastroenterology*, 95, 651, 1988.

246. **Granger, D. N., Zimmerman, B. J., Sekizuka, E., and Grisham, M. B.,** Intestinal microvascular exchange in the rat during luminal perfusion with formyl-methionyl-leucyl-phenylalanine, *Gastroenterology*, 94, 673, 1988.

247. **Hobson, C. H., Butt, T. J., Ferry, D. M., Hunter, J., Chadwick, V. S., and Broom, M. F.,** Enterohepatic circulation of bacterial chemotactic peptide in rats with experimental colitis, *Gastroenterology*, 94, 1006, 1988.

248. **Woodhouse, A. F., Myers, D. B., Broom, M. F., Hobson, C. H., and Chadwick, V. S.,** Intestinal absorption, metabolism and effects of bacterial chemotactic peptides in rat intestine, *J. Gastroenterol. Hepatol.*, 2, 35, 1987.

249. **Ferry, D. M., Butt, T. J., Broom, M. F., Hunter, J., and Chadwick, V. S.,** Bacterial chemotactic oligopeptides and the intestinal mucosal barrier, *Gastroenterology*, 97, 61, 1989.

# Chapter 11

# Role of Probiotics

*Gerald W. Tannock*

## CONTENTS

I. The Normal Microflora of the Digestive Tract .................................................................. 257
   A. Influences of the Normal Microflora on the Host ............................................... 259
   B. Modification of the Normal Microflora ............................................................. 259
II. A Critical View of Current Probiotics ......................................................................... 260
III. Promising Developments in Probiotic Research ......................................................... 263
   A. Fermented Milk Products for Lactose-Intolerant Subjects ................................ 263
   B. Avirulent *Clostridium difficile* Strains as Probiotics ......................................... 264
   C. Influence of Lactobacilli on Enzyme Activities in the Large Bowel ................. 264
IV. Molecular Genetics and the Development of Probiotics ........................................... 265
V. Development of Efficacious Probiotics Through Gene Technology ......................... 267
VI. Detection of Probiotic Strains in Digestive Tract Samples ...................................... 267
VII. Conclusions .................................................................................................................. 269
References .............................................................................................................................. 269

## I. THE NORMAL MICROFLORA OF THE DIGESTIVE TRACT

The human digestive tract is exposed to numerous microorganisms throughout the life of the host. Many of these microbes have only a brief, transient existence in the digestive tract (environmental microbes) whereas others, through their adhesive, invasive, or toxigenic abilities persist for days or weeks (pathogens) before succumbing to the host's defense mechanisms. The digestive tract is the habitat for other microbial types that are detected consistently. This collection of microbes, mainly bacteria, that permanently colonize the digestive tract is known as the normal microflora.

The normal microflora of the human digestive tract is large and complex being composed of several hundred different species (Tables 1 and 2). Because host factors restrict the extent of microbial colonization, only certain regions of the human digestive tract harbor a normal microflora.[1] Until recently, the human stomach was considered not to be colonized by microbes in the case of healthy subjects, but the long-term association of *Helicobacter pylori* with the mucosal surface of the antrum of the stomach of many human subjects, in the absence of overt symptoms of disease, has lead Lee and colleagues to describe these bacteria as "almost normal microflora." *Helicobacter pylori* is involved in the etiology of at least two distinct pathologies (i.e., chronic active gastritis, peptic ulcers), which argues for the bacteria to be considered pathogenic rather than as a member of the normal microflora. Further knowledge of these microbes, and their relationship with the human host, is required before their exact ecological status can be determined.[2] Other bacterial types (Gram-positive cocci and bacilli) can be detected in gastric contents from healthy subjects, but never in numbers greater than about ten thousand per ml of aspirate. The duodenum and jejunum likewise harbor few microbial cells, but bacterial numbers are markedly higher in ileal contents (about $10^8$ per ml). Both Gram-positive and Gram-negative bacteria are present in this region. Extensive microbial colonization of the human digestive tract is primarily restricted to the large bowel (colon) where bacterial populations of $10^{10}$ per gram (wet weight) of contents are detectable. Obligately anaerobic species are numerically predominant, facultative anaerobes are nevertheless common, and both Gram-positive and Gram-negative genera are represented (Tables 1 and 2).[1]

Three host factors appear to be largely responsible for the distinctive distribution of the normal microflora of the human digestive tract.

1. Nature of the gastrointestinal mucosa. A microflora containing lactic acid bacteria is not present in the gastric region of humans, unlike mice, rats, pigs, and fowl. The forestomach of mice and rats, the pars oesophagea of the porcine stomach, and the fowl crop are lined by a stratified squamous epithelium.

**Table 1**   The 25 Most Prevalent Bacterial Species in the Feces of Human Subjects Consuming a Western-style Diet

1. *Bacteroides vulgatus*
2. *Bacteroides fragilis* sp.
3. *Bacteroides fragilis*
4. *Bacteroides thetaiotaomicron*
5. *Peptostreptococcus micros*
6. *Bacillus* sp.
7. *Bifidobacterium adolescentis* D
8. *Eubacterium aerofaciens*
9. *Bifidobacterium infantis*
10. *Ruminococcus albus*
11. *Bacteroides distasonis*
12. *Peptostreptococcus intermedius*
13. *Peptostreptococcus* sp. 2
14. *Peptostreptococcus productus*
15. *Eubacterium lentum*
16. Facultatively anaerobic streptococci
17. *Fusobacterium russii*
18. *Bifidobacterium adolescentis* A
19. *Bifidobacterium adolescentis* C
20. *Bacteroides clostridiiformis* ssp. *clostridiiformis*
21. *Peptococcus prevotii*
22. *Bifidobacterium infantis* ssp. *liberorum*
23. *Clostridium indolis*
24. *Enterococcus faecium*
25. *Bifidobacterium longum* ssp. *longum*

From Finegold, S. M., Attebery, H. R., and Sutter, V. L., *Am. J. Clin. Nutr.*, 27, 1456, 1974.

**Table 2**   Bacterial Species (Including Facultative Anaerobes) that are Commonly Detected as Members of the Fecal Microflora of Humans

*Bacteroides vulgatus*
*Bacteroides fragilis*
*Bacteroides thetaiotaomicron*
*Bacteroides distasonis*
*Bacteroides clostridiiformis*

*Bifidobacterium adolescentis*
*Bifidobacterium infantis*
*Bifidobacterium longum*

*Peptostreptococcus micros*
*Peptostreptococcus intermedius*
*Peptostreptococcus productus*
*Peptostreptococcus prevotii*

*Ruminococcus albus*

*Eubacterium aerofaciens*
*Eubacterium lentum*

*Fusobacterium russii*

*Enterococcus faecium*
*Enterococcus faecalis*

*Veillonella* sp.

*Lactobacillus acidophilus*
*Lactobacillus fermentum*
*Lactobacillus plantarum*

*Clostridium indolis*
*Clostridium perfringens*
*Clostridium ramosum*

*Escherichia coli*

Lactobacilli and, in the case of piglets, certain streptococci can adhere to the surface of these stratified squamous epithelia. Replication of attached cells forms a layer of bacteria on the epithelial surface.[3] The human gastrointestinal tract, except for the distal anal canal, does not contain a stratified squamous epithelium, so epithelial-associated lactic acid bacteria are not present in the human stomach.

2. Gastric acidity. Secretion of HCl and the associated acidic pH of gastric contents effectively prevents colonization of the human stomach. Patients with achlorhydria harbor at least 100 times more microbes in their stomach contents as compared to healthy subjects.[4]

3. Peristalsis. The propulsion of intestinal contents through the intestinal tract by peristalsis produces a "wash-out" effect that prevents the accumulation of significant numbers of microbes in the duodenum and jejunum. Patients with pathologies resulting in areas of intestinal stasis contain numerous microbes in these regions, usually bacterial types considered to be inhabitants of the large bowel. Colonization of the small bowel by these microbes can lead to malabsorptive conditions.[4] Interestingly, subjects in Third World countries usually have larger microbial populations in the small bowel than is encountered in individuals in other parts of the world. This may be due to the degree of microbial contamination of food, water, and environment in developing countries or, more likely, the effect of malnutrition on host physiology.[5]

It is clear, therefore, that care must be taken in assessing the role of the normal microflora of the digestive tract with regard to the well being of the human host. Marked differences in the distribution of the microflora occur between different animal species and possibly under different environmental conditions. Generalizations regarding the human microflora and extrapolations of observations from one animal species to another can sometimes be made, but with caution.

**Table 3**  Some Microflora-Associated Characteristics of Animals[a]

| Activity or substance | Comments |
|---|---|
| β-D-Glucuronidase | Absent in germ-free intestine |
| Short chain fatty acids (propionic, butyric) | Absent in germ-free intestine |
| Coprostanol | Absent in germ-free intestine |
| β-Aspartyl-glycine | Degraded in conventional intestine |
| Secondary bile acids | Absent in germ-free intestine |
| Urobilinogens | Absent in germ-free intestine |
| Trypsin | Inactivated in conventional cecum |
| Epithelial cell turnover rate | Slower in germ-free animals |
| Intestinal transit time | Slower in germ-free animals |
| Appearance of cecum (rodents only) | Distended in germ-free animals |
| Presence of neutrophils and lymphocytes in intestinal mucosa | Scarce in germ-free animals |
| Peyer's patches and mesenteric lymph nodes | Poorly demarcated in germ-free animals |
| Immunoglobulins in blood | Low concentrations in germ-free animals |

[a]  Table from References 6, 41, and 72.

## A. INFLUENCES OF THE NORMAL MICROFLORA ON THE HOST

The collective metabolic activities of the normal microflora of the large bowel have been likened to that of the liver, which is considered to be the most metabolically diverse and active human organ. The influence of a collection of at least 40 commonly detected, and up to 400 to 500, bacterial species on the intestinal ecosystem is likely to be appreciable. Microbially produced substances, absorbed into the blood or lymphatic circulations, could influence host factors even in sites remote from the digestive tract. Comparisons of the characteristics of animals harboring a normal microflora (conventional animals) with those of identical species and strain, but raised in the absence of microbes (germ free) have shown that such influences do occur.[6] Microflora-associated characteristics, examples of which are summarized in Table 3, involve the biochemistry, physiology, anatomy, and immunology of the host.

## B. MODIFICATION OF THE NORMAL MICROFLORA

The microflora of the large bowel of humans, despite the connotation "normal," is essentially a collection of parasites that are well adapted for life in the intestinal tract. As is the case with all forms of life, the activities of the microbes are aimed at exploitation of their environment so as to permit their cellular replication and hence continuation of the species. "Normal" as used in relation to the microflora denotes that it is commonly encountered in clinically healthy subjects, and not that the microflora necessarily confers normality on the host. In fact, the microflora has many undesirable attributes not the least of which is its involvement in post-surgical sepsis and urinary tract infection. Most pathologies associated with the microflora, however, result from the escape of microbes from the large bowel to usually sterile regions of the host as a result of conditions that predispose to this spread.[7,8]

The effect of the metabolic activities of the normal microflora on the human host has nevertheless long been of interest. In the early years of this century, Metchnikoff proposed that humans would benefit from the surgical removal of the colon so as to eliminate a reservoir of bacterial metabolites, produced by proteolytic (putrefactive) activities, that are toxic to human tissues.[9] Although bacterial metabolites which are potentially harmful to tissues are absorbed from the intestinal tract into the blood circulation, a normal functioning liver can detoxify these microbial metabolites so that systemic tissues are not affected.[4] Nevertheless, Metchnikoff's ideas on intestinal putrefaction and fermentation have been taken up and broadened by subsequent workers so that a widespread concept exists that the development of a lactic acid bacterial microflora in the colon is beneficial to the host. The concept of the desirability of modifying the microflora with lactic acid bacteria was initially due to the observation that the naturally occurring fermentation of milk by these microorganisms prevented the growth of nonacid-tolerant, proteolytic species. If lactic fermentation prevented the putrefaction of milk, it was reasoned that it should have a similar effect in the digestive tract. The implantation of lactobacilli in the gastrointestinal tract, therefore,

**Table 4** Microorganisms
Commonly Used in Probiotics

---

*Products for consumption by humans*

    *Lactobacillus acidophilus*
    *Lactobacillus casei* ssp. *rhamnosus*
    *Lactobacillus casei* Shirota strain
    *Lactobacillus delbrueckii* ssp. *bulgaricus*
    *Bifidobacterium adolescentis*
    *Bifidobacterium bifidum*
    *Bifidobacterium breve*
    *Bifidobacterium longum*
    *Bifidobacterium infantis*
    *Streptococcus salivarius* ssp. *thermophilus*

*Products for consumption by farm animals*

    *Lactobacillus acidophilus*
    *Lactobacillus casei*
    *Lactobacillus plantarum*
    *Lactobacillus delbrueckii* ssp. *bulgaricus*
    *Bifidobacterium bifidum*
    *Bacillus subtilis*
    *Streptococcus salivarius* ssp. *thermophilus*
    *Pediococcus pentosaceus*
    *Enterococcus faecium*
    *Saccharomyces cerevisiae*
    *Aspergillus oryzae*
    *Torulopsis* spp.

---

would suppress the growth of "putrefactive" bacteria, thus reducing the amount of toxic substances generated in the intestine. Subsequently widened to include modification of the microflora as a means of increasing nonspecific resistance to intestinal infection, the concept of the derivation of a beneficial intestinal microflora continues to attract strong interest.[10] Many commercial products that contain living microbes of intestinal origin are destined for consumption with the aim of improving "intestinal balance" are available. These products are termed "probiotics."[11]

Lactic acid-producing bacteria are a popular choice of microbes for use in intestinal microflora modifications (Table 4). This choice is based partly on the historical concept discussed above. Other reasons for using these bacteria are: (a) lactococci and lactobacilli have long been used in the production of foods containing fermented milk (cheese, yogurt) without harm to the consumer; therefore, by a rather doubtful logic, Gram-positive, lactic acid-producing bacteria can be safely consumed and (b), the large scale culture of lactic acid bacteria is possible and already developed by the dairy industry. Members of the genus *Lactobacillus*, many species of which are intestinal inhabitants, have been obvious choices for use in probiotics because they combine both lactic acid-producing and intestinal colonization attributes. The lactobacilli can be accepted as suitable candidates for probiotics in the case of pigs and fowl because the bacteria colonize the proximal regions of the digestive tract in large numbers in these animals. Moreover, they are present throughout the remainder of the gastrointestinal tract of the host. Their use in probiotics for human consumption could be considered curious because they are not numerically dominant in the digestive tract and are absent from the microflora of about 25% of human subjects.[12] On the other hand, members of the genus *Bifidobacterium*, when present in the intestinal tract, are a consistently more numerous bacterial group in feces. Bifidobacteria are obligate anaerobes that have a fermentative metabolism that produces high amounts of acetic and lactic acid. The increasing use of these organisms in probiotics presumably reflects their numerical predominance over lactobacilli.

## II. A CRITICAL VIEW OF CURRENT PROBIOTICS

The human digestive tract is colonized in early life by microorganisms. Delays in colonization may be produced by cesarean delivery but, even so, the acquisition of a normal microflora proceeds in a predictable manner until the mature microflora detected in adults is achieved.[13] Since the acquisition and maintenance of the microflora is achieved naturally without human intervention, the need for consumption of microbes as dietary supplements must surely be questioned. After all, why should the introduction into the digestive tract of bacteria contained in capsules, or yogurts, be more satisfactory than natural exposure to these bacteria, considering that they have the same source in nature? It can be argued that the bacteria contained in probiotics have special properties that will affect the well being of the host. This type of argument cannot withstand scientific testing, however, because the bacterial strains used in current probiotics have been chosen largely on the basis of their amenability to industrial-scale cultivation and survival ability during storage. Most, if not all, of the characteristics of these strains have been determined under *in vitro* conditions, and scientifically valid field trials of efficacy are rarely made.

Probiotics for use with farm animals may have validity in providing an appropriate intestinal microflora since, under modern intensive farming conditions, young animals may have little, if any, contact with adult animals and are raised under clean conditions relative to the farmyard. The opportunities for the acquisition of an intestinal microflora may be decreased under these circumstances.[11] This condition is of consequence in young animals with an already high susceptibility to infection by intestinal pathogens since the microflora confers a degree of nonspecific resistance to some infectious diseases. Studies relating to the nonspecific resistance of chickens to *Salmonella* colonization, however, show that the resistance is mediated by a collection of bacteria (mostly obligate anaerobes), and not by single bacterial types.[14] Stressful conditions are known to alter the microflora, including a decrease in the *Lactobacillus* population, of the digestive tract of farm and experimental animals.[15] It is unlikely, however, that the administration of lactic acid bacteria in times of stress could lead to recolonization of the tract: if conditions in the intestine were unsuitable for the replication of indigenous lactobacilli, they would be equally unsuitable for colonization by the probiotic strain. Taking all these factors into consideration, it is unlikely that humans can benefit from the administration of probiotics as currently formulated. Could a physician with clear conscience prescribe the administration of a relatively poorly defined, Gram-positive bacterial strain of intestinal origin to human neonates that have been delivered by cesarean section? How could one evaluate the well being of neonates or stressed adults with or without probiotic treatment?

Perhaps because of these considerations, use of probiotic preparations on the advice of a medical practitioner is largely limited to the consumption of yogurt after treatment with an antibiotic administered by the oral route. Antibiotic administration in this way can result in alterations to the normal microflora of human subjects[16] with associated changes in bowel habit. General practitioners sometimes advise patients to consume yogurts to "restore the microbial balance" in their intestinal tract. Often, according to anecdotal reports, the diarrhea accompanying or following antibiotic usage ceases about the same time as the consumption of fermented milk occurs. The bacteria utilized commonly in the manufacture of yogurt (*Streptococcus salivarius* ssp. *thermophilus*, *Lactobacillus delbrueckii* ssp. *bulgaricus*) apparently do not colonize the gastrointestinal tract.[17] Hence, any amelioration of antibiotic-associated diarrhea must be due to a placebo effect, coincidental resolution of the problem as the microflora reconstitutes naturally, or because the yogurt bolus tends to have a constipating effect as does kaolin (a fine white clay), which was at one time a popular remedy for stopping diarrhea.

Probiotics containing lactic acid bacteria are claimed to produce beneficial effects on the host because they stimulate the defensive mechanisms of the animal's body so that they are "primed" and ready to quickly destroy invading pathogens.[6] The cell wall components of lactic acid-producing bacteria do stimulate the immunological system. Peptidoglycan from *Bifidobacterium longum* and *Bifidobacterium thermophilum*, and cell wall material from *Lactobacillus casei*, *Lactobacillus delbrueckii* ssp. *bulgaricus*, and *Lactobacillus plantarum* have been reported to influence certain immunological parameters of various animal species (Table 5). The major effect exerted by cell wall material seems to be on macrophages. These are mononuclear cells that live free in tissues or are attached to walls of blood sinuses. They have different names and functions depending on the tissue in which they reside. For example, those in the liver are called Kupffer cells, those in bone marrow are osteoclasts, and those in the lungs are alveolar macrophages. Macrophages can ingest and destroy microbial cells and other "foreign" cells or materials, or damaged body components, contributing greatly to the inflammatory response through the release of interleukin-1 and tissue necrosis factor. Both of these substances enhance neutrophil activities. Like neutrophils, the cytoplasm of macrophages contains membrane-bounded sacs (lysosomes) that contain hydrolytic enzymes, cationic antibacterial proteins, and molecules that catalyze the formation of the bactericidal superoxide anion from oxygen. Phagocytosed bacteria are contained within vacuoles (phagosomes) in the macrophage cytoplasm. Lysosomes migrate to, and fuse with, the phagosome thereby liberating their contents. The bactericidal and digestive processes of the lysosomal

**Table 5**  Examples of Published Papers Reporting
Stimulation of the Immune System by Lactic Acid Bacteria

| Subject | Reference |
|---------|-----------|
| Adjuvant effect of *Lactobacillus plantarum* | 73 |
| Antitumor effect of *Lactobacillus casei* | 74 |
| Resistance to *Mycobacterium* infection in mice using *L. casei* | 75 |
| Resistance to *Pseudomonas* infection in mice using *L. casei* | 76 |
| Survival of mice inoculated with *Listeria monocytogenes* using *L. casei* | 77 |
| Antitumor effect of *Bifidobacterium infantis* | 78 |
| Antitumor effect of yogurt | 79 |
| Adjuvant effect of fermented milk (*L. acidophilus, L. casei*) | 80 |

contents kill and destroy the bacteria. Unlike neutrophils, macrophages continue to differentiate after they have left the bone marrow and, under conditions of appropriate stimulation, become "activated." Macrophages that have been activated phagocytose more vigorously, take up more oxygen, and secrete larger amounts of hydrolytic enzymes. In general, activated macrophages are better able to kill microbial cells. Material from *Lactobacillus casei*, for example, activates macrophages so that they have increased oxygen radical production (superoxide anion) and increased lysosomal enzyme activity. They also have increased phagocytic ability and are induced to release colony-stimulating factor (stimulation of proliferation and differentiation of granulocytes and macrophages).[18,19] As well as their role in nonspecific resistance, macrophages participate in inducing specific immune responses by stimulating the development of lymphocytes. Macrophages respond to chemical signals sent by lymphocytes and are, in turn, stimulated to differentiate into activated macrophages. Therefore, there is interaction between the constitutive and inducible defense systems.

While many reports record the stimulation by lactic acid bacteria, or their products, of immunological parameters to statistically significant degrees in experimental studies, the biological significance of the results needs always to be considered. In other words, would the degree of stimulation of the immunological mechanisms obtained be of any consequence to the well being of the animal? To be of biological significance, immunostimulation would need to render the animal refractory to infection when exposed to a pathogen in the long term (the same effect as a vaccine used to immunize against a specific infection), or would shorten the period of the illness, or decrease mortality. Furthermore, the immunostimulatory effect must be produced under conditions that resemble those in real life. In general, studies reported so far in the literature do not meet these criteria. For example, the antitumor effects that have been observed in animals inoculated by parenteral routes, with lactobacilli or bifidobacterial products, were obtained using high dosages of bacterial cells over a protracted period [e.g., 10 mg of bacteria (about $10^{10}$ bacterial cells) per kg body weight given on five occasions] and bear no relationship to the consumption of fermented milk products. While translocation of bacteria from the intestinal tract is known to occur under certain circumstances,[20] its magnitude is not equivalent to the parenteral injection of animals with large doses of microbial cells. Experiments need to mimic the situation where small amounts of bacterial components would pass from the gastrointestinal lumen into the lymphatic or blood circulation of the host. In addition, survival times of experimental animals must be presented realistically. A small proportion of treated animals may live longer than untreated controls, but comparisons should be made so that an impression of the fate of the majority of experimental subjects is obtained. The biological significance of the degree of even transient resistance produced against pathogenic bacteria by injection of animals with lactobacilli cannot be realistically assessed on the basis of the reported studies. Parenteral inoculation routes have been used which bypass the normal barriers to infection that would be encountered by the pathogen.

Lactic acid-producing bacteria are not, of course, the only bacteria that activate macrophages in this way. Any cell with different antigenicity to that of the host will stimulate some part of the animal's immune system. The activation of macrophages by the B.C.G. (vaccine) strain of *Mycobacterium bovis*, and by "*Corynebacterium parvum*" (actually *Propionibacterium acnes*, a member of the skin microflora) has been thoroughly studied.[21] The use of bacteria or bacterial compounds as nonspecific stimulators of the immunological system must be approached with caution, because in some subjects components of the immune system that are activated and respond to bacterial substances may also react with host components, thereby producing an autoimmune disease. Mycobacteria, for example, are being investigated for their possible involvement in causing rheumatoid arthritis in humans.[22] Cell wall material from *Bifidobacterium breve*, *Bifidobacterium adolescentis*, *Eubacterium aerofaciens*, and *L. casei* have been reported to produce moderate to severe arthritis in rats inoculated intraperitoneally with the bacterial components.[23,24] A safer approach to the controlled modulation of immunological mechanisms than whole bacterial cells or their cell wall components appears to be the use of purified cytokines (i.e., interleukin-1, interleukin-2, gamma interferon) obtained through recombinant DNA technology.[25] Probiotics need to be derived on a rational basis so that they too can enter the armamentarium of the medical practitioner on the same footing as pharmaceutical products.

## III. PROMISING DEVELOPMENTS IN PROBIOTIC RESEARCH

Probiotics only have promise when predictable outcomes resulting from their use can be measured. Probiotic research in three areas appears to have potential for humans in this respect.

### A. FERMENTED MILK PRODUCTS FOR LACTOSE-INTOLERANT SUBJECTS

The disaccharide lactose, which constitutes about 5% of cow's milk, is hydrolyzed to glucose and galactose moieties by lactase (a β-galactosidase) associated with the brush border of enterocytes lining the small intestinal tract of children. These saccharides can be absorbed and metabolized. Congenital deficiency in lactase is extremely rare, but late-onset hypolactasia, developing during childhood, is common in many regions of the world, including Asian, Arab, and African populations (80 to 100% of individuals) and also occurs in caucasians, but at a lower incidence (25%).[26] Deficiency in the production of lactase in the intestinal tract results in passage of dietary lactose in an unaltered state to the lower small bowel and the colon. These sites harbor an extensive microflora (about $10^8$ bacteria per ml ileal contents; $10^{10}$ per gram of colon contents), several microbial components of which produce β-galactosidase that enables them to hydrolyze the lactose. Products of the hydrolysis are fermented by the bacteria, with production of gases (including $H_2$) and organic acids which give rise to the symptoms of abdominal discomfort and osmotic diarrhea that are experienced by lactose-intolerant individuals.

Clinically, lactose intolerance is defined as the occurrence of gastrointestinal symptoms after administration of a single test dose of 50 g of lactose in aqueous solution. More specifically, lactose intolerant individuals have a breath $H_2$ concentration over 20 ppm during an 8 h period following a 20 g ingestion.[27] This breath test provides a standard, internationally recognized, noninvasive clinical assay by which lactose intolerance can be diagnosed, and the effects of fermented milk products on lactose-intolerant individuals can be evaluated under conditions that exclude a placebo effect. Studies have demonstrated that lactose-intolerant individuals can tolerate fermented milk (yogurt) better than base milk containing the same amount of lactose.[28] Patient perceptions correlate with $H_2$ breath test results. It appears that although the lactic acid bacteria themselves cannot replicate in the small bowel, β-galactosidases produced by *S. salivarius* ssp. *thermophilus* and *L. delbrueckii* ssp. *bulgaricus* during production of the yogurt can pass through the stomach (the enzyme is apparently intracellular). The β-galactosidases are inactive at the acid pH of the final yogurt product and do not hydrolyze lactose until the digesta reaches the lower small intestine.[28] By this time, the pH of the digesta, even containing acidic substances such as yogurt, has reached near neutrality, which is optimal for β-galactosidases produced by streptococci and lactobacilli.[27,28] Additionally, the bacterial cells may lyze as they pass through the small bowel.

On the basis of these studies, therefore, it can be proposed that yogurts prepared using lactic acid bacteria that form β-galactosidases in large amounts, which are active at acid pH, would enable lactose-intolerant individuals to ingest more milk-based food. With associated importance in nutrition, relating to protein and calcium requirements, it could thus be an aid to the dietary supplementation of malnourished populations in Africa and other countries where lactose intolerance is common. Yogurts containing β-galactosidases, with these properties, would pass through the stomach and begin to hydrolyze lactose as soon as the pH of the duodenal contents became slightly more alkaline than the yogurt. Lactose hydrolysis would therefore occur in the proximal small bowel rather than in the distal bowel and would mean that the products of hydrolysis would be absorbed from the intestine before they reached regions of the digestive tract colonized by a microflora. Additionally, the absence in humans of a microflora in the proximal intestinal tract would ensure that microbial metabolism of the products of lactose hydrolysis would not cause problems in the duodenum and jejunum of the subjects. Research aimed at screening lactic acid bacteria for production of β-galactosidases has indicated that lactococci do not produce β-galactosidase (they take up lactose by the phosphoenolpyruvate-dependent transferase system and hydrolysis is carried out by phospho-β-galactosidase) but that lactobacilli, except for *Lactobacillus casei* do produce this enzyme.[29] Some work on the β-galactosidase activity of "dairy strains" of *Lactobacillus helveticus* and *L. acidophilus* has been done, but strains of lactic acid bacteria that inhabit the digestive tract need to be examined further. Admittedly, strains isolated from the digestive tract may not have other qualities necessary for the production of a quality yogurt, but molecular biological methods could be used in the long term to incorporate the appropriate β-galactosidase-encoding gene into a strain of lactic acid bacterium already used commercially.

## B. AVIRULENT *CLOSTRIDIUM DIFFICILE* STRAINS AS PROBIOTICS

Pseudomembranous colitis, associated with the oral administration of ampicillin, clindamycin, cephalothin, or other antibiotics, is characterized by watery, nonbloody diarrhea and abdominal pain. Sigmoidoscopy demonstrates white-yellow areas (plaques) on the surface of the colonic mucosa and, in severe cases, the plaques coalesce to form a sheet composed of fibrin, mucins, and leukocytes (a pseudomembrane) on the epithelial surface. In some cases, the disease can contribute directly to death of the patient. The signs and symptoms of pseudomembranous colitis, and in many cases of the less severe disease, antibiotic-associated colitis, are due to proliferation of an obligately anaerobic bacterium, *Clostridium difficile*, in the large bowel. Attaining populations of about $10^6$ per gram of large bowel contents, the clostridia secrete two toxins (A and B).[30] Toxin A is an enterotoxin since it causes fluid accumulation in ileal loops formed experimentally in rabbits. It is a 308 kD protein that also has cytotoxic activity since it causes a cytopathic effect on cultured mammalian cells. Toxin B is a cytotoxic 207 kD protein that is lethal to experimental animals when administered with a sublethal dose of toxin A.[31] It has been proposed that toxins A and B act synergistically: the former damaging the epithelium, thus allowing toxin B access to the underlying tissue. Both toxins are important in the etiology of disease as demonstrated by the observation that hamsters require immunization against both toxins to be protected from ileocecitis caused by *C. difficile*.[32]

The proliferation of *C. difficile* in the large bowel results from suppression of other members of the normal microflora during prolonged antibiotic therapy. The absence of these bacteria removes, from the ecosystem, the mechanisms by which *C. difficile* populations are regulated. In the antibiotic-altered ecosystem, *C. difficile* replicates to levels where its toxins reach pathological concentrations. *Clostridium difficile* is present in the feces of about 3% of healthy subjects but is easily transmitted between patients in hospital wards. Results from *in vitro* investigations using fecal homogenates inoculated with *C. difficile* suggest the involvement of obligately anaerobic bacteria as the members of the microflora that normally regulate *C. difficile*. The species that achieve this effect have not, however, been conclusively identified. Current research is aimed at identifying the bacteria with regulatory characteristics so that they can be used in probiotics in prophylaxis against pseudomembranous colitis.[33] In other research, nontoxigenic (avirulent) strains of *C. difficile* are being investigated in this role because they would presumably occupy the same ecological niche as virulent strains and, if of superior colonizing ability, could exclude the toxin producing strains from the large bowel.[34] Probiotics containing obligately anaerobic bacteria might need to be administered in the form of enemas to ensure that the bacteria arrive at the appropriate anatomical location in a viable state. Obtaining informed consent from patients for the administration of a fecal emulsion or culture of fecal bacteria by enema may be a task that would tax even Demosthenes.

## C. INFLUENCE OF LACTOBACILLI ON ENZYME ACTIVITIES IN THE LARGE BOWEL

Cancer of the large bowel is a major cause of mortality in countries such as New Zealand, the United Kingdom, and the United States.[35] Epidemiological evidence suggests that incidence of the disease is strongly influenced by lifestyle, in particular by the type of diet. Inhabitants of countries in which cancer of the colon is relatively common tend to consume more fat (meat) in their diet than do other population groups. In some cases, populations with a higher incidence of the disease also consume less fiber.[35] Although dietary composition does not markedly affect the composition of the intestinal microflora in terms of bacterial species present,[36] diet has been shown to influence enzyme activities that are of microbial origin.[37] Cancer of the colon is likely to be multifactorial in etiology, but bacterial enzymes could be of significance in the formation and release of carcinogens (mutagens) in the large bowel. Azoreductase activity has received attention, for example, because these enzymes, which are produced by several bacterial species comprising the normal microflora, catalyze the cleavage of azo bonds in dyes used in the food industry as coloring agents. There has been concern that these azoreductases can mediate the formation of toxic amines (some of which are carcinogens) in the intestinal ecosystem.[38] β-glucuronidase, which is produced by *Escherichia coli* and obligate anaerobes such as bacteroides and clostridia, catalyze the cleavage of glucuronic acid molecules from glucuronides entering the intestinal tract in bile. This bacterial enzyme activity could be significant for reactivation of potentially toxic molecules, including carcinogens or co-carcinogens, that had been detoxified by the formation of glucuronides in the liver.[39] While the significance of these, and other bacterial enzymes, in the causation of colon cancer is not clear, efforts at modifying the levels of their activity by the use of a *Lactobacillus* strain administered to experimental subjects in milk have been reported. β-glucuronidase, nitroreductase, and azoreductase activities were monitored in the feces of human subjects during a regimen that included a four-week period in which they consumed their usual diet, four weeks in which their diet was supplemented with low fat milk, return for four weeks to their usual unsupplemented diet, four weeks of diet supplemented with milk containing *L. acidophilus* (about $10^9$ lactobacilli per ml), and four weeks of their usual unsupplemented diet. During the period where the diet was supplemented with *L. acidophilus*, a two- to fourfold reduction was recorded in the activity of the three enzymes.[40]

Unfortunately, the status of the lactobacilli of the subjects in this study before, during, and after administration of the *Lactobacillus* strain was not determined. This makes reconciliation of the observed changes in enzyme activities with that of the presence of the ingested species difficult. Did the "probiotic" strain colonize the individuals? Did it survive transit through the digestive tract? These questions cannot be answered on the basis of the reported findings. Some degree of support for these experimental results is available from studies using mice that do not have lactobacilli as members of their gastrointestinal microflora. These animals harbor a complex microflora in their digestive tracts which is functionally equivalent to conventional mice, as judged by the determination of 26 microflora-associated characteristics, but are not host to lactobacilli. Maintained in isolators by germ-free technological methods, the colony of mice provides a microbiologically constant model system with which to determine the influences of lactobacilli when resident in the digestive tract of the host.[41] Comparison of the biochemistry of the large bowel content of *Lactobacillus*-free mice compared with that of mice with an identical microflora, but to which three *Lactobacillus* strains had been added, showed that the lactobacilli influenced enzymatic activities. Azoreductase activity was 31% lower in animals colonized by lactobacilli as compared to those that were *Lactobacillus*-free.[42] Male *Lactobacillus*-free mice had 52% more β-glucuronidase activity in the cecum than female mice. The enzyme activity was reduced compared to that observed in females when lactobacilli colonized the digestive tract of the male mice.[43] The mechanisms by which lactobacilli mediate these phenomena and the significance of the altered biochemistry to the animal host is presently unknown. Collectively, the results from experiments with human subjects and with mice indicate that further research into this area is justified.

## IV. MOLECULAR GENETICS AND THE DEVELOPMENT OF PROBIOTICS

The development and production of scientifically valid probiotics is a difficult task given that it can be speculated that a wide range of properties should be present in the "ideal" probiotic strain:

1. The microorganism should persist in the digestive tract for a period sufficient to exert the desired effect on the host. The ability of the bacteria to adhere to, or associate with, mucosal surfaces could be important in this respect. Bacterial growth rate and efficient utilization of nutrients are also thought to be important colonization attributes.
2. The microbial strain should produce a substance inhibitory to digestive tract pathogens, or stimulate host immunity, so as to increase the resistance of the animal to intestinal infection.
3. The microbe should contribute to the nutrition of the host animal by synthesizing essential nutrients that become more readily available to the host, and/or by digesting dietary substances that the host is physiologically ill equipped to utilize.
4. The microbial strain must be suitable for cultivation under large-scale industrial conditions, and must be amenable to preservation for storage, prior to retailing and use in the field.
5. The organism must be avirulent and devoid of metabolic characteristics that could compromise the health of the host.
6. All of the above properties must be stable characteristics.

The detection of such a prototype strain in nature would require a large and expensive screening program and, indeed, would be unlikely to succeed. The alternative is to derive a microbial strain with appropriate characteristics by means of genetic manipulation. An organism known to colonize the digestive tract, for example, could be modified so that genes encoding undesirable attributes were inactivated, while those encoding characteristics desirable in a probiotic culture could be introduced using recombinant DNA technology.[44,45]

Recombinant DNA technology is well developed for use with *Escherichia coli* but knowledge of the genetics and molecular biology of other species comprising the normal microflora has accumulated at a relatively slow rate due to the relatively small number of research groups pursuing this kind of work. Using lactobacilli as an example, work has progressed slowly because, although the dairy industry is greatly interested in lactic acid bacteria, most of the research in this respect has been directed toward lactococci. Additionally, technical difficulties are greater when working at the molecular level with Gram-positive bacteria as compared to *Escherichia coli*. The basic tools required for the genetic manipulation of some gastrointestinal inhabitants are, however, available. Using lactobacilli as an example, plasmids with potential use as cloning vectors have been derived and the three "classical" methods of introducing DNA molecules into bacterial cells (i.e., conjugation, transduction, and transformation) have been demonstrated using lactobacilli.[46-48] Of particular significance has been the development of electrotransformation (electroporation) of Gram-positive bacteria as described in Chapter 2. This technique enables many species of Gram-positive bacteria, such as the lactobacilli for which transformation methods were not previously available, to be genetically modified.[48] At least 13 *Lactobacillus* genes have been cloned in *E. coli* hosts but the genetic modification of these species for biotechnological purposes is still in its infancy.[44] Nevertheless, Scheirlinck and colleagues[49] have demonstrated that the derivation of recombinant *Lactobacillus* strains with a stable, novel phenotype is feasible. In their experiments, a strain of *Lactobacillus plantarum* was electrotransformed with plasmids containing an α-amylase gene from *Bacillus stearothermophilus* and an endoglucanase gene from *Clostridium thermocellum*. Both genes were expressed in the *Lactobacillus* host and the active enzymes were excreted by the bacterial cells into the culture medium. The plasmids, however, were of low stability in the *Lactobacillus* host. Further plasmids were constructed that contained fragments of *Lactobacillus* chromosome, an origin of replication recognized only by Gram-negative bacteria, and the α-amylase and endoglucanase genes. Following electrotransformation of the *L. plantarum* strain with this type of plasmid, the plasmid DNA constructs were inserted into the *Lactobacillus* chromosome by means of single homologous recombination events. The transformed lactobacilli secreted endoglucanase, but α-amylase was secreted poorly.

Stability of newly acquired characteristics by recombinant bacteria would be critical to the successful development of a novel probiotic strain by genetic modification. The introduced DNA must be maintained by the bacterial host and transmitted to its progeny under industrial conditions and in the gastrointestinal environment. Recombinant plasmids capable of autonomous replication in the bacterial cytoplasm will probably not prove to be a satisfactory means of deriving genetically modified strains for use as probiotics. These plasmids are frequently not maintained by bacterial cells, even under laboratory conditions, and are likely to be quickly lost under the harsh conditions prevailing in natural habitats.[50] Retention of recombinant plasmids would be encouraged if a genetic determinant essential to the survival of the bacterium was incorporated into the recombinant DNA molecule.

A strong argument against the use of autonomously replicating, recombinant plasmids to modify gastrointestinal bacteria is the evidence that interspecies and intergeneric transfer of DNA is common.[51-53] The potential exists, therefore, for recombinant plasmids to be transferred widely within the normal microflora of the digestive tract. This possibility is unlikely to please scientific or governmental agencies concerned with the regulation of the release of genetically modified microbes. The insertion of novel genetic determinants into the chromosome of the recipient bacterial strain would overcome these problems; the newly acquired DNA would be:

1. replicated as part of the bacterial chromosome (stable mainentance);
2. inherited by the recipient's progeny; and
3. no more likely to be transferred to other bacterial types than any other gene on the chromosome.

## V. DEVELOPMENT OF EFFICACIOUS PROBIOTICS THROUGH GENE TECHNOLOGY

Advances in microecological knowledge of the digestive tract and the molecular biology of the intestinal inhabitants permits the derivation of probiotics that will have specific, scientifically measurable effects on the host. The derivation of probiotic strains by molecular genetic technologies will result in a conceptual change regarding probiotics. The concept will change from the use of strains selected from nature and administered to animals with the rather vague aim of improving "intestinal balance," to one in which members of the normal microflora of the digestive tract are genetically modified so that when they colonize the animal recipient they become an internal source of molecules with specific immunological or pharmacological activity. The types of probiotics that could be developed by using molecular genetics are as follows:

1. Immunizing strains. A member of the microflora could be genetically modified so that its cells synthesized an immunogen characteristic of a particular intestinal pathogen. Colonization of the digestive tract by such a strain could result in a continuous exposure of the intestinal mucosa to the immunogen so that secretory IgA antibodies would be synthesized by the host. This process could result in immunity of the host to the specific intestinal pathogen since the antibodies would prevent binding of the pathogenic cells or toxins to the intestinal epithelial surface. The genetically modified microbe could be transmitted from mother to infant, thus perpetuating the cycle of immunization without further attention from a medical practitioner. Genetic modification of bacteria or viruses so they synthesize a variety of immunogens has already been achieved using avirulent *Salmonella* strains[54] and vaccinia virus,[55] but the use of members of the normal microflora may be aesthetically more pleasing to the general public.

2. Delivery strains. Biotechnological research has led to the derivation of genetically modified bacteria and yeasts that can be cultivated in large volumes under industrial conditions. The modified organisms synthesize large amounts of substances of medical or industrial significance, which are subsequently harvested and purified from the cultures. It may be possible to derive strains of gastrointestinal, or other microbes, that similarly synthesize novel substances following genetic modification, but that can be used to deliver molecules with biological activity to particular regions of the intestinal tract, without the need to industrially purify the beneficial substance in question. For example, a microbial strain whose cells lyze under specific conditions of surfactant concentration and/or pH might be derived for incorporation into foods so that microbial cells would deliver a product to a specific level of the intestinal tract.

## VI. DETECTION OF PROBIOTIC STRAINS IN DIGESTIVE TRACT SAMPLES

In instances other than that of the delivery strains mentioned in the previous section, probiotic organisms would appear to require colonization attributes to be effective. It is difficult to envisage how probiotics could influence host biochemistry, or other factors, if they could not metabolize and function for a relatively prolonged period in the digestive tract. Methods that permit the rapid and accurate detection and enumeration of specific bacterial strains are therefore required during developmental research and field trials to confirm that bacteria administered as probiotics survive and replicate in appropriate regions of the digestive tract.

Colony morphology as a means of recognizing a specific strain of microbe has been used in at least one study involving a *Lactobacillus* strain.[56] Given the diversity of colonial forms of lactobacilli and other members of the normal microflora encountered in cultures from digestive tract samples, it is doubtful that this is a realistic proposition in most instances. Phenotypic differentiation on the basis of metabolic properties of strains is logistically impossible since large numbers of isolates from gastrointestinal tract samples would have to be examined using several tests. Although serological recognition of strains is somewhat simpler, it again involves the testing of many individual isolates. The major difficulty with serological recognition is that an antiserum which will react only with the bacterial strain of interest must first be derived.

Molecular biological techniques in which bacterial strains are characterized by genotypic traits rather than phenotype now permit the reliable differentiation of bacterial strains belonging to diverse species. Techniques that may be useful with respect to ecological studies of the normal microflora include:

1. Plasmid profiling. The collection (number and size) of plasmids harbored by a bacterial strain, referred to as the plasmid profile, can be used to differentiate it from other similar strains. Plasmid profiling has

been used to distinguish strains of bacteria in medical, industrial, and environmental investigations as well as to study colonization of the digestive tract of piglets and of human infants by normal microflora strains.[57,58] Plasmid profile instability (i.e., loss of a plasmid from the collection; acquisition of a plasmid by conjugation or other mechanisms) has been considered to be a disadvantage of this method, but, at least in the case of lactobacilli, indigenous plasmid profiles are as stable as chromosomal "fingerprints" derived from restriction endonuclease digest analysis.[59] An obvious disadvantage of plasmid profiling, however, is that not all bacterial strains harbor plasmids.

2. Ribosomal RNA (rRNA) as a broad-spectrum probe for strain differentiation. Ribotyping is a method by which restriction fragment length polymorphism of DNA can be detected in bacterial strains of the same or different species of a particular genus.[60] The method uses ribosomal RNA sequences as the basis for a broad-spectrum probe for strain differentiation. DNA extracted from bacterial isolates is digested with appropriate restriction endonucleases, the resulting fragments of DNA are separated in an agarose electrophoretic gel, transferred to a hybridization membrane, and probed with a radiolabeled nucleotide sequence derived from that of *E. coli* ribosomal RNA. Because bacteria have multiple copies of rRNA operons in their chromosome, several fragments containing rRNA gene sequences are observed after hybridization with the labeled probe. Ribosomal RNA sequences from *E. coli* can be used as a probe for any bacterial species because highly conserved rRNA sequences are available. Fragment length polymorphism revealed by comparison of the hybridization patterns permits differentiation between bacterial strains.

   Ribotyping of lactobacilli of gastrointestinal tract origin has recently been demonstrated to be a practical and sensitive means of differentiating between strains. Although minor changes in ribotype occur over time when lactobacilli are inhabiting the digestive tract, the use of two restriction endonucleases with markedly different recognition sequences to generate ribotypes reduces the possibility of mistaking the variant as a novel strain.[59]

3. Pulsed field gel electrophoresis (PFGE). Chromosomal DNA is digested by restriction endonucleases chosen on the basis of the mol% G+C content of the DNA of the bacterial species and on the recognition sequence of the enzyme. Endonucleases that produce a relatively small number of DNA fragments are desired so that an easily recognizable pattern can be visualized after PFGE.[61] The pattern of fragments generated can be characteristic of a strain.[62] Pulsed field gel electrophoresis separates large DNA fragments (as large as 10 megabases) by exposing them to alternating, perpendicular electrical fields. The alternating field forces the fragments to change orientation rather than to migrate immediately after the field is changed from one direction to another. The rate of reorientation is size dependent so larger molecules change direction more slowly than small ones. The pulse time (the time spent in a field of particular direction) is varied (between 1 sec and 1.5 h) and this dictates the DNA class size which spends most of the time reorientating rather than migrating. The DNA molecules are therefore separated by the retardation of net movement rather than by sieving. Preliminary studies in my laboratory have demonstrated that strains of intestinal lactobacilli can be characterized by this method (Heng and Tannock, unpublished).

4. Repetitive extragenic palindromic (REP) elements/enterobacterial repetitive intergenic consensus (ERIC) sequences/polymerase chain reaction (PCR). Repetitive nucleotide sequences dispersed randomly in the DNA of Gram-negative bacteria have been detected. Two classes of dispersed sequences exist: REPs and ERIC sequences. Oligonucleotide primers that can be used with the PCR to amplify these dispersed sequences and hence produce a characteristic strain fingerprint composed of PCR products of different molecular weights have been reported. The method is apparently rapid and sensitive, requiring a single set of two REP and two ERIC primers for PCR amplification, and agarose gel electrophoresis to detect the products (i.e., to obtain the "fingerprint").[63]

5. Nucleic acid probes. Cloned DNA can be used to prepare labeled probes to detect specific nucleotide sequences in DNA liberated from microbial cells. Probes labeled radioactively or nonradioactively (i.e., biotin, digoxigenin, acridinium) can be used in hybridization studies with chromosomal or plasmid DNA as target immobilized on nylon membranes. Such probes can detect species and even specific strains of bacteria cultured on selective media on membranes. Derivation of specific probes by random cloning and testing of fragments of DNA generated by restriction endonuclease digestion of bacterial genomes, or the use of whole plasmids as probes, has been described for some intestinal inhabitants.[64-68] Synthetic oligonucleotide probes that have ribosomal RNA sequences as their target have also been derived and may be useful in the detection and enumeration of bacterial strains.[69,70]

## VII. CONCLUSIONS

Increasing knowledge of the molecular biology of microbial cells, including those of members of the normal microflora, permits the development of scientifically valid probiotics. Strains for use in microbial interference, immunization, and delivery of pharmacological substances are likely developments in the near future. Since these developments depend on the derivation and intentional release of genetically modified bacteria into the environment, public fears of the outcomes of molecular genetic research must be allayed. The acceptance of the use of recombinant microbes to prepare foods (e.g., chymosin from *E. coli*; the mycoprotein retailed as "Quorn;" genetically modified yeast for bread making) provides a source of optimism for the use of genetically modified microbes in the development of probiotics. Genetically modified probiotic strains will doubtless need to be assessed by appropriate regulatory agencies before they are released for commercial use. It is important, therefore, that the microecology of gastrointestinal inhabitants be intensively studied in parallel to investigations of their molecular biology. A balanced, detailed knowledge will be needed of their colonization attributes, their activities in the digestive tract ecosystem and resulting influences on the host of members of the normal microflora will be essential if sound, scientifically based decisions concerning the development and regulation of this new generation of probiotic organisms are to be made.

## REFERENCES

1. **Savage, D. C.,** Microbial ecology of the gastrointestinal tract, *Ann. Rev. Microbiol.,* 31, 107, 1977.
2. **Lee, A., Fox, J., and Hazell, S.,** Pathogenesis of *Helicobacter pylori*: a perspective, *Infect. Immun.,* 61, 1601, 1993.
3. **Tannock, G. W.,** The microecology of lactobacilli inhabiting the gastrointestinal tract, *Adv. Microb. Ecol.,* 11, 147, 1990.
4. **Draser, B. S. and Hill, M. J.,** *Human Intestinal Flora,* Academic Press, London, 1974.
5. **Gracey, M.,** The contaminated small bowel syndrome, in *Human Intestinal Microflora in Health and Disease,* Hentges, D. J., Ed., Academic Press, New York, 1983, chap. 19.
6. **Gordon, H. A. and Pesti, L.,** The gnotobiotic animal as a tool in the study of host microbial relationships, *Bact. Rev.,* 35, 390, 1971.
7. **Finegold, S. M.,** *Anaerobic Bacteria in Human Disease,* Academic Press, New York, 1977.
8. **Plorde, J. J.,** Urinary tract infections, in *Medical Microbiology,* Sherris, J. C., Ed., Elsevier, New York, 1990, chap. 66.
9. **Metchnikoff, E.,** *The Prolongation of Life. Optimistic Studies,* William Heinemann, London, 1907.
10. **Tannock, G. W.,** Control of gastrointestinal pathogens by normal flora, in *Current Perspectives in Microbial Ecology,* Klug, M. A. and Reddy, C. R., Eds., American Society for Microbiology, Washington, D.C., 1984, 374.
11. **Fuller, R.,** Probiotics in man and animals, *J. Appl. Bact.,* 66, 365, 1989.
12. **Finegold, S. M., Sutter, V. L., and Mathisen, G. E.,** Normal indigenous intestinal flora, in *Human Intestinal Microflora in Health and Disease,* Hentges, D. J., Ed., Academic Press, New York, 1983, chap. 1.
13. **Tannock, G. W.,** The acquisition of the normal microflora of the gastrointestinal tract, in *Human Health: The Contribution of Microorganisms,* Gibson, S. A. W., Ed., Springer-Verlag, London, 1994, chap. 1.
14. **Schneitz, C., Hakkinene, M., Nutio, L., Nurmi, E., and Mead, G.,** Droplet application for protecting chicks against *Salmonella* colonization by competitive exclusion, *Vet. Res.,* 126, 510, 1990.
15. **Tannock, G. W.,** Effect of dietary and environmental stress on the gastrointestinal microbiota, in *Human Intestinal Microflora in Health and Disease,* Hentges, D. J., Ed., Academic Press, New York, 1983, chap. 20.
16. **Nord, C. E. and Edlund, C.,** Ecological effects of antimicrobial agents on the human intestinal microflora, *Microb. Ecol. Health Dis.,* 4, 193, 1991.
17. **Anon.,** The influence of feeding cultured products on the microbial ecology of the gut, *Int. Dairy Fed. Bull.,* 159, 5, 1983.
18. **Hashimoto, S., Nomoto, K., Matsuzala, T., Yokokura, T., and Mutai, M.,** Oxygen radical production by peritoneal macrophages and Kupffer cells elicited with *Lactobacillus casei, Infect. Immun.,* 44, 61, 1984.
19. **Nanno, M., Shimizu, T., Mike, A., Ohwaki, M., and Mutai, M.,** Role of macrophages in serum colony-stimulating factor induction by *Lactobacillus casei* in mice, *Infect. Immun.,* 56, 357, 1988.
20. **Berg, R. D.,** Translocation of indigenous bacteria from the intestinal tract, in *Human Intestinal Microflora in Health and Disease,* Hentges, D. J., Ed., Academic Press, New York, 1983, chap. 15.
21. **Nathan, C. F.,** Secretion of oxygen mediators: role in effector functions of activated macrophages, *Fed. Proc.,* 41, 2206, 1982.
22. **Holoshitz, J., Koning, F., Coligan, J. E., de Bruyn, J., and Strober, S.,** Isolation of CD4-CD8-mycobacteria-reactive T-lymphocyte clones from rheumatoid arthritis synovial fluid, *Nature,* 339, 226, 1989.
23. **Lehman, T. J., Cremer, M. A., Walker, S. M., and Dillon, A. M.,** The role of humoral immunity in *Lactobacillus casei* cell wall induced arthritis, *J. Rheumatol.,* 14, 415, 1987.

24. **Severijnen, A. J., van Kleef, R., Hazenberg, M. P., and van de Merwe, J.,** Cell wall fragments from major residents of the human intestinal flora induce chronic arthritis in rats, *J. Rheumatol.,* 16, 1061, 1989.

25. **Heath, A. W. and Playfield, J. H. L.,** Cytokines as immunological adjuvants, *Vaccine,* 10, 427, 1992.

26. **Gracey, M.,** Sugar intolerance, in *Diarrhea,* Gracey, M., Ed., CRC Press, Boca Raton, FL, 1991, chap. 13.

27. **Kolars, J. C., Levitt, M. D., Aouji, M., and Savaiano, D. A.,** Yogurt — an autodigesting source of lactose, *N. Engl. J. Med.,* 310, 1, 1984.

28. **Pochart, P., Dewit, O., Desjeux, J.-F., and Bourlioux, P.,** Viable starter culture, β-galactosidase activity, and lactose in duodenum after yoghurt ingestion in lactose-deficient humans, *Am. J. Clin. Nutr.,* 49, 828, 1989.

29. **Premi, P., Sandine, W. E., and Elliker, P. R.,** Lactose-hydrolyzing enzymes of *Lactobacillus* species, *Appl. Microbiol.,* 24, 51, 1972.

30. **Bartlett, J. G.,** Pseudomembranous Colitis, in *Human Intestinal Microflora in Health and Disease,* Hentges, D. J., Ed., Academic Press, New York, 1983, chap. 17.

31. **Wren, B. W.,** Bacterial enterotoxin interactions, in *Molecular Biology of Bacterial Infection,* Hormaeche, C. E., Penn, C. W., and Symth, C. J., Eds., Cambridge University Press, Cambridge, 1992, 127.

32. **Lyerly, D. M., Saum, K. E., MacDonald, D. K., and Wilkins, T. D.,** Effects of *Clostridium difficile* toxins given intragastrically to animals, *Infect. Immun.,* 47, 349, 1985.

33. **Borriello, S. P.,** The influence of the normal flora on *Clostridium difficile* colonisation of the gut, *Ann. Med.,* 22, 61, 1990.

34. **Borriello, S. P.,** Pathogenesis of *Clostridium difficile* infection of the gut, *J. Med. Microbiol.,* 33, 207, 1990.

35. **Walker, A. R. P.,** Colon cancer and diet, with special reference to intakes of fat and fiber, *Am. J. Clin. Nutr.,* 29, 1417, 1976.

36. **Finegold, S. M. and Sutter, V. L.,** Fecal flora in different populations with special reference to diet, *Am. J. Clin. Nutr.,* 31, S116, 1978.

37. **Goldin, B., Dwyer, J., Gorbach, S. L., Gordon, W., and Swenson, L.,** Influence of diet and age on fecal enzymes, *Am. J. Clin. Nutr.,* 31, S136, 1978.

38. **Rowland, I.,** The influence of the gut microflora on food toxicity, *Proc. Nutr. Soc.,* 40, 67, 1981.

39. **Drasar, B. S. and Barrow, P. A.,** Benefit and mischief from the intestinal microflora, in *Intestinal Microbiology,* American Society for Microbiology, Washington, D.C., 1985, chap. 4.

40. **Goldin, B. R., Swenson, L., Dwyer, J., Sexton, M., and Gorbach, S. L.,** Effect of diet and *Lactobacillus acidophilus* supplements on human fecal enzymes, *J. Natl. Cancer Inst.,* 64, 255, 1980.

41. **Tannock, G. W., Crichton, C., Welling, G. W., Koopman, J. P., and Midtvedt, T.,** Reconstitution of the gastrointestinal microflora of *Lactobacillus*-free mice, *Appl. Environ. Microbiol.,* 54, 2971, 1988.

42. **McConnell, M. A. and Tannock, G. W.,** Lactobacilli and azoreductase activity in the murine cecum, *Appl. Environ. Microbiol.,* 57, 3664, 1991.

43. **McConnell, M. A. and Tannock, G. W.,** A note on lactobacilli and β-glucuronidase activity in the intestinal contents of mice, *J. Appl. Bacteriol.,* 74, 649, 1993.

44. **Tannock, G. W.,** Genetic manipulation of gut microorganisms, in *Probiotics. The Scientific Basis,* Fuller, R., Ed., Chapman and Hall, London, 1992, chap. 8.

45. **Tannock, G. W.,** Molecular genetics, a new tool for investigating the microbial ecology of the gastrointestinal tract?, *Microb. Ecol.,* 15, 239, 1988.

46. **Raya, R. R., Kleeman, E. G., Luchansky, J. B., and Klaenhammer, T. R.,** Characterization of the temprate bacteriophage øadh and plasmid transduction in *Lactobacillus acidophilus* ADH, *Appl. Environ. Microbiol.,* 55, 2206, 1989.

47. **Chassy, B. M.,** Prospects for the genetic manipulation of lactobacilli, *FEMS Microbiol. Rev.,* 46, 297, 1987.

48. **Luchansky, J. B., Muriana, P. M., and Klaenhammer, T. R.,** Application of electroporation for transfer of plasmid DNA to *Lactobacillus, Lactococcus, Listeria, Pediococcus, Bacillus, Staphylococcus, Enterococcus* and *Propionibacterium, Mol. Microbiol.,* 2, 637, 1988.

49. **Scheirlinck, J. B., Mahillon, J., Joos, H., Dhaese, P., and Michiels, F.,** Integration and expression of α-amylase and endoglucanase genes in *Lactobacillus plantarum* chromosome, *Appl. Environ. Microbiol.,* 55, 2130, 1989.

50. **Davies, F. L. and Gasson, M. J.,** Reviews of the progress of dairy science, genetics of lactic acid bacteria, *J. Dairy Sci.,* 48, 263, 1987.

51. **Trieu-Cuot, P., Arthur, M., and Courvalin, P.,** Origin, evolution and dissemination of antibiotic resistance genes, *Microbiol. Sci.,* 4, 263, 1981.

52. **Roberts, M. C. and Hillier, S. L.,** Genetic basis of tetracycline resistance in urogenital bacteria, *Antimicrob. Agents Chemother.,* 34, 261, 1990.

53. **McConnell, M. A., Mercer, A. A., and Tannock, G. W.,** Transfer of plasmid pAMβ1 between members of the normal microflora inhabiting the murine anterior digestive tract and modification of the plasmid in a *Lactobacillus reuteri* host, *Microbiol. Ecol. Health Dis.,* 4, 343, 1991.

54. **Cardenas, L. and Clements, J. D.,** Oral immunization using live attenuated *Salmonella* spp. as carriers of foreign antigens, *Clin. Microbiol. Rev.,* 5, 328, 1992.

55. **Tartagelia, J., Pincus, S., and Paoletti, E.,** Poxvirus-based vectors as vaccine candidates, *Crit. Rev. Immunol.,* 10, 13, 1990.

56. **Saxelin, M., Elo, S., Salminen, S., and Vapaatalo, H.,** Dose response colonisation of faeces after oral administration of *Lactobacillus casei* strain *GG, Microb. Ecol. Health Dis.,* 4, 209, 1991.

57. **Tannock, G. W., Fuller, R., and Pedersen, K.,** *Lactobacillus* succession in the piglet digestive tract demonstrated by plasmid profiling, *Appl. Environ. Microbiol.,* 56, 1310, 1990.

58. **Tannock, G. W., Fuller, R., Smith, S. L., and Hall, M. A.,** Plasmid profiling of members of the family *Enterobacteriaceae,* lactobacilli, and bifidobacteria to study the transmission of bacteria from mother to infant, *J. Clin. Microbiol.,* 28, 1225, 1990.

59. **Rodtong, S. and Tannock, G. W.,** Differentiation of *Lactobacillus* strains by ribotyping, *Appl. Environ. Microbiol.,* 59, 3480, 1993.

60. **Stull, T. L., LiPuma, J. J., and Edlind, T. D.,** A broad-spectrum probe for molecular epidemiology of bacteria: ribosomal RNA, *J. Infect. Dis.,* 157, 280, 1988.

61. **Gardiner, K.,** Pulsed field gel electrophoresis, *Anal. Chem.,* 63, 658, 1991.

62. **Howard, P. J., Harsono, K. D., and Luchansky, J. B.,** Differentiation of *Listeria monocytogenes, Listeria innocua, Listeria ivanovii* and *Listeria seeligeri* by pulsed-field gel electrophoresis, *Appl. Environ. Microbiol.,* 58, 709, 1992.

63. **Versalovic, J., Koeuth, T., and Lupski, J. R.,** Distribution of repetitive DNA sequences in eubacteria and application to fingerprinting of bacterial genomes, *Nucleic Acids Res.,* 19, 6823, 1991.

64. **Kuritza, A. P., Shaughnessy, P., and Salyers, A. A.,** Enumeration of polysaccharide-degrading *Bacteroides* species in human feces by using species-specific DNA probes, *Appl. Environ. Microbiol.,* 51, 385, 1986.

65. **Attwood, G. T., Lockington, R. A., Xue, G.-P., and Brooker, J. D.,** Use of a unique gene sequence as a probe to enumerate a strain of *Bacteroides ruminicola* introduced into the rumen, *Appl. Environ. Microbiol.,* 54, 534, 1988.

66. **Hensiek, R., Krupp, G., and Stackenbrandt, E.,** Development of diagnostic oligonucleotide probes for four *Lactobacillus* species occurring in the intestinal tract, *System. Appl. Microbiol.,* 15, 123, 1992.

67. **Tannock, G. W.,** Biotin-labeled plasmid DNA probes for detection of epithelium-associated strains of lactobacilli, *Appl. Environ. Microbiol.,* 55, 461, 1989.

68. **Tannock, G. W., McConnell, M. A., and Fuller, R.,** A note on the use of a plasmid as a DNA probe in the detection of a *Lactobacillus fermentum* strain in porcine stomach contents, *J. Appl. Bacteriol.,* 73, 60, 1992.

69. **Frothingham, R., Duncan, A. J., and Wilson, K. H.,** Ribosomal DNA sequences of bifidobacteria: implications for sequence-based identification of the human colonic flora, *Microb. Ecol. Health Dis.,* 6, 23, 1993.

70. **Yamamoto, T., Morotomi, M., and Tanaka, R.,** Species-specific oligonucleotide probes for five *Bifidobacterium* species detected in human intestinal microflora, *Appl. Environ. Microbiol.,* 58, 4076, 1992.

71. **Finegold, S. M., Attebery, H. R., and Sutter, V. L.,** Effect of diet on human fecal flora; comparison of Japanese and American diets, *Am. J. Clin. Nutr.,* 27, 1456, 1974.

72. **Luckey, T. D.,** *Germfree Life and Gnotobiology,* Academic Press, New York, 1963.

73. **Bloksma, N., de Heer, E., van Dijk, H., and Willers, J. M.,** Adjuvancy of lactobacilli. I. Differential effects of viable and killed bacteria, *Clin. Exp. Immunol.,* 37, 367, 1979.

74. **Kato, I., Kobayashi, S., Yokokura, T., and Mutai, M.,** Antitumor activity of *Lactobacillus casei* in mice, *Gann.,* 72, 517, 1981.

75. **Saito, H., Tomioko, H., and Nagashima, K.,** Protective and therapeutic efficacy of *Lactobacillus casei* against experimental murine infections due to *Mycobacterium fortuitum* complex, *J. Gen. Microbiol.,* 133, 2843, 1987.

76. **Saito, H., Watanabe, T., and Horikawa, Y.,** Effects of *Lactobacillus casei* on *Pseudomonas aeruginosa* infection in normal and dexamethasone-treated mice, *Microbiol. Immunol.,* 30, 249, 1986.

77. **Sato, K.,** Enhancement of host resistance against *Listeria* infection by *Lactobacillus casei*: role of macrophages, *Infect. Immun.,* 44, 445, 1984.

78. **Kowhi, Y., Imai, K., Tamura, Z., and Hashimoto, Y.,** Antitumor effect of *Bifidobacterium infantis* in mice, *Gann.,* 69, 613, 1978.

79. **Reddy, G. V., Shahani, K. M., and Banerjee, M. R.,** Inhibitory effect of yoghurt on Ehrlich ascites tumor-cell proliferation, *J. Natl. Cancer Inst.,* 50, 815, 1973.

80. **Perdigon, G., de Macias, M. E. N., Alvarez, S., Oliver, G., and de Ruiz Holgado, A. P.,** Systemic augmentation of the immune response in mice by feeding fermented milks with *Lactobacillus casei* and *Lactobacillus acidophilus, Immunology,* 63, 17, 1973.

# INDEX

## A

AB blood group antigens, see Blood group antigens

Absorption, nutrient
  carbohydrate malabsorption, 68–71, see also Resistant starches
  SCFAs and, 111–112

Acabose, 147

Acetate, 102, 103
  amino acid fermentation products, 88, 89
  carbohydrate fermentation end products, 62
  carbohydrate substrates, 104–105
  and fatty acid oxidation, 116
  portal blood, 109
  stoichiometry, 110

Acetogenesis
  hydrogen consumption, 137
  hydrogen disposal, 145–146

N-Acetyl derivatives of putrescine and cadaverine, 91

Acetylglucosaminidases, Clostridium, 214

N-Acetyl muramyl-L-alanyl-D-isoglutamine (MDP), 244

Achlorhydria, 202

Acidaminococcus, 89, 93

Acid protease inhibitors, 87

Acid proteases, 79, 80

Acquisition of microbiota, 2, 9, 11, 13, see also Colonization

Acrylate pathway, 89

Actinomyces, diet and, 14

Actinomycetaceae, 28

Acylneuramininate lyase, 189

Adhesion, 4, 184
  colonization and, 7
  in inflammatory bowel disease, 237–238
  pathogens, 202

Adhesion index, 237–238

Aeromonas
  diarrhea, 202, 211–212
  hydrophila, 86, 211, 237
  in inflammatory bowel disease, 237
  sobria, 211

Aflatoxins, 159, 168

Agammaglobulinemia, 11

Agar surface matings, 36

Agmatine, 90

AIDS
  mycobacterial infections, 210
  and Mycobacterium avium infection, 239

Albumin
  amino acid fermentation products, 88
  proteolysis, 78
    Bacteroides fragilis, 85
    small intestine and fecal fractions, 80

Alcaligenes eutrophus, 39

Alcohols, 116
  carbohydrate fermentation end products, 62
  measurement methods, 184

Allochthonous populations, 2

Amides, nitroso compound formation, 162

Amines
  amino acid fermentation, 90–91
  gases, 140
  measurement methods, 184
  nitroso compound formation, 162

Amino acid decarboxylation, 90–91

Amino acids
  fermentation
    amines, 90–91
    ammonia, 89–90
    gas metabolism, 94
    organic N-containing compounds, 87–88
    phenols and indoles, 91–94
    SCFAs, 88–89
  mucin, 177, 178
    analytical methods, 183–184
    utilization studies, 186
  and SCFA production, 110
  tumor promoter synthesis, 164

2-Amino-3-methyl-3H-imidazo[4,5-f]quinoline (IQ), 156, 161–162, 168

Aminopeptidase, 78

Ammonia, 71, 89
  and clostridial proteases, 86
  diet and, 93
  gas metabolism, 140
  Helicobacter pylori mucus penetration, 185
  measurement methods, 184
  as nitrogen source, 85
  and portal systemic encephalopathy, 68, 90, 147
  protein and amino acid metabolites, 164

α-Amylase gene transfer, 266

Amylase-resistant starch, see Starch, resistant

Anaerobes
  acquisition of microbiota, 13
  molecular genetics, see Molecular genetics

Anaerobic conditions, 6

Anaerobic continuous-flow culture model, 4

Anaerobic electoporation, 40, 42

Anaerobic streptococci
  diet and, 14
  normal microflora, 258
  in ulcerative colitis, 233

Anaerobiospirillum, 16

Anaerorhabdus (Bacteroides) furcosus, 29

Animal models, 4–5, see also Germ-free animals
  advantages and disadvantages, 157, 158

and inflammation, 244
inflammatory bowel disease, 242
Anthraquinones, 160
Antibiotic resistance plasmids, 31–33, see also Molecular
   genetics
Antibiotics
   *Clostridium difficile* overgrowth, 202
   culture conditions, 6
   ecological shifts, 62
   and L forms, 238
   pseudomembranous colitis, 12
   and pseudomembranous colitis, 215–216, 264
   and SCFAs, 105
   suppression of microbiota, 17
Antimicrobials, nitro compounds, 161
Antineutrophil cytoplasmic antibodies (ANCA), 229
Apparent total *N*-nitroso compound (ATNC) analysis, 162
Arabinoglactan, and SFCAs, 103, 104
*Arachnia-propionibacterium*, 14
Aromatic amino acids
   mucin, 178
   phenol and indole formation, 69, 91–94, 164
Aromatic nitro compounds, 161
Arthropathies, 11–12, 228, 243–244, 262–263
Arylamidase activity, *Bacteroides fragilis* protease, 85
Ascorbic acid, 12
A secretors, mucin utilization, 186
Aspartic protease inhibitors, 87
Aspartic proteases, 79, 80
*Aspergillus*
   *niger*, 69
   *oryzae*, 260
A + T rich regions, 37
Autoantibodies, inflammatory bowel disease, 229
Autochthonous populations, 2–3
Autolysis, 87
Azo compounds
   activation to toxic, carcinogenic, or mutagenic
      derivatives, 160–61
   carbohydrate metabolites, 69
   proteolysis, 78, 80, 85
   toxicology, 160–161
Azoreductase, 156, 265

**B**

Bacillary dysentery, 207–209
*Bacillus*
   diet and, 14
   molecular genetics, cloning vectors, 44
   normal microflora, 258
   probiotics, 260
   *stearothermophilus*, 48, 266
   *subtilis*, 30, 38, 260
Bacterial cell wall, see Cell wall
Bacterial infections, see Infections and diarrhea

Bacterial overgrowth
   bacterial proteases in, 84
   and breath hydrogen, 147
   hydrogen excretion measurements, 147
   indican as indicator in, 94
Bacterial proteases, oteases, bacterial
Bacterial reductases, 161
Bacteriocin, 30
Bacteriophages, see Transduction
*Bacteroides*
   acquisition of microbiota, 13
   amino acid fermentation
      amine formation, 90
      phenol and indole formation, 91, 93
   carbohydrate metabolism, 70, 71
      fructo-oligosaccharide and, 69
      mean numbers and fermentation end products, 62
      mucin degradation, 65
      nonstarch polysaccharides, 67
   culture conditions, 6
   diet and, 14
   fecapentaene production, 163
   in inflammatory bowel disease
      Crohns' disease, 233, 239
      mucosal, 236
      after surgery, 235
      ulcerative colitis, 233
   molecular genetics, 26
      classification, 28–29
      cloned genes, 49
      cloning vectors, 43, 45, 46
      gene expression, 47
      gene transfer systems and methods, 40
      plasmid conjugation, 38, 39
      plasmids, 31, 32
      release of genetically engineered bacteria, 48
      restriction barriers, 38
      transposons and insertion sequences, 33–34
   mucin utilization, 186
   normal microflora, 258
   proteolysis, 84–85
*Bacteroides* species
   *(Porphyromonas) asaccharolyticus*, 28
   *capillosus*, 29
   *clostridiformis*, 258
   *coagulan*s, 29, 93
   *distasonis*
      diet and, 14
      normal microflora, 258
      proteolysis, 84
   *eggerthii*
      molecular genetics, cloning vectors, 43
      phenol and indole formation, 93
   *(Porphyromonas) endodontalis*, 28
   *fragilis*, 28, 32
      acquisition of microbiota, 13

amino acid fermentation, amine formation, 91
carbohydrate metabolism, nonstarch polysaccharides, 67
diarrhea, 213–214
diet and, 14
enumeration and isolation, 83, 84
formyl peptides, 245
in inflammatory bowel disease, 242
  Crohns' disease, 234
  after surgery, 235
molecular genetics
  cloned genes, 49
  cloning vectors, 43, 46
  gene transfer systems and methods, 40
  mobile genetic elements, 33, 34
  plasmids, 31
  species differences in restriction and
    modification systems, 38
  normal microflora, 258
  proteolysis, 84–85
*(Anaerorhabdus) furcosus*, 29
*(Porphyromonas) gingivalis*, 28
*(Mitsuokella) multiacidus*, 29, 169
*ovatus*, 31
  amino acid fermentation, phenol and indole formation,
    93
  diet and, 14
  molecular genetics
    cloned genes, 49
    cloning vectors, 43
    gene transfer systems and methods, 40
    mobile genetic elements, 33, 34
  proteolysis, 84
  SCFA production, 106
*(Tissierella) preacutus*, 29
*putredinis*, 29, 93
*ruminicola*, 31, 49
*splanchnicus*, 84, 85, 93
*thetaiotaomicron*
  amino acid fermentation, phenol and indole formation,
    93
  chondroitin lyase enzymes, 36–37
  detection of, 29
  diet and, 14
  in inflammatory bowel disease, L-forms, 238
  molecular genetics
    cloned genes, 49
    mobile genetic elements, 34
  mucin degradation, 189
  normal microflora, 258
  plasmids, 31
  proteolysis, 84
*uniformis*, 32
  amino acid fermentation, phenol and indole formation,
    93
  molecular genetics
    cloned genes, 49

electoporation, 42
  gene transfer systems and methods, 40
  mobile genetic elements, 33, 34
*vulgatus*
  acquisition of microbiota, 13
  detection of, 29
  diet and, 14
  in inflammatory bowel disease, 234, 242
  normal microflora, 258
  plasmids, 31
  proteolysis, 84
B antigen secretors, mucin utilization, 186
Batch cultures, 4
B cells, in inflammatory bowel disease, 242
Beaded mucin model, 177
Benzo[a]pyrene, 164–165, 167
Bicarbonate, SCFAs and, 112
*Bifidobacterium*
  acquisition of microbiota, 13
  amino acid fermentation, 91, 93
  amino acid release, 88
  and arthritis, 263
  carbohydrate metabolism, 62, 68–71
  diet and, 14
  in inflammatory bowel disease
    Crohns' disease, 233
    after surgery, 235
  molecular genetics, 26
    classification, 28
    plamids, 30
    whole chomosome analysis, 27
  mucin utilization, 65, 185, 186
  in probiotics, 260
*Bifidobacterium* species
  *adolescentis*, 14
    and arthritis, 263
    normal microflora, 258
    in probiotics, 260
  *bifidum*, 30
    in probiotics, 260
    supplementation of bottle-fed infants, 13
  *breve*, 30
    and arthritis, 263
    in probiotics, 260
    SCFA production, 107
  *dentium*, 30
  *infantis*, 14
    normal microflora, 258
    in probiotics, 260
  *infantis* ssp. *liberorum*, 258
  *longum*, 14, 30
    and hepatoma incidence, 169
    normal microflora, 258
    in probiotics, 260, 261
  *longum* ssp. *longum*, 258
  *pseudocatenulatum*, 30

*thermophilum*, 261
Bifidus factor, 66
Bile
    elimination of drugs and xenobiotics, 156, 157
    formyl peptide excretion, 247
    and *Salmonella typhi*, 207
Bile acid dehydroxylation, 8
Bile acids
    in germ-free animals, 259
    nitrate reduction, 69
    tumor promoter synthesis, 163–164
Bilirubin sulfate, 189
Bilirubin transformation, 8
Biotechnology, see also Molecular genetics
    cloned genes of intestinal anerobes, 49–50
    probiotics, 265–268
    release of genetically engineered bacteria, 48, 50
Birth, see Acquisition of microbiota
Blood group antigens, 64–65, 176
    in inflammatory bowel disease, 188
    mucin, 64–65
    mucin utilization, 186
    salivary mucin glycoprotein, 66
Blood group antigen secretors, mucin utilization, 186
Botulism, infant, 12, 215
Bovine serum albumin
    amino acid fermentation products, 88
    proteolysis, 78
        *Bacteroides fragilis*, 85
        small intestine and fecal fractions, 80
Bran, and SFCAs, 103–105
Branched chain fatty acids, 88, 89, 110
Breast feeding, 12–13
Breath, pulmonary gas excretion, see also Hydrogen, breath
    measurement methods, 133–135
    methane, 140, 142
Brooke ileostomy, 235
*Brucella*, 240
Brush border enzymes, proteolysis, 84
Butylamine, 91
Butyrate, 102, 103
    amino acid fermentation products, 88, 89
    carbohydrate fermentation end products, 62
    carbohydrate substrates, 104–105
    portal blood, 109
    stoichiometry, 110
    and ulcerative colitis, 148
*Butyrivibrio*, 48, 93

**C**

Cadaverine, 90, 91
*Campylobacter*
    culture conditions, 6
    diarrhea, 202, 212–213
    in inflammatory bowel disease, 236, 237, 240
*Campylobacter* species
    *cinaedi*, 212, 213
    *coli*, 212, 213
    *fennelliae*, 212, 213
    *fetus* ssp. *fetus*, 213
    *hyointestinalis*, 213
    *jejuni*, 237
    *laridis*, 213
    *upsaliensis*, 213
Cancer, see also Carcinogenesis
    indoles and skatoles in, 94
    methane and, 148
    mucus in, 177
*Candida*, 62
    *albicans*, 14, 237
Caproic acid, 102, 103
Carageenan, 148
Carbohydrate metabolism
    and amino acid fermentation products, 88
    carbohydrate utilization, 62–63
    dietary carbohydrates, 66–68
    endogenous carbohydrates, 64–66
    fermentation products, see also Short chain fatty acids
    malabsorbed carbohydrates, bifidogenic potential, 68–71
        fructo-oligosaccharides and inulin, 69
        isomalto-oligosaccharides, 70
        raffinose, 69
        *trans*-galactosylated oligosaccharides, 70–71
    nature of colonic microflora, 61–62
    and proteolysis, 83
    sources of carbohydrate in colon, 63–64, see also Mucin; Mucus glycoproteins, host
Carbohydrates, cellular, 64–66
    in inflammatory bowel disease, 243–244
    mucin utilization, 185–186
    sources of carbohydrate in colon, 63–64
Carbohydrates, dietary, 66–68, see also Starch, resistant
    and amino acid fermentation, 93, 94
    and composition of microbiota, 9
    gas production, see Gas metabolism
    malabsorption
        bifidogenic potential, 68–71
        and breath hydrogen, 146–147
    and SCFAs, 104–105
Carbon dioxide
    amino acid decarboxylation and, 90
    carbohydrate fermentation end products, 62
    flatus composition, 131
    methanogenesis, 137
    production in large intestine, 139–140
    SCFA energy metabolism, 113
Carbon dioxide, culture conditions, 6
Carboxylic acids
    carbohydrate fermentation end products, 62–63

SCFAs, see Short chain fatty acids
Carboxymethylcellulose, 67, 104
Carcinogenesis
    carcinogen binding by microflora, 168
    flavonoids and, 167
    lactobacilli and, 264–265
    methane and colorectal cancer, 148
    microflora role in etiology of cancer, 156
    modification by microflora, 168–170
    phenols as cocarcinogens, 94
    probiotics and, 262
    SCFAs and, 114–116
    tumor promoter synthesis, 163–164
Carcinogens, 8, see also Toxicology
    activation of chemicals, 158–162
    binding by microflora, 168
    synthesis of, 159, 162–163
Carrageenans, 189
Casein
    amino acid fermentation products, 88
    as nitrogen source, 85
    proteolysis, 78
        activities of small intestine and fecal fractions, 80
        *Bacteroides fragilis*, 85
Cecal flora, 4
Celiac sprue, 139
Cell differentiation, SCFAs and, 114–115
Cell mediated immunity, 241, see also Immunodeficiency
        states
Cell metabolism and growth, see also Epithelium;
        Mucosa
    SCFAs in, 112–116
    tumorigenesis, see Cancer; Carcinogenesis
Celluloses, 67, 104, see also Fiber, dietary
Cell wall deficient (L) forms, in inflammatory bowel
        disease, 238
Cell walls
    glycoconjugates, 65
    and inflammation, 263
    in inflammatory bowel disease, 243–244
    lactobacilli, 261
C + G rich regions, 37
CheckProbe program, 7
Chemical-induced diarrhea, 10
Chemically modified sugars, 68
Chemical transformation systems, 40
Chemoattractants, 245–246
Chemostats, see *In vitro* systems
Chenodeoxycholic acid, 164
*Chlamydia*, 236, 237
    *trachomatis*, 23, 241
Cholesterol conversion to coprostanol, 8
Cholic acid, 164
Choline sulfate, 189
Chomosome analysis, 27
Chondroitin lyase, 36–37

Chondroitin sulfate, 186
Chromosomal mapping, 37
Chymostatin, 79–82
Chymotrypsin
    bacterial, 81, 85
    inhibitors, 79–82
    pancreatic, 77
*Citrobacter*, 237
Classification
    genetic methods, 26–27
    phylogenetic analysis, 27–29
Cloning
    analysis of genetic determinants, 34–37
    probiotics, 263, 265–267
Cloning vectors, 47–48
*Clostridium*
    amino acid fermentation, 88–91, 93
    A + T rich regions, 37
    carbohydrate metabolism, 62, 70, 71
    diarrhea, 202, 214–216
    diet and, 14
    molecular genetics
        classification, 28
        cloned genes, 49–50
        gene expression, 47
        gene transfer systems and methods, 40
        plasmids, 30–32
        transposons, 34
        whole chomosome analysis, 27
    normal microflora, 258
    proteolysis, 86
    in ulcerative colitis, 233
    virulence determinants, 35–36
*Clostridium* species
    *acetobutylicum*, 36, 38, 39
    *bifermentans*, 28
        amino acid fermentation, phenol and indole
            formation, 92, 93
        enumeration and isolation, 84
        molecular genetics, cloned genes, 49
        proteolysis, 86
    *botulinum*
        diarrhea, 215
        infant botulism, 12, 215
        proteases, 86
    *cadaveris*, 93
    *difficile*, 28, 202
        colonization by, 62
        diarrhea, 215–216
        in inflammatory bowel disease, 237
        molecular genetics
            cloned genes, 49
            mobile genetic elements, 34–35
        probiotics, 264
        pseudomembranous colitis, 12
        treatment with microbial preparations, 17

*indolis*, 258
*malenominatum*, 93
*paraputrificum*, 169
*perfringens*, 5, 14, 28, 31
    acquisition of microbiota, 13
    amino acid fermentation, 91
    bifidobacteria and, 69
    diarrhea, 214–215
    enumeration and isolation, 83
    mapping, 37
    molecular genetics
        cloned genes, 49
        cloning vectors, 42, 43, 45
        electoporation, 41
        gene expression, 47
        gene transfer systems and methods, 40
        mobile genetic elements, 33, 34
        virulence determinants, 35–36
    normal microflora, 258
    proteolysis, 86
    SCFA production, 106, 107
    transposons, 36
    virulence determinants, 35–36
*ramosum*, 8, 14
    normal microflora, 258
    proteases, 87
*septicum*
    diarrhea, 214
    molecular genetics, cloned genes, 50
    proteases, 87
*sorelli*, 92
*sporogenes*, 28
    amino acid fermentation, 92
    proteases, 87
*thermocellum*, 48, 266
Co-carcinogens, phenols, 94
Colectomy, in inflammatory bowel disease, 234–236
Colitis, see also Infections and diarrhea; Inflammatory
        bowel disease
    nonspecific, 236
    segmental, 212
    ulcerative, See Inflammatory bowel disease; Ulcerative
        colitis
Collagen
    amino acid fermentation products, 88
    proteolysis
        *Bacteroides fragilis*, 85
        clostridial protease activity, 86
        small intestine and fecal fractions, 80
Collagenases, 87
Colon, structure of, 3
Colon cancer, see Cancer; Carcinogenesis
Colonization, see also Acquisition of microbiota
    adhesion and, 7
    *Lactobacillus delbruckei* ssp. *bulgaricus*, 261

micronutrients and, 12
    significance of microbe presence, 5
    transient species, 5
Colostomies, SFCAs, 102, 103
Colostrum, 12
Complement activation, inflammatory bowel disease, 233
Conformational changes, and proteolysis, 78
Conjugation, metabolic, 164–168
    drugs and xenobiotics, 156, 157
    phenol and indoles, 92, 94
Conjugation, plasmid, 30, 31, 36, 38–39
    probiotic development, 266
    systems and methods, 40
Continuous-culture fermenters, 4, 5
Cooked food mutagens, 159, 168, see also Food preparation
        and processing
*Coprococcus*, 16, 229, 234, 241
    *comes*, 233, 241
Coprostanol
    cholesterol conversion to, 8
    in germ-free animals, 259
*Corynebacterium parvum*, 262
CPN50 (*Clostridium*), 37
Cresols, 91, 92, 164
Crohn's disease, see Inflammatory bowel disease
Cryptic plasmids, 44
*Cryptosporidium*, 236
Crypts, microhabitats, 10
Culture of bacteria, 6–7, see also Growth conditions; *In
        vitro* systems
Cyanogenic glycosides, 159
Cycasin, 159, 160
Cysteine, 84
    metalloprotease inhibition, 79–82
    mucin utilization studies, 186
Cysteine proteases, 79, 80, 84
Cystic fibrosis, 17
Cytochromes P450–dependent mixed function oxidases,
        148, 157, 167, 168
Cytokines, 9, see also Inflammatory mediators
Cytotoxic compounds, metabolic
    ammonia, 89–90
    carbohydrate metabolites, 68
Cytotoxins, microbial, see also Toxins, microbial
    *Aeromonas*, 211
    *Clostridium*, 214
    pathogens, 202

**D**

Darmbrand, 215
Decarboxylation of amino acids, 90–91
Demethylation of mercury, 165–166
Deoxycholic acid, 164
*Desulfobacter*, 144

*Desulfobulbus*, 144
*Desulfomonas*, 144
*Desulfotomaculum*, 28, 144
*Desulfovibrio*, 32, 37, 39, 144
   *desulfuricans*, 39, 50, 118
   *fructosovorans*, 40, 42
   gene transfer systems and methods, 40
   and SCFAs, 116
   SP 8031, 39
   taxonomic divergence, 39
   *vulgaris*, 32, 39, 50
Detoxification, 156, 157, 159
   gut microflora and, 165–168
   phenol and indoles, 92, 94
Dextran sulfate, 148
Diabetic food products, 68
Diamine oxidases, 91
Diarrhea, 10, 17, see also Infections and diarrhea
   in inflammatory bowel disease, 228
   new species isolated from patients with, 16
   and SCFAs, 103, 105–106
Diazotized protein derivatives, 78, 80
Diet, 17
   amine formation, 91
   and amino acid fermentation, 93, 94
   and bacterial populations, 14
   ecosystem components, 9–10
   interactions within ecosystem, 11
   nitroso compound formation, 162
   and phenol and indoles, 93
   and proteolysis, 81
   and SCFAs, 104–106
   toxic metabolites
      lignans and phytoestrogens, 166–167
      plant, 158–160
Dietary carbohydrates
Diet therapy, in inflammatory bowel disease, 229, 232
Digestive enzymes, 10, 139
   nitrogen sources, 76
   proteolytic, 77, 78, 215
Dimethylamine, 90, 91, 94
4-Dimethylaminoazobenzene, 167
1-Dimethylhydrazine, 164
Dimethylsulfide, 140
Direct Black 38, 160–161
Disaccharides, mucin breakdown and, 66
Dissimilatory nitrate reduction, 145
Dissimilatory sulfate reduction, 145–146
Disulfide bridges, mucin, 178
Dithioerythritol, 85
Dithiothreitol, 85
Diversion colitis, SCFAs in, 110
DNA
   genetic engineering, see Molecular genetics
   rolling circle replication, 42

DNA adducts, 159
DNA composition, 26
Drugs
   enterohepatic circulation, 165
   and mucin, 177, 179
Dry weight, 7

**E**

Ecological principles, 2–3, 16–17, see also Microbial
      ecology
EDTA, 79–82, 84
Elastin, 84, 85
Elastinal, 79, 80
Electron donors/acceptors, amino acids, 88
Electrophoresis, 27, 29, 37, 268
Electroporation, 36, 42–47
   probiotic development, 266
   systems and methods, 40
Elemental diet
   and composition of microbiota, 9
   in inflammatory bowel disease, 229, 232
Encephalopathy, portal-systemic, 68, 90
Endogenous sources
   carbohydrates, 64–66, see also Mucin; Mucus
      glycoproteins, host
   proteins, peptides, and amino acids, 10, 76–78
Endoglucanase gene transfer, 266
Endopeptidase inhibitors, 79, 80, 82
Endopeptidases, pancreatic, 77
Endotoxin
Energy metabolism
   acetate in, 116
   SCFAs and, 113
Enteritis, see Infections and diarrhea
Enteritis necroticans, 214, 215
Enteroaggregative *Escherichia coli* (EAggEC), 203, 205,
      206
Enterobacteria
   amino acid fermentation, 90
   carbohydrate metabolism, 70, 71
   and inflammation, 244
   in inflammatory bowel disease, 237
*Enterococcus*
   acquisition of microbiota, 13
   in inflammatory bowel disease, translocation across
      intestine, 239
   molecular genetics, cloning vectors, 44
   proteases, 87
*Enterococcus* species
   *faecalis*, 2, 14
      cloning vectors, 44
      formyl peptides, 245
      germ-free animal colonization, 5
      and hepatoma incidence, 169

in inflammatory bowel disease, 241
L-forms, 238
translocation across intestine, 240
ulcerative colitis, 233
molecular genetics
plasmid pAM, 38
transposons, 36
normal microflora, 258
*faecium*
normal microflora, 258
in probiotics, 260
proteases, 87
Enterohemorrhagic *Escherichia coli* (EHEC), 205, 206
Enterohepatic circulation, 157, 159, 164–165
formyl peptides and, 247
SCFAs in, 109
Enteroinvasive *Escherichia coli* (EIEC), 205, 206
Enteropathogenic *Escherichia coli* (EPEC), 203–205
Enterotoxigenic *Escherichia coli* (ETEC), 205–207
Enterotoxins, see also Toxins, microbial; specific agents
*Aeromonas*, 211
diarrheal disease, 202
virulence determinants, 35–36
Enumeration, 6, 7
molecular genetics, 29
proteolytic bacteria, 83–84
Environmental factors, and intestinal flora in
inflammatory bowel disease, 230
Enzyme activity, microflora-associated characteristics, 8
Enzymes
mucin utilization, 185, 186
nitrogen sources, 76
as protein source, 10
Enzymes, microbial
activity absent in germ-free animals, 259
toxic products, 156
Enzymes, pancreatic, see Digestive enzymes; Pancreatic
enzymes
Epithelial cell associated antigens (ECAC), 229
Epithelium, see also Mucosa
ammonia and, 89–90
butyrate and, 112–116
in germ-free animals, 259
microhabitats, 10
mucous layer, see Mucus glycoproteins, host
and normal microflora, 258
pathogenicity in diarrheal diseases, see Infections and
diarrhea; specific agents
turnover of, 66
Erythritol, 68
*Escherichia coli*, 2, 14, 37
acquisition of microbiota, 13
adherence to mucosa, 184
conjugation systems, 40
diarrhea, 202–206
formyl peptides, 245, 246
genetic engineering, 38

germ-free animal colonization, 5
glucuronidase, 265
and hepatoma incidence, 169
in inflammatory bowel disease, 237, 238
L-forms, 238
after surgery, 235
translocation across intestine, 239
ulcerative colitis, 233
molecular genetics
cloning vectors, 44, 45
gene expression, 47
plasmid conjugation, 39
normal microflora, 258
plasmids, 37
somatic O-antigen, 65
transfer of genetic material, 33, 34
Estrogenic compounds, 159, 166–167
Ethanol
acetate source, 116
carbohydrate fermentation end products, 62
measurement methods, 184
4–Ethylphenol, 91, 92
*Eubacterium*, 28
carbohydrate metabolism, 62
diet and, 14
in inflammatory bowel disease, 241
Crohns' disease, 233, 234
Crohns' disease patients, 229
mucosal, 236
after surgery, 235
molecular genetics, cloned genes, 50
quinoline (IQ) production, 161–162
unusual and new species, 16
*Eubacterium* species
*aerofaciens*, 12, 14
and arthritis, 263
normal microflora, 258
*contortum*, 14, 241
*cylindroides*, 14
*lentum*, 14
amino acid fermentation products, 89
normal microflora, 258
*limosum*, 146
*rectale*, 14
*tenue*, 93
European Molecular Biology Laboratory (EMBL) library,
28
Experimental methods, 3–9

**F**

Facultative organisms, 61
Fasting
fatty acid production, 111, 116
and mucin, 177, 179
Fat, dietary
bile acid secretion, 163–164

and carcinogenesis, 169
and composition of microbiota, 9
Fatty acids, see also Short chain fatty acids
  amino acid fermentation, 88
  antagonism of *Clostridium*, 214
  phenol substituted, 91
Fecal ammonia, 89, 93
Fecal fractions
  proteolytic activity, 79–83
  study methods, 13–14
Fecal populations, 13–14
Fecapentaenes, 159, 163, 164
Feces
  elimination of drugs and xenobiotics, 156, 157
  proteolysis, 78
Fermentation, see also Carbohydrate metabolism;
        Proteolysis and amino acid fermentation;
        Short chain fatty acids
  gas production, see Gas metabolism
  mucin utilization studies, 186
  sugars, concurrent with amino acids, 88
Fermentation products, see also Short chain fatty acids
  carbohydrates, 62–63
  mucin, analytical methods, 184
  and pathogen establishment, 203
Fermenter systems, 4, see also *In vitro* systems
Fiber, dietary, 9
  carbohydrate fermentation end products, 64
  and SFCAs, 103–105
*Fibrobacter*, 48
Flagellar (H) antigens, 203
Flatulence, 147
Flatus
  composition of, 131
  hydrogen and methane excretion, 133, 134
  measurement of, 132–133
  methane excretion, 140
  trace gases, 140
Flavonoids, 159, 160, 167–168
Flavonols, 160
Food additives, 67, 189
Food preparation and processing
  cooked food mutagens, 159, 168
  quinoline (IQ) production, 161–162
  resistant starches, 67–68
Formate, 62, 102, 103
Formula vs. breast-fed infants, 12–13
Formyl peptides, in inflammatory bowel disease, 228, 244–247
Fructans, and SFCAs, 103
Fructo-oligosaccharides, bifidogenic potential, 69
Fucose residue hydrolysis, 186
Fucosidases, 185, 186
Functional studies, 8
*Fusobacterium*
  amino acid fermentation products, 89
  in Crohns' disease, 229, 233

diet and, 14
*Fusobacterium* species
  *naviforme*, 93
  *necrophorum*, 87
  *nucleatum*, 88, 93
  *russii*, 258
  *varium*, 93

**G**

Galactosamines, mucin, 64–65, 186
Galactosaminidases, mucin utilization, 185, 186
Galactose, mucin, 64–65, 186
Galactosidases, 64–66, 70
  *Clostridium*, 214
  mucin utilization, 185, 186
  probiotics, 263
*trans*-Galactosylated oligosaccharides, bifidogenic
        potential, 70–71
Gas-liquid chromatography, 184
Gas metabolism
  amino acid decarboxylation and, 90
  amino acid fermentation, 94
  carbohydrate fermentation end products, 62
  clinical applications, 146–148
  hydrogen disposal mechanisms, 145–146
  methanogenesis, 140–142
  microflora-associated characteristics, 8
  mucin utilization studies, 186
  and pathogen establishment, 203
  production in large intestine, 136–140
    carbon dioxide, 139–140
    hydrogen, 136–139
    trace gases, 140
  SCFAs and, 113, 116
  study methods, 132–136, 184
    *in vitro*, 135–136
    *in vivo*, 132–135
  sulfate reduction, dissimilatory, 142–145
Gastric acidity, and normal microflora, 258
Gastroenteritis, 139
G + C ratios, 26–28
Gelatin, 85, 86
Gelatinases, 87, 214
Gene expression
  genetically engineered organisms, 48–51
  SCFAs and, 114–115
Genetic engineering
  molecular genetics, 37–51, see also Molecular genetics
  probiotics, 265–268
Genetic factors, host
  in inflammatory bowel disease, 233
  and intestinal bacteria, 10, 228–229
Genetics, microbial, see Molecular genetics
Gene transfer systems, 37–51
  cloning vectors, 47–48
  conjugation, 38–39

electroporation, 42–47
gene expression, 48–51
plasmid conjugation, 38–39
probiotics, 265–268
protoplast transformation, 41–42
release of genetically engineered bacteria, 51
transduction, 39–41
Genome size, 37
Genotoxins
IQ metabolites, 162
secondary bile acids, 164
Geographic factors in inflammatory bowel disease, 230
Germ-free animals, 2, 5
advantages and disadvantages, 157, 158
enzymes and metabolic activity absent in, 259
pancreatic protease activity in, 77
tumor incidence, 168, 169
Giardiasis, 139
Globular protein proteolysis, 78, 85
Glucuronidases, 8, 71, 156
biliary conjugates, toxic, 164–165
in germ-free animals, 259
*Lactobacillus* and, 265
Glucuronides, toxic metabolites, 68–69
Glycoproteins, see also Mucin; Mucus glycoproteins, host
metabolism of, 65–66
nitrogen sources, 76
and peptic ulceration, 187
sulfated, 189
Glycosaminoglycans, sulfated substrates, 189
Glycosidases, 8, 64–66, 70, 71
plant glycosides as substrates, 156
sulfatase and, 189
Glycosides, see also Mucin; Mucus glycoproteins, host
activation to toxic, carcinogenic, or mutagenic
derivatives, 158–160
mucin, 64–65
toxic metabolites, 68–69
Glycosidic bonds, mucin, 176
Glycosphingolipids, 66
Glycosulfatase, 189
Gram-negative bacteria
diet and, 14
glycoconjugates, 65
Gram-positive organisms, cloning vectors, 43–44
Growth conditions
amino acid decarboxylation, 91
and clostridial proteases, 86
and intracellular proteases, 84–85
mucin utilization studies, 186, 187
SCFA production, 107
Growth cycle, and intracellular proteases, 84–85
Growth factors
butyric acid and, 115
immune system, see Immune mediators
Guar, and SFCAs, 103, 104

Gut associated immune system, 9, 11

**H**

Habitat, 2
*Hafnia*, 237
H (flagellar) antigens, 203
Heat labile haemolysin, 209
*Helicobacter pylori*, 6, 76, 185
as "almost normal microflora,", 257
formyl peptides, 245
mucus and, 186–188
penetration of mucin, 184
Hemicelluloses, 67
Hemolysins
*Aeromonas*, 211
*Clostridium*, 214
*Vibrio*, 209
virulence determinants, 35–36
Hemorrhagic colitis, 206
Hepatic coma (portal-systemic encephalopathy), 68, 90, 147
Hepatomas, 159, 168, 169
Heterocyclic amines, see IQ
Heterocyclic nitro compounds, 161
High performance anion-exchange chromatography with
pulsed amperometric detection (HPAEC-PAD),
180, 183
Histamine, 90, 91
HLA antigens, in inflammatory bowel disease, 229
HLA-B27/β-microglobin transgenic rat, 242
Hormones, 159
enterohepatic circulation, 165
and mucin, 179
Host factors
autochthonous populations, 2
ecosystem components, 9
inflammatory bowel disease, 233
interactions within ecosystem, 11
and intestinal bacteria, 10, 228–229
and normal microflora, 257–258
Host mucus glycoproteins, see Mucin; Mucus glycoproteins,
host
Human volunteers, 157, 158
Hyaluronidases, *Clostridium*, 214
Hydrogen
acetogenesis, 145–146
amino acid fermentation, 88, 94
carbohydrate fermentation end products, 62
flatus composition, 131
mucin utilization studies, 186
nitrate reduction, 145
pulmonary excretion, 133–135
SCFAs and, 116
and sulfate reducers, 142, 143
Hydrogen, breath, 133–135, 142
clinical applications, 146–147

production in large intestine, 136–139
Hydrogen sulfide
  and inflammatory bowel disease, 189
  mucin utilization studies, 186
  and pathogen establishment, 203
  and ulcerative colitis, 148

# I

Identification of microbiota, 6, 29
Ileal protease activity, 79, 80
Ileo-anal pouch, 235–236
Ileocecal valve, 236
Ileostomy fluid
  carbohydrate fermentation end products, 63
  dietary carbohydrates, 67
  nitrogen-containing materials, 75–76
  proteolytic activities, 77–81
  SCFA production, 108
Ileostomy reservoir, in inflammatory bowel disease, 234–235
Immune mediators
  in diarrheal disease, 206
  endotoxin activation of, 243
  and mucin, 179
  probiotics and, 261
Immune system, 9, 11
  bacterial products and, 244
  in germ-free animals, 259
  host factors, 9, 11
  in inflammatory bowel disease, 240–241
  lactic acid bacteria and, 261, 262
Immunodeficiency states, 11
  and diarrheal disease, 206–207
  mycobacterial infections, 210, 239
  and translocation of bacteria across intestine, 23
Immunoglobulins
  and bacterial translocation, 240
  in breast milk, 12
  in germ-free animals, 259
  host factors, 9, 11
IncP trasnsfer system, 37
Indican, 94
Indigenous populations, 2–3
Indoles, 94
  amino acid fermentation, 91–94
  in cancer patients, 94
  carbohydrate metabolites, 69
Infant botulism, 12, 215
Infant formulas, lactulose in, 68
Infants, colonization of gut, see Colonization
Infections and diarrhea
  *Aeromonas*, 211–212
  *Bacteroides fragilis*, 213–214
  *Campylobacter*, 212–213
  clostridia, 214–216

  *Escherichia coli*, 203–206
  *Mycobacterium*, 209–210
  *Plesiomonas*, 212
  *Salmonella*, 206–207
  *Shigella*, 207–209
  *Vibrio*, 209
  *Yersinia*, 210–211
Infective doses, 17, 208–209
Inflammation, 11, see also Infections and diarrhea
  bacterial components inducing, 243–247
  cell wall materials and, 263
  and mucin, 177, 179
  SCFAs in, 116–119
Inflammatory bowel disease
  animal models, 242
  definitions and clinical overview, 227–228
  environmental factors and intestinal microflora, 230
  genetics and intestinal bacteria, 228–229
  immunological evidence for bacterial involvement in, 240–241
  indole and phenol detoxification, 94
  inflammatory products of intestinal bacteria, 243–247
    endotoxin, 243
    formyl peptides, 245–247
    muramyl peptides, 244–245
    peptidoglycan-polysaccharide compexes, 243–244
  luminal bacteria in, 230–236
  metabolic activity of commensal bacteria in, 242
  mucosal bacteria in, 236–239
  mucus in, 177, 188–189
  SCFAs in, 116–119
  sulfide and ulcerative colitis, 148
  sulfur metabolism in ulcerative colitis, 117–118
  translocation of bacteria across intestine in, 239–240
  treatment approaches, advances in, 229
Inflammatory mediators
  in diarrheal disease, 206
  endotoxin activation of, 243
  and mucin, 179
  probiotics and, 261
Insertional mutagenesis, 36–37
Insertion sequences
  *Bacteroides*, 33–34
  lactobacilli, 32–33
*In situ* studies, 4
Interspecies restriction barriers, 38
Interspecific gene transfer, 266
Intestinal motility, see Motility, gut
Intravenous feeding, 64
Inulin
  bifidogenic potential, 69
  and SFCAs, 103, 104
Invasion
  by *Clostridium*, 214
  by *Helicobacter pylori*, 185
*In vitro* studies, 4–5

advantages and disadvantages, 157, 158
carbohydrate fermentation end products, 63
gas production, 135–136
mucin study methods, 179–184
mucin utilization, 186, 187
proteolytic systems, 78
SCFA production, 107
*In vivo* studies, 4–5, see also Animal models; Germ-free
      animals
advantages and disadvantages, 157, 158
gas metabolism, 132–135
Iodoacetate, 79–82
IQ (2-amino-3-methyl-3*H*-imidazo[4,5-*f*]quinoline), 156
activation to toxic, carcinogenic, or mutagenic
      derivatives, 161–162
binding by intestinal bacteria, 168
Isobutyrate, 102, 103
Isolation, proteolytic bacteria, 83–84
Isomalto-oligosaccharides, bifidogenic potential, 70
Isovaleric acid, 102, 103, 110

## J

Jejuno-ileal motility, ileo-anal pouch and, 235–236
Jejunostomates, carbohydrate fermentation end products,
      63
Johne's disease, 239
Joint inflammation, 11–12
cell wall components and, 262–263
in inflammatory bowel disease, 228, 243–244

## K

Kaempferol, 159, 160
Keto acids, amino acid fermentation, 88
*Klebsiella*, in inflammatory bowel disease, 235, 236, 237
*Klebsiella pneumoniae*, 32
formyl peptides, 245
in inflammatory bowel disease, 237
Kock pouch ileostomy, 235

## L

*Lachnospira*, 93
Lactase deficiency, 139, 147
Lactate, 102, 103
carbohydrate fermentation end products, 62
fermentation of, 103
and pathogen establishment, 203
Lactic acid bacteria
and immune system, 261, 262
modification of, 259, 260
normal microflora, 258
Lactitol, 68, 147
*Lactobacillus*
amino acid fermentation

amine formation, 90
phenol and indole formation, 91, 93
carbohydrate metabolism, 70, 71
fructo-oligosaccharide and, 69
mean numbers and fermentation end products, 62
diet and, 14
in inflammatory bowel disease
Crohns' disease, 234
ulcerative colitis, 233
molecular genetics, 26, 266
cloned genes, 50
cloning vectors, 43, 44
electoporation, 41–42
gene transfer systems and methods, 40
plasmids, 30, 31
protoplast transformation, 41
transposons and insertion sequences, 32–33
normal microflora, 258
probiotics, 260, 263–265
proteases, 87
treatment of *Clostridium difficile* infections, 17
*Lactobacillus* species
*acidophilus*, 14
and hepatoma incidence, 169
molecular genetics
electoporation, 41–42
gene transfer systems and methods, 40
plasmids, 31, 38
protoplast transformation, 41
normal microflora, 258
probiotics, 260, 262, 264, 265
*casei*, 32
and arthritis, 263
molecular genetics, protoplast transformation, 41
in probiotics, 260–262, 264
*curvatus*, 36, 48
*delbruckei* ssp. *bulgaricus*, 260, 261
*fermentum*
molecular genetics
cloned genes, 50
gene transfer systems and methods, 40
normal microflora, 258
*helveticus*, 264
*pentosus*, 31, 43
*plantarum*
molecular genetics, 266
cloned genes, 50
cloning vectors, 43, 46
gene expression, 48
gene transfer systems and methods, 40
plasmid pAMβ1, 38
Tn*919*, 36
normal microflora, 258
probiotics, 260, 261, 262
*reuteri*, 30
*salivarius*, 39, 40
*Lactococcus lactis*, 38, 43, 44, 45

Lactoferrin, 9
Lactose, 10, 111
Lactose intolerance, 260–262, 264
Lactulose, 68, 90, 147
   gas production, 138
   SCFA production, 108
Lambda bacteriophage, 39, 44
Leucine aminipeptidase, 78
*Leuconostic mesenteroides* Tn*919*, 36
Lewis antigens, 186, 188
L-forms of mucosal bacteria, 238
Lignans, 159, 166–167
Lipid, dietary, see Fat, dietary
Lipids, sulfated, 189
*Listeria*
   in inflammatory bowel disease, 239, 240
   molecular genetics, cloning vectors, 44
   *monocytogenes*, 262
Lithocholic acid, 164
Liver
   conjugation of xenobiotics, 156, 157
   enterohepatic circulation, see Enterohepatic circulation
   in inflammatory bowel disease, 228
   portal-systemic encephalopathy, 68, 90, 147
   propionate uptake, 110
   in *Salmonella* diseases, 207
Liver tumors, 168, 169
Low-calorie food products, 68
Low-molecular weight carbohydrate metabolism, 68
Luminal bacteria
   in inflammatory bowel disease, 230–236
   microhabitats, 10
   populations, 14–15
Lymphocytes
   butyric acid and, 115
   in germ-free animals, 259
   in inflammatory bowel disease, 242
Lysis, bacterial, protease release, 84, 87

**M**

Macrophages
   bacteria surviving in, 239
   endotoxin activation of, 243
   inflammatory mediator induction by cell wall components, 244
   muramyl peptides and, 244
   probiotic activation, 262
Malabsorption
   carbohydrates, bifidogenic potential, 68–71
   gas production, see Gas metabolism
Malnutrition, and mucin, 177, see also Fasting; Nutritional status; Starvation
Maltitol, 68
Mapping, physical, 37, 45
MDP (N-acetyl muramyl-L-alanyl-D-isoglutamine), 244
Mean transit time, SCFA production, 107

Media, culture, 6, see also *In vitro* systems; Nutrient limitation
*Megasphaera*, 93
   *elsdenii*, 50, 88
Mercaptoacetate, 117
Mercury, 159
   detoxification, 165–166
   resistance plasmids, 31
Metabolism, host
   epithelial cells, ammonia and, 90
   modification by microflora, 168–170
   modification of, 159
Metabolism, microbial, see also Carbohydrate metabolism; Proteolysis and amino acid fermentation; Toxicology
   activity absent in germ-free animals, 259
   gas production, see Gas metabolism
   genetic classification, 28
   in inflammatory bowel disease, 242
   microflora-associated characteristics, 8
   mucin utilization, 185, 186
Metalloproteases, 79, 80
   clostridial protease activity, 86
   mucin utilization studies, 186
   *Vibrio cholerae*, 185
Meterorism, 147
Methane, see also Methanogenesis
   and colorectal cancer, 148
   flatus composition, 131
   gas metabolism, 140–142
   mucin utilization studies, 186
   pulmonary excretion, 133, 134
Methanethiol, 140
*Methanobacterium thermautotrophicum*, 32
*Methanobrevibacter*, 33, 34
   *smithii*, 137, 140
*Methanococcus*, 42
   *voltae*
      cloning vectors, 45, 46
      molecular genetics
         electoporation, 42
         gene transfer systems and methods, 40
Methanogenesis, 140–142, see also Methane
   and cancer, 148
   hydrogen consumption, 137
   SCFAs and, 116
Methanogens, 16, see also specific organisms
   molecular genetics, cloning vectors, 43, 45
   mucin utilization studies, 186
   and sulfate reducers, 142, 143
*Methanosphaera stadtmaniae*, 140
Methemoglobinemia, 161
Methylazoxymethanol, 164
Methylcelluloses, 67
Methylmalonaciduria, 110
Methylmercury detoxification, 159, 165–166
N-Methyl-N-nitrosourea (MNU), 164

Methyl sulfides, 140
Methylureas, nitroso compound formation, 162
Microbial ecology
    components, 9–12
        dietary contributions, 9–10
        diversity within ecosystem, 10
        host factors, 9
        interactions within ecosystem, 10–12
        microbiota, 10
    ecosystem, 2–9
        description, 3
        ecological principles, 2–3
        *in situ* studies, 4
        *in vitro* and *in vivo* models, 4–5
        sample analysis, 5–8
        study methods, 3–9
    microbiota, 12–16
        acquisition, 12–13
        adult profile, 13–16
        unusual components, 16
    stability of ecosystem, 16–17
Microbiota
    acquisition of, 12–13
    adult profile, 13–16
        fecal population, 13–14
        luminal populations, 14–15
        mucosal-associated populations, 15–16
    ecosystem components, 10
    and SCFAs, 106–107
Microflora-associated characteristics (MACs), functional
        studies, 8
Microhabitats, 6, 10
*Mitsuokella* (*Bacteroides*) *multiacidus*, 29
Mixed function oxidases, 148, 157, 167, 168
Mobile genetic elements, 29–35
Molecular genetics, 7–8
    bacteroides
        plasmids, 32
        transposons and insertion sequences, 33–34
    bifidobacteria, plasmids, 30
    cloning and analysis, 34–37
        cloning strategies, 34–35
        physical mapping, 37
        transposons and targeted insertional mutagenesis,
            36–37
        virulence determinants, 35–36
    clostridia
        plasmids, 30–32
        transposons, 34
    genetic diversity, 26–29
        identification and classification methods, 26–27
        identification and enumeration of organisms, 29
        phylogenetic analysis, 27–29
    genetic manipulation, 37–51
        cloning vectors, 47–48
        electroporation, 42–47

gene expression, 48–51
        plasmid conjugation, 38–39
        protoplast transformation, 41–42
        release of genetically engineered bacteria, 51
        transduction, 39–41
    lactobacilli
        plasmids, 30
        transposons and insertion sequences, 33–34
    methanogens, plasmids, 32
    plasmids and mobile genetic elements, 29–34
    probiotics, 265–266
    SCFAs and, 114–115
    sulfate-reducing bacteria, 32
Monoamine oxidases, 91
*Morganella*, 235
Morin, 167
Motility, gut
    ileo-anal pouch and, 235–236
    toxins and, 203
Motility, microbial, 184
    *Campylobacter*, 212–213
    flagellar antigens, 203
Mucin, see also Mucus glycoproteins, host
    *Clostridium* and, 214
    metabolism of, 64–65
        analytical methods, 180–184
        degradation of, pathological implications, 184–186
    microflora-associated characteristics, 8
    nitrogen sources, 76
    proteolysis, 84
Mucin oligosaccharide degrading (MOD) bacteria, 185
Mucopolysaccharides, nitrogen sources, 76
Mucoprotective index, 189
Mucosa, see also Epithelium; Mucus glycoproteins, host
    adhesion to, 4
    carboydrates, 66
    culture conditions, 6
    formyl peptides and, 247
    host factors, 9, 11
    *in vitro* models and, 5
    microhabitats, 10
    and normal microflora, 257–258
    pancreatic protease binding, 77
    pathogenicity in diarrheal diseases, see Infections and
            diarrhea; specific agents
    pathogens and, 202, 203
    SCFAs and, 113–114
Mucosal-associated microbes, 15–16
    enumeration, 7
    in inflammatory bowel disease, 236–239
    nutrient deprivation and, 17
Mucus gel, microhabitats, 10
Mucus glycoproteins, host
    bacterial metabolism of, analytical techniques, 180–184
        fermentation products, 184
        proteins and amino acids, 183–184

sialic acids, 183
  sugars, 181–183
degradation of, 184–189
  and inflammatory bowel disease, 188–189
  mucin metabolism, 184–186
  peptic ulceration and *Helicobacter pylori*, 186–188
factors affecting integrity of mucus gel layer, 175–180
  intracellular organization and biosynthesis, 175–176
  structure and function of secretions, 176–177
  turnover, 177–180
Multistage continuous culture fermenters, 4
Muramyl peptides, in inflammatory bowel disease, 244–245
Mutagenesis, insertional, 36–37
Mutagens, 8
  activation of chemicals, 158–162
  carbohydrate metabolites, 68
  cooked food, 159
  flavonoids and, 167
  nitrosamines, 91
*Mycobacterium*, 262–263
  N-acetyl muramyl-L-alanyl-D-isoglutamine (MDP), 244
  diarrhea, 209–210
  and inflammation, 244
  in inflammatory bowel disease, 238–239
    Crohns' disease, 238
    immunological evidence, 241–242
    L-forms, 238
  probiotics and, 262
*Mycobacterium* species
  *avium*, 238, 239
  *avium-intracellulare* (MAI), 210
  *bovis* (BCG strain), 262
  *kansasii*, 238
  *leprae*, 209
  *paratuberculosis*, 238, 239, 241
  *tuberculosis*, 209–210
Myricetin, 167

**N**

Necrotic eneritis, 214, 215
Neonates, see Acquisition of microbiota
Neuraminidase
  *Clostridium*, 214
  virulence determinants, 35
Niche, 2
Nitrate reductase, 71, 156
Nitrate reduction, see also Ammonia
  hydrogen disposal, 145
  to nitroso compunds, 162–163
Nitrates
  nitrite formation, 69
  reduction to nitroso compounds, 162–163
Nitrite reductase, 156
Nitrites, 69, 71, 156, 161
  nitrosamine formation, 91

nitrosation reactions, 94
Nitro compounds
  activation to toxic, carcinogenic, or mutagenic
    derivatives, 161
  carbohydrate metabolites, 69
Nitrogen, 140
Nitrogenase, *Desulfovibrio* genes, 32
Nitrogen limitation, amine formation, 91
Nitrogen metabolism, see also Ammonia; Proteolysis and
    amino acid fermentation
  *Lactobacillus* and, 265
Nitrogen source
  and intracellular proteases, 85
  peptides, 85, 88
Nitroreductase, 8, 156, 265
Nitrosamines, 91
Nitrosation of dimethylamine, 94
Nitroso compounds, 94, 162–163
  protein and amino acid metabolites, 164
  synthesis of, 159
Nonspecific colitis, 236
Nonstarch polysaccharides
  carbohydrate fermentation end products, 64
  metabolism of, 67
  and SFCAs, 103, 104
Normal microflora, 257–260
Nutrient deprivation, 17
Nutrient limitation
  and amine formation, 92
  and clostridial proteases, 86
  mucin utilization studies, 186, 187
Nutrients, microflora as source of, 155
Nutritional status
  fatty acid production, 111, 116
  and mucin, 177, 179
  and proteolysis, 81

**O**

O-antigens, 65
Ofloxacin, 17
Oral streptococci, mucin utilization, 186
Organic acids
  carbohydrate fermentation end products, 62
  SCFAs, see Short chain fatty acids
Organic N-containing compounds, amino acid fermentation,
    87–88
O (somatic) antigens, 203
Outer membrane proteins, 7, 210, 211
Ovalbumin, 78, 80
Oxidation-reduction reactions, amino acid fermentation, 88

**P**

Palatinose, 68
Pancreatic amylase-resistant starch, see Starch, resistant

Pancreatic enzymes, 76, 215
    bacterial degradation of, 78
    binding in mucosa, 77
    as protein source, 10
Pancreatic insufficiency, 139
Paratyphoid fever, 207
Parenteral feeding, 64
Pathogens, 3, 10, see also Infections and diarrhea
    in inflammatory bowel disease, 236, 237
    probiotics and, 262
Pectin, 69, 104
*Pediococcus*, 44
    *pentosaceus*, 260
Pepsin, 189
Pepstatin A, 80, 81, 82
Peptic ulceration, see also *Helicobacter pylori*
    mucus and, 186–188
Peptide hormones, and mucin, 179
Peptide hydrolases, clostridial, 86
Peptides
    in inflammatory bowel disease, 244–247
    as nitrogen source, 85, 88
Peptidoglycan-polysaccharide complexes, in inflammatory
        bowel disease, 243–244
*Peptococcus*
    diet and, 14
    *prevotii*, 258
Peptone, as nitrogen source, 85
*Peptostreptococcus*, 28
    amino acid fermentation products, 89
    carbohydrate metabolism, 62
    diet and, 14
    and inflammation, 244
    in inflammatory bowel disease, 241
        Crohns' disease, 229, 234
        mucosal, 236
        after surgery, 235
    normal microflora, 258
*Peptostreptococcus* species
    *asaccharolyticus*, 93
    *intermedius*, 258
    *micros*, 258
    *productus*, 14, 146, 241
Peristalsis, 258, see also Transit/retention time
pH
    amine formation, 90
    amino acid fermentation and, 94
    carbohydrate metabolism and, 71
    proteolysis, 78, 79, 81, 83, 85
Phages, see Transduction
Phenols
    amino acid fermentation, 91–94
    carbohydrate metabolites, 69
    measurement methods, 184
    protein and amino acid metabolites, 164
Phenol-substituted fatty acids, 89

Phenylalanine, phenol and indole formation, 91
Phenylethylamine, 91
Phenylpropionate, 92
Pheromones, 245
Phospholipase A, 209
Phospholipase C, 35
Phylogenetic analysis, 26–29
Physical mapping, 37, 45
Phytoestrogens, 166–167
Phytosterols, 159
Pig-Bel, 215
Piperidine, 90, 91
Plant glycosides, 158–160
Plasmid conjugation, 30, 31, 38–39
Plasmid profiling, 267
Plasmids, 29–34
    *Bacteroides*, 29, 32
    clostridia, 30–32
    cryptic, 44
    detection of probiotic strains, 267
    lactobacilli and bifidobacilli, 30
    methanogens and sulfate-reducing bacteria, 32
    probiotic development, 266
    stability of, 38
*Plesiomonas*
    diarrhea, 212
    *shigelloides*, 212
PMSF, 79–82, 84
Pneumatosis cystoides intestinalis, 147–148
Polycyclic aromatic hydrocarbons, 161
Polydextrose, 69
Polyethylene glycol, 40, 41
Polymerase chain reaction (PCR), 7, 26, 28
    detection of probiotic strains, 268
    nested, 239
Polyols, 68
Polysaccharides, cellular, in inflammatory bowel disease,
        243–244
*Porphyromonas*, 28
    *asaccharolytica*, 93
    *gingivalis*, 84
Portal blood, see also Enterohepatic circulation
    SCFAs in, 109, 113
Portal-systemic encephalopathy, 68, 90, 147
Postnatal acquisition of microbiota, see Colonization
Pouchitis, 110, 235
*Prevotella*, 189
    *melaninogenica*, 87, 89
    *ruminicola*, 87
Probiotics, role of
    critical view of, 260–263
    detection in digestive tract samples, 267
    developments in research, 263–265
    gene technology, 267
    molecular genetics, 265–267
    normal microflora, 257–260

and host, 259
    modification of, 259–260
Proctitis, 16
Promoters, tumor, see Tumor promoters
Pronase, 185
Propionate, 102, 103
    amino acid fermentation products, 88, 89
    carbohydrate fermentation end products, 62
    carbohydrate substrates, 104–105
    portal blood, 109
    stoichiometry, 110
*Propionibacterium*
    *acnes*, 87, 93, 262
    diet and, 14
    *jensenii*, 40
    molecular genetics, 44
Propylamine, 91
Proteases
    in inflammatory bowel disease, 189
    mucin breakdown, 184, 185
    mucin utilization studies, 186
Proteases, bacterial
    cell lysis and, 81
    nitrogen sources, 76
    release in gut, 77
Proteases, pancreatic
    bacterial degradation of, 78
    binding in mucosa, 77
    clostridial toxin modification, 215
Protein
    mucin, analytical methods, 183–184
    as nitrogen source, 85
    sources of, 10
Protein, dietary, 9
    amine formation, 91
    and amino acid fermentation, 93, 94
    SCFA production, 110
    toxic metabolites, 164
    tumor promoter synthesis, 164
Protein depolymerization, 88
Protein fermentation
    amino acid fermentation products, 88
    gas production, see Gas metabolism
    SCFAs, see Short chain fatty acids
Proteolysis
    *Clostridium*, 214
    in germ-free animals, 259
    inactivation of, 8
Proteolysis and amino acid fermentation
    amino acid fermentation, 87–94
        amines, 90–91
        ammonia, 89–90
        gas metabolism, 94
        organic N-containing compounds, 87–88
        phenols and indoles, 91–94
        SCFAs, 88–89

in large intestine, 77–83
    characterization of activity in large intestine, 79–83
    comparison of activities, large vs. small intestine, 78–79
microflora, 83–87
    *Bacteroides fragilis*, 84–85
    *Bacteroides splanchnicus*, 85
    *Clostridium bifermentans*, 86
    *Clostridium perfringens*, 86
    isolation and enumeration, 83–84
    miscellaneous bacteria, 86
*Proteus*, in inflammatory bowel disease, 237
    after surgery, 235
    ulcerative colitis, 233
*Proteus vulgaris*
    formyl peptides, 245
    germ-free animal colonization, 5
Protoplast fusion, 40
Protoplast transformation, 40, 41–42
*Providencia*, 235
Pseudomembranous colitis, 215–216
    *Clostridium difficile* overgrowth, 202
    probiotics, 264
*Pseudomonas*
    *aeruginosa*, 86, 237, 238
    in inflammatory bowel disease, 236, 237
        L-forms, 238
        translocation across intestine, 239
    *maltophila* L-forms, 238
    probiotics and, 262
Pulsed field gel electrophoresis (PFGE), 27, 29, 37, 268
Putrescine, 90, 91
Pyrrolidine, 90, 91

**Q**

Quercetin, 159, 160, 167
Quinolines, see IQ
Quinolone antibiotics, 17

**R**

Raffinose, 10, 68
    bifidogenic potential, 69
    gas production, 138
Rare cutters, 37
Rectal proctitis, 16
Redox potential, culture, 6
Redox reactions, amino acid fermentation, 88
Reductases, bacteria, 161
Regional distribution, phenol and indoles, 92
*res*, 31
Residence time, see Transit/residence time
Resistant starch, see also Starch, resistant
    gas production, 138
    metabolism of, 67–68

and SFCAs, 103
Restriction barriers, 38
Restriction enzyme analysis, 7, 28
Restriction enzymes, rare cutters, 37
Restriction fragment length polymorphisms (RFLP), 27
Retention time, see Transit/residence time
Ribosomal RNA, 7, 26–29
    *Clostridium perfringens*, 35
    detection of probiotic strains, 268
Ribotyping, 268
Rolling circle method of replication, 42
Roll-tube culture, 6
Rumen organisms, 76
*Ruminococcus*, 48, 93
    *albus*, 50, 258
    carbohydrate metabolism, mean numbers and
        fermentation end products, 62
    diet and, 14
    *gnavus*, 185, 186
    mucin degradation, 65
    *torques*, 65, 185, 186
Rutin, 167

**S**

Saccharolytic activity, clostridial, 86
*Saccharomyces cerevisiae*, 260
*Salmonella*, 261
    diarrhea, 202, 206–207
    *enteriditis*, 206
    in inflammatory bowel disease, 236, 239
    *paratyphi*, 206
    release of genetically engineered bacteria, 48
    *typhi*, 206, 207
    *typhimurium*, 5, 237
Sample analysis, 5–8
*Sarcina*, 28, 93
SCFAs, see Short chain fatty acids
Secondary bile acids
    in germ-free animals, 259
    tumor promoter synthesis, 163–164
Secretory IgA
    and bacterial adhesion, 240
    in breast milk, 12
    host factors, 9, 11
Segmental colitis, 212
Selective media, 6
*Selenomonas*, 28
Sequence homologies, 26, 27
Serine, mucin utilization studies, 186
Serine protease inhibitors, 87
Serine proteases, 79, 80
    bacterial, 81
    *Bacteroides fragilis*, 84
Sex pheromones, 245
*Shigella*
    *boydii*, 208

diarrhea, 202, 207–209
    *dysenteriae*, 208
    *flexneri*, 208
    in inflammatory bowel disease, 236, 237
    *sonnei*, 208
Short bowel syndrome, carbohydrate fermentation end
        products, 63
Short chain fatty acids, 159
    absorption, 111–112
    amino acid fermentation, 88–89
    carbohydrate fermentation end products, 62
    cell metabolism and growth, 112–116
        differentiation, gene expression, and colon cancer,
            114–116
        energy metabolism, 113
        mucosal growth, 113–114
    clinical conditions, 116–119
        diversion colitis, 119
        pouchitis, 119
        ulcerative colitis, 116–119
    in germ-free animals, 259
    measurement methods, 184
    metabolism of, 116
    microflora-associated characteristics, 8
    mucin utilization, 186
    occurrence in hindgut, 102–108
        antibiotics, 105
        diet, 106
        microflora, 106–107
        stool output and diarrhea, 105–106
        substrates, 102–105
        transit time, 107–108
    and pathogen establishment, 203
    production rates, 108–111
        amino acids as substrates, 110
        portal blood studies, 108–109
        stable isotope studies, 110–111
        stoichiometry, 109–110
    protective effects, 168
    surgical diversion/ileostomy reservoirs and, 234
    and ulcerative colitis, 148
Shuttle vectors, 38
Sialic acids
    in inflammatory bowel disease, 188
    mucin, 64, 184, 185
Sialidases, 185
Sickle cell anemia, 206
Skatoles, in cancer patients, 94
Small intestine, proteolysis in, 78–80
Sodium, SCFAs and, 112
Somatic (O) antigens, 65, 203
Sorbitol, 68
Soybean trypsin inhibitor, 79–82
Species differences, restriction and genetic modification
        systems, 38
Spheroplasts, 238
Spirochetes, 15, 16

Stability of ecosystem, 16–17
Stachyose, 68, 138
*Staphylococcus*
   *aureus*, 245
   *carnosus*, 48
   cloning vectors, 44
   *hyicus*, 48
Starch
   bifidobacteria effects, 69
   resistant, 9–10, see also Resistant starch
Starvation
   fatty acid production, 111, 115
   and mucin, 177, 179
Steric effects, in proteolysis, 78
Steroid sulfate, 189
Stickland reactions, 94
Stool output, see also Transit/residence time
   and SCFAs, 105–106
*Streptococcus*
   acquisition of microbiota, 13
   amino acid fermentation, 90
   anaerobic, 14
      normal microflora, 258
      in ulcerative colitis, 233
   carbohydrate metabolism, 62
   diet and, 14
   Group A, 244
   molecular genetics, 38, 40
   mucin utilization, 186
*Streptococcus* species
   *bovis*, 48
   *faecalis*, 40
   *mitis*, 186
   *salivarius* ssp. *thermophilus*, 260, 261
   *sanguis*, 36
*Streptomyces alboniger*, 42, 46
Succinate, 62
Sugars, 10
   chemically modified, 68
   fermentation, see also Carbohydrate metabolism
      concurrent with amino acids, 88
   mucin, 178, see also Mucin; Mucus glycoproteins, host
      analytical methods, 181–183
      utilization of, 185–186
Sulfatase, mucin degradation, 189
Sulfated substrates, and inflammatory bowel disease, 188
Sulfate-reducing bacteria
   and inflammatory bowel disease, 189
   *in vitro* studies of gas production, 135–136
   molecular genetics
      classification, 28
      plasmids, 32
   mucin utilization studies, 186
   and SCFAs, 116
Sulfate reduction, dissimilatory, 145–146
Sulfide
   measurement methods, 184

   trace gases, 140
   and ulcerative colitis, 117–118, 148
Surface-associated populations, nutrient deprivation and, 17
Surgery, in inflammatory bowel disease, 230–232, 234–236

## T

Targeted insertional mutagenesis, 36–37
Taxonomy, 26–29
T cells, 242
Tertiary structure
   mucin, 177, 187
   protein, 78
Thermostable membrane damaging protein (TDH), 209
Thimerosal, 79–82, 84
Thiol protease inhibitors, 79, 80, 82, 86, 87
Thiol proteases, 79, 80, 84–86
*Tissierella (Bacteroides) preacutus*, 29
TLCK, 80
TLPK, 80
*Torulopsis*, 260
Toxic metabolites, 8
Toxicology
   activation of chemicals to toxic, mutagenic, or
      carcinogenic derivatives, 158–162
      azo compounds, 160–161
      IQ, 161–162
      nitro compounds, 161
      plant glycosides, 158–160
   bacterial metabolism, 156–157
   biliary conjugates and enterohepatic circulation, 164–165
   carcinogen synthesis, 162–163
   detoxification and protective effects of microflora,
      165–168
   methods for study of, 157–158
   microflora-host interactions, 155–156
   modification of metabolism and tumor incidence by
      microflora, 168–170
   tumor promoter synthesis, 163–164
Toxins, metabolic products
   ammonia, 89–90
   carbohydrate metabolites, 68
   detoxification, 148, 156, 157, 164–168
   in inflammatory bowel disease, 243
Toxins, microbial, 155, see also specific agents
   clostridial
      plasmid-encoded, 31
      proteases, 86
   diarrheal disease, 202, see also specific agents
   ecological factors, 12
   *Escherichia coli*, 205, 206
   and mucin, 179, 186
   virulence determinants, 35–36
Trace gases, production in large intestine, 140
Transamination, amino acid fermentation, 88
Transduction, 39–41
   probiotic development, 266

systems and methods, 40
Transformation, 38
   probiotic development, 266
   protoplast, 41–42
   systems and methods, 40
Transgenic rat, inflammatory bowel disease model, 242
Transient species, 2, 5
Transit/residence time, 4
   and amino acid fermentation products, 88
   carbohydrate fermentation end products, 62–63
   gas, 131
   in germ-free animals, 259
   ileo-anal pouch and, 236
   measurement of, breath hydrogen, 146–147
   and proteolysis, 83
   and proteolytic activity, 77, 78
   and SCFAs, 105–108
Translocation of bacteria across intestine, 239–240
Transposon insertional mutagenesis, 37
Transposons
   *Bacteroides*, 33–34
   clostridia, 34
   genetic analysis, 36–37
   lactobacilli, 32–33
Traveler's diarrhea, 17
Trypsin
   in germ-free animals, 259
   inactivation of, 8
   mucosal barrier breakdown, 185
Trypsin, bacterial, 81, 85
Trypsin, pancreatic, binding in mucosa, 77
Trypsin inhibitors, 79, 80, 82
Tryptophan, 91
Tumorigenicity, probiotics and, 262
Tumor inducers, flavonoids and, 167
Tumor promoter synthesis, 159, 163–164
Twins, 10, 229, 233
Typhoid fever, 207
Tyramine, 90, 91
Tyrosine, 91, 93

**U**

Ulcerative colitis, see also Inflammatory bowel disease
   indole and phenol detoxification, 94
   SCFAs in, 116–119, 168
   sulfur metabolism in, 117–118, 148
Unsaturated fatty acids, amino acid fermentation, 88
Urea, and mucosal barrier, 187

**V**

Valerate, 102, 103
Vancomycin, 6
Vectors, cloning, 38, 47–48
Vegetarian diet, bacterial populations with, 14
*Veillonella*, 69, 93, 258
*Vibrio*, 202, 209
   *alginolyticus*, 86, 209
   *cholerae*, 209
      infective dose, 209
      proteases, 185
   *fluvialis*, 209
   *hollisae*, 209
   *mimicus*, 209
   *parahemolyticus*, 209
   *vulnificus*, 209
Viellonella, 14
Virulence, 202
Virulence determinants, 35–36
   *Escherichia coli*, 205, 206
   *Salmonella*, 207
Vitamins, 155

**W**

Wash-out effect, 258
Whipple's disease, 210
Whole chomosome analysis, 27
Wood pigeon strain *M. avium*, 239

**X**

Xenobiotic detoxification, 156, 157
Xylitol, 68

**Y**

Yeasts
   diet and, 14
   in inflammatory bowel disease, 237
   probiotics, 260
*Yersinia*, 202, 210–211, 236
   *enterocolitica*, 210
      adherence to mucosa, 184–185
      in inflammatory bowel disease, 237, 241
   *pestis*, 210
   *pseudotuberculosis*, 210, 240